设施园艺科技创新进展

——2018 第五届中国·寿光国际设施园艺研讨会论文集

Advances in Innovation Technology of Protected Horticulture

—— Proceedings of 2018 the Fifith International Forum on Protected Horticulture (Shouguang ·China)

● 杨其长 （日）古在丰树（Kozai Toyoki） （荷）伯特（Gerard P.A. Bot） 主编

Edited by Yang Qichang Kozai Toyoki (Japan) Gerard P.A. Bot (The Netherlands)

中国农业科学技术出版社

China Agricultural Science and Technology Press

图书在版编目（CIP）数据

设施园艺科技创新进展：2018 第五届中国·寿光国际设施园艺研讨会论文集 / 杨其长，（日）古在丰树（Kozai Toyoki），（荷）伯特（Gerard P.A. Bot）主编 .—北京：中国农业科学技术出版社，2018. 4

ISBN 978-7-5116-3587-7

Ⅰ. ①设… Ⅱ. ①杨… ②古… ③伯… Ⅲ. ①园艺—设施农业—国际学术会议—文集 Ⅳ. ①S62-53

中国版本图书馆 CIP 数据核字（2018）第 054765 号

责任编辑 张孝安 崔改泵
责任校对 李向荣
出 版 者 中国农业科学技术出版社
北京市中关村南大街12号 邮编：100081
电 话 （010）82109708（编辑室）（010）82109702（发行部）
（010）82109709（读者服务部）
传 真 （010）82106650
网 址 http: // www.castp.cn
经 销 者 各地新华书店
印 刷 者 北京建宏印刷有限公司
开 本 710mm × 1 000mm 1/16
印 张 29.5 彩插 2面
字 数 679千字
版 次 2018年4月第1版 2018年4月第1次印刷
定 价 98.00元

2018第五届中国·寿光国际设施园艺研讨会组织委员会名单

组委会主席：

吴孔明　中国工程院院士　中国农业科学院副院长

赵绪春　山东省寿光市委副书记　市政府市长

组委会副主席：

孙修炜　中共寿光市委副书记

张燕卿　中国农业科学院农业环境与可持续发展研究所所长　研究员

组委会成员：

王丽君　山东省寿光市人民政府副市长

贡锡锋　中国农业科学院国际合作局局长

王述民　中国农业科学院成果转化局局长

郝志强　中国农业科学院农业环境与可持续发展研究所党委书记　研究员

杨其长　中国农业科学院农业环境与可持续发展研究所研究员

王启龙　山东省寿光市蔬菜高科技示范园管理处主任

陈青云　中国农业大学教授　中国农业工程学会设施园艺工程专委会主任

蒋卫杰　中国农业科学院蔬菜花卉研究所研究员　中国园艺学会设施园艺分会会长

王乐义　山东省寿光市三元朱村党支部书记

隋申利　中国（寿光）国际蔬菜科技博览会组委会办公室常务副主任

赵利华　山东省寿光市蔬菜高科技示范园管理处副主任

程瑞锋　中国农业科学院农业环境与可持续发展研究所环境工程室主任

Organizing Committee of 2018 the Fifth International Forum on Protected Horticulture（Shouguang.China）

Chairman of the Organizing Committee

Wu Kongming（Vice President of Chinese Agricultural Academy of Sciences）

Zhao Xuchun（Mayor of The People's Government of Shouguang City）

Vice-Chairman of the Organizing Committee

Sun Xiuwei（Vice Secretary of The People's Government of Shouguang City）

Zhang Yanqing（Director of Institute of Environment and Sustainable Development in Agri\ culture, CAAS）

Members of the organizing committee

Wang Lijun（Vice mayor of The People's Government of Shouguang City）

Gong Xifeng（Director of the Bureau of International Cooperation, CAAS）

Wang Shumin（Director of the Bureau of Transforming Scientific Results into Practice, CAAS）

Hao Zhiqiang（Party secretary of Institute of Environment and Sustainable Development in Agriculture, CAAS）

Yang Qichang（Professor of Institute of Environment and Sustainable Development in Agriculture, CAAS）

Wang Qilong（Director of International Vegetable Science and Technology Expo Organizing Committee Office in Shouguang, Shandong）

Chen Qingyun（Professor of China Agricultural University, Director of Chinese Society of Agricultural Engineering Horticultural Engineering of Special Committees）

Jiang Weijie（Professor of Institute of Vegetables and Flowers, Chinese Academy of Agricultural Sciences, Director of Chinese Society of Horticultural Science, Horticultural Engineering）

Wang Ley（Secretary of Sanyuanzhu Village in Shouguang City, Shandong Province）

Sui Shenli（Vice- Director of International Vegetable Science and Technology Expo Organizing Committee Office in Shouguang, Shandong）

Zhao Lihua（Vice-Director of Vegetable Hi-Tech Demonstration Park Office in Shouguang, Shandong）

Cheng Ruifeng（Director of Research Laboratory of Environmental Engineering in Institute of Environment and Sustainable Development in Agriculture, CAAS）

2018第五届中国·寿光国际设施园艺研讨会学术委员会名单

学术委员会主席：

Kozai Toyoki教授	日本千叶大学　日本植物工厂联盟主席
杨其长研究员	中国农业科学院农业环境与可持续发展研究所
Gerard P. A. Bot教授	荷兰瓦赫宁根大学

学术委员会成员（按姓氏字母顺序）：

艾希珍教授	山东农业大学
Bournet Pierre-Emmanuel	法国国立路桥大学
白义奎教授	沈阳农业大学
别之龙教授	华中农业大学
Katsoulas Nikolaos教授	希腊塞萨洛尼基大学
崔瑾教授	南京农业大学
崔世茂教授	内蒙古农业大学
陈超教授	北京工业大学
陈青云教授	中国农业大学
方炜教授	台湾大学
郭世荣教授	南京农业大学
郭文忠教授	北京市农林科学院
黄丹枫教授	上海交通大学
Hikosaka Shoko教授	日本千叶大学
Hayashi Eri部长	日本植物工厂联盟国际部
贺超兴研究员	中国农业科学院蔬菜花卉研究所
郝秀明教授	加拿大温室与加工作物研究中心
蒋卫杰教授	中国农业科学院蔬菜花卉研究所
李天来院士	沈阳农业大学
李亚灵教授	山西农业大学
李建设教授	宁夏大学
李建明教授	西北农林科技大学
刘厚诚教授	华南农业大学
刘红教授	北京航空航天大学
刘士哲教授	华南农业大学
刘滨疆研究员	大连市农业机械化研究所
陆春贵教授	英国诺丁汉大学
罗卫红教授	南京农业大学
毛罕平教授	江苏大学

马承伟教授	中国农业大学
钮根花教授	美国德州农工大学
乔晓军研究员	北京市农林科学院
宋卫堂教授	中国农业大学
孙忠富研究员	中国农业科学院农业环境与可持续发展研究所
孙治强教授	河南农业大学
史庆华教授	山东农业大学
Takayama Kotato教授	日本爱媛大学
Tom Dueck教授	荷兰瓦赫宁根大学
佟国红教授	沈阳农业大学
仝宇欣副研究员	中国农业科学院农业环境与可持续发展研究所
王秀峰教授	山东农业大学
魏珉教授	山东农业大学
温祥珍教授	山西农业大学
武占会研究员	北京市农林科学院
须晖教授	沈阳农业大学
肖玉兰教授	首都师范大学
喻景权教授	浙江大学
于贤昌教授	中国农业科学院蔬菜花卉研究所
张志斌研究员	中国农业科学院蔬菜花卉研究所
张振贤教授	中国农业大学
郑宏飞教授	北京理工大学
周长吉研究员	农业部规划设计研究院
邹志荣教授	西北农林科技大学
赵淑梅副教授	中国农业大学

Scientific Committee of 2018 the Fifth International Forum on Protected Horticulture（Shouguang.China）

Chairman of the Scientific Committee

Kozai Toyoki, Professor	Chiba University in Japan　Plant Factory Association in Japan
Yang Qichang, Professor	Institute of Environment and Sustainable Development in Agriculture, CAAS
Gerard P. A. Bot, Professor	Wageningen University

Member of Scientific Committee

Ai Xizhen, Professor	Shandong Agricultural University
Bournet Pierre-Emmanuel, Professor	Agrocampus Ouest
Bai Yikui, Professor	Shenyang Agricultural University
Bie Zhilong, Professor	Huazhong Agricultural University
Katsoulas Nikolaos, Professor	University of Thessaloniki
Cui Jin, Professor	Nanjing Agricultural University
Cui Shimao, Professor	Agricultural University of The Inner Mongol
Chen Chao, Professor	Beijing University of Technology
Chen Qingyun, Professor	China Agricultural University
Fang Wei, Professor	National Taiwan University
Guo Shirong, Professor	Nanjing Agricultural University
Guo Wenzhong, Professor	Beijing Academy of Agriculture and forestry Sciences
Huang Danfeng, Professor	Shanghai Jiao Tong University
Hikosaka Shoko, Professor	Chiba University in Japan
Hayashi Eri	Plant Factory Association in Japan, Division Head of International Relations
He Chaoxing, Professor	Institute of Vegetables and Flowers, CAAS
Hao Xiuming, Professor	Canada Greenhouse and Processing Crops Research Center
Jiang Weijie, Professor	Institute of Vegetables and Flowers, CAAS
Li Tianlai, Academician	Shenyang Agricultural University
Li Yaling, Professor	Shanxi Agricultural University
Li Jianshe, Professor	Ningxia University
Li Jianming, Professor	Northwest Agriculture and Forestry University of Science and Technology

前　言

PREFACE

自2009年以来，借助中国（寿光）国际蔬菜科技博览会平台，中国农业科学院和寿光市人民政府先后于2009年、2011年、2013年、2015年在博览会期间连续举办了四届“中国·寿光国际设施园艺高层学术论坛”（High-level International Forum on Protected Horticulture，HIFPH）。论坛主要以“低碳、节能、高效、安全”为主题，汇聚了国内外数十位知名设施园艺专家以及数百位参会代表，就设施园艺科技进展、节能与新能源利用、温室结构工程、环境模拟与优化控制、LED光源、设施高效栽培、植物工厂技术等专题进行深入研讨，论坛取得了圆满成功。作为论坛的成果之一，四届论坛还汇集与会专家的200余篇论文，正式编辑出版了四本论文专集，受到业内同行的广泛关注。

近年来，设施园艺产业发展迅速,为改善城乡居民的生活质量、增加农民收入做出了重要贡献。但是随着耕地的不断减少、化石能源的日益紧缺、劳动力成本的持续上升以及人们对食品安全的高度关注，设施园艺产业也面临着诸多亟待解决的难题。如何利用现代科技成果解决设施园艺生产中面临的资源、环境与可持续发展问题，是摆在世界设施园艺专家面前的重大课题。为此，本届论坛选择以“节能、高效、绿色、智能”为主题，邀请了来自美国、荷兰、日本、法国、希腊以及国内的40余位知名专家作大会主题报告和专题报告，并围绕设施结构工程、环境模拟与调控、高效栽培、节能与新能源利用、LED光源、新型材料与装备、太空农业、物联网技术以及植物工厂等热点内容进行交流与研讨，探讨实现设施园艺绿色安全、提质增效的技术途径，并汇集与会专家的50余篇论文，正式编辑出版。

在论坛组织过程中，得到了中国农业科学院、山东省寿光市人民政府、荷兰瓦赫宁根大学（Wageningen University）、日本千叶大学（Chiba University）、山东省农业大学、山东省农业科学院、中国园艺学会设施园艺分会、中国农业工程学会设施园艺工程专业委员会等单位的大力支持，在此表示衷心感谢。本书由中国农业科学院科技创新工程（农业环境与可持续发展研究所“设施植物工程”团队）和“863”课题（2013AA103004）资助出版。

由于时间仓促，论文集难免会有错漏之处，恳请各位同仁和读者批评指正。

编　者

2018年3月

目　录

CONTENTS

综　述
REVIEW PAPER

设施园艺工程技术
PROTECTED HORTICULTURE ENGINEERING TECHNOLOGY

设施栽培理论和技术
ROTECTED CULTIVATION THEORY AND TECHNOLOGY

综　述

REVIEW PAPER

Towards Sustainable Smart Plant Factories with LEDs, Artificial Intelligence and Phenotyping Unit

Kozai Toyoki

(*Japan Plant Factory Association*, *Kashiwa-no-ha*, *Kashiwa*, *Chiba* 277-0882, *Japan*)

Abstract: The benefits, unsolved problems and challenges for plant factories with artificial lighting (PFALs) are discussed. The remarkable benefits are high resource use efficiency, high annual productivity per unit land area, and production of high-quality plants without using pesticides. Major unsolved problems are high initial investment, electricity cost and labor cost. A major challenge for the next-generation smart PFAL is the introduction of advanced technologies such as artificial intelligence with the use of big data, genomics and phenomics (or methodologies and protocols for noninvasive measurement of plant-specific traits related to plant structure and function). Finally, the concept and methods of plant cohort research for production in PFALs are discussed. In plant cohort research, the time courses of plant traits (i.e., phenomes) are measured for plant individuals together with the data on environments, human and machine interventions, and resource inputs and outputs. The components of the plant cohort research system for a PFAL include the cultivation system module (CSM), phenotyping unit, database or data warehouse and application software for the collection, analysis and visualization of the data. Possible applications of the plant cohort research system are very briefly discussed.

Key words: Artificial intelligence (AI); Big data; Light emitting diodes (LEDs); Phenotyping

1 Introduction

We are facing a trilemma in which there are three almost equally undesirable alternatives:①shortage and/or unstable supply of food; ②shortage of resources; and ③degradation of the environment. This trilemma is occurring at the global as well as local and national level amid an increasing urban population and a decreasing and/or aging agricultural population.

To help solve this trilemma, transdisciplinary methodologies based on new concepts need to be developed by which the yield and quality of food are substantially improved with less resource consumption and environmental degradation compared to current plant production systems. Plant factories with artificial lighting (PFALs) are one such system expected to achieve this mission. The benefits of the PFAL include high resource use efficiency, high annual productivity per unit land area, and production of high-quality plants without using pesticides (Kozai et al., 2015). Current major problems to be solved are the high initial investment and the electricity and labor costs.

The next-generation 'smart' plant factories with artificial lighting (smart PFALs) are expected to help solve food, resource and environment issues concurrently, by significantly reducing the initial and operation costs. This article is a reduced version of Kozai (2018a) and Kozai (2018b).

2 Potential and Actualized Benefits of the PFAL (Kozai, 2018a; Kozai, 2018b)

The PFALs that have actualized most of the potential benefits described below are making a profit and expanding their production capacity; however, the number of such profitable PFALs is currently limited. To actualize the potential benefits of the PFAL, the concepts behind the benefits and the methodology for their actualization must be thoroughly understood before designing and operating the PFAL. At the same time, the vision, mission and goals of the PFAL design, operation and business model must be clearly established and shared with the team members.

(1) High Resource Use Efficiency (RUE) which reduces resource consumption and waste, and thus lowers production costs. RUE is defined as the amount of resource fixed or utilized in the plants (F) divided by the amount of resource supplied to the PFAL (S) (Kozai et al., 2015). Essential resources to be supplied regularly to grow plants in the PFAL are light energy, CO_2, water, fertilizer (nutrients), seeds/transplants and labor only. The light energy, CO_2, water and fertilizer (nutrients) are essential resources for seeds/transplants to grow by photosynthesis (Kozai, 2013).

The first, Water use efficiency (WUE) is the amount of water fixed or held in the plants divided by the net amount of water supplied to the culture beds and absorbed by the plant roots. In the airtight PFAL, almost all the transpired water is condensed and collected at the cooling panels of the air conditioners and returned to the nutrient solution tank. Then, the net consumption of water is the difference in amount between the irrigated water and the water returned to the nutrient solution tank. WUE for the PFAL is around 0.95 (= the fresh weight of plants divided by the difference in weight between the irrigated water and the water returned to the nutrient solution tank) (Kozai, 2013; Kozai et al., 2015). This high WUE (95% water saving) of the PFAL is a big advantage in arid regions and other water-scarce areas.

The second, CO_2 and fertilizer efficiencies of the PFAL are also relatively high (0.80 ~ 0.90) compared with those of CO_2-enriched and soil-cultivation greenhouses with ventilators closed (0.5–0.6 for both) (Kozai et al., 2015).

The third, Light energy use efficiency of the PFAL, however, is still very low (0.032 ~ 0.043), although it is higher than that of the greenhouse (0.017) (Kozai et al., 2015). Use efficiencies of electric energy, light energy, space and labor in the current PFALs need to be considerably improved in the next-generation PFALs through the application of LEDs, intelligent lighting systems, better environmental control, and the introduction of new cultivars that grow well under low photosynthetic photon flux densities (PPFD).

(2) High annual productivity per unit land area. Over 100-fold annual productivity per unit

land area can be achieved without the use of pesticides, compared with the annual productivity per unit land area in open fields, mainly due to the use of multilayers (10 tiers on average), shortened cultivation period (often by half) by optimal environmental control, high land area use efficiency (no vacant cultivation space throughout the year), high planting density and virtually no damage by weather and pest insects. It would be interesting to estimate the maximum annual productivity of the PFAL under the optimal environment using a simulation model.

(3) Land area use efficiency is defined as ($A_u \times n \times N$) divided by ($365 \times A_t$) where A_u is the area of a unit cultivation space, n is the number of units of cultivation space in the PFAL, N is the average number of days per year during which the unit cultivation space is occupied by plants under cultivation, and A_t is the land area occupied by the floor of the PFAL. The unit cultivation space can be a cultivation panel, a tier or a rack consisting of more than one tier.

(4) Major components of the production cost in the PFAL are electricity, labor and depreciation of initial investment (the sum of these three cost components account for 75% ~ 80% of the total production cost). Thus, electric energy productivity (kg of produce per kWh of electricity consumption), labor productivity (kg of produce per labor hour), and space productivity (kg of produce per floor area or cultivation area) are important indices for analyzing and improving the productivity of the PFAL.

(5) High weight percentage of marketable parts over the whole plant biomass. In other words, a low percentage of trimmed/damaged plant parts as waste over the whole plant biomass. The percentage can be increased by the proper environmental control method, cultivation system and cultivar selection. Currently, the fresh weight percentages of marketable leafy parts and trimmed leafy or root parts of leaf lettuce plants are, respectively, estimated to be 77% ~ 80% and 20% ~ 23% in most PFALs in Japan. Leafy parts of root crops such as carrot, turnip and radish need to be edible, tasty, nutritious and presentable to improve the weight percentage of the marketable parts of root crops.

(6) High-quality plants can be produced as scheduled by proper environmental control, cultivar selection, and cultivation system (Kozai et al., 2016). Shape/appearance, taste, and mouth sensation, as well as composition/contents of functional components such as vitamins, polyphenols and minerals can be controlled. However, most of such factors are still controlled by trial-and-error based on past experiences. On the other hand, researchers are systematically conducting a series of experiments to produce consistent functional components.

(7) High controllability of plant environment. Controlled aerial environmental factors include PPFD, VPD (water vapor pressure deficit), air temperature, CO_2 concentration, light quality (spectral distribution), lighting cycle (photoperiod/dark period) and air current speed. Controlled hydroponic culture factors include strength, composition, temperature, pH (potential of hydrogen), dissolved O_2 concentration and flow rate of the nutrient solution.

(8) High reproducibility and predictability of yield and quality. Because of the high

controllability of the environment all year round, scheduled and/or on-demand plant production is possible regardless of the weather. The quality of produce and the yield can be controlled by controlling the environment. For example, mouth sensation, taste, color and flavor of lettuce can be finely controlled to suit its use in salads, sandwiches or hamburgers by environmental control.

(9) High traceability throughout the supply chain of PFAL industry, which enables a high level of risk management.

(10) High adaptability for the location (near or in food/meal delivery shops, etc.). The PFAL can be built without any problems in shaded areas, on contaminated or infertile soil, and in vacant rooms/buildings/land in urban areas. The PFAL can also be built in very cold, arid or hot areas. For example, there are no heating costs for the PFAL even when the outside air temperature is below −40℃ because the walls and floor are thermally well insulated and heat is generated by the lamps in the culture room. The PFAL is most suited to urban areas where the production site is close to the consumption site (local production for local consumption). This saves fuel, time and labor for transportation of fresh produce, and creates job opportunities for handicapped, elderly and young people in or near their residential area.

(11) High controllability of sanitary conditions. Because of this high controllability, pesticide-free and other contaminant-free plants are produced. Global GAP (Good Agricultural Practice) and/or HACCP (Hazard Analysis and Critical Control Point) can be introduced relatively easily, and so a high level of risk management can be achieved.

(12) Long shelf life due to low CFU (colony forming units of microorganisms) per gram, which decreases the amount of vegetable garbage or loss at home and in shops. The shelf life is estimated to be around two times longer for lettuce plants grown in the PFAL than for those grown in the field. Due to this advantage, the market price of PFAL-grown vegetables is often 20% ~ 30% higher than that of field-grown and greenhouse-grown vegetables.

(13) No need to wash or cook before serving, if packed in a sealed package after harvest in the culture room. This reduces the consumption of water for washing, electricity/city gas/fuel for boiling and stir-frying, and labor for washing and cooking. On the other hand, when eating fresh vegetables, the CFU per gram of vegetable needs to be lower than around 300.

(14) Easy measurement of hourly and/or daily rates of resource supply, production and waste. The RUEs can be estimated online. Based on the estimation, the production costs can be predicted and subsequently reduced using the data on RUEs.

(15) Stepwise improvement of the RUEs, productivity and economic value of the plants is possible by visualizing the flow of energy, substances and workers in the production process and other related costs/sales using the measured data. To do so, models of plant growth, energy/substance balance and production process scheduling are necessary.

(16) Light and safe work under comfortable air temperature and moderate air movement. There are still some problems to be solved to further improve the working environment both for

large-scale PFALs with automation and small-scale PFALs in order to increase job opportunities.

（17）Design and environment control are simpler in the PFAL than in the greenhouse due to its airtightness, high thermal insulation of walls and floor and no solar light transmission to the cultivation space. Global standardization of the PFAL design （except for building design）is easier than that of the greenhouse design.

（18）A small PFAL with a floor area of 0.1 ~ 10 m^2 is a wonderful way to learn the principles of life science, engineering and technology at home, school or a community center, especially when the PFAL is connected via the Internet to other small PFALs and a PFAL database for the exchange of information and opinions （Harper and Siller 2015）.

3 Current Unsolved Problems of PFALs（Kozai, 2018a, 2018b）.

3.1 Actions Required for Solving the Problems

（1）Drastic reduction in initial investment and operation costs. The operation cost per kg of fresh produce needs to be reduced by 30% ~ 50%, and the current initial cost per annual production capacity needs to be reduced by about 30% by the year 2020—2022, compared with the costs in 2017.

（2）Sustainable production. Improvement of the cultivation system and its operation to reduce, recycle and reuse resources, and use natural energy is essential. Energy-autonomous PFALs need to be designed, operated and commercialized.

（3）Advanced technologies including artificial intelligence （AI）, big data, the Internet of Things （IoT）, bioinformatics, genomics and phenomics need to be introduced to improve the resource use efficiency and cost performance of the PFAL. Phenomics is an emerging research field along with the methodologies and protocols for the noninvasive measurement of cellular- to canopy-level plant-specific traits related to plant structure and function. Flexible robotic automation needs to be introduced to reduce the amount of heavy, dangerous, simple and/or troublesome manual work.

（4）Medicinal plants for high-quality health care and cosmetics products need to be produced at low cost. Genetically engineered plants for production of pharmaceuticals such as vaccines for influenza and other viruses need to be produced in a specially designed PFAL.

（5）Worldwide active organizations of plant factory industries and academics need to be established for better global communication and information sharing.

（6）Organic hydroponic systems for PFALs that are easy to handle and economically viable need to be developed. Symbiosis of plants with microorganisms will benefit plant growth in the PFAL. Organic fertilizer can be produced from fish waste, vegetable garbage, mushroom waste and other types of biomass.

3.2 Some Specific Technical Problems

（1）Efficient use of white LEDs, which emit a significant amount of green light

（20% ~ 40% of total light energy）. The optimal spectral distribution of white LEDs to meet a specific requirement is still unknown.

（2）Green light effect on photosynthesis, growth, development, secondary metabolite production, disease resistance and human health in the PFAL has become an emerging research topic in relation to white LEDs.

（3）Net photosynthesis, transpiration and dark respiration of plants in the PFAL can be continuously measured. Efficient methods for utilizing this data need to be developed.

（4）Energy and mass （substance）balance in the PFAL can be continuously measured. Efficient methods for utilizing this data need to be developed.

（5）Resource use efficiency （RUE）and cost performance of the PFAL can be measured, visualized and controlled. Efficient methods for utilizing this data need to be developed.

（6）Algae growth inhibition in hydroponics. Also, the occurrence of intumescence （or edema）and tip burn symptoms on the leaves of leafy vegetables needs to be inhibited by proper environmental control and cultivar selection.

（7）Microbiological ecosystems in the culture beds are currently unknown and uncontrolled. Organic acids produced by plant roots, dead roots, dead and living algae and many kinds of microorganisms including pathogens should be present in the culture beds. Beneficial and stable microbiological ecosystems need to be established.

3.3 Actions Required for Enhancing PFAL R&D and Business

（1）Rational, powerful and clear messages on the vision, mission and goals of the PFAL. People are increasingly interested in the potential benefits of the PFAL, and are expecting further progress in R&D on smart PFALs.

（2）Open database and open business planning and management system.

（3）Human resource development for PFAL managers and workers. Human resource development programs for capacity building of PFAL managers are crucially needed. Well-edited books, manuals and guidelines need to be published. Software/hardware systems for managing the complicated cause-and-effect relationships in the PFAL need to be developed.

（4）Tools, facilities and guidelines for worker safety, labor saving and quality operations. Compact and safe systems for efficient seeding, transplanting, harvesting, transporting and packaging.

（5）Software with database for minimizing the electricity costs for lighting and air conditioning under a given lighting schedule.

（6）Software with database for 'smart' environmental control for production of targeted functional components of a plant species under economic, botanical and engineering constraints. A computer-assisted support system for sensing, data analysis, control, visualization and decision-making needs to be developed.

（7）Well-designed floor plan and equipment layout to maximize labor and space productivity.

（8）Marketing for creating new markets for health care. New products that do not compete with currently used products are necessary.

（9）Understanding and anticipation by local residents regarding the potential of the PFAL. Actual working/virtual models showing people the future are necessary.

（10）Breeding the plant cultivars for the PFAL. The characteristics of plants suited to cultivation in the PFAL are: ①fast growth under relatively low PPFD，high CO_2 concentration and high planting density，②fast growth under low stresses of water，temperature and pest insects/pathogens，③fast growth without physiological disorders，④secondary metabolite production sensitive to environmental conditions or stresses，and ⑤high economic value per kg of produce due to qualitative plant traits. Molecular breeding can be a powerful tool.

（11）Guidelines and manuals for sanitary control，food and worker safety，and LED lighting.

（12）Standardization of terminology and units for the basic properties of light，lamps and nutrient solution.

（13）Standardization of PFAL components.

3.4 Challenges for the Smart PFAL

Challenges for developing the software/hardware units or systems to be implemented in the smart PFAL as the next-generation PFAL include:

（1）The PFAL as an essential unit to be integrated with other biological systems to improve the sustainability of a building or a city.

（2）The PFAL for large-scale production and breeding of high-wire tomatoes and other fruit-vegetables and berries such as strawberry and blueberry.

（3）Cultivation system module（or unit）as a minimum component of the plant cultivation space in the PFAL，which is easily connected to other basic module units to make a larger PFAL.

（4）Hydroponic system without the use of substrate（supports）and a nutrient solution circulation unit without the drainage of nutrient solution from the culture beds. Then，the total volume of nutrient solution in the culture beds，piping and nutrient solution unit is greatly reduced.

（5）Phenotyping unit for continuous and nondestructive（or noninvasive）measurement of plant traits such as fresh weight，leaf area，number of leaves，leaf angle，three-dimensional plant community architecture，leaf surface temperature，optical properties and chemical components of the plants and physiological disorders such as tip burn and intumescence（or edema）of the leaves. The measured data on plant traits is used as input data for the phenome-genome-environment-human/machine interventions model to determine the setpoints of environmental factors and/or the selection of elite plants for breeding.

（6）Periodic movement of plants due to circadian rhythms（biological clock），water stress，air current patterns，etc.，and their effects on hormonal balance，photosynthesis，transpiration and growth of plants.

（7）Smart LED lighting unit for maximizing the cost performance（product of the unit economic value and the yield of produce divided by the operating cost）by time-dependent control of light environment factors such as light quality，photosynthetic photon flux density（PPFD），lighting cycle（photo-/dark period）and lighting direction.

（8）Ion concentration control unit for the hydroponic system. The concentration of each major ion type（NO_3^-、K^+、Mg^{2+}、Ca^{2+}、Na^+、NH_4^+、Mg^{2+}、Cl^-、PO_4^{3-} and SO_4^{2-}）in the nutrient solution is separately measured or estimated and controlled.

（9）Software unit for discriminating the effects of spatial variations of the environment on the spatial variations of individual plant growth from the effects of genetic variations among the plants on the spatial variations of individual plant growth. The spatial variation of plant growth is obtained from the data measured by the phenotyping unit.

（10）Hardware/software unit for minimizing the spatial variations of the air temperature，vapor pressure deficit（VPD）and air current speed by controlling the spatial distributions of PPFD and air current speed under the given LED lighting system and three-dimensional plant canopy architecture. The spatial variations of the environment are obtained from the distributed environmental sensors.

（11）Software unit for automatically determining the setpoints of environmental factors to meet the objectives of PFAL operation under given constraints，using the phenome data and other data.

（12）Deep learning unit for searching for a function（G，E，M）or relationship among the phenome，genome and environment datasets. P = function（G，E，M）where P is phenome data，G is genome data，E is environment data and M is management（human and machine interventions）data. Using the big datasets of P，G，E and M，the function（G，E，M）is found by deep learning. In the PFAL with a controlled environment，the datasets of P，E and M can be collected relatively accurately and easily. Thus，the genome dataset is known，and the function（G，E，M）can be found relatively easily by deep learning.

（13）Software/hardware unit for searching for DNA expressions/markers driven by environmental changes，using the genome，phenome and environment datasets，and for determining the setpoints of environmental factors using the above dataset. Deep learning unit using the phenome，genome and environment datasets would become a powerful breeding tool.

（14）Integration of the deep learning model with mechanistic and multivariate statistical models.

（15）Speed breeding Watson et al.（2017）proposed a method of ‘speed breeding’ which greatly shortens generation time and accelerates breeding and research programs. They envisage great potential for integration speed breeding with other crop breeding technologies，including high-throughput genotyping，genome editing and genomic selection，for accelerating the rate of crop improvement.

（16）Dual （virtual/actual）PFAL，a pair of virtual and actual PFALs. The virtual PFAL placed in the cloud is used to simulate the actual PFAL output using the data input to the actual PFAL. The parameter values in the virtual PFAL are adjusted automatically by using the input and output data of the actual PFAL. The virtual PFAL can be used for training，self-learning，education，fun，and research and development of the PFAL.

4 Plant Cohort Research and Its Application （Kozai，2018b）

4.1 Introduction

The phenotyping unit mentioned above is used to reveal the plant trait dynamics in a noninvasive manner. Plant cohort research for PFALs in Japan started at the end of 2017 as part of a research project on phenotyping- and artificial intelligence （AI）-based environmental control and breeding for the PFAL. A special feature of the plant cohort research is that the lifecycle phenome history of an individual plant and its related data are continuously measured and analyzed for all the plants in the PFAL.

Using recent advanced technologies related to phenotyping and big data mining，cohort research can bc conducted for plant production and breeding in the PFAL. Plant cohort means a relatively large group of plant individuals having a statistical factor （such as cultivar，planting density and cultivation system）in common with the plants in a PFAL for commercial plant production and/or breeding. The plant cohort research can be conducted for multiple purposes in a PFAL. For example，plant cohort research aiming to prove yield and quality by environmental control and breeding can be conducted concurrently using the same set of big data.

4.2 Causes of the Variation in Plant Traits and Separation of the Causes

The variations in the traits of plants sown at the same time and place are attributable to four factors: ①environment，②genome （species，cultivar，DNA markers），③human intervention，and ④machine intervention. Human and/or machine intervention，which can cause a positive or negative change in the plant traits，includes seed priming，seed coating，seed selection，seeding，transplanting or spacing，harvesting，cleaning/washing，and maintenance. A physiological disorder of a plant is not considered to be caused by a disease but instead is considered as a plant trait. Variations in plant traits that cannot be explained by environmental variations or human and/or machine interventions are possibly due to the genetic variations of seeds，which should be random in terms of the position of the plants over the trays. Plants with an extreme genetic variation can be used for breeding purposes.

4.3 Application Software for the Plant Cohort Research

To make full use of the big data obtained，application software such as the ones listed below needs to be developed. Easiest but useful software to develop include: ①Three-dimensional structure of plant canopy，②Vertical distribution of leaf area，number of leaves and PPFD in the plant canopy，③Ratio of photosynthetic photons received by plants to those emitted by lamps，

④Vertical distribution of air current and diffusion coefficient in the plant canopy, ⑤Vertical distribution of net photosynthetic and transpiration rates, ⑥Alarm system for malfunctions, accidents and damage of plants and equipment, ⑦Alarm system for malfunctions, accidents and damage of plants and equipment, ⑧Tracking the position of plant individuals in the PFAL, ⑨Tracking the growth of plant individuals in the PFAL, ⑩Averaged growth rates of leaf area, height, fresh weight and their variances, ⑪Histograms of fresh weight of produce, trimmed leaves and roots, and their causes, ⑫Hourly resource inputs (supply rates), their cost and resource use efficiencies, ⑬Motion analysis of workers, equipment and tools, and motion improvements, and ⑭Location (s) of tipburn spots on leaves, and its causes and solution. Software with higher function than above to be developed include: ①Plant growth curves under various conditions, ②Parameterization of mechanistic, statistic and AI models using actual and predicted resource inputs and output rates, ③Determination of the dates of first and second transplanting to maximize the cost performance of the PFAL, ④Hourly resource output (production) rates of produce, waste, heat energy, etc., ⑤Prediction of physiological disorders, spread of insects and microorganisms, ⑥Vertical distribution of specific secondary metabolites in the plant canopy, ⑦Costs and benefits relationships under various growth curves, ⑧Plant production schedule to maximize the cost performance of the PFAL, and ⑨Prediction of plant canopy structure and secondary metabolite production.

Plant cohort research is a new area in PFAL research, and few experimental results have been obtained so far. However, such research has the potential to reveal the relationships between plant phenomes, genomic traits, resource inputs/outputs and human/machine interventions, and to improve plant productivity in terms of electric energy, cultivation area and labor hours.

5 Conclusion

To actualize the potential benefits of the PFAL, a considerable amount of systematic research, development and marketing with the appropriate vision, mission, strategy and methodologies are necessary. On the other hand, actualization of the potential benefits is relatively easy. Because, the energy and material balance and the plant-environment relationship in the PFAL are much simpler than in the greenhouse, and the plant environment is no affected by the weather outside. Thus, the methods for actualizing the benefits are relatively straightforward. Plant cohort research for plant production in the PFALs will be a useful tool for production and breeding of plants in the PFALs.

6 Acknowledgements

This article is partially based on results obtained from a project commissioned by the New Energy and Industrial Technology Development Organization (NEDO). The authors express their deep appreciation to all the members, especially to Na Lu, Rikuo Hasegawa, Osamu

Nunomura, Tomomi Nozaki, Yumiko Amagai, Eri Hayashi, Toru Maruo and Toshitaka Yamaguchi, of Chiba University and/or the Japan Plant Factory Association for their technical and administrative support.

References

Harper C and M Siller. 2015. OpenAg: a globally distributed network of food computing. Pervasive Computer., 14（4）: 24–27.

Kozai, T. 2013. Resource use efficiency of closed plant production system with artificial light: Concept, estimation and application to plant factory. Proc Jpn Acad Ser B, 89: 447-461.

Kozai T, G Niu and M Takagaki. 2015. Plant factory: an indoor vertical farming system for efficient quality food production. Academic Press. 405 pages.

Kozai T, K Fujiwara and E Runkle. 2016. LED lighting for urban agriculture. Springer. 454 pages.

Kozai, T. 2018a. Benefits, problems and challenges of plant factories with artificial lighting（PFALs）: a short review. Acta Horticulturae（GreenSys 2017, Beijing, China）（in press）.

Kozai T.（ed.）. 2018b. Smart Plant Factory, Springer（in press）.

Watson A, Ghosh S, Williams M J., et al., 2018. Speed breeding is a powerful tool to accelerate crop research and breeding. Nature Plants, （4）: 23-29.

Advances and Challenges in CFD Modelling of Crop-climate Interaction in Greenhouses

P. E. Bournet[1, a]

(1. *Ephor*, *Agrocampus*, *Quest Sfr Quasav*, *Fr Irstv*, 49000, *Angers*, *France*)

Abstract: From a bibliographic analysis of recent CFD studies on greenhouses, it can be seenthat the publications in that field still increase. Among the corresponding published papers however, only half consider the cropwhile only 10% include a real modelling of the crop interaction with local environment.Since the pioneering work of Boulard and Wang (2002) , substantial advances have been reported concerning the modelling of crop interaction with local environment. Studies have progressively included the mechanical strain exerted by the crop on the flow, together with the heat and water vapor exchanges at the level of the leaves. A particular attention has been paid to the radiative transfer modelling, improving the calculation of the distributed radiation inside the crop. Then, authors included specific submodels to take account of water restriction or condensation process. Recent studies considered the CO_2 balance inside the greenhouse, together with the photosynthesis activity.The present paper investigates the fundamentals of the recently developed models used in CFD to assess the crop impact on local climate: crop submodel, radiative submodel, water balance submodel, condensation submodel and CO_2-photosynthesis submodel. Then, an overview of the main conclusions of the corresponding studies is provided, stressing the main advances. Finally, considerationson the current challenges in CFD modelling of crop impact on climate in greenhouses are provided.

Key words: Greenhouse; CFD; Modelling; Plant; Microclimate; Radiation; Condensation; Temperature; Humidity; CO_2; Photosynthesis

1 Introduction

Assessing climate and plant characteristics in greenhouse is of prime interest not only to improve the control of plant growth, but also to improve the design and equipment of greenhouses. In the last decades, thanks to the development of computer facilities and software, it became possible to simulate greenhouse climate characteristics, which offered new opportunities to curtailtime developments and reduce costs. Among the available modelling tools, CFD (Computational Fluid Dynamics) technology has significantly contributed to improve efficiency andcredibility of research results in numerous fields of engineering. Simulation technologies makes it possible to apply environmental conditions with a full consideration of diverse factors.

CFD is a computer-based technique aiming atcharacterizing, interpreting, and quantifying flow phenomena by solvingconservation equations (or extended conservation equations)

E-mail: pierre-emmanuel.bournet@agrocampus-ouest.fr

explaining the phenomena. Governing equations of fluid dynamics are non-linear partial differential equations which cannot be solved using analytical techniques. Thanks to the fast development of computation technology since the 1 990s, it is now possible to describe fluid phenomena by solvingNavier-Stokes equations in volumes surrounding three dimensional shapes and for unsteady conditions.

In the last 3 decades, the CFD tool was widely applied to the greenhouse system, mainly to study the ventilation process in the first stage, and then to assess plant-climate interactions. Based on a short bibliographic investigation, an analysis of the last trends of the peer reviewed studies devoted to greenhouse system modelling with CFD is provided. Then, fundamentals of modelling of crop interaction with local flow and climate are given. A focus on radiative transfers, condensation process and water balance is also proposed. Recent developments concerning the modelling of CO_2concentration and photosynthesis activity are then analyzed. Finally, an overview of corresponding papers and main results is provided, and challenges for the future are addressed.

2 Current Evolution of Published Researches

In order to analyse the current evolution of the published papers on CFD studies in the field of greenhouse climate, a search was performed on the Scopus^{TM}database. The following keywords were retained: (i) greenhouse, and CFD, and not climate change and (ii) greenhouse, and CFD, and crop, and not climate change. Only papers for which the wordgreenhouse was in the title and/or abstract were retained. A period extending only from 2000 to 2017 was considered as very few papers were published on CFD greenhouse issues before. The corresponding histograms are presented in Figure 1 which shows both the number of peer reviewed published papers (distinguishing articles from proceedings) and the number of citations per year. CFD published studies considerably grew from 2000 to 2008 but then tended to stagnate and re-increased in 2017. Proceeding papers remain quite numerous in that field of research (51% proceeding papers *vs* 49% articles for greenhouse + CFD; and 57% proceeding papers *vs* 43% articles for greenhouse + CFD + crop). The papers including a true crop model (drag effect + energy and water balance) are nevertheless quite few and represent only 10% of the total published items.

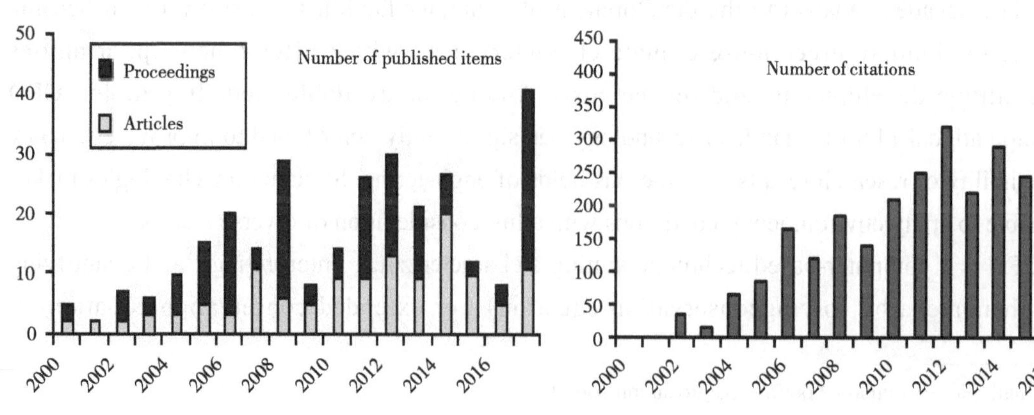

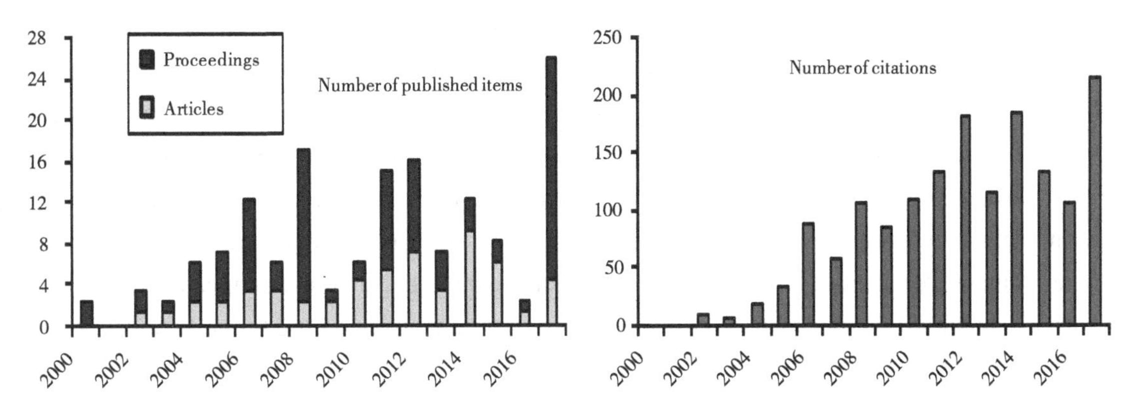

Figure 1 Published items and citation in each year: A: topic=greenhouse and CFD, period 2000—2017 (295 papers/145 articles or reviews) . B: topic=Greenhouse and CFD and crop, (149papers/64 articles or reviews) .

3 Fundamentals of Crop Modelling

3.1 Crop Submodel

The crop interacts with the local ambient conditions in two ways: first it creates a pressure drop and second it exchanges heat and water vapor with the surrounding air. Water vapor fluxes are related to the transpiration process which itself depends on solar radiation, CO_2 concentration, relative humidity and temperatures of the greenhouse air and plant leaves (Stanghellini, 1987) . Basically, the canopy absorbs a part of the incident radiation and if the net absorbed radiation is greater than the latent heat demand, the plant evacuates the excess heat by increasing its temperature while conversely, if the net radiation is lower than the latent heat demand, the plant decreases its temperature.

In CFD studies, the pressure drop is calculated by assimilating the crop to a porous medium in which the Darcy–Forchheimer equation is solved (Molina-Aiz et al., 2004; Teitel et al., 2008) . It is generally assumed that pressure forces cause the main part of the total canopy drag, meaning that the viscous resistance of the crop may be neglected. The sink of momentum due to the drag effect of the crop is then expressed as a function of the squared air velocity. The crop also creates sinks or sources of heat and water vapor, which have to be included as additional terms (sensible and latent heat flux) in the conservation equations for heat and mass of water vapor. The latent heat flux or transpiration is generally obtained directly (so-called direct method) from the local water vapor concentration in the leaves and neighboring air and from the aerodynamic and stomatal resistances (Boulard and Wang, 2002) . These latter parameters are calculated from local conditions: temperature, radiation, relative humidity and/or air velocity. Alternatively, transpiration may be estimated from the Penman-Monteith model as a function of the absorbed net radiation, vapor pressure deficit and aerodynamic and stomatal resistances (Bouhoun Ali et al., 2016a, b, 2017) . In both methods, the knowledge of the radiation distribution inside the canopy is required. The sensible heat flux and therefore the leaf temperature are then calculated from the resolution of an energy balance over each cell of the crop domain. Recently, Bouhoun Ali et al. (2016a, b, 2017) included the possibility to take account of the crop behavior under suboptimal water inputs (all previous studies had been conducted for well-watered plants). Indeed, water restriction may lead to stomatal partial closing and transpiration rate reductions. In their model, they defined a multiplicative function of the water matric potential in the expression of the stomatal resistance, and developed a specific submodel of the water balance over the substrate, linking the water matric potential to the water content according to the Van Genuchten model (van Genuchten, 1980).

As the crop submodel requires the knowledge of the water distribution inside the canopy, a species model in which air and water vapor are treated independently must be activated. This model solves the transport equation for the mass fraction of water vapor (Majdoubi et al., 2009). The latent heat flux from the crop is included as a source term inside this equation.

3.2 Radiative Submodel

Thc distribution of radiation inside the crop intervenes in the energy balance over the leaves. The determination of radiation distribution was a great issue in the literature in the pastand is still a major concern. In most CFD studies, this energy balance over the canopy only involves short wave length radiations as the authors assume that long wave length radiation may be neglected given the temperature distribution inside the canopy is almost uniform. In their study, Boulard and Wang (2002) determined the distribution of solar radiation within a greenhouse tunnel taking account of the path of the sun, greenhouse geometry, cover transmittance and sky conditions. Then, from the calculated incident solar radiation at the top of the crop, and considering a given extinction coefficient, they applied the Beer' s law (radiation decay along the optical path) to estimate the vertical distribution of short wave length radiations inside the crop. Such a method was used by Fatnassi et al. (2003, 2006), Bartzanas et al. (2004), Roy and Boulard (2005) or Majdoubi et al. (2009). These authors however did not explicitly solve the radiative transfers within the crop, andconsequently did not take account of the directional nature of the solar radiation. Fidaros et al. (2010), Nebbali et al. (2012) and Pouillard et al. (2012) were among the first to solve the Radiative Transfer Equation (RTE) inside the crop and thereforeto calculate the solar radiation distribution – inferred from the luminance - inside tomato stands. They defined the absorptivity of plants and conducted unsteady calculation from which they succeeded in increasing the realism of simulations. This approach was also implemented by Morille et al. (2014), Bournet et al. (2017) and Bouhoun Ali et al. (2017).

3.3 Condensation Submodel

To date CFDstudies of greenhouse climate including water vaportransport and changes of phase remain scarce. In their pioneering work on that topic, Tong et al. (2009) took account of condensation by estimating the corresponding latent heat flux from a formulaprovided by Garzoli (1985). More recently, a specific subroutine (Bell, 2003) was implemented in greenhouse CFD studies to take account of the condensationalong the surfaces (Piscia et al., 2012a). The water uptake from the ambient air (which occurs where the local temperature goes below the dew point) wasexpressed as a mass flux sink term in the water vapor conservationequation according to the equation provided by Bird et al. (1960). In their study, Piscia et al. (2012a) however omitted the associated heat flux along the walls which could be obtained by multiplying the mass flux by the latent heat of vaporization, as done by Bouhoun Ali et al. (2014). These source terms were finally applied to the volume along the inner roof or walls surfaces where the condensation phenomenon is assumed to occur. The thickness of this volume was calculated according to the Nusselt theory for flat plates (Bouhoun Ali et al., 2014).

3.4 CO_2-photosynthesis Submodel

A first attempt to include CO_2 cycle and photosynthesis in CFD modelling approach was undertaken by Roy et al. (2014). The photosynthesis process was taken into account by considering the absorption or production of CO_2 of plants as a sink or source term in the transport equation of CO_2, following the formalism proposed by Thornley (1976). The production of CO_2 consists in the maintenance and growth respiration. It was estimated for the case of tomato to 22% of the raw photosynthesis consumption. A similar approach was implemented by Molina-Aiz et al. (2017) who computed the volumetric net photosynthesis as difference between canopy photosynthesis and crop respiration on the basis of the photosynthesis model of Acock et al. (1976) modified by Nederhoff and Vegter (1994). Boulard et al. (2017) used the model of Thornley (1976) for gross photosysnthesis combined with the growth and development model for greenhouse tomato crops proposed by Heuvelink (1999) to calculatethe maintenance and growth respirationswhich were then subtracted from gross photosynthesis to obtain tomato net photosynthesis.

4 Overview of Studies Using a Crop Submodel

In CFD studies, the inclusion of crop effects requires the development of a user-defined-function (UDF). Boulard and Wang (2002) were the first to implement this model for lettuces. It was then applied by Bartzanas et al. (2004), Fatnassi et al. (2003, 2015), Majdoubi et al. (2009, 2016), Fidaros et al. (2010), Nebbali et al. (2012) and Wang et al. (2013) to tomatoes, by Fatnassi et al. (2006) to roses, by Kichah et al. (2011, 2012), Morille et al. (2012), Bournet et al. (2017) and Bouhoun Ali et al. (2014, 2016, 2017) to New Guinea Impatiens, and by Chen et al. (2015) to Begonia (Table 1).

Table 1 Characteristics of the CFD studies including a crop submodel.

Authors	Greenhouse	Dimen-sion	Crop model characteristics (and other submodels)	Validation
Boulard and Wang (2002)	Tunnel	3D Steady	Crop submodel (porous medium + sensible and latent heat flux) Specific model for radiation distribution	Transmittance Air velocity Temperature Transpiration flux
Fatnassi et al. (2003)	Moroccan type	3D Steady	Crop submodel (porous medium + sensible and latent heat flux) Specific model for radiation distribution	Ventilation rate
Bartzanas et al. (2004)	Tunnel	2D/3D Steady	Crop submodel (porous medium + sensible and latent heat flux)	Air velocity Ventilation rate Temperature
Roy and Boulard (2005)	Tunnel	3D Steady	Crop submodel (porous medium + sensible and latent heat flux)	
Fatnassi et al. (2006)	Multi span		Crop submodel (porous medium + sensible and latent heat flux) Specific model for radiation distribution	Ventilation rate
Majdoubi et al. (2009)	Canary type greenhouse	3D Steady	Crop submodel (porous medium + sensible and latent heat flux) Specific model for radiation distribution	Temperature Relative Humidity
Tong et al. (2009)	Chinese greenhouse	2D Unsteady	No crop Condensation	Air temperature
Nebbali et al. (2012)	Tunnel	3D Unsteady	Crop submodel (porous medium + sensible and latent heat flux) Resolution of RTE distinguishing short and long wave radiation	no
Piscia et al. (2012a, b)	4-span plastic greenhouse	3D Unsteady	No crop Resolution of RTE distinguishing short and long wave radiation Condensation	Air temperature Roof temperature Humidity ratio
Pouillard et al. (2012)	Energy saving closed greenhouse	3D	Crop submodel (porous medium + sensible and latent heat flux) Resolution of RTE distinguishing short and long wave radiation	Air temperature Leaf temperature Temperature Saturated humidity at leaf temperature Air humidity

continued

Authors	Greenhouse	Dimen-sion	Crop model characteristics (and other submodels)	Validation
Tamimi et al. (2013)	Arch type greenhouse	3D steady	Crop submodel (porous medium + sensible and latent heat flux) Resolution of RTE Discrete phase model for cooling of fogging system	Air temperature Air humidity
Roy et al. (2014)	Venlo glasshouse	2D Unsteady	Crop submodel (porous medium + sensible and latent heat flux) Resolution of RTE distinguishing short and long wave radiation Submodel for the absorption or production of CO_2	CO_2 rates in the crop CO_2 net consumption in the crop
Fatnassi et al. (2015)	Photovoltaic greenhouse Venlo greenhouse Asymmetric greenhouse	3D Steady	Crop submodel (porous medium + sensible and latent heat flux) Solar load model : solar ray tracing algorithm and radiation model called surface-to-surface (S2S)	no
Majdoubi et al. (2016)	Canarian greenhouse	3D Steady	Crop submodel (porous medium + sensible and latent heat flux) Specific model for radiation distribution	Air temperature Air humidity
Bouhoun Ali et al. (2016)	Venlo Greenhouse	2D Unsteady	Crop submodel (porous medium + sensible and latent heat flux) Resolution of RTE distinguishing short and long wave radiation Submodel for water balance	Air temperature Leaf temperature Matric potential Stomatal resistance Air humidity Transpiration rate
Bournet et al. (2017)	Venlo Greenhouse	2D Unsteady	Crop submodel (porous medium + sensible and latent heat flux) Resolution of RTE distinguishing short and long wave radiation	Air temperature Leaf temperature Air humidity Transpiration rate Ground surface temperature
Bouhoun Ali et al. (2017)	Venlo Greenhouse	2D Unsteady	Crop submodel (porous medium + sensible and latent heat flux) Resolution of RTE distinguishing short and long wave radiation Submodel for water balance	Air temperature Leaf temperature Matric potential Stomatal resistance Air humidity Transpiration rate

continued

Authors	Greenhouse	Dimen-sion	Crop model characteristics (and other submodels)	Validation
Molina-Aiz et al. (2017)	Almeria type	2D Steady	No crop submodel: constant source of water vapour + constant source of heat Values of radiative fluxes inferred from measurements Submodel for the absorption or production of CO_2	Air velocity Air temperature Air humidity CO_2 concentration
Boulard et al. (2017)	6-span glasshouse	3D Unsteady	Crop submodel (porous medium + sensible and latent heat flux) Resolution of RTE distinguishing short and long wave radiation Submodel for the absorption or production of CO_2	Air temperature Leaf temperature Saturated humidity at leaf temperature Air humidity Shortwave radiation Air speed Crop transpiration CO_2 concentration

4.1 Results from the crop-submodel and combination with a radiative submodel

The pioneering work of Boulard and Wang (2002) mainly aimed at validating the crop submodel inside a greenhouse tunnel on the basis of air velocity, temperature and transpiration flux measurements. Following studies were mostly conducted for 3D geometries and under steady state conditions. The authors mainly tested the impact of a variation of outside climatic conditions, of different vent shapes and combinations and/or of different facilities such as nets. Validation was undertaken on the basis of air velocity, air temperature, leaf temperature, humidity and/or ventilation rate. Fatnassi et al. (2003, 2006) focused on the impact of insect nets on the inside climate considering a plastic tomato greenhouse and a multi-span rose greenhouse. They evidenced the significant temperature and humidity increase caused by the nets and concluded that arrangements such as wiser orientations of the roof vents and the addition of side openings could reduce this effect. For a Canarian greenhouse, and considering a wind direction perpendicular to the side and roof openings, Majdoubi et al. (2009) also showed that the insect screen significantly reduced inside air velocity and increased inside temperature and humidity, especially in the vicinity of the crop. Considering a tunnel greenhouse Bartzanas et al. (2004) tested four different vent configurations and analysed their impact on the climate homogeneity inside the greenhouse. Roy and Boulard (2005) found important differences in the temperature and humidity patterns according to different wind incidences (0°, 45°and 90°) . Kichah et al. (2011, 2012) considered a calculation domain restricted to the crop and the air in the vicinity of the plants.

For a 34h period, they showed the ability of the model to predict both the absorbed radiation inside the canopy and the transpiration rate. Their model however did not include the directional characteristics of solar radiation inside the canopy. Fatnassi et al. (2015) focussed on the influence of solar panels on the solar distribution inside two types of greenhouses (asymmetric and Venlo).

During the last few years most studies focused on the evolution of the microclimate inside the crop and solved the RTE not only inside the greenhouse, but also inside the crop itself by specifying an extinction coefficient (Fidaros et al., 2010; Nebbali et al., 2012; Pouillard et al., 2012; Bouhoun Ali et al., 2017). Fidaros et al. (2010) simulated the climate inside a ventilated arch type tunnel greenhouse during a solar day taking into account the optical properties of plants only in the photosynthetic active radiation (PAR) band. Nebbali (2008) and Nebbali et al. (2012) considered a tunnel type greenhouse and investigated the heterogeneity of the crop transpiration together with its evolution all day long. They got very realistic results. Pouillard et al. (2012) considered a closed energy saving greenhouse and established that two thirds of the captured radiative energy were transferred to latent heat thus increasing air humidity while the remaining part contributed to greenhouse air warm up and heat accumulation just below the roof. Chen et al. (2015) implemented a coupling between a CFD model and an energy prediction model (EPM) with the aim to increase energy savingsand system performance. They activated a crop submodel adapted to a begonia crop and showed that the CFD–EPM-based method could improve system performance with more accurate temperature, more rapid responses and lower energy consumption. Bournet et al. (2017) carried out simulations at a daily timescale including sun path, ground conduction and crop interaction. They validated their results on seven different variables (temperature and humidity both above and inside the canopy, leaf temperature, transpiration rate and ground temperature) with good accuracy.Bouhoun Ali et al. (2017) considered potted New Guinea Impatiens under well-watered and water restriction conditions. CFD simulations showed the ability of the model to correctly predict transpiration, air and leaf temperatures, air humidity and matric potential inside the greenhouse for both water regimes. Then testing different irrigation scenarios, they showed that the water supply could be reduced by 20% without significantly impacting the transpiration rate. Also very recently, the CFD tool was adapted to new crop systems such as plant factories (Zhang et al., 2016). CFD Studies however remain scarce on that topic.

4.2 Results from the Condensation Submodel

Assessing condensation requires the development of a specific User Defined Function.Tong et al. (2009) implemented a simplified approach in which they did not explicitly solve the equation for water vapour mass fraction transport, but estimated the latent heat flux due to condensation from an empirical formula, assuming a constant value for the inside surface heat transfer coefficient. They validated their model against ground and air temperature data inside a Chinese solar greenhouse. Piscia et al. (2012a, b) analyzed the condensation process during night-time inside a four span plastic greenhouse, considering the crop as a constant source of water vapor, thus neglecting the interaction of the crop with the ambient climatic conditions. In particular, they showed that the roof was the coolest surface in the greenhouse, and therefore the sink for the water vapor produced by the crop. Following the work of Piscia et al. (2012a), Farber et al. (2014, 2017) focused on the transient and spatially varying condensation rate for a tomato crop, assuming a constant transpiration rate. They validated their model against the data obtained by Piscia et al. (2012a). They also provided a map of the condensation rate distribution over the walls and roofs showing its spatial heterogeneity. More recently, Bouhoun Ali et al. (2014, 2017) conducted a CFD study for an Impatiens New Guinea crop grown inside a glasshouse. In their study, they took account of the evolving transpiration rate of the crop, by activating a specific crop submodel. Their results revealed the ability of the model to predict the air, wall and leaf temperatures together with the relative humidity and transpiration rate. They also tested the influence of five parameters: outside temperature, wind velocity, sky temperature, emissivity of the glass and ground heat flux on the response of the model and concluded that all these parameters may affect wall and roof temperature and therefore condensation. They evidenced that low outside temperature or sky temperature may enhance condensation because they involve an important decrease of the cover temperature.

4.3 Results from the CO_2 Submodel

Assessing the time evolution of the distribution of the CO_2 concentration in a cropped greenhouse, Roy et al. (2014) correctly predicted the evolution of CO_2 concentration both when vents were open or closed. In particular, they could predict the stratification of the CO_2 concentration when vents were closed. Boulard et al. (2017) highlighted the heterogeneity of canopy temperature, transpiration and carbon dioxide fluxes and their daily evolution. After validating their numerical results with respect to dynamic and distributed measurements, their global canopyclimate model was used to virtually test biological and technological system adaptations that significantly improve its functioning, such as crop Leaf Area Density (LAD) development and the positioning of the air suction inlets. Improved designs and cultural practices were suggested and further integration of this simulation tool for improving the climate of closed greenhouses and growth chambers is expected. Molina-Aiz et al. (2017) also validated their results of net photosynthesis against measurements in an experimental greenhouse and values reported in the literature.

5 Conclusion

Recent developments in CFD made it possible to considerably improve the realism of simulation of climate distribution inside greenhouses. During the last fifteen years, a particular attention was paid to the development of specific submodels taking account of the interaction of plants with local climate. These submodels were progressively included to predict the mechanical strain exerted by plants on to the flow, the sensible and latent heat fluxes together with radiation distribution inside the canopy and condensation process. More recently photosynthesis and CO_2 concentration were included in CFD models. These advances offer new opportunities towards an improvement of greenhouse design, climate control and crop management.

Still, the realism of physical processes included in CFD models will have to be improved to get better and more reliable results. Complementary submodels will have to be developed such as comprehensive models describing bio-chemi-physical phenomena like heat and mass exchanges and yield of primary and secondary metabolites. Probably there will be an interest in the future also for modelling the effect of supplemental lighting, LED lighting, lighting fixture designs and their effect on crop and greenhouse climate.

Concerning validation of CFD models, there is also a need to define a reliable protocol: which variables to consider ? How many ? Which criteria to conclude about quality of the model to predict the climate ? Another challenge in the future will be to cope with calculation power limitation. From that point of view, improving the coupling between CFD and global models to fasten calculations could be reached by combining the use of BES (Building Energy Simulations tools) with CFD simulation tools.

Finally connecting CFD results to a Decision Making system will be a challenge in the future. The question is how to make the tool easy to use and helpful to decision makers. Indeed, the use of CFD tool as a device to improve greenhouse design, performance, sustainability and to optimize control (smart greenhouses) remains in the heart of future CFD development strategies.

References

Acock, B., Hand, D.W., Thornley, J.H.M., et al. 1976. Photosynthesis in stands of green peppers. Anapplication of empirical and mechanistic models to controlled-environment. Ann. Bot. (Lond.), 40 (6) : 1 293–1 307. http: //dx.doi.org/10.1093/oxfordjournals.aob.a085250

Bartzanas, T., Boulard, T. and Kittas, C.2004. Effect of vent arrangement on windward ventilation of a tunnel greenhouse. Biosystems Engineering, 88, 479-490http: //dx.doi.org/10.1016/j.biosystemseng.2003.10.006

Bell, B. 2003. Application Brief: Film Condensation of Water Vapor. Lebanon, New Hampshire: Fluent, Inc.

Bird, R.B., Stewart, W.E., and Lightfoot, E.N. 1960. Transport Phenomena, New York: John Wiley & Sons.

Bouhoun Ali, H., Bournet, P.E., Danjou V. et al. 2014. CFD simulations of the night-time condensation inside a closed glasshouse: Sensitivity analysis to outside external conditions, heating and glass properties, Biosystems Engineering, 127, 159–175http: //dx.doi.org/10.1016/j.biosystemseng.2014.08.017.

Bouhoun Ali H., Bournet P.E., Cannavo P. and Chantoiseau E. 2016a. CFD analysis of Impatiens crop irrigation scenarios inside a greenhouse, CIGR-Ageng International Conference of Agricultural Engineering, Aarhus, Denmark June 26-29, 8p.

Bouhoun Ali H., Bournet P.E., Cannavo P. and Chantoiseau E. 2016b. CFD simulation of greenhouse microclimate and crop transpiration under water restriction, CIGR-Ageng International Conference of Agricultural Engineering, Aarhus, Denmark June 26-29, 8p.

Bouhoun Ali H., Bournet P.E., Cannavo P. and Chantoiseau E. 2017. Development of a CFD crop submodel for simulating microclimate and transpiration of ornamental plants grown in a greenhouse under water restriction, Computers and Electronics in Agriculture, in press. http: //dx.doi.org/10.1016/j.compag.2017.06.021.

Boulard, T. and Wang, S. 2002. Experimental and numerical studies on the heterogeneity of crop transpiration in a plastic tunnel. Computers and Electronics in Agriculture, 173, 190-34http://dx.doi.org/10.1016/S0168-1699(01)00186-7.

Boulard, T. Jean-Claude Roy, J.C., Pouillard, et al., 2017. Modelling of micrometeorology, canopy transpiration and photosynthesis in a closed greenhouse using computational fluid dynamics, Biosystems Engineering, 158, 110-133http://dx.doi.org/10.1016/j.biosystemseng.2017.04.001.

Bournet P.E., Morille B. andMigeon C. 2017. CFD prediction of the daytime climate evolution inside a greenhouse, taking account of the crop interaction, sun path and ground conduction, Greensys 2015 Evora July 19-23rd, ActaHorticulturae, 1170, 61-70http://dx.doi.org/10.17660/ActaHortic.2017.1170.6.

Chen, J., Xu, F., Tan, D., Shen. et al., 2015. A control method for agricultural greenhouses heating based on computational fluid dynamics and energy prediction model. Appl. Energy, 141, 106–118http://dx.doi.org/10.1016/j.apenergy.2014.12.026.

Farber, P., Decker, N., Farber, K.. et al., 2014. A computational fluid dynamics investigation of condensation in greenhouses. Paper presented at: 28th European Simulation and Modelling Conference - ESM '2014 (Porto, Portugal), 382–386.

Farber, K., Farber, P., Gräbel, J. et al., 2017. A computational fluid dynamics investigation of flow, solar radiation, heat transfer, transpiration and condensation in a greenhouse. ActaHorticulturae, 1170, 45-52http://dx.doi.org/10.17660/ActaHortic.2017.1170.4.

Fatnassi H., Boulard T. and Bouirden, L. 2003. Simulation of climatic conditions in full-scale greenhouse fitted with insect-proof screens. Agricultural and Forest Meteorology, 118, 97–111http://dx.doi.org/10.1016/s0168-1923(03)00071-6.

Fatnassi, H., Boulard, T., Poncet. et al., 2006. Optimisation of greenhouse insect screening with computational fluid dynamics. Biosystems Engineering, 93, 301–312 http://dx.doi.org/10.1016/j.biosystemseng.2005.11.014.

Fatnassi H., Poncet C., Bazzano M. et al., 2015. A numerical simulation of the photovoltaic greenhouse microclimate, Solar Energy, 120, 575–584http://dx.doi.org/10.1016/j.solener.2015.07.019.

Fidaros D.K., Baxevanou C.A., Bartzanas, T. et al., 2010. Numerical simulation of thermal behavior of a ventilated arc greenhouse during a solar day. Renewable Energy, 35, 1380-1386http://dx.doi.org/10.1016/j.renene.2009.11.013.

Garzoli, K.V.1985. A simple greenhouse climate model. ActaHortic. 174, 393–400http://dx.doi.org/10.17660/ActaHortic.1985.174.52.

Heuvelink. 1999. Evaluation of a dynamic simulation model for tomato crop growth and development. Annals of Botany, 83, 413-422http://dx.doi.org/10.1006/anbo.1998.0832.

Kichah A., Bournet, P.E., Migeon, C. et al., 2011. Experimental and Numerical Study of Heat and Mass Transfer Occurring at Plant Level inside a Greenhouse, ActaHorticulturae, 893, 621-628http://dx.doi.org/10.17660/ActaHortic.2011.893.65.

Kichah, A., Bournet, P.E., Migeon, C. et al., 2012. Measurement and CFD simulation of microclimate characteristics and transpiration of an Impatiens pot plant crop in a greenhouse. Biosystems Engineering, 112(1), 22-34http://dx.doi.org/10.1016/j.biosystemseng.2012.04.006.

Majdoubi, H., Boulard, T., Fatnassi, H. et al., 2009. Airflow and microclimate patterns in a one-hectare Canary type greenhouse: an experimental and CFD assisted study. Agricultural and Forest Meteorology, 149(6-7): 1050-1062http://dx.doi.org/10.1016/j.agrformet.2009.01.002.

Majdoubi H., Boulard T., Fatnassi H. et al., 2016. Canary greenhouse CFD nocturnal climate simulation. Open Journal of Fluid Dynamics, 6(2), 88-100http://dx.doi.org/10.4236/ojfd.2016.62008.

Molina-Aiz, F.D., Valera, D.L. and Alvarez, A.J. 2004. Measurement and simulation of climate inside Almeria-type greenhouses using computational fluid dynamics. Agricultural and Forest Meteorology 125: 33-51http://dx.doi.org/10.1016/j.agrformet.2004.03.009.

Molina-Aiz, F.D., Norton, T., López, A. et al., 2017. Using Computational Fluid Dynamics to analyse the CO_2 transfer in naturally ventilated greenhouses. ActaHorticulturae. 1182, 283-292http://dx.doi.org/10.17660/ActaHortic.2017.1182.34.

Morille, B., Bournet, P.E. and Migeon, C. 2014. Analysis of the time evolution of the intercepted radiation inside a crop using CFD. ActaHorticulturae, 1037, 1017-1026http://dx.doi.org/10.17660/ActaHortic.2014.1037.134.

Nebbali, R. 2008. Modelling of the dynamics of the distributed internal climate in a greenhouse. PhD Thesis, Université de Franche-

Comté, France.

Nebbali, R., Roy, J.C. andBoulard, T. 2012. Dynamic simulation of the distributed radiative and convective climate within a cropped greenhouse. Renewable Energy, 43, 111-129http: //dx.doi.org/10.1016/j.renene.2011.12.003.

Nederhoff, E.M. and Vegter, J.G. 1994. Canopy photosynthesis of tomato, cucumber and sweet pepper ingreenhouses: measurements compared to models. Ann. Bot. (Lond.) 73 (4), 421–427 http: //dx.doi.org/10.1006/anbo.1994.1052.

Piscia, D., Montero, J.I., Baeza, E. et al., 2012a. A CFD greenhouse night-time condensation model. Biosystems Engineering, 111, 141-154http: //dx.doi.org/10.1016/j.biosystemseng.2011.11.006.

Piscia, D., Montero, J. I., and Flores, J. 2012b. Predicting night-time condensation in a multi-span greenhouse using computational fluid dynamic simulations. ActaHorticulturae, 927, 627-634http: //dx.doi.org/10.17660/ActaHortic.2012.927.77.

Pouillard, J.B., Boulard, T., Fatnassi, H., et al., 2012. Preliminary experimental and CFD results on airflow and microclimate patterns in a closed greenhouse. ActaHorticulturae, 952, 191-200 http: //dx.doi.org/10.17660/ActaHortic.2012.952.23.

Roy, J.C. and Boulard, T. 2005. CFD prediction of the natural ventilation in a tunnel-type greenhouse: influence of wind direction and sensibility to turbulence models. ActaHorticulturae, 691, 457–464 http: //dx.doi.org/10.17660/ActaHortic.2005.691.55.

Roy, J.C., Pouillard, J.B., Boulard, T., Fatnassi, et al., 2014. Experimental and CFD Results on the CO_2 Distribution in a Semi Closed Greenhouse. ActaHorticulturae. 1037, 993-1000http: //dx.doi.org/10.17660/ActaHortic.2014.1037.131.

Stanghellini, C. (1987). Transpiration of greenhouse crops: an aid to climate management. PhD Thesis, Agricultural University of Wageningen, The Netherlands

Tamimi, E., M. Kacira, C. Choi, L. An. 2013. Analysis of climate uniformity in a naturally ventilated greenhouse equipped with high pressure fogging system. Transactions of ASABE, 56 (3): 1241-1254.http: //dx.doi.org/10.13031/trans.56.9985.

Teitel, M., Ziskind, G., Liran, O., et al., 2008. Effect of wind direction on greenhouse ventilation rate, airflow patterns and temperature distributions. Biosystems Engineering, 101, 351-369 http: //dx.doi.org/10.1016/j.biosystemseng.2008.09.004.

Thornley, J.H.M. 1976. Mathematical Models in Plant Physiology, London, UK: Academic Press, pp.331.

Tong G., Christopher D. M. and Li, B. 2009. Numerical modelling of temperature variations in a Chinese solar greenhouse. Computers and Electronics in Agriculture, 68, 129-139 http: //dx.doi.org/10.1016/j.compag.2009.05.004.

Van Genuchten, M.T., 1980. A closed-form equation for predicting the hydraulic conductivity of unsaturated soils. Soil Sci. Soc. Am. J. 44, 892–898http: //dx.doi.org/10.2136/sssaj1980.03615995004400050002x.

Wang, X., Luo, J., and Li, X. 2013. CFD based study of heterogeneous microclimate in a typical Chinese greenhouse in central China. Journal of Integrative Agriculture, 12 (5) 914–923http: //dx.doi.org/10.1016/S2095-3119 (13) 60309-3.

Zhang, Y., Kacira, M. and An, L. 2016. A CFD study on improving air flow uniformity in indoor plant factory system, Biosystems Engineering, 147, 193-205http: //dx.doi.org/10.1016/j.biosystemseng.2016.04.012.

Manipulating Light Environment for Enhancing Growth and Nutritional Quality of Basil Plants under Indoor Controlled Environment

Dou Haijie, Niu Genhua*, Gu Mengmeng, Joseph Masabni

(1. *Department of Horticultural Sciences, Texas A&M University, College Station, TX* 77843, *USA*; 2. T*exas A&M AgriLife Research and Extension Center at El Paso*, T*exas A&M University System*, 1380 *A&M Circle, El Paso, TX* 79927, *USA gniu@ag.tamu.edu*; 3. *Department of Horticultural Sciences, Texas A&M AgriLife Extension Service, College Station, TX* 77843, *USA*; 4. *Texas A&M AgriLife Research and Extension Center, Texas A&M University System*, 1710 *FM* 3053 *N, Overton, TX* 75684, *USA*)

Abstract: Indoor vertical farming, also known as "plant factory", is a highly controlled environmental system for plant production that utilizes multi-layer culture shelves with artificial lighting. Due to increasing urban populations and diminishing resources such as water and arable land, indoor vertical farming emerged as an alternative and supplementary food crop growing system to conventional field-based system. However, due to its high capital and operational costs, indoor vertical farming is currently limited to the production of high-value, quick-turning crops such as culinary and medicinal herbs, microgreens, baby greens, and leafy greens such as lettuce and transplants. Basil (*Ocimum basilicum*) is one of the most popular herbs in the United States, and its production under indoor vertical farming is gaining more popularity owing to a stable supply with high nutritional quality. This paper reviews the latest research on the effect of light conditions on basil growth and development and presents some of our latest research results on the effects of daily light integral, light quality (various combination of red and blue light vs. fluorescent and white light emitting diode) , and ultraviolet (UV) light as a supplemental light before harvest on basil growth and nutritional quality. Our results demonstrate a great potential to enhance growth and nutritional quality of basil (as a model crop) through manipulation of the light environment under indoor controlled environment.

Key words: Daily light integral; Light quality; Plant Factory; *Ocimum basilicum*; Supplemental UV-B radiation

1 Introduction

Basil (*Ocimum basilicum*) is called the "king of herbs" or the "royal herb" and is widely used as a culinary herb and medicinal plant due to its specific aromatic flavor and relatively high content of phenolic compound (Chiang et al., 2005; Makri and Kintzios, 2008) . To achieve a stable and reliable supply, basil production in indoor vertical farming is gaining more popularity (Kozai, 2013; Kruma et al., 2008; Liaros et al., 2016; Saha et al., 2016) . However, the commercial application of indoor farms is limited by high levels of construction and operation costs, and artificial lighting is the costliest (or largest) operation-factor with about 80% of total electricity consumption (Kozai, 2013). Meanwhile, light is one of the most important environmental factors affecting plant growth and development and regulating plant behavior depending on light quantity, quality, direction, and duration (Chang et al., 2008; Dou et al., 2017; Shafiee-Hajiabad et al., 2016). Therefore, characterizing the effects of lighting conditions on the growth and nutritional quality of basil plants in indoor vertical farming is one of the most important goals in the future.

2 PPFD and Photoperiod

Under given environmental conditions, plant growth responds almost linearly to increasing photosynthetic photon flux density (PPFD) before reaching a light saturation point. Photoperiod also influences leaf expansion, crop yield, and nutritional content accumulation of plants (Beaman et al., 2009; Shiga et al., 2009). Basil originated in tropical and subtropical regions and is adapted to moderate to high PPFD and long day irradiation (Pushpangadan and George, 2012). A 16-h photoperiod is used in most studies in controlled environment basil cultivation with artificial lighting (Amaki et al., 2011; Piovene et al., 2015).

Daily light integral (DLI, the product of PPFD and photoperiod) represents the total photosynthetic photon flux radiated by a light source in 24 hours, and usually has a linear relationship with crop yield. The target DLI for leafy crops in indoor vertical farming is recommended as 12~17 $mol \cdot m^{-2} \cdot d^{-1}$ for enhancing growth and energy conservation (Albright et al., 2000; Kozai et al., 2015). A few studies investigated the effects of DLIs ranging from 13.5 to 34.6 $mol \cdot m^{-2} \cdot d^{-1}$ on basil growth and development (Beaman et al., 2009; Chang et al., 2008), but no study has determined the optimum DLI between 12 and 17 $mol \cdot m^{-2} \cdot d^{-1}$ to minimize the energy cost while maintaining a high yield.

Table 1 The net photosynthetic rate, transpiration rate, and stomatal conductance of green basil 'Improved Genovese Compact' grown for 21 days at different daily light integrals (DLI) in indoor controlled environment. (Dou et al., 2018).

Treatment	Net Photosynthetic Rate ($mmol \cdot m^{-2} \cdot s^{-1}$)		Transpiration Rate ($mmol \cdot m^{-2} \cdot s^{-1}$)		Stomatal Conductance ($mmol \cdot m^{-2} \cdot s^{-1}$)	
DLI 9.3	6.1	c^x	1.26	c	86	b
DLI 11.5	7.8	bc	1.43	bc	106	b
DLI 12.9	11.5	a	2.24	a	194	a
DLI 16.5	10.6	a	2.01	a	172	a
DLI 17.8	10.4	ab	1.85	ab	142	ab

x Means followed by the same letters are not significantly different within a column, according to Student's t mean comparison ($P<0.05$).

Table 2 Plant leaf area, specific leaf area, and shoot FW of green basil 'Improved Genovese Compact' grown for 21 days at different DLIs in indoor controlled environment. (Dou et al., 2018) .

Treatment	Leaf Area (cm^2)		Specific Leaf Area[y] ($cm^2 \cdot g^{-1}$, DW)		Shoot FW (g)	
DLI 9.3	406	b[x]	518	a	13.1	c
DLI 11.5	454	b	480	ab	15.7	b
DLI 12.9	560	a	462	b	20.2	a
DLI 16.5	609	a	389	c	23.4	a
DLI 17.8	614	a	398	c	23.3	a

x Means followed by the same letters are not significantly different within a column, according to Student's t mean comparison ($P<0.05$) .

y Specific leaf area= leaf area per unit leaf dry weight.

An experiment was conducted at Texas AgriLife Research and Extension Center at El Paso to determine the minimum DLI between 9~18 $mol \cdot m^{-2} \cdot d^{-1}$ (five DLI levels of 9.3, 11.5, 12.9, 16.5, and 17.8 $mol \cdot m^{-2} \cdot d^{-1}$ with the same 16-h photoperiod provided by Philips fluorescent lamps) for sweet basil production with comparable yield and nutritional values in indoor vertical farming (Dou et al., 2018) . Results indicated that basil plants grown under higher DLIs of 12.9, 16.5 or 17.8 $mol \cdot m^{-2} \cdot d^{-1}$ had higher net photosynthetic rate, transpiration rate, and stomatal conductance, compared to those under lower DLIs of 9.3 and 11.5 $mol \cdot m^{-2} \cdot d^{-1}$ (Table 1). Under lower DLIs, basil plants maximize light-harvesting capacity by increasing light harvesting chlorophyll-protein complex in photosystem II (Kitajima and Hogan, 2003; Sarijeva et al., 2007) , which contains the majority of Chl b, and consequently resulting in a higher Chl a+b content and lower Chl a/b ratio (Figure 1) . Higher photosynthesis under DLIs of 12.9, 16.5 and 17.8 $mol \cdot m^{-2} \cdot d^{-1}$ resulted in larger and thicker leaves of basil plants, and the shoot fresh weight(FW) was 54.2%, 78.6%, and 77.9%, respectively, higher than that at DLI of 9.3 $mol \cdot m^{-2} \cdot d^{-1}$(Table 2). The amounts of total anthocyanin, phenolics, and flavonoids per plant were also positively correlated to DLIs (Figure 2) . Combining results of growth, yield, and nutritional quality of sweet basil, a DLI of 12.9 $mol \cdot m^{-2} \cdot d^{-1}$ is suggested for sweet basil commercial production in indoor vertical farming to minimize the energy cost while maintaining a high yield and nutritional quality.

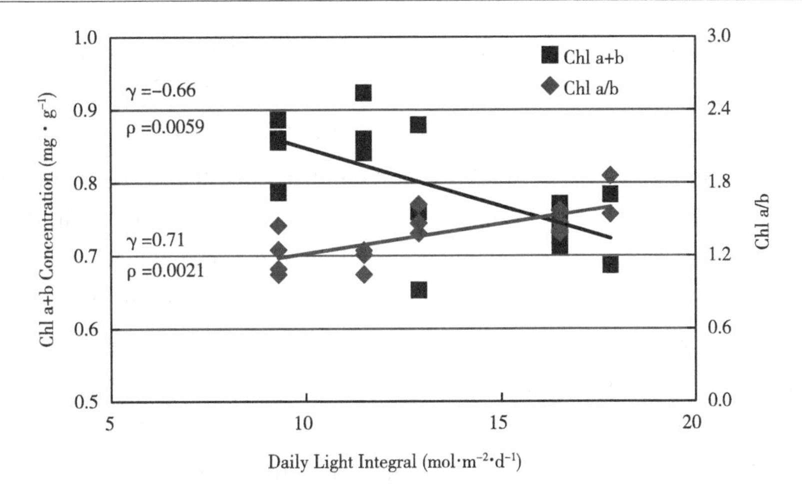

Figure 1　Correlation between Chl a+b concentration per leaf fresh weight and Chl a/b ratios with DLI of green basil 'Improved Genovese Compact' grown for 21 days at different DLIs in indoor controlled environment. (Dou et al., 2018).

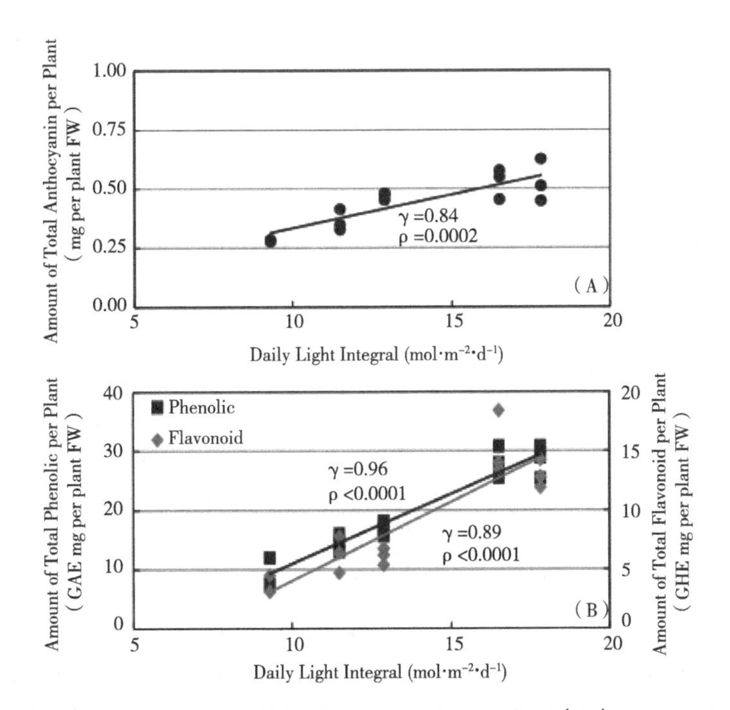

Figure 2　Correlation between amount of total anthocyanin per plant (A), amount of total phenolic per plant, and amount of total flavonoid per plant (B) with DLIs of green basil 'Improved Genovese Compact' grown for 21 days at different DLIs in indoor controlled environment. (Dou et al., 2018).

3　Light Quality

Plants sense and respond to a broad range of light spectra from ultraviolet (UV) to far-red regions, and light quality or light spectrum wavelength significantly affects plant growth, development, morphology, and secondary metabolism (Amaki et al., 2011; Brazaitytė et al., 2013; Bugbee, 2016; Dou et al., 2017) . The development of LED technology provided researchers opportunities to regulate crop yield and nutritional quality using different light wavelength. The high efficiency of red and blue light on plant growth is easily understood since they perfectly fit the absorption peak of chlorophylls and combining red and blue light is more effective than monochromatic red or blue light for plant growth due to better excitation of photoreceptors including phytochromes, cryptochromes, and phototropins (Dong et al., 2013; Sabzalian et al., 2014) . Most previous research only examined the effects of monochromatic red or blue light, and few studies investigated the effects of combined red and blue light, and for those who did so, the blue light proportion (BP) threshold was not determined. Basil shoot FW increased by 214% when BP increased from 15% to 59% under combined red and blue light at 200 $\mu mol \cdot m^{-2} \cdot s^{-1}$ with a 16-h photoperiod (Piovene et al., 2015) . However, 'Wala' basil plants grown under white fluorescent lamps (FLs) with lower BP (red:blue:far-red or R:B:FR at 24%:8%:2%) had higher plant height and greater shoot FW compared with plants grown under white LEDs with higher BP (R:B:FR at 14%:16%:6%) at 160 $\mu mol \cdot m^{-2} \cdot s^{-1}$ with a 16-h photoperiod, and no differences in leaf area or photosynthetic rate were observed (Fraszczak et al., 2014) .

To investigate the optimal BP in combined red and blue light for basil production, green basil 'Improved Genovese Compact' and purple basil 'Red Rubin' were treated with six light spectrum combinations including white fluorescent light (WFL) , white LED (WLED) , and four combinations of red and blue LEDs (R:B=1.18, BP 46%, R1B1; R:B=3.44, BP 23%, R3B1; R:B=6.39, BP 14%, R6B1; R:B=7.84, BP 11%, R8B1) at 224 $\mu mol \cdot m^{-2} \cdot s^{-1}$ with a 16-h photoperiod. Results were comparable to Fraszczak et al. (2014) where the shoot FW of green and purple basil both increased as BP decreased, and was the lowest under full spectrum treatments of WFL and WLED. The shoot DW of purple basil had a similar trend as shoot FW, whereas shoot DW of green basil showed no difference among treatments. The anthocyanin, total phenolic, and flavonoid concentration of green basil were the highest under treatment R3B1. The anthocyanin concentration of green basil was enhanced when BP increased, and no differences were found among treatments WFL, WLED, R1B1, and R3B1. As key enzymes in the synthesis of anthocyanins, the expression of phenylalanine ammonia-lyase, chalcone synthase, and dihydroflavonol 4-reductase increased under blue light, resulting in an enhanced accumulation of anthocyanin in lettuce (*Lactuca sativa*) , *Salvia miltiorrhiza*, and *Gerbera hybrid* (Li, 2010; Meng et al., 2004) . Red light benefited phenolic compounds accumulation, especially rosmarinic acid, the main component of phenolic acids in basil plants, compared with other light spectra (Samuoliene et al., 2016; Shiga et al., 2009) . Continuous red and white light irradiation at 100 $\mu mol \cdot m^{-2} \cdot s^{-1}$ induced rosmarinic acid accumulation of basil plants up to 6 $mg \cdot g^{-1}$ FW within 14 days, whereas blue light irradiation only induced rosmarinic acid accumulation up to 3 $mg \cdot g^{-1}$ FW (Shiga et al., 2009) .

4 Supplemental UV Light Radiation

UV light is commonly considered a stress factor to plant growth, but some studies indicated that supplemental UV radiation induced the synthesis of phytochemicals, which provide protection against potential UV damage, including UV-B absorbing anthocyanins, flavonoids, and antioxidants such as ascorbate, carotenoids, glutathione and a broad range of other metabolites (Bantis et al., 2016; Sakalauskaite et al., 2013). Supplemental UV-B radiation at 2.5 mmol·m^{-2}·s^{-1} for 1 h or 2 h per day significantly increased the content of total phenolic compounds, anthocyanin concentrations, and antioxidant activity in basil plants without suppressing biomass accumulation, and the 1-h UV-B treatment was more efficient for anthocyanin accumulation than the 2-h treatment (Sakalauskaite et al., 2012; Sakalauskaitė et al., 2013). Similarly, supplemental UV-A radiation to white LEDs enhanced the antioxidant properties of 'Genovese' basil microgreen (Brazaityte et al., 2016).

Effects of supplemental UV-B radiation and PPFD on the growth and nutritional quality of basil 'Improved Genovese Compact' and purple basil 'Red Rubin' were evaluated. Uniform basil seedlings were treated with two PPFD levels (160 and 224 mmol·m^{-2}·s^{-1}) after transplanting, and at five or two days before harvest, basil plants were applied with one of five supplemental UV-B radiation levels (no UV-B radiation, CNT; UV-B radiation of 1 h per day for 2 days, 1H2D; 1 h per day for 5 days, 1H5D; 2 h per day for 2 days, 2H2D; and 2 h per day for 5 days, 2H5D). Results showed that 2H5D treatment significantly decreased the net photosynthetic rate, transpiration rate, and stomatal conductance of green and purple basil plants by 68%/70%, 55%/68%, and 65%/76%, respectively, and PPFD had no effects. Shoot FW of green and purple basil were both decreased by supplemental UV-B and lower PPFD treatments, and no interactive effects were observed. The anthocyanin, total phenolics, and flavonoid concentration of green basil under supplemental UV-B treatments were enhanced by 18%~22%, 35%~126%, and 80%~169%, respectively, whereas anthocyanin concentration of purple basil showed no difference, and phenolic and flavonoid concentration of purple basil were slightly decreased under 1H2D treatment. Considering the yield reduction and nutritional content enhancement of basil plants by supplemental UV-B radiation, one hour for 2 or 5 days UV-B treatments before harvest at PPFD of 224 mmol·m^{-2}·s^{-1} was recommended for green basil production.

5 Conclusions

It is apparent that lighting conditions including PPFD, photoperiod, light quality, and supplemental UV-B radiation could be used to regulate the yield and nutritional quality of basil plants in indoor vertical farming. A DLI of 12.9 mol·m^{-2}·d^{-1} with a 16-h photoperiod, combinations of red and blue LED with blue light proportion of 23%, and 1 h supplemental UV-B radiation for 2 days or 5 days before harvest significantly increased the shoot FW and enhanced the anthocyanin, total phenolic, and flavonoid concentration of green basil 'Improved Genovese Compact'. The anthocyanin, total phenolic, and flavonoid concentration of purple basil 'Red Rubin' showed no difference or were slightly increased under supplemental UV-B treatments due to higher phytochemical concentrations comparing to green basil.

A single lighting solution will not fit all plant species for optimal yields. The effects of photosynthetically less-efficient light, namely, cyan, green, yellow, orange, and far-red on plant growth and development are still unknown. Therefore, more research on light environment need to be done on various species to better understand the effects of light quality on plant growth, development and nutritional quality to guide indoor vertical production of basil and other leafy greens and herbs.

6 Acknowledgements

This research is partially supported by the United State Department of Agriculture, National Institute of Food and Agriculture Hatch project TEX090450 and Texas A&M AgriLife Research.

References

Albright, L., Both, A.J. and Chiu, A. 2000. Controlling greenhouse light to a consistent daily integral. Transactions of the ASAE 43, 421.

Amaki, W., Yamazaki, N., Ichimura, M. and Watanabe, H. 2011. Effects of light quality on the growth and essential oil content in sweet basil. Acta Hort. 907, 91-94.

Bantis, F., Ouzounis, T. and Radoglou, K. 2016. Artificial LED lighting enhances growth characteristics and total phenolic content of *Ocimum basilicum*, but variably affects transplant success. Scientia Hort. 198, 277-283.

Beaman, A.R., Gladon, R.J. and Schrader, J.A. 2009. Sweet basil requires an irradiance of 500 mmol·m^{-2}·s^{-1} for greatest edible biomass production. HortScience 44, 64-67.

Brazaitytė, A., Virsilė, A., Samuolienė, G., et al., 2016. In Light quality: Growth and nutritional value of microgreens under indoor and greenhouse conditions, VIII Intl. Symp. Light Hort. 277-284.

Bugbee, B. 2016. In Toward an optimal spectral quality for plant growth and development: The importance of radiation capture, VIII Intl. Symp. Light Hort. 1-12.

Chang, X., Alderson, P.G. and Wright, C.J. 2008. Solar irradiance level alters the growth of basil (*Ocimum basilicum* L.) and its content of volatile oils. Envir. Expt. Botany 63, 216-223.

Chiang, L.C., Ng, L.T., Cheng, P.W., et al., 2005. Antiviral activities of extracts and selected pure constituents of *Ocimum basilicum*. Clinical Expt. Pharma. Physiol. 32, 811-816.

Dong, J.Z., Lei, C., Zheng, X.J. et al., 2013. Light wavelengths regulate growth and active components of *Cordyceps militaris* fruit bodies. J. Food Biochem. 37, 578-584.

Dou, H., Niu, G., Gu, M., et al., 2017. Effects of light quality on growth and phytonutrient accumulation of herbs under controlled environments. Horticulturae, 3, 36.

Dou, H., Niu, G., Gu, M., et al., 2018. Responses of sweet basil to different DLIs in photosynthesis, morphology, yield, and nutritional quality. HortScience (in press) .

Frąszczak, B., Golcz, A., Zawirska-Wojtasiak, et al., 2014. Growth rate of sweet basil and lemon balm plants grown under fluorescent lamps and LED modules. Acta Sci. Pol. Hortorum Cultus 13, 3-13.

Li, Q. 2010. Effects of light quality on growth and phytochemical accumulation of lettuce and salvia miltiorrhiza bunge. Northwest A&F

University, Shanxi, China, 2010.

Liaros, S., Botsis, K. and Xydis, G. 2016. Technoeconomic evaluation of urban plant factories: The case of basil (*Ocimum basilicum*). Sci. Total Envir. 554, 218-227.

Makri, O., and Kintzios, S. 2008. *Ocimum* sp. (basil): Botany, cultivation, pharmaceutical properties, and biotechnology. J. Herbs, Spices Med. Plants 13, 123-150.

Meng, X., T. Xing, and X. Wang. The role of light in the regulation of anthocyanin accumulation in *Gerbera hybrida*. Plant Growth Regulation 44, 243-250.

Kitajima, K., and Hogan, K. 2003. Increases of chlorophyll a/b ratios during acclimation of tropical woody seedlings to nitrogen limitation and high light. Plant, Cell & Environ. 26, 857-865.

Kozai, T. 2013. Resource use efficiency of closed plant production system with artificial light: Concept, estimation and application to plant factory. Proceedings of the Japan Academy. Series B, Physical and Biol. Sci. 89, 447.

Kozai, T., Niu, G. and Takagaki, M. 2015. Plant factory: An indoor vertical farming system for efficient quality food production. Academic Press: San Diego, USA.

Kruma, Z., Andjelkovic, M., Verhe, R. aet al., 2008. Phenolic compounds in basil, oregano and thyme. Foodbalt 5, 99-103.

Piovene, C., Orsini, F., Bosi, S., et al., 2015. Optimal red: Blue ratio in led lighting for nutraceutical indoor horticulture. Scientia Hort. 193, 202-208.

Pushpangadan, P., and George, V. 2012. Basil. In Handbook of herbs and spices, 2nd Edition ed.; Peter, K.V., Ed.

Sabzalian, M.R., Heydarizadeh, P., Zahedi, M., et al., 2014. High performance of vegetables, flowers, and medicinal plants in a red-blue LED incubator for indoor plant production. Agron. Sustainable Dev. 34, 879-886.

Saha, S., Monroe, A. and Day, M.R. 2016. Growth, yield, plant quality and nutrition of basil (*Ocimum basilicum* L.) under soilless agricultural systems. Annals Agri. Sci. 61, 181-186.

Sakalauskaitė, J., Viškelis, P., Duchovskis, P., et al., 2012. Supplementary UV-B irradiation effects on basil (*Ocimum basilicum* L.) growth and phytochemical properties. J. Food Agri. Envir. 10, 342-346.

Sakalauskaitė, J., Viskelis, P., Dambrauskienė, E., et al., 2013. The effects of different UV-B radiation intensities on morphological and biochemical characteristics in *Ocimum basilicum* L. J. Sci. Food Agri. 93, 1266-1271.

Samuoliene, G., Brazaityte, A., Virsile, A., et al., 2016. Red light-dose or wavelength-dependent photoresponse of antioxidants in herb microgreens. PLOS One 11, e0163405.

Sarijeva, G., Knapp, M. and Lichtenthaler, H.K. 2007. Differences in photosynthetic activity, chlorophyll and carotenoid levels, and in chlorophyll fluorescence parameters in green sun and shade leaves of ginkgo and fagus. J. Plant Physiol. 164, 950-955.

Shafiee-Hajiabad, M., Novak, J. and Honermeier, B. 2016. Content and composition of essential oil of four *Origanum vulgare* L. Accessions under reduced and normal light intensity conditions. J. Appl. Botany Food Quality 89.

Shiga, T., Shoji, K., Shimada, H., et al., 2009. Effect of light quality on rosmarinic acid content and antioxidant activity of sweet basil, *Ocimum basilicum* L. Plant Biotech. 26, 255-259.

植物工厂人工补光技术现状与发展趋势

赵静[1, 2]，周增产[1, 2]，卜云龙[1, 2]，兰立波[1, 2]，姚涛[1, 2]，李秀刚[1, 2]

（1. 北京京鹏环球科技股份有限公司，北京　100094；
2. 北京市植物工厂工程技术研究中心，北京　100094）

摘要：植物工厂的主要特征之一就是利用全人工光源实现光环境的智能控制。基于植物生长发育合理需求的电光源产品及其智能光控设备的研发与应用是今后植物工厂及栽培方法革新的核心内容。本文阐述了植物工厂中常见的补光照明方法设备的特性与优点，介绍了典型作物补光系统的设计思路，并对LED光源在植物工厂中的应用发展前景进行趋势分析。

关键词：植物工厂；光环境调控；补光装置；LED光源

Current Situation and Trends of Artificial Lighting Technology in Plant Factory

Zhao Jing[1, 2], Zhou Zengchan[1, 2], Bu Yunlong[1, 2], Lan Libo[1, 2], Yao Tao[1, 2], Li Xiugang[1, 2]

（1. *Beijing Kingpeng International Hi-tech Corporation*, *Beijing*　100094;
2. *Beijing Plant Engineering Technology Research Center*, *Beijing*　100094）

Abstract: One of the main features of the plant factory is the use of artificial intelligent light source light environment control. Therefore, the development of intelligent optical control technology is the core content of plant factory development. This paper introduces the characteristics of lighting system and common equipment in plant factory. This article introduces the design ideas of the typical light crops system, and the application of LED light source in plant factory was prospected.

Key words: Plant factory; Light environment regulation; Lighting device; LED light source

1　引言

植物工厂是设施农业的最高级发展阶段，它很好地将现代工业、生物科技、营养液栽培和信息技术等相结合，对设施内环境因子实施高精度控制，具有全封闭、对周围环境要求低，缩短植物收获期，节水节肥、无农药生产、不向外排放废物等优点[1]，单位土地利用效率是露地生产的40～108倍，其中智能化人工光源及其光环境调控对其生产效率起到决定性作用[2]。

光作为重要的物理环境因子，对植物的生长发育和物质代谢均起到关键的调控作用。“植物工厂的主要特征之一就是全人工光源并实现光环境的智能调控”已经成为业界的普遍共识[3]。

项目资助：科技部科技伙伴计划资助（KY201702008）

2　植物对光照的需求

植物的生命活动离不开光照，植物通过叶绿素在光能作用下将CO_2和水合成为糖和淀粉等碳水化合物并释放出氧气的生理过程称为光合作用。光是植物光合作用的唯一能量来源，光照强度、光质（光谱）及光的周期性变化对作物的生长发育具有深刻影响，其中，以光照强度对植物的光合作用影响最大[4]。

2.1　光强对作物的影响

光照的强弱一方面影响着光合强度，同时还能改变作物的形态，如开花、节间长短、茎的粗细及叶片的大小与薄厚。植物对光照强度的要求可分为喜光型、喜中光型、耐弱光型植物。蔬菜多数属于喜光型植物，其光补偿点和光饱和点均比较高，在人工光植物工厂中作物对光照强度的相关要求是选择人工光源的重要依据，了解不同植物的光照需求对设计人工光源、提高系统的生产性能都极为必要。

2.2　光质对作物的影响

光质（光谱）分布对植物光合作用和形态建成同样具有重要影响。光是辐射的一部分，而辐射是一种电磁波。电磁波具有波的特性与量子（粒子）特性。光的量子称为光子（photon），在园艺领域亦称为光量子。波长范围为300 ~ 800nm的辐射称为植物的生理有效辐射（physiologically active radiation）；而波长范围400 ~ 700nm的辐射称为植物的光合有效辐射（photosynthetically active radiation，PAR），如图1所示。

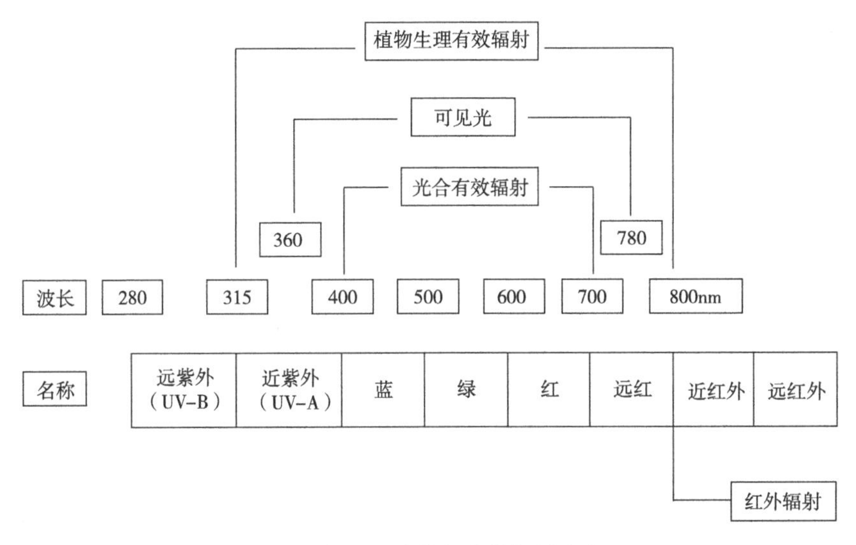

图 1　与植物生长相关的辐射分类

植物光合作用最重要的两种色素是叶绿素（Chlorophyll）和胡萝卜素（Carotenes）。图2为各光合色素的光谱吸收图谱，其中叶绿素吸收光谱集中在红蓝波段，照明补光系统是根据作物的光谱需求设计相关的照明系统进行人工补光并促进植物的光合作用。

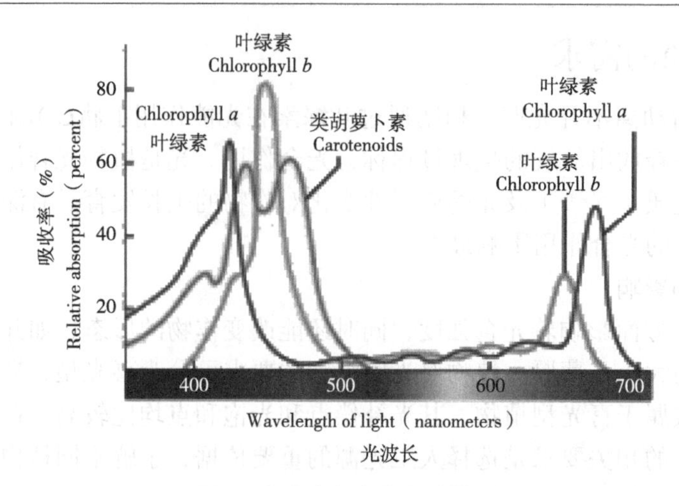

图 2 各光合色素的光谱吸收

2.3 光周期对作物的影响

植物的光合作用和光形态建成与日长（或光期时间）之间的相互关系称为植物的光周性。光周性与光照时数密切相关，光照时数是指作物被光照射的时间。不同的作物，完成光周期需要一定的光照时数才能开花结实。按照光周期的不同，可分为长日照作物，如白菜等，在其生育的某一阶段需要12～14h以上的光照时数；短日照作物，如洋葱、大豆等，需要12～14h以下的光照时数；中日照作物，如黄光、番茄、辣椒等，在较长或者较短的日照下，都能开花结实。

2.4 光照强度表述方式

环境三要素中，光照强度是选择人工光源的重要依据，目前对光照强度有多种表述方法，主要包括如下3种。

光照度（illumination）是指受照平面上接受的光通量面密度（单位面积的光通量），单位：勒克斯（lx）。

光合有效辐射照度PAR，单位：W/m^2。

光合有效光量子流密度PPFD或PPF即单位时间、单位面积上到达或通过的光合有效辐射的光量子数，单位$\mu mol \cdot m^{-2} \cdot s^{-1}$。主要指与光合作用直接有关的400～700nm的光照强度。也是植物生产领域最常用的光照强度指标。

3 典型补光系统的光源分析

人工补光是通过安装补光系统，提高目标区域内的光照强度或延长光照时间从而实现植物对光的需求。一般来说，补光系统包括补光设备、电路及其控制系统。补光光源主要包括白炽灯、荧光灯、金属卤化物灯、高压钠灯及LED等几种常用类型。由于白炽灯电光效率低，光合能效低等缺点，已经被市场淘汰了，故本文不做详细分析。

3.1 补光光源——荧光灯

属于低压气体放电灯的类型。玻璃管内充有水银蒸汽或惰性气体，管内壁涂有荧光粉，光色随管内所涂荧光材料的不同而异。荧光灯光谱性能好，发光效率较高，功率较小，寿命长（12 000h），成本相对较低。因荧光灯自身发热量较小，可以贴近植物进行照明，适用于立体栽培，但荧光灯光谱布局不合理，目前国际上比较常用的方法是增设反光罩，尽量增加栽培区作物的有效光源成分（图3）[5]。目前，日本adv-agri公司还开发了新型补光光源HEFL，HEFL实际上属于荧光灯范畴，是冷阴极荧光灯（CCFL）及外部电极荧光灯（EEFL）的总称，是一种混合电极荧光灯。HEFL灯管极细，直径仅4mm左右，长度按照栽培需要可从450mm到1 200mm，色别可选择白色、红色和蓝色，算是常规荧光灯的改进版。

图 3 荧光灯下作物生长

3.2 补光光源——金属卤化物灯

在高压水银灯的基础上，通过在放电管内添加各种金属卤化物（溴化锡、碘化钠等）而形成的可激发不同元素产生不同波长的一种高强度放电灯[5]。卤化灯发光效率较高、功率大、光色好、寿命较高、光谱大。但由于发光效率低于高压钠灯，寿命也比高压钠灯短，目前仅在少数植物工厂使用（图4）。

图 4 高压钠灯（左）与卤素灯（右）下玫瑰生长

3.3 补光光源——高压钠灯

属于高压气体放电灯类型（图5）。高压钠灯是在放电管内充高压钠蒸汽，并添加少量氙（Xe）和汞灯金属的卤化物的一种高效灯[5]。一方面，由于其具有较高的电光转换效率同时兼顾较低制造成本，所以目前农业设施补光应用中高压钠灯是最广泛的，但由于其光谱存在光合效率低的缺点，造成能效较低的短板。另一方面，高压钠灯所发出的光谱成分主要集中在黄橙光波段，缺少植物生长必需的红色和蓝色光谱。

图5 高压钠灯雾培叶菜生长

3.4 补光光源一发光二极管（LED）

LED作为新一代光源，具有更高的电光转换效率，光谱可调、光合效率高等诸多优点。LED能够发出植物生长所需要的单色光，与普通荧光灯等补光光源相比，LED具有节能、环保、寿命长、单色光、冷光源等优点（图6）[5]。随着LED的电光效率进一步提升，规模效应产生的成本下降，LED将成为农业设施补光的主流设备。

图6 LED补光灯下作物生长

通过比较可清晰了解不同补光光源的特性如下表所示[5]。

表　各类人工光源的特性指标比较

人工光源	功率（W）	发光效率〔（lm/W）〕	可见光比（%）	使用寿命（h）
荧光灯	45	100	34	12 000
金属卤化物灯	400	110	30	6 000
高压钠灯	360	125	32	12 000
LED灯	0.04	20	90	50 000

4　移动式补光装置

光照强弱与作物的生长密切相关，植物工厂中多采用立体栽培，但受栽培架结构所限，层架间的光照、温度分布不均，这样会影响作物的产量，收获期也不同步。

4.1　立体可移动补光装置

北京京鹏环球科技股份有限公司早在2010年已成功开发手动升降补光装置，其原理是通过摇动摇柄转动小型卷膜器带动传动轴及固定在其上面的绕线器转动，实现收放钢丝绳的目的，吊挂补光灯钢丝绳通过多组换向轮与升降器的绕线轮连接，从而达到了调整补光灯高低的效果[6]。2017年，京鹏公司设计开发了新型移动式补光装置，该装置可以根据作物生长需求实时自动调节补光高度。该调节装置现安装在3层式光源升降式立体栽培架上，其中顶层为光照情况最好的层级，故配置高压钠灯；中间层及底层配置可升降调节系统，为LED灯具，可根据光传感器的检测信号，自动调节补光灯高度，为作物提供合适的光照环境（图7）[7]。

图 7　京鹏自动升降补光装置

4.2　水平可移动补光装置

相对于为立体栽培量身订制的移动式补光装置（图8），荷兰开发了一种可作水平移动的补光灯装置同样为植物补光方式拓宽了新思路。为了免于阳光下补光灯的阴影对植物生长的影响，可将补光灯沿水平方向通过伸缩式滑道推向支架两边，让阳光全面照射到植物上；没有太阳光的阴雨天，可将补光灯推向支架中间，使补光灯的光照均匀地给植物补光；通过支架上的滑道水平方向移动补光灯，避免了频繁地拆装和搬动补光灯，降低了员工的劳动强度，有效提高了工作效率。

图 8　荷兰水平移动补光装置

5　典型作物补光系统设计思路

从移动式补光装置的设计不难看出，植物工厂补光系统的设计通常以不同作物生育期的光强、光质和光周期参数以及末端调控特殊手段最为设计的核心内容，依赖智能控制系统来执行实施，达到节能高产的终极目标。人工光植物工厂中光环境的调控更为必要，光源质量直接影响着植物的生产效率及其品质。渡边博之采用水冷模板LED光源栽培植物工厂内生菜、芹菜等蔬菜，能利用效率较高，能明显缩短生产周期[1]。Okamoto等[9]使用超高亮度LED红光：蓝光=2：1培育莴苣获得成功。Yanagi等[10]研究表明，单色蓝光LED处理下莴苣干物重较纯红光或红蓝光组合下小，植株显得更加矮壮和健康。Nichols等[11]研究发现传统人工光源产热较多，LED补充照明，将电能高效转变为有效光合辐射。研究发现，LED单一或组合光源作为植物工厂内菠菜[12]、萝卜[13]、甜菜[14]等蔬菜的光源，提高光合效率，进植株生长，控其形态建成。LED的光谱域宽在 ± 20nm左右，完全能够满足植物对单色光的需求，红光、蓝光及其PPFD实现单独控制[10]。刘水丽[15]研究表明，人工光照处理相比，LED光源下植株生长速率高，面积大，片数多，片生长速度快，根较深。张欢等[16]研究表明，光有利于萝卜芽苗菜生长，高产量，改善部分营养品质。陈晓丽等[17]研究了LED组合光谱对水培生菜矿物质吸收的影响，结果表明，叶绿素生理吸收波峰（峰值450、660nm）对应的单一或组合光谱均可增强水培生菜根对Na、Fe、Mn、Cu、Mo元素的吸收能力，单一红光光谱的促进作用最为显著。周华等[18]在江西省科学院植物工厂内研究不同波长LED光源对生菜生长和品质的影响，结果表明红光能提高地上生物量，光能矮化植物，高生菜维生素C和粗蛋白质含量，合光质（R：B：UV-B=20：5：1）可降低生菜叶面积，高维生素C、粗蛋白质和粗纤维含量。

目前，有关叶菜的补光设计构建已逐渐成熟。举例来说，叶菜可分为苗期、生长中期、生长后期和末端处置4个阶段；果菜可分为苗期、营养生长期、开花阶段、采收阶段。从补光光强属性来说，苗期光强应略低，应在60 ~ 200μmol·m^{-2}·s^{-1}，随后逐渐增大。叶菜最高可至100 ~ 200μmol·m^{-2}·s^{-1}，果菜可达300 ~ 500μmol·m^{-2}·s^{-1}，以保障各生育期植物光合作用对光强的要求，实现高产的需要；对光质而言，红蓝比例至关重要。苗期为了增加苗质量，防止徒长，一般把红蓝比例设置在较低的水平［1 ~ 2:1］，随后逐渐降低，以满足植物光形态建成的需求，可将叶菜红蓝光比设置在（3 ~ 6）:1之间[8]。对光周期而言，与光强类似，应呈现随生育期延长而增加的趋势，以使得叶菜有更多地光合时间进行光合作用。果菜的补光设计会更加复杂，除上述基本规律外，应重点关注开花期光环境设置中的光周期设置，必须促进蔬菜的开花结果，以免适得其反。值得提出的是，光配方应包括末端处置光环境设置内容，比如通过连续光大幅提高水培叶菜苗菜的产量和品质，或者通过UV处理显著提高芽菜和叶菜（尤其是紫叶和红叶生菜的营养品质）。

6 结束语

植物工厂被认为是21世纪解决世界资源、人口和环境问题的重要途径，是未来高科技工程中实现食物自给的重要途径[3]。本文阐述了植物工厂中常见的补光照明方法设备的特性与优点，介绍了典型作物补光系统的设计思路，通过比较不难发现，为应对连阴天、雾霾等恶劣天气导致的弱光寡照逆境，同时保障设施作物高产稳产，LED光源装备最为符合当前发展趋势。未来植物工厂的发展方向应侧重新型高精度、低成本传感器和远程可控的、可调光谱的补光照明装置系统及专家化控制系统。同时，未来植物工厂将向低成本化、智能化、自适应化不断发展。

参考文献

[1] 徐圆圆，覃仪，吕蔓芳，等. LED光源在植物工厂中的应用［J］.现代农业科技，2016（6）：161-162+170.
[2] 郑盛华，覃志豪，王志丹.我国现代设施农业发展趋势及关键技术［J］.农业经济，2015（4）：62-63.
[3] 周小丽，刘木清.植物工厂中的关键技术［J］.中国照明电器，2016（12）：1-4.
[4] 钟培芳，陈年来.光照强度对园艺植物光合作用影响的研究进展［J］.甘肃农业大学学报，2008（5）：104-109.
[5] 杨其长，魏玲玲，刘文科，等. 植物工厂系统与实践［M］. 北京：化学工业出版社，2012. 35-48.
[6] 龙智强，周增产，卜云龙，等. 植物工厂人工环境控制栽培室的设计研究［J］. 北方园艺，2010（15）：85-88.
[7] 赵静，周增产，卜云龙，等. 植物工厂自动立体栽培系统的研究开发［J］.农业工程，2018（1）：18-21.
[8] 刘文科，刘义飞.设施园艺LED光源设计与应用策略探讨［J］.中国照明电器，2017（11）：24-27.
[9] Okamoto K，Anagi T，Ondo S.Growth and morphogenesis of lettuce seedlings raised under different combinations of red and blue light［J］.Acta Horticulturac，1997，435：149-157.
[10] Yanagi T，Kamoto K，Akita S.Effects of blue and blue/red lights of two different PPF levels on growth and morphogenesis of lettuceplants［J］.Acta Horticulturac，1996，400：117-122.
[11] Nichols M，Hristie C B.Towards a sustainable “greenhouse” vegetable factory［J］.Acta Horticulturac，2002，578：153-156.
[12] Yanagi T，Kamoto K.Utilization of super -bright light emittingdiodes as an artificial light source for plant growth［J］.Acta Horticulturac，1997，18：223-228.
[13] Yorio N C，Oins G D，Agie H R，et al.Improving spinach，radish，and lettuce growth under red light -emittting diodes（LEDs）with blue light supplementation［J］.Hort Science，2001，36（2）：380-383.
[14] Shin K S，Urthy H N，Eo J W，et al.Induction of betala in pigmentation in hairy roots of red beet under different radiation sources［J］.Biologia Plantarum，2003，47（1）：149-152.
[15] 刘水丽.人工光源在闭锁式植物工厂中的应用研究［D］.北京：中国农业科学院，2007.
[16] 张欢，徐志刚，崔瑾.不同光质对萝卜芽苗菜生长和营养品质的影响［J］.中国蔬菜，2009（10）：28-32.
[17] 陈晓丽，郭文忠，薛绪掌.LED 组合光谱对水培生菜矿物质吸收的影响［J］.光谱学与光谱分析，2014（5）：1 394-1 397.
[18] 周华，刘淑娟，王碧琴.不同波长 LED 光 源对生菜生长和品质的影响［J］.江苏农业学报，2015（2）：429-433.

设施园艺工程技术

PROTECTED HORTICULTURE

ENGINEERING TECHNOLOGY

植物无糖培养技术在苹果工厂化育苗中的应用

杨成贺[1]，姜世豪[1]，党康[1]，肖玉兰[2]

（1. 上海离草科技有限公司，上海　201203；2. 浙江清华长三角研究院，杭州　314006）

摘要：以苹果品种m9T337试管苗为材料，应用植物无糖培养微繁殖技术于组培生根壮苗阶段，与传统的有糖培养相比，应用植物无糖培养微繁殖技术可以有效提高苹果组培苗植株长势和根系的发育质量，生根时间和种苗繁育周期明显缩短。实践证明：植物无糖培养微繁殖技术在苹果组培工厂化育苗中反馈效果很好，在未来的种苗工厂化生产中将有着广阔的应用前景。

关键词：无糖培养微繁殖；苹果无糖苗；植株长势；根系发育；培养时间

Application of Sugar-free Micropropagation System on Industrial-scale Production in Apple Plantlets

Yang Chenghe[1]，Jiang Shihao[1]，Dang Kang[1]，Xiao Yulan[2]

（1. *Shanghai LiCao Biotech Co.，Ltd. Shanghai*　201203；
2. *Yangtze Delta Region Institute of Tsinghua University，Hang zhou*　314006）

Abstract：Apple ‘m9T337’ was used as the explants and cultured with sugar-free micropropation system and conventional or sugar-containing micropropation system in the rooting and acclimatization stages. Compared with sugar-containing micropropagation，sugar-free micropropagation showed better growth with good quality，developed rooting system and shortened culture period. Practice has proved that sugar-free micropropagation system has a higher effective on industrial-scale production of apple plantlets，and there will have a wide application prospect in the future in industrial micropropation.

Key words：Sugar-free micropropagation；Apple plantlet；Rooting；Survival percent；Culture period

1　引言

苹果脱毒种苗在栽培生产中能够有效的增强树势，提高果树的抗逆性，提高苹果的产量和品质，因而在种苗市场上的需求量巨大。目前在苹果脱毒苗的生产中主要依靠传统的有糖培养手段获得，但在苹果脱毒苗在组培工厂化生产中存在生根慢、长势弱、缓苗期长、成苗率不高等问题，这些问题制约了苹果脱毒苗的规模化和产业化生产。

植物无糖培养微繁殖（sugar-free micropropagation）技术，又称光自养微繁殖（photoautotrophic micropropagation）技术，是指容器中的小植株在人工光照下，吸收CO_2进行光合作用，以完全自养的方式进行生长繁殖。该技术自20世纪80年代末，世界著

杨成贺，男，硕士，主要从事植物无糖培养技术和无糖培养系统的研究。E-mail：445761383@qq.com

名的环境调控专家古在丰树教授提出以来，受到了全球学者的广泛关注。以往三十余年的大量研究表明[1][2]，相比传统的植物有糖培养，应用植物无糖培养微繁殖技术进行组培苗的繁育，能够有效的促进组培试管苗的生长和发育，组培污染率大幅度降低，整个培养周期也能得到明显的缩短。

试验通过应用植物无糖培养微繁殖技术于苹果组培苗的生根壮苗阶段，并与苹果有糖培养作比较，旨在解决苹果脱毒种苗组培工厂化生产中在生根壮苗阶段存在的问题，也为其他作物的无糖组培工厂化育苗提供借鉴与参考。

2 试验材料与方法

2.1 试验材料

苹果矮化砧木品种M9T337增殖培养30d的繁殖瓶苗为供试材料，选取平均高度为3.04cm，叶片数9.6个，鲜重108.4mg，干重13.5mg的单株试管苗作为生根培养的接种材料。

2.2 试验方法

无糖培养处理采用上海离草科技有限公司开发生产的植物无糖培养系统，系统由培养容器、空气过滤装置、自然换气装置和强制换气装置组成，相关设备安装于培养室组培架上，培养室独立配备CO_2供给系统和补光灯。

接种容器尺寸为28cm（长）×22cm（宽）×15cm（高）的长方形底无糖培养盒，每盒加入含水量35%（重量比）的无菌蛭石350g，无糖液体培养基800ml，培养基为1/2MS培养基，不加糖、琼脂、激素和有机物，pH值为5.8。每盒接种苹果试管苗150株，设置3次重复处理，合计接种6盒。

培养条件为温度25℃±2℃，光照强度为100lx±18$\mu mol \cdot m^{-2} \cdot s^{-1}$，光照时数14h/d，$CO_2$浓度为800～1 000$\mu mol \cdot mol^{-1}$，培养室通气时间与光照时间相同，培养时间30d。

有糖培养处理的培养基为1/2MS+IAA0.3mg/L+琼脂4.2g/L+蔗糖30g/L，pH值为5.8，接种瓶为7cm（直径）×9cm（高）的圆形底组培瓶，每瓶接种试管苗8株，设置3次重复处理，合计接种114瓶。培养条件光照强度为50lx±18$\mu mol \cdot m^{-2} \cdot s^{-1}$，其他条件与无糖培养处理相同，培养时间40d。

在无糖培养30d和有糖培养40d时，每个处理分别随机抽取30株生根苗，测定植株的高度、叶片数、鲜重、干重和单株根数，随后两个处理的所有生根苗移到温室进行炼苗和移栽，分别统计成活率。

3 结果与分析

3.1 苹果无糖培养苗和有糖培养苗的植株长势对比

由表1可知，苹果无糖培养苗（30d）的平均株高为7.68cm，单株叶片数为14.8个，单株鲜重为720.1mg，单株干重为81.4mg；苹果有糖培养苗（40d）的平均株高为3.59cm，单株叶片数为10.4个，单株鲜重为220.6mg，单株干重为26.3mg。苹果无糖培养苗单株的株高、叶片数、鲜重、干重分别是有糖培养苗的2.1倍、1.4倍、3.3倍和3.1倍，苹果无糖培养苗的长势明显优于有糖培养苗。

表1　不同培养方式对苹果试管苗生长的影响

Table 1　Effects of different culture methods on the growth of apple plantlets in vitro

项目	株高（cm）	叶片数（个）	植株鲜重（mg）	植株干重（mg）	单株根数（条）	生根率（%）
无糖培养微繁殖（30d）	7.68	14.8	720.1	81.4	8.2	94.67
有糖培养微繁殖（40d）	3.59	10.4	220.6	26.3	6.4	90.02

3.2　苹果无糖培养苗和有糖培养苗的生根质量对比

无糖培养的苹果苗30d时，生根率为94.67%，平均单株根数为8.2条（表1）；有糖培养40d时苹果苗生根率为90.02%，平均单株根数为6.4条（表1）。无糖培养苗的根部无愈伤组织，根系白而粗壮，不同植株间根系的生长发育差异较小；有糖培养苗根部有较为明显的愈伤组织形成，根系呈淡黄色，不同植株间根系的发育差异较大（图1）。

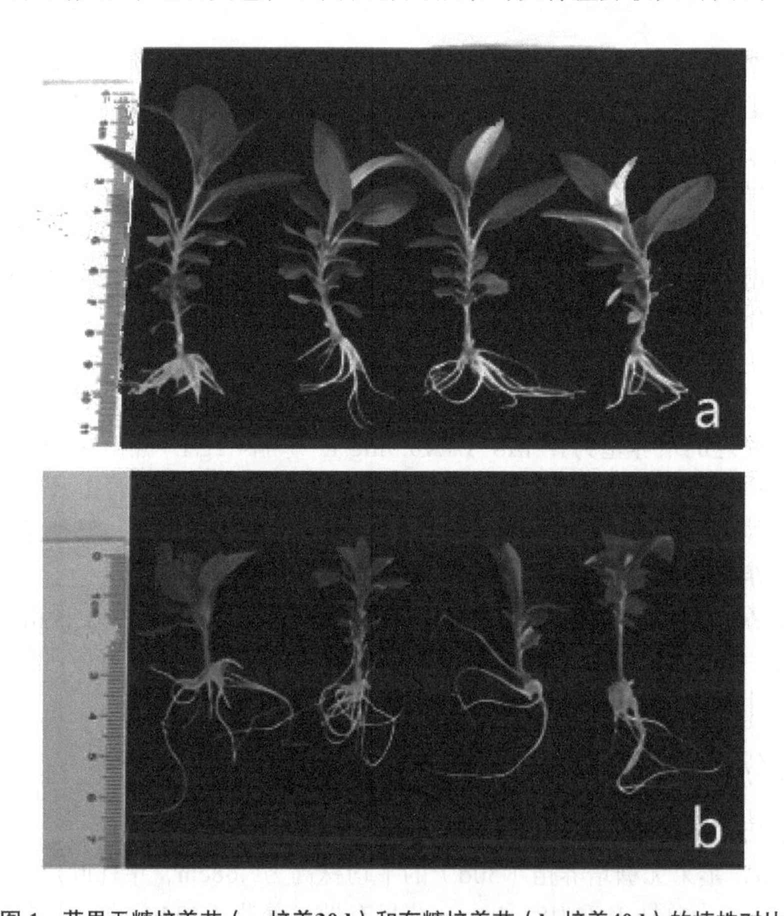

图1　苹果无糖培养苗（a. 培养30d）和有糖培养苗（b. 培养40d）的植株对比

Figure 1　Comparison apple plantlets with sugar-free micropropagation（a. 30d）and sugar-containing micropropagation（b. 40d）

3.3 苹果无糖培养苗和有糖培养苗的温室过渡移苗程序对比

苹果苗无糖培养至30d时即可移植到温室炼苗，把培养容器开盖置于温室中，进行种苗的外界环境适应，5d后就可开始进行苹果无糖培养苗移栽工作。种苗可带蛭石直接从培养容器中栽种至育苗杯中，浇透水，遮阴缓苗处理3d后就可除去遮阴进行种苗的正常管理工作。从试管苗生根培养到温室苗的正常管理，苹果无糖培养苗时间为38d，移栽成活率为83.22%（表2）。

表 2　不同培养方式的苹果组培苗的培养时间和移栽成活率

Table 2　Culture time and transplant survival percents on tissue cultured apple plantlets in different culture methods

项目	培养容器内生根培养天数（d）	移栽前过渡天数（d）	温室移栽后缓苗天数（d）	合计培养天数（d）	移栽前生根苗数量（株）	移栽后成苗数量（株）	成活率（%）
无糖培养微繁殖	30	5	3	38	852	709	83.22
有糖培养微繁殖	40	14	7	61	821	483	58.83

苹果有糖苗的温室过渡在生根培养40d后进行，组培瓶苗不开盖置于温室中过渡适应7d，然后进行时间为7d的开盖缓苗处理，最后进行苹果有糖培养苗的移栽工作。种苗移栽时为保证成活率，先进行根系的琼脂清洗处理工作；然后栽种在育苗杯中，浇透水，遮阴缓苗处理7d后，可进行温室种苗的正常管理工作。从试管苗生根培养到温室苗的正常管理，苹果有糖培养苗的时间为61d，移栽成活率为58.83%，明显低于无糖培养处理（表2）。

4 结论与讨论

4.1 关于苹果无糖培养苗的植株长势

传统的植物有糖培养中，小植物主要依靠糖作为碳源，植株以异养的方式获取植株生长的养分。但在植物组织培养中，培养基中糖是造成污染发生的主要原因之一。为了减少微生物的污染，在有糖培养的过程中，不得不减少与外界的气体交换，使用封闭的、小的培养容器以减少微生物感染的机率。培养容器内的环境不便于调控，使小植株生长在透气性差且高温高湿的环境中，易导致植株长势差，严重时会导致植株生长发育不良甚至死亡。

植物无糖培养微繁殖技术的碳源来自于空气中的CO_2，培养基中不添加糖，这就等于没有了微生物萌发的温床，因此污染发生的机率大大降低，大空间的培养容器可以用于无糖培养使用，便于培养容器内部的环境调控[5]。通过环境调控保持小植株适宜生长的温度和湿度，提高培养容器内部的光照强度和CO_2浓度，可以极大的促进小植株光合作用，使植株的干物质积累和生长发育速度加快，无糖培养苗在植株长势方面优于有糖培养苗[6]。肖玉兰等[3]对比非洲菊的无糖培养苗和有糖培养苗，无糖苗的植株在长势和干物质积累明显优于有糖苗；和世平[4]在对培养半夏无糖培养研究时发现半夏无糖组培苗的株高显著高于有糖组培苗。在本试验中，苹果无糖培养苗的长势显著优于有糖培养，也进一步证实了无糖培养在木本植物的培养上也很有优势。

4.2 关于苹果无糖培养苗的根系发育

传统植物有糖培养中，常常使用凝胶状的物质作为培养基质，如琼脂、卡那胶等，这些材料的透气性极差，进行灭菌后，培养基中的O_2已完全散失，不利于植株根系的发育，且发育的根系细弱，移栽时易受到损伤。

植物无糖培养可以采用多孔的无机物质作为生根培养基质，如蛭石、珍珠岩、砂等，相比有糖培养使用的培养基质，这些材料有较高的空气扩散系数，透气性较好，大大改善了植株的生根环境，有助于根系对养分的吸收，促使根系早生快发，使小植株根系发育良好[7]。在本试验中，苹果苗无糖培养处理的生根率比有糖培养提高了4.65%，生根质量也明显提高，这与采用多孔的无机材料——蛭石作为培养基质也的极大的关系。多孔培养基质利于提高植株的生根质量，这在桉树、甘薯、咖啡、山竹果等多种作物中均得到验证[1]，祁永琼[8]在仙客来的无糖组培中使用蛭石和珍珠岩作为栽培基质，与仙客来使用琼脂作为培养基质有糖培养比较，无糖培养苗的根系发育和生根率均优于有糖培养苗。

4.3 关于苹果无糖苗的室外过渡程序

本试验苹果无糖培养苗所需的培养时间比有糖培养减少了23d，而最终种苗的成活率还提高了24.39%，其原因是：植物有糖培养容器内的小植株生长的环境与外界环境差异较大，高温高湿的环境造成小植株的组织结构和气孔功能异常，向瓶外过渡时，植株需要进行营养方式的转变（异养→自养），同时小植株需要逐渐恢复正常的组织结构和气孔功能以适应外界环境条件，因此为保证移苗成活率，这个过程需要繁琐的炼苗程序和较长的时间。

植物无糖培养微繁殖中，通过环境调控使小植株以完全光合自养的方式获取生长所需养分，因此小植株具有较高的光合能力，培养容器内的环境与外界高度相似，因而小植株对外界环境具有较好的适应能力，在向室外进行过渡移苗时，小植株能够保持正常的组织结构和气孔功能，且无需进行营养方式的转变，这就大大缩短了炼苗时间，同时保证了良好的移苗成活率。相比植物有糖培养，在组培工厂化生产中应用无糖培养微繁殖技术，能够有效的缩短种苗的繁育周期，简化移苗过渡程序。

4.4 关于植物无糖培养微繁殖技术在工厂化育苗中的适用性

应用植物无糖培养微繁殖技术，培养容器内部的环境调控极为关键，关于植物无糖培养装置的开发国内外已有不少试验和研究[9]。但以往的无糖培养装置多为大体积的箱式培养系统，应用该类系统对培养室改造较大，前期的接种操作和后期的培养管理相对不便，不便于进行种苗的规模化和产业化生产，加之昂贵的设备价格，使得植物无糖培养微繁殖技术和相关设备难以在组培生产中大面积推广应用。

近两年，上海离草科技有限公司开发了可以用于生产的植物无糖组培系统，适用于组培试管苗的增殖和生根壮苗阶段，该系统完成了培养容器和相关装置的优化，便于组培生产企业的使用操作，在种苗生产企业的实际应用中，能够充分发挥植物无糖培养微繁殖技术的优势。

陕西青美生物科技有限公司应用该系统进行苹果的组培脱毒苗规模化生产，该公司已投入使用的无糖培养室面积为300m^2，当月可完成生产苹果无糖组培苗约52万株。相比于传统的有糖组培手段，应用无糖快繁系统的培养室可培养种苗数量提高了1.4倍，培养室单位面积的利用率提高，种苗的繁育周期缩短了一半，种苗移栽后的成活率也有了很好的保证（表3）。植物无糖培养技术在苹果上的成功应用，也会为其他作物的组培工厂化育苗提供良好的借鉴和参考。相信在未来的种苗的工厂化生产中，植物无糖培养微繁殖技术将会有着非常广阔的应用前景（图2）。

表 3　应用无糖培养微繁殖技术和有糖培养微繁殖技术进行苹果苗工厂化生产的各项数据对比（该数据由陕西青美生物科技有限公司提供）

Table 3　Comparison of apple plantlet production cultured by sugar-free micropropation and sugar-containing micropropation（data from Shaanxi Qingmei Biotechnology Co.，Ltd.）

项目	无糖培养微繁殖	有糖培养微繁殖
培养容器尺寸及类型	28cm（长）×22cm（宽）×15cm（高）的长方形底培养盒	9cm（直径）×14.5cm（高）的圆底培养瓶
（尺寸为150cm长×60cm宽×30cm高）单层培养架可放置培养容器数量（个）	10	70
单个容器接种数量（株）	200	12
单层培养架培养植株数量（株）	2 000	840
生根+炼苗时间（d）	40	80
室外过渡前生根率（%）	98	98
室外过渡缓苗后成活率（%）	80	50
80d出苗数量（株）	3 136	411
80d出苗比例	7.5	1

图 2　植物无糖培养系统在苹果组培工厂化育苗生产中的应用（陕西青美生物科技有限公司）

（a）无糖培养室；（b）、（c）苹果无糖苗的温室过渡炼苗；（d）温室移栽缓苗后的苹果无糖苗

Figure 2　Application of sugar-free micropropagation system on a large scale production in Apple plantlets（Shanxi Qingmei Biotechnology Co.，Ltd.）

（a）Sugar-free micropropation system's room；（b）、（c）Environment adapting in greenhouse of apple plantlets cultured by sugar-free micropropagation；（d）Apple plants cultured by sugar-free micropropagation after transplanted in greenhouse

参考文献

［1］肖玉兰. 植物无糖组培快繁工厂化生产技术［M］. 昆明：云南科技出版社，2003.

［2］屈云慧，熊丽，吴丽芳，等. 无糖组培技术的应用及发展前景［J］.中国种业，2003（12）：17-18.

［3］肖玉兰，张立力，张光怡，等. 非洲菊无糖组织培养技术的应用研究［J］.园艺学报，1998（04）：97-99.

［4］和世平，王荔，陈疏影，等. 半夏无糖组培苗营养生长和光合生理对增施CO_2的响应［J］.云南农业大学学报，2009，24（02）：204-209.

［5］张美君. 微繁殖中的环境控制和培养容器［A］. 中国农业科学院、山东省寿光市人民政府.

［6］设施园艺研究新进展——2009中国·寿光国际设施园艺高层学术论坛论文集［C］.中国农业科学院、山东省寿光市人民政府，2009：8.

［7］Fujiwara K. S. Kira and T. Kozai. Contribution of photosynthesis to dry weight increase of in vitro potato cultures under different CO_2 concentrations［J］. ActaHorticulturae，1995，393：119-126.

［8］Afreen ZobayedF，Zobayed SMA，Kubota C，et al.Supporting material affects the growth and development of in vitro sweet potato plantlets cultured photoautotrophically.［J］.In Vitro Cellular and Development Biology. Plant：Journal of the Tissue Culture Association，1999（6）：470-474.

［9］祁永琼，许邦丽，褚素贞，等. 仙客来组织培养技术的研究［J］. 云南农业科技，2008（05）：13-15.

［10］Xiao Y，Niu G，Kozai T. Development and application of photoautotrophic micropropagation plant system［J］. Plant Cell Tissue & Organ Culture，2011，105（2）：149-158.

Modelling and Experimental Validation of the Thermal Performance of an Active Solar Heat Storage-Release System in a Chinese Solar Greenhouse

Lu Wei, Zhang Yi, Fang Hui, Wu Gang, Ke Xinglin, Wei Xiaoran, He Yongkang, Zhan Zhengpeng, Zhang Chen, Yang Qichang*, Zhou Bo

(1. *Institute of Environment and Sustainable Development in Agriculture*, *Chinese Academy of Agricultural Sciences*, *Beijing* 100081; 2. *Institute of Environment and Sustainable Development in Agriculture*, *Chinese Academy of Agricultural Sciences*, 12 *Southern Street of Zhongguancun*, *Beijing* 100081 *yangqichang@caas.cn*; 3. *School of Animal*, *Rural and Environmental Sciences*, *Nottingham Trent University*, *Brackenhurst Campus*, *Southwell*, *Nottinghamshire*, *NG25 0QF*, *UK*, *zhonghua.bian@ntu.ac.uk*)

Abstract: An active solar heat storage-release (AHS) system that stores solar energy in a water storage tank can supplement heat to raise the air temperature in Chinese solar greenhouses (CSGs) during cold winter nights. To quantify such heat transfer processes and to improve the performance of AHS systems, a tank temperature model was developed. The model was calibrated and validated, and it can predict water temperatures in the storage tank with an average accuracy of 0.4℃. The model was used to determine the required solar collector area and storage tank volume under various desired air temperatures in different greenhouses. In three cases, greenhouses with a ground area of 272 m^2 and roughly 62 m^2 of solar collectors were installed to maintain temperatures above 12℃ under Beijing's climate conditions. To facilitate a one-degree increase in the air temperature set-point, approximately 2 m^2 of additional solar collectors and 0.1 m^3 of additional storage tanks are needed.

Key words: Solar Energy; Greenhouse; Model; Proper design

Nomenclature			
A	area, m^2	g	experimental greenhouse with AHS
C	specific heat, $J \cdot kg^{-3} \cdot k^{-1}$	g	inlet
CH	cost of heating, US$	k	fan-coil unit number, k= 1, 2, ⋯ k
Co	energy conversion efficiency, %		outlet
CV	specific calorific value of the energy source	out	outdoor
D	total heating time, d	p	pump
E	energy, J	plate	heat collection-heat release plate
F	the view factor, dimensionless		release
I	solar radiation intensity,	rg	reference greenhouse without AHS
k	heat transfer coefficient, $W \cdot m^{-2} \cdot K^{-1}$	s	solar energy
Q	heat flux, W	t	time

continued

Nomenclature			
T	temperature,	tank	AHS storage tank
v	volumetric flow rate of water,	s	water
V	volume of water in the storage tank,	x	specific heating system
W	the electric power of water pump W		time interval
a	solar radiation absorptivity		
η	heat collecting efficiency	*Abbreviations*	
ρ	density, $kg \cdot m^{-3}$	AHS	active solar heat storage-release
σ	the Stefan-Boltzmann constant, $W \cdot m^{-2} \cdot K^{-4}$	CFH	coal-fired heating system
		COP	coefficient of performance
Subscripts		CSG	Chinese solar greenhouse
b	insulating blanket covering the south-facing roof of	CV	calorific value
	a greenhouse	EHS	electric heating system
C	collection	GFH	gas-fired heating system
con	consumption	PE	polyethylene
e	extra	PVC	polyvinyl chloride

1 Introduction

The use of greenhouses in agricultural production has increased dramatically over the last three decades. The primary purpose of a greenhouse is to extend the cultivation season and to achieve high yields of quality produce. These aims are only possible when the indoor temperature can be controlled within an optimal range. Heating is an important measure for maintaining greenhouse air temperatures under cold weather conditions (Hassanien, Li, & Lin, 2016; Sethi & Sharma, 2008). To reduce energy consumption (particularly the use of fossil fuels) from greenhouse heating and to increase the sustainability of protected horticulture, considerable attention has been devoted to novel and renewable sources of energy as alternative means for heating greenhouses. Moreover, the efficient development of economical heat storage systems and related devices is as important as the development of novel and renewable energy sources from the perspective of energy conservation (Öztürk & Başçetinçelik, 2003). Solar energy, an abundant and clean energy source, is an attractive and less expensive alternative to conventional fuels for enclosed heating applications (Öztürk, 2005). Based on the working principle, solar heating can be classified into passive or active heating. A traditional Chinese solar greenhouse is a passive solar heating system in which little human intervention is imposed on the heating process. All heat transfer processes in such a greenhouse are self-regulated. Compared to passive heating, active heating utilizes mechanical devices (circulation pumps, fans, etc.) and materials (heat media such as refrigerants, water, air, gravel, etc.) to alter heat transfer and capture processes to increase energy utilization. This approach allows for more control over the level and timing of heating in comparison to passive heating systems at the cost of higher up-front operating expenses (Bot, 2001; Grafiadellis, 1985; Hazami, Kooli, Lazaar, Farhat, & Belghith, 2005; Kumari, Tiwari, & Sodha, 2006; Mavrogianopoulos & Kyritsis, 1993; Santamouris, Argiriou, & Vallindras, 1994; Sethi & Sharma, 2007).

Chinese solar greenhouses (CSGs) are primarily located in northern China from 32 to 41N latitude. Most CSGs function via mechanisms of passive energy storage and release, wherein northern walls intercept radiation and store solar energy during the day and release it at night. Through this passive energy transfer and high insulation at night, CSGs will be frost free even during cold nights with an outdoor temperature of −20℃, but night greenhouse air temperature will decrease to a level of approximately 6~10℃. (Chai, Ma, & Ni, 2012; Chen, 1994; Jinquan, 1996; Tong, Christopher, Li, & Wang, 2013; Wang et al., 2014). For the high tomato production in the Netherlands night temperature should not be lower than 18℃ (Vogelezang, 1993; Bakker et al., 1995), but most of Chinese growers recommend 8℃ as the critical night temperature for tomato production in CSGs without any auxiliary heating systems (Ma, C., & Miao, X., 2005; JiaDianJiaJu, 2016). With the limited CSG climate control capabilities, it is impossible to satisfy temperature requirements for optimal fruit, vegetable and flower production, especially during cold winter nights (Bartzanas, Tchamitchian, & Kittas, 2005; Gao, Liang, & Duan, 2003; Ma & Miao, 2005; Wang, Ma, Chai, & Kong, 2007). To improve collection and release outcomes, the researchers studied many heating systems in CSGs, but their expensive price and limited ability for heating make them difficult for practical application (Liu et al., 2003; Wang, 2004; Bai et al., 2006; Wang et al., 2007; Zhang et al., 2009). A low-price, active heat storage-release (AHS) system was developed, and the thermal performance of CSGs utilizing this system was studied. The main components of an AHS system are solar collectors. Water, the heat-transfer medium, circulates between the solar collectors and a water storage tank via a pump. AHS systems have been experimentally tested and have been shown to be effective at increasing night time air temperatures (Fang et al., 2013, 2015; Hui, Qichang, & Ji, 2008; Liang, Fang, Yang, Zhang, & Sun, 2013; Zhang, Yang, & Fang, 2012).

In principle, AHS systems can be applied in any greenhouse, but the current design method has not been formalized and is still based on rules of thumb. An accurate description of the thermal performance of an AHS system to facilitate the proper design of the collecting and release capacity of AHS systems in relation to the greenhouse temperature requirements is still missing. Therefore, the goal of this study was to develop and validate a numerical model for AHS systems in connection to the thermal behaviour of CSGs. The model was calibrated and verified with experimental data from functioning Chinese solar greenhouses and used to design AHS capacity. While the Northern

Chinese continental climate only allows greenhouse production in winter emphasis is on this season.

2 Greenhouse and AHS

An experiment was conducted in two identical CSGs in the Shunyi District of Beijing (latitude 40°26′ N, longitude 116°52′ E) to determine the effects of AHS. The CSGs were east-west oriented, 34 m long and 8 m wide. Their ridge height was 3.8 m, and the height of the northern wall was 2.6 m. In one CSG (the experimental greenhouse), an AHS system was installed. The other CSG without AHS served as a reference greenhouse. The CSGs were positioned 8 m apart in the north-south direction. A full-size tomato crop (Solanum lycopersicum, cv. 'Ruifen882'; Rijk Zwaan, De Lier, The Netherlands) was grown during the experimental period. Some specifications on the CSGs are given in Table 1.

Table 1 Characteristics of the tested CSGs.

Section	Construction and coverage materials	Surface area, m^2
North wall	Two 10-mm-thick cement boards with 100-mm polystyrene foam in between	85
West and east end walls	240-mm clay brick + 100-mm polystyrene foam + 120-mm clay brick + 10-mm tiles	46
North roof	Two 20-mm fibre cement boards with a 150-mm polystyrene board in between	51
South roof	0.08-mm layer of transparent PVC film covered with a 20-mm PE foam thermal blanket at night	277
Floor	Bare soil	272
Framework structure	Steel skeleton	

The AHS system (Figure 1) consisted of three components:①solar collectors including one layer of 20mm polystyrene insulation board as a base and two layers of 0.2mm black polyethylene film with a 3mm gap allowing water to flow through; ②a heat storage tank with 12cm water-resistant clay brick walls externally insulated with 10cm polystyrene board; and ③a plumbing system with a water circulation pump at a rated power level of 0.75 kW. The plumbing system connected the solar collector and the storage tank.

Nineteen solar collectors were installed on the interior surface of the north wall with a flow rate of $0.3m^3 \cdot h^{-1}$ through each solar collector. The solar collectors were 2 m high and 1.22 m wide and were positioned 0.4 m above the floor and spaced 0.20 m horizontally. The solar collectors were connected in parallel and shared an inflow pipe and return pipe. Circulating water was fed into the inlets from above and was gravity-drained to the return pipe. The solar collectors covered 46 m^2, accounting for 54% of the north wall area. Water was circulated through rigid insulated PVC pipes. An $8m^3$ capacity storage tank was buried underground on the east side of the greenhouse.

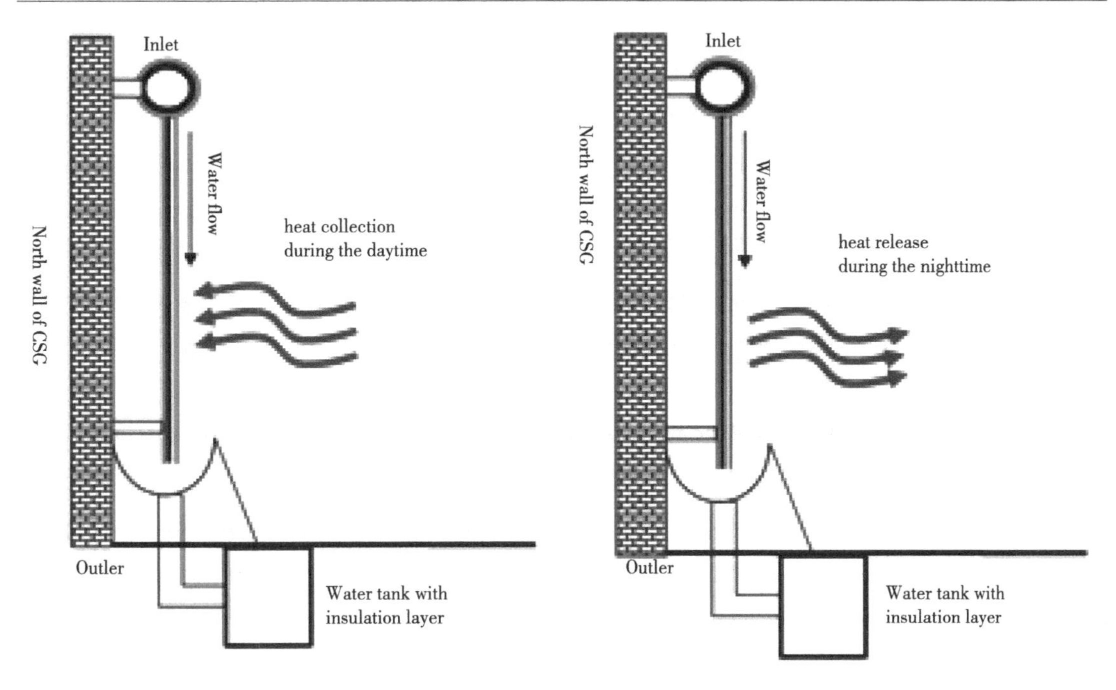

Figure 1　Schematic diagram of the AHS system and its two running modes - heat collection and release.

3　Measurements

Temperatures were measured using T-type thermocouples (manufacturer-claimed accuracy of ±0.2℃). Three air temperature sensors protected from solar radiation were placed in the middle of the greenhouse at a height of 1.5 m, positioned 2 m, 4 m and 6 m from the north wall (Figure 2). Three anti-rust processed water temperature sensing thermocouples were placed in the water tank, with one in the centre, one at the inlet and one at the outlet. The solar irradiation normal to the north wall was measured using a pyranometer (model CMP3, Kipp & Zonen, Delft, The Netherlands; measurement range: 0 to 2 000$W \cdot m^{-2}$; spectral range: 300 to 2 800nm; manufacturer-stated accuracy: ±0.5%). The pyranometer was installed vertically along the south-north centreline of the north wall surface at a height of 1.5 m. All temperatures and solar irradiation levels were recorded using a CR1000 data logger at intervals of 10 min (CR1000, Campbell Scientific Inc., UT, USA. Temperature: measurement range: −25 to 50℃, manufacturer-stated accuracy: ±0.3℃. Solar radiation: measurement range: 0 to 2, 000 $W \cdot m^{-2}$, manufacturer-stated accuracy: 0.5%). Outdoor air temperatures were measured using environment sensors (HOBO U14-001, Onset Computer Corporation, USA; accuracy: temperature 0.2℃, RH ±2.5 %). An electric power meter (T8, Northmeter Co., Ltd., Shenzhen, China) was used to measure the electricity consumption of the water pump, and a vortex flowmeter (LWGY, D & F Company, China; measurement range: 1.5 to 15 $m^3 \cdot h^{-1}$) measured the flow rate in the main circuit pipe.

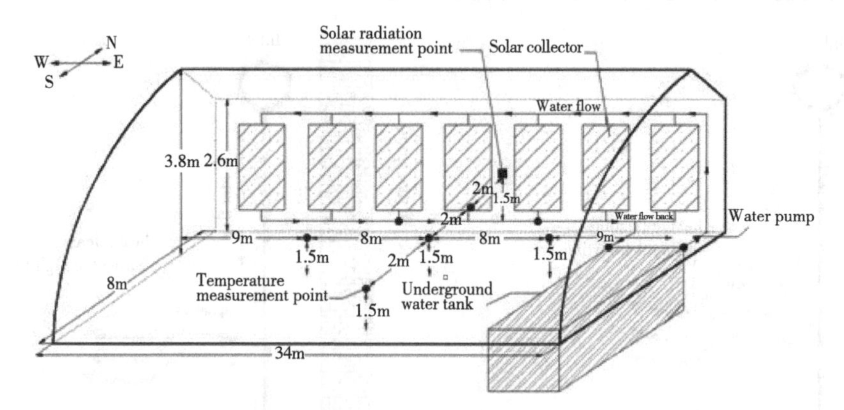

Figure 2 Schematic diagram of the AHS system and positions of measurement instruments in the experimental CSG.

The thermocouples were calibrated using a low-temperature tank（DC-300，Ningbotianheng Instrument Factory，Ningbo，China）at temperatures of 0℃，10℃，20℃，30℃，and 40℃. The HOBO environment sensors were calibrated using an Assmann-aspirated psychrometer（Y-5011，Yoshino，Co.，Tokyo，Japan）at air temperatures of −20℃，−10℃，0℃，10℃，20℃，30℃ and 40℃.

4 Modelling

Solar collectors absorb solar energy and transfer it into the water flowing through them. Some heat energy is lost to the greenhouse. The temperature difference between the inlet and outlet of a solar collector is small compared to the temperature difference that can be realized in an infinitely long collector. Moreover，the volume of the collector is negligible compared to that of the storage tank.

The amount of collected heat at a specific moment is given by

$$Q_c = \rho_w C_w v_w \left(\overline{T}_o - \overline{T}_i\right) \quad (1)$$

where is the instantaneous heat collected，W；is the density of water，kg m^{-3}；is the specific heat of water，J kg^{-1} K^{-1}；is the water flow rate of the AHS system，m^3 s^{-1}；and and are the temperatures at the inlet and outlet of the AHS system，respectively，K.

The total amount of energy E_c（J）collected during time period τ is

$$E_c = \sum_\tau Q_c \cdot \tau \quad (2)$$

The instantaneously released energy Q_r（W）is

$$Q_r = \rho_w v_w C_w \left(\overline{T}_i - \overline{T}_o\right) \quad (3)$$

The total heat E_r（J）released during period τ is

$$E_r = \sum_\tau Q_r \cdot \tau \quad (4)$$

The instantaneous heat collection efficiency of the AHS system is defined as the ratio of instantaneously collected heat energy to available solar radiation:

$$\eta_c = \frac{Q_c}{A_c I_c} = \frac{\rho_w v_w C_w \left(\overline{T}_i - \overline{T}_o\right)}{A_c I_c} \quad (5)$$

Where is the effective heat collecting area of the AHS, m^2; and is the solar radiation intensity normal to the surface of the north wall, $W\cdot m^{-2}$. As the collected energy is equal to the absorbed radiation energy subtracted from the heat lost from the collectors to the greenhouse (Goldberg & Klein, 1980), equation 5 can be written as

$$\eta_c = \frac{aA_cI_c - k_cA_c\left(\bar{T}_{plate} - \bar{T}_g\right)}{A_cI_c} = a - \frac{k_c\left(\bar{T}_{plate} - \bar{T}_g\right)}{I_c} \quad (6)$$

Where is the absorptivity of the solar collectors, dimensionless; is the overall heat transfer coefficient between the collectors and air, $W\cdot m^{-2}\cdot K^{-1}$; is the mean plate temperature, K; and heat losses from the solar collectors include convective heat loss and radiation loss. To simplify the model, the radiation heat loss was combined with the convective heat loss, and the total heat loss is denoted by one term. Furthermore, the surrounding surface temperature was assumed to be equal to the greenhouse air temperature (Incropera et al., 2007).

We can define an average efficiency over period τ as

$$\eta_c = \frac{E_c}{E_s} = \frac{E_c}{\sum_{\tau} A_cI_c\tau} \quad (7)$$

Where E_s is the total solar energy (J) reached during period τ.

The energy balance of the storage tank is given by the inflow and outflow of water and by the heat loss:

$$V\rho_wC_w\frac{d\bar{T}_{\tan k}}{dt} = v_w\rho_wC_w\left(\bar{T}_o - \bar{T}_{\tan k}\right) \quad (8)$$

Where is the water volume of the storage tank, m^3; is the mean temperature from the inlet of the tank to the outlet of the solar collectors, K; and is the mean temperature of the water tank, K. The heat loss from the well-insulated water tank was neglected in the present study.

Using the Euler forward method, the tank temperature at time t+Δt can be calculated from the temperature at time *t:*

$$\bar{T}_{\tan k,t+\Delta t} - \bar{T}_{\tan k,t} = \frac{v_w}{V}\left(\bar{T}_o - \bar{T}_{\tan k,t}\right)\Delta t \quad (9)$$

The energy balance of the storage tank is associated with the collector based on the following equation:

$$V\rho_wC_w\frac{d\bar{T}_{\tan k}}{dt} = aA_cI_c - k_cA_c\left(\bar{T}_{plate} - \bar{T}_g\right) \quad (10)$$

Using the Euler forward method, equation 10 is discretized as

$$V\rho_wC_w\left(\bar{T}_{\tan k,t+\Delta t} - \bar{T}_{\tan k,t}\right) = aA_cI_c\Delta t - k_cA_c\left(\bar{T}_{plate,t} - \bar{T}_{g,t}\right)\Delta t \quad (11)$$

The mean plate temperature is given by

$$\bar{T}_{plate,t} = \frac{\bar{T}_o + \bar{T}_{\tan k,t}}{2} \quad (12)$$

The substitution of in equation 11 and the substitution of in equation 9 produce the following equation:

$$\bar{T}_{\tan k,t+\Delta t}=\frac{2v_w}{2Vv_w\rho_w C_w+k_c A_c V}\left[\left(V\rho_w C_w-k_c A_c\Delta t+\frac{k_c A_c V}{2v_w}\right)\bar{T}_{\tan k,t}+aA_c I_c\Delta t+k_c A_c\bar{T}_g\Delta t\right] \quad (13)$$

T_{tank} can thus be calculated from the temperature of the previous time step.

5 Economic performance of the AHS system

The coefficient of performance (COP) is the ratio of the energy released from the AHS system E_r to the energy consumption E_{con} of the system (so of the water pump) (Sun et al., 2013) :

$$COP=\frac{E_r}{E_{con}} \quad (14)$$

Power consumption of the water pump is the only energy consumption of AHS.

$$E_{con}=W_p\tau_p \quad (15)$$

Where is the electric power of the water pump, W; is the water pump total run time, s.

The economic performance of the AHS system was analysed and compared with three conventional heating systems that are popular in greenhouse heating in Beijing based on the same energy input as AHS and the local energy prices (electricity, coal and natural gas) obtained during the experimental period. The heating costs of AHS and other heating systems were calculated using Eqs. (16) - (18) (Chai et al., 2012; Liu et al., 2003) :

$$E_x=\frac{E_r}{Co_x CV_x} \quad (16)$$

$$CH_x=E_x P_x \quad (17)$$

$$CH_{z,d,A}=\frac{CH_x}{A_g D} \quad (18)$$

Where is the equivalent energy consumed by heating system x to provide the primary energy in the experimental greenhouse in kWh for AHS and EHS, kg for CFH, and m^{-3} for GFH; is the energy conversion efficiency of heating system x, %; is the specific calorific value of the energy source consumed by heating system x, in kJ $kW\cdot h^{-1}$ for electricity, $kJ\cdot kg^{-1}$ for coal and kJ m^{-3} for natural gas; is the unit cost of the energy source in heating system x, in US\$ kWh^{-1} for electricity, US\$$\cdot kg^{-1}$ for coal and US\$$\cdot m^{-3}$ for natural gas; is the cost of heating in US\$; is the cost of heating per square metre greenhouse area per day in US\$$\cdot m^{-2}\cdot d^{-1}$; A_g is the experimental greenhouse area in m^2; and D is total heating time, d.

6 Environmental conditions

The experiment was performed from 1 December, 2014 to 13 January, 2015. Over this period, data for twenty days (21 December 2014 to 9 January 2015) were selected to illustrate the behaviour of the AHS system. A day was defined as running from 8:00 AM to 8:00 AM of the next day. Variations in outdoor air temperature, solar irradiation on the surface of the north wall and greenhouse air temperatures with and without the AHS are shown in Figure 3 (a-b). These twenty days included twelve sunny days and eight cloudy or hazy and foggy days with low temperatures, which is normal for the Beijing area during this time of year. The maximum outdoor air temperature reached 13.5℃ over the twenty days.

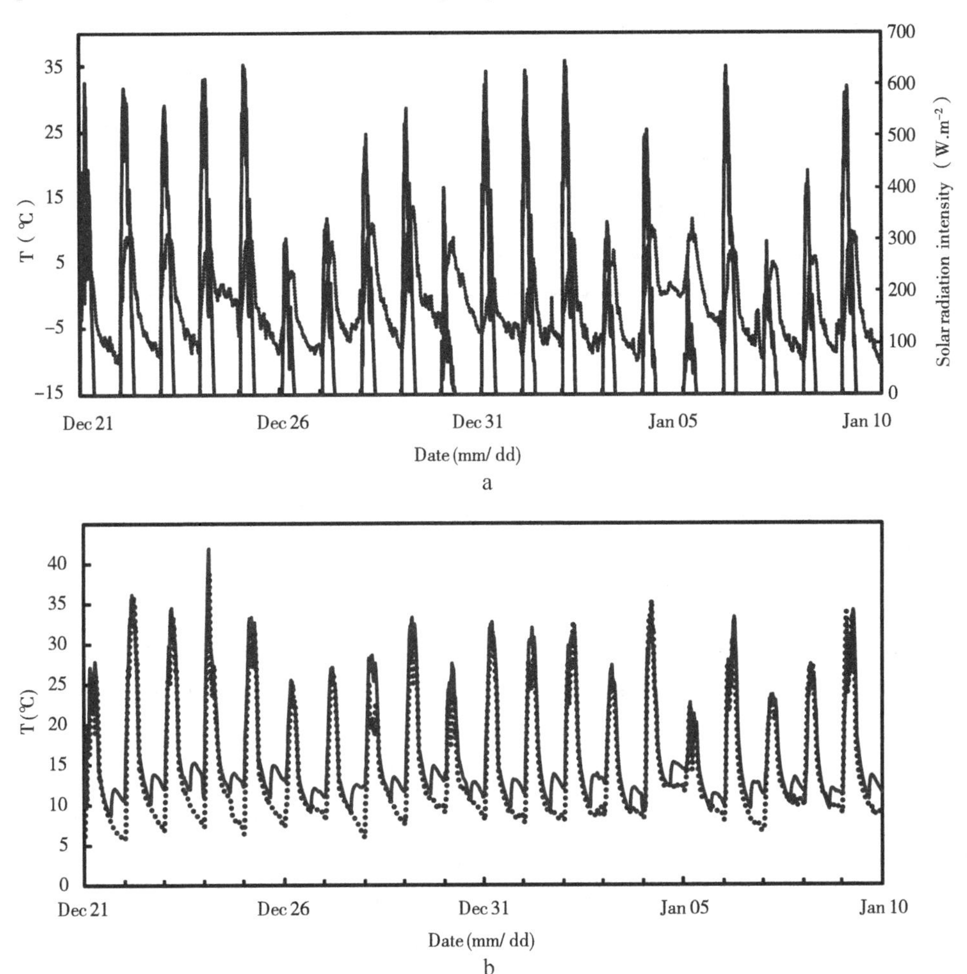

Figure 3 (a-b) Environmental conditions and air temperatures in the experimental and reference greenhouses from 21 December 2014 to 9 January 2015. (a), ——: solar radiation intensity (at the vertical north wall); ········: outdoor air temperature. (b), ——:indoor air temperature in the experimental greenhouse; ········: indoor air temperature in the reference greenhouse.

The daily peak solar radiation on the north wall surface ranged from 267 to 643 $W \cdot m^{-2}$ over the twenty days. CSGs with south-facing transparent covers exhibited a high transmissivity to direct radiation. However, the greenhouse transmissivity to diffuse radiation was low due to the presence of opaque north walls and roofs (Gao et al., 2003). At the lowest solar altitude reached during the year (in Beijing on 22 December at solar noon: 26.5°), the indoor irradiation on the vertical surface was roughly 85% of that on the outdoor horizontal surface for the observed days. The daily peak air temperatures in the reference greenhouses were not significantly different from those of the experimental greenhouse, although the ventilation rates in the two greenhouses may have been different over the ventilating period at around noon. This trend suggests that the peak air temperature is primarily determined by the available solar irradiation.

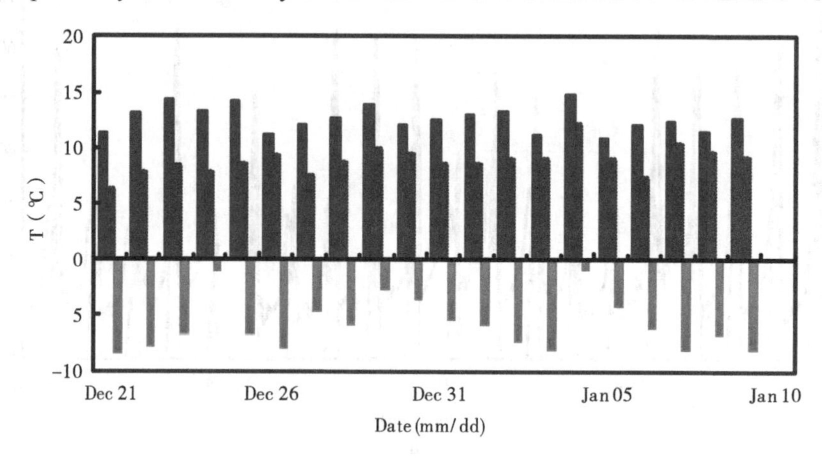

Figure 4　The average air temperature in the CSG with and without AHS and outdoor air temperature for the twenty nights. ■: the average air temperature in the CSG with AHS; ■: the average air temperature in the CSG without AHS; ■: the average outdoor air temperature.

External thermal blankets in the two greenhouses were retracted at 8:00 AM and covered the roof at 4:00 PM. The AHS system was switched on for heating at 0:00 AM at night. The air temperature in the experimental greenhouse increased and then remained stable while the air temperature continued to decrease in the reference greenhouse. Figure 4 presents the average air temperatures in the experimental and reference CSGs, the temperature differences between them and the average outdoor air temperature for the twenty nights. The resulting average air temperature in the greenhouse with the AHS system ranged from 10.9℃ to 14.8℃, with a mean value of 12.6℃ (Table 2) over the twenty nights. The average air temperature in the reference greenhouse ranged from 6.4℃ to 12.2℃, with a mean value of 8.9℃. During heat releasing periods from 0:00 to 8:00, the average outdoor air temperature ranged from −8.3℃ to −0.8℃, with a mean value of −5.7℃. The average indoor-outdoor temperature difference ranged from 14.3℃ to 21.0℃ for the experimental greenhouse and from 8.9℃ to 17.1℃ for the reference greenhouse over the twenty nights. On average, the AHS system increased night-time air temperatures by 3.7℃ (Table 2). Despite the fact that the reference greenhouse was not equipped with an AHS system, the mean indoor air temperature was still approximately 14.6℃ higher than the outdoor temperature because of the insulation of the enclosure and the heat released from the north wall and soil（Li，Bai，& Zhang，2010；Tong，Christopher，& Li，2009；Tong，Wang，Bai，& Liu，2003）.

Table 2　The average air temperature in the CSG with AHS，；the average air temperature in the CSG without AHS，；the average outdoor air temperature，；and the difference between and，and，and for the twenty nights.

Period	$\overline{T}_g$	$\overline{T}_{rg}$	$\overline{T}_{out}$	$\overline{T}_g-\overline{T}_{rg}$	$\overline{T}_g-\overline{T}_{out}$	$\overline{T}_{rg}-\overline{T}_{out}$
	℃	℃	℃	℃	℃	
21 December，2014-9 January，2015	12.6	8.9	−5.7	3.7	18.3	14.6

7　Technical Performance of the AHS System

Figure 5 presents the water temperatures reached in the storage tank over the twenty days. The water temperature is an important indicator of how well a system performs；thus，we can interpret the performance of the AHS system from the tank temperature. The AHS system was automatically turned off at 16:00. It can be observed that daily peak temperatures occurred at approximately 15:00，after which the water temperature started to decrease. This trend suggests that the energy gain was offset by heat losses from the storage tank and that the energy content of the tank started to decrease. Heat losses from the tank occurred due to conduction through the tank walls and/or convection（sensible and latent heat）from the water surface. The former is undesirable；the latter（heat loss through convection），although unintentional and uncontrolled，may not be counterproductive because the heat is lost to the greenhouse air. At 0:00，the AHS system was turned on and started to release energy，causing the water tank temperature to decrease more rapidly in a range from 32.6℃ to 19.2℃ over the twenty nights. The tank temperature increase during the twenty days ranged from 19.2℃ to 36.2℃ and was affected by the available solar energy when the water temperature（Figure 5）and solar radiation（Figure 3a）patterns differed.

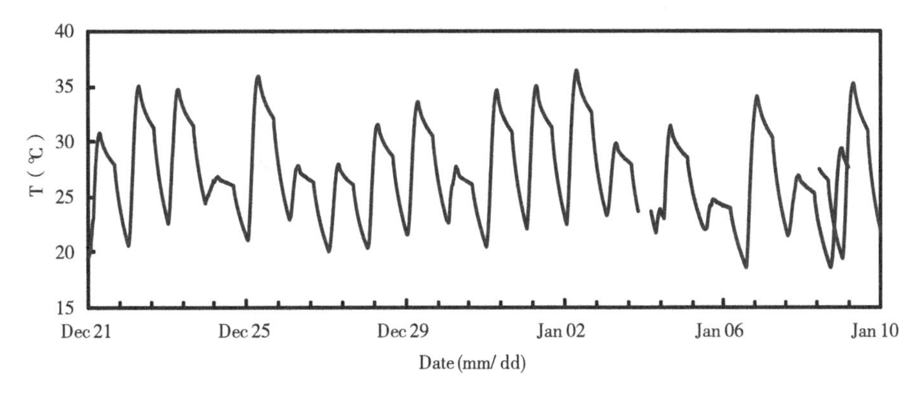

Figure 5　Storage tank temperatures from 21 December 2014 to 9 January 2015.

The performance of the AHS system can also be evaluated by the heat collecting efficiency, utilization efficiency and coefficient of performance (COP) . The results for the twenty-day period are presented in Table 3. The heat collecting efficiency ranged from 41% to 75% over the twenty days. This range may be dependent on outdoor wind speed and ventilation levels in both greenhouses, as openings in the covers were manually adjusted each day and indoor air flow affected the heat loss from solar collectors (Bot, 1983; Incropera et al., 2007; Iordanau, 2009) . Over the entire measurement period, the average heat collection efficiency level of the AHS system reached 62% as the AHS structure was improved and the internal water flow became more uniform, producing an average heat collection efficiency of 1.2 ~ 1.6 times the values reported in previous studies. (Fang et al., 2015; Wang et al., 2007; Zhang et al., 2012) . The utilization efficiency, defined as , ranged from 51% to 198%. A value greater than 100% indicates that the energy released exceeded the energy stored, suggesting that the energy stored from previous days was used.

Table 3 Accumulated solar radiation E_S, collected and released energy E_C and E_r, energy consumed by the water pump for the AHS system over the twenty-day period.

Day	E_S	E_C	Er	Econ	Collection Efficiency	Utilization Efficiency	Coefficient of performance
	MJ	MJ	MJ	MJ	E_C/E_S	Er/E_C	$Er/Econ$
21 December, 2014	338	202	143	41	0.60	0.71	3.5
22 December, 2014	429	273	170	41	0.64	0.62	4.2
23 December, 2014	378	229	170	41	0.61	0.74	4.2
24 December, 2014	221	144	113	41	0.65	0.78	2.8
25 December, 2014	465	349	177	41	0.75	0.51	4.4
26 December, 2014	174	89	115	41	0.51	1.29	2.8
27 December, 2014	229	160	114	41	0.70	0.71	2.8
28 December, 2014	332	244	136	41	0.73	0.56	3.4
29 December, 2014	389	269	151	41	0.69	0.56	3.7
30 December, 2014	144	99	115	41	0.69	1.16	2.8
31 December, 2014	442	303	171	41	0.69	0.56	4.2
1 January, 2015	410	280	174	41	0.68	0.62	4.3
2 January, 2015	444	296	178	41	0.67	0.60	4.4
3 January, 2015	222	123	336	41	0.55	1.18	3.6
4 January, 2015	313	154	101	35	0.49	0.66	2.9
5 January, 2015	105	43	94	27	0.41	1.98	3.1
6 January, 2015	460	297	155	41	0.65	0.52	3.8
7 January, 2015	175	80	112	41	0.46	1.40	2.8
8 January, 2015	277	179	124	41	0.65	0.69	3.1
9 January, 2015	434	284	130	41	0.65	0.46	3.2

Energy lost from the greenhouse is balanced by energy released from the AHS system, uncovered areas of wall, and ground soil. To estimate their relative contributions, the energy lost from the greenhouse to the environment was calculated by applying a constant value (1.85 $W \cdot m^{-2} \cdot k^{-1}$) calculated from the thermal resistance of the insulating blanket (2-cm PE foam in this study) and from convective heat transfer coefficients for the interior and exterior surfaces of the south-facing roof. The resulting heat flows are shown in Figure 6 for four characteristic nights. It is evident that the heat released from the AHS decreased at night due to decreasing water temperatures and that heat losses fluctuated around a constant level. The heat released from uncovered walls and ground soil is the difference between the two lines, corresponding to 51%, 39%, 53% and 44% for the four nights. The COP calculated from equation 15 ranged from 2.8 to 4.4 over the twenty days (see Table 3) . It is dependent mainly on the energy consumption of the water pump that is linked to the flow resistance of the equipment and connecting pipes so to the plumbing of the system. This was done better than previous studies for AHS system with the attention to the pressure losses. But in practical application the plumbing still must be designed more carefully to minimise the demanded energy for pumping.

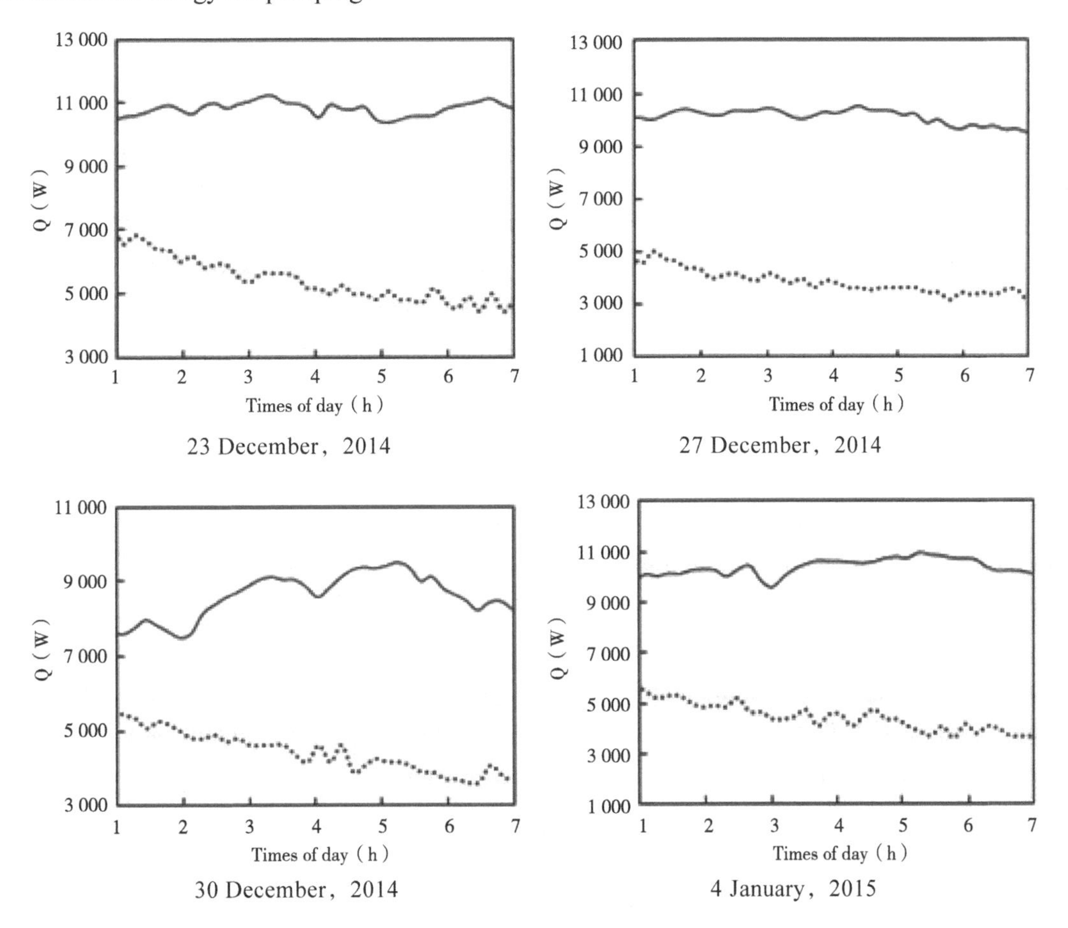

Figure 6 Heat released from the AHS and estimated heat losses from the CSG to the environment for four different nights. ——: the estimated heat losses from the CSG to the environment; ········: the heat release from AHS.

8 Economic Performance of the AHS System

The energy provided by the AHS system was 6114 MJ in the experimental greenhouse from 1 December 2014 to 13 January 2015. Four hundred and eighty-five kWh (1747 MJ) of electric energy were consumed by the AHS system during the experimental period (Table 4) . The peak price and valley price of electricity were 0.226 US$ $kW{\cdot}h^{-1}$ and 0.064 US$ $kW{\cdot}h^{-1}$ during the experimental period in Beijing, and an electricity cost of 0.145 US$ kWh^{-1} was used for calculation in this study because the AHS system was working for approximately 8 hours in the daytime and night-time. The daily heating cost per square metre of greenhouse floor was 0.0059 US$$\cdot m^{-2}\cdot d^{-1}$ in the greenhouse with AHS. Table 4 also illustrates the economic performance comparison results among AHS, EHS, CFH and GFH during the test. The EHS would require an equivalent of 1788 kWh of electricity, the CFH would consume 348 kg of standard coal, and the GFH would expend 203 m^3 of natural gas (249 kg at 288 K and 1 atm) to provide the same quantity of heat as the AHS system released to the greenhouse. Take the heating cost for AHS system as 100; then, the relative heating cost for EHS, CFH and GFH were 368.7, 71.8 and 153.6, respectively. The heating cost of AHS was lower than EHS and GFH by 267.8% and 53.6%, respectively, but was higher than CFH by 28.2%.

In this paper, in Beijing in 2014, the initial total investment in the AHS system was approximately 9.19 US$$\cdot m^{-2}$ per unit of floor area. With an expected 15% increase of tomato production over the reference greenhouse (JiaDianJiaJu, 2013; Zhang et al., 2012) , the payback period was calculated as approximately 5 years. In addition, a grower was paid about a salary of approximately 0.0119 US$$\cdot m^{-2}$ d^{-1} to perform maintenance work on the AHS system during the test, and the depreciation cost could be calculated as approximately 0.00252·US$$\cdot m^{-2}\cdot d^{-1}$, with 10 years as the designed life. The initial investments of EHS, CFH and GFH were 10.318 US$$\cdot m^{-2}$, 8.544 US$$\cdot m^{-2}$ and 8.705 US$$\cdot m^{-2}$, respectively (Ma, Zou, & Lu, 2015) . Taking the labour cost, maintenance cost, and depreciation cot etc. into consideration, and assuming 15 years as the same designated life for EHS, CFH and GFH, the total daily costs of AHS, EHS, CFH and GFH were 0.02 US$$\cdot m^{-2}\cdot d^{-1}$, 0.035 US$$\cdot m^{-2}\cdot d^{-1}$, 0.018 US$$\cdot m^{-2}\cdot d^{-1}$ and 0.022 US$$\cdot m^{-2}\cdot d^{-1}$, respectively.

The heating cost of AHS varies with the fluctuation of energy prices, which were primarily determined by supply and demand, along with regulations and transportation costs. Although the heating cost of AHS was higher than CFH, the benefits of non-renewable energy saving and low CO_2 emission made the AHS a better choice for greenhouse heating in Northern China (for example, in Beijing) , which is confronted by a substantial air-pollution problem, especially in winter. Moreover, the economic comparison of AHS and other conventional heating systems should develop a more complete investigation and analysis based on the investment cost, operational cost, maintenance cost and labour cost associated with practical greenhouse heating, instead of reference to residential buildings.

Table 4 Comparison of energy sources and greenhouse heating costs during the experimental period in Beijing.

Thermo-economic parameters	AHS	EHS	CFH	GFH
Total energy consumption[a]	485 kW·h	1788 kW·h	348 kg	203 m^3
Caloric value (Cv_k), kJ	3600 $kW·h^{-1}$	3600 $kW·h^{-1}$	29306 kg^{-1}	37590 m^{-3}
Energy conversion efficiency (C_k) %[b]	351	95	60	80
Energy price (P_k), US$[c]	0.145 $kW·h^{-1}$	0.145 $kW·h^{-1}$	0.146 kg^{-1}	0.532 m^{-3}
Unit heating cost, US$·$m^{-2}$·$d^{-1}$	0.0059	0.0217	0.0042	0.0090
Relative heating cost, %	100	368.7	71.8	153.6

a Based on 6114 MJ actually provided by the AHS to CSG from 1 December 2014 to 13 January 2015.

b The C_k values are from Ma et al. (2015) .

c Based on mean exchange rate of 1.000 US$ to 6.203 RMB from 1 December 2014 to 13 January 2015 (People’s Bank of China, 2015)

9 Parameter Estimation

Using equation 6, the absorptivity and heat transfer coefficient for the solar collectors and greenhouse air during heat collection were determined by linear regression from the plotted Energy Collecting Efficiency as function of ($T_{plate}-T_g$) / I_c (Figure 7) . The resulting absorptivity of the solar collectors is 0.79 and is 14.7 W·m^{-2}·k^{-1}. The value of the heat transfer coefficient (14.7 W m^{-2} k^{-1}) is deemed reasonable for heat transfer due to convection and thermal radiation in a greenhouse with modest natural ventilation (Bot, 1983) .

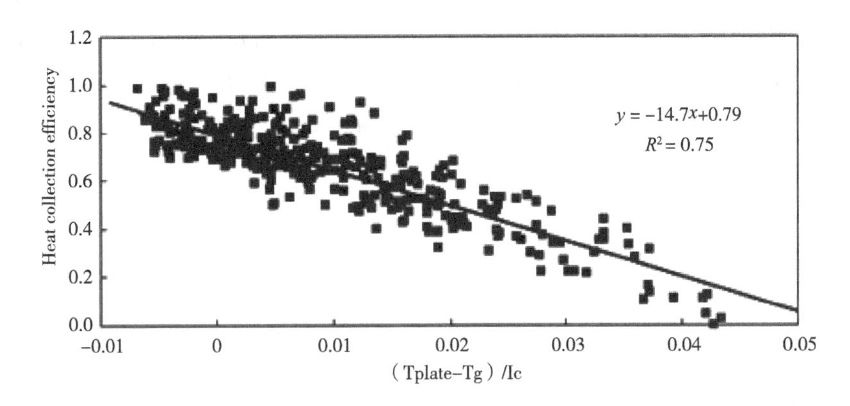

Figure 7 Average heat collecting efficiency (equation 6) of the AHS system as affected by the ratio of the temperature difference between the collectors and the indoor air over the solar radiation intensity. Each data point represents a 10-min average. ■: the data were obtained from 12 December 2014 to 9 January 2015.

In Figure 8, the released heat is plotted (night periods) against the solar collector-greenhouse temperature difference. In some studies, is used for natural convection, and the equation fits our data well. Thus, the resulting heat transfer coefficient equals , in agreement with natural convection heat transfer (Bot, 1983; Fang et al., 2015). In this study, the average was 7.1

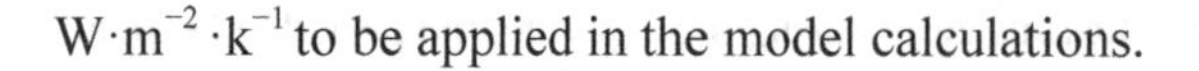

$W \cdot m^{-2} \cdot k^{-1}$ to be applied in the model calculations.

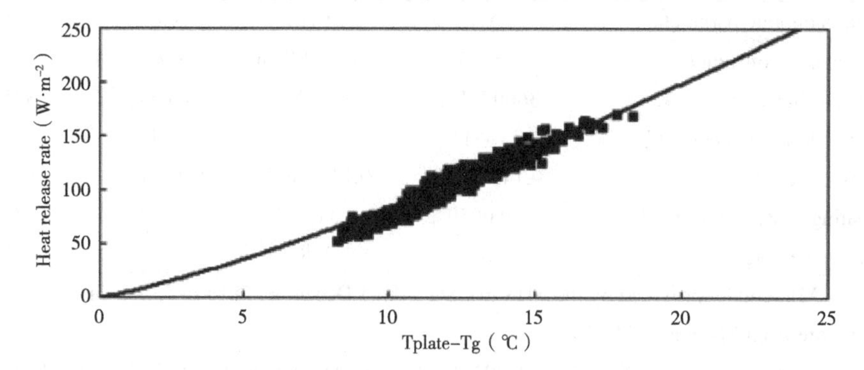

Figure 8　Average heat release rate of the AHS system as affected by the temperature difference between the collectors and the indoor air. Each data point represents a 10-min average. ■ : the data were obtained from 12 December 2014 to 9 January 2015.

10　Storage Tank Temperature Model Validation

The storage tank temperature model presented in equation 13 was validated with six days of data. The parameter was taken as 7.1 $W \cdot m^{-2} \cdot k^{-1}$ for the night and 14.7 $W \cdot m^{-2} \cdot k^{-1}$ for the day; both values were obtained from the experiment. A good fit was obtained between the measured and calculated water temperatures（Figure 9）. Small discrepancies can be observed between 13:00 and 14:00, possibly because of a shadow over the radiation sensor（see Figure 3a）, which caused an underestimation of solar radiation. The model was validated using data for another six days. The calculated water temperature is in good agreement with the measured water temperature. The average error is 0.4℃, and the average percentage error is 1.5%. Thus, the model can be used to accurately predict the storage tank temperature.

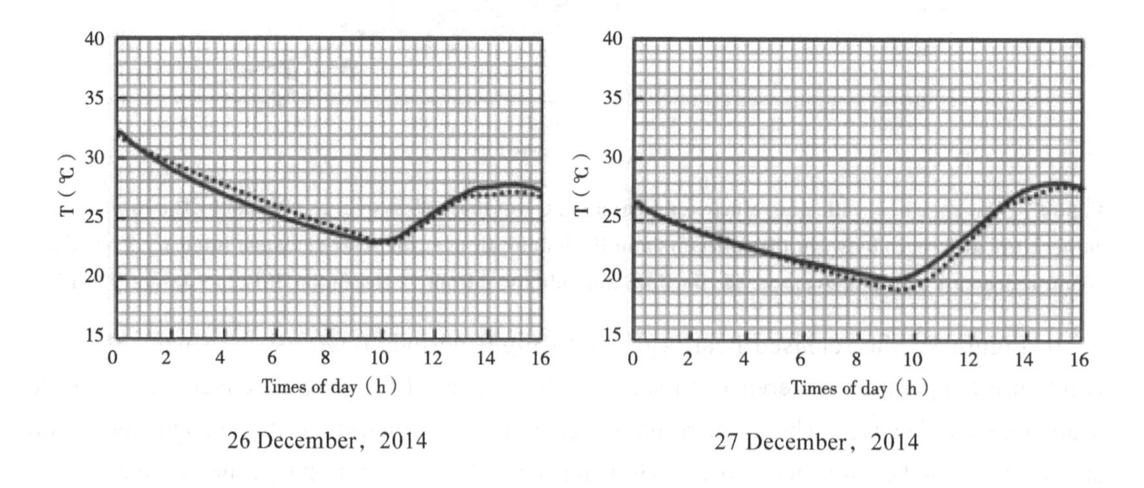

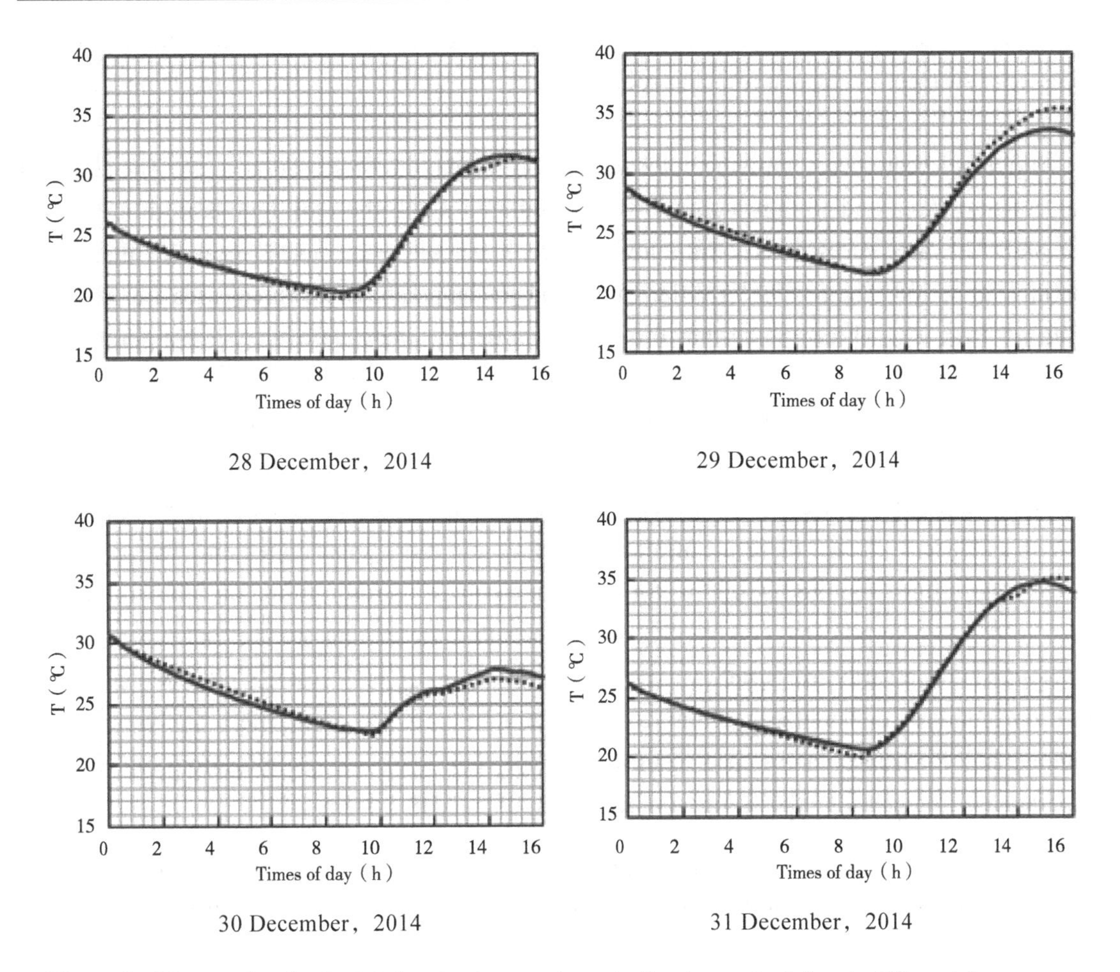

Figure 9 Measured and calculated water temperatures in the storage tank for six different days. ——: measured water tank temperature; ········: calculated water tank temperature.

11 Determination of solar collector area and storage tank volume

The water tank temperature model can be used to determine the solar collector area A_c and storage tank volume V. In this study, three cases of CSG were analysed to make the determination of A_c and V for AHS system. Case-I is the experimental greenhouse with AHS, Case-II is the reference greenhouse without AHS, and Case-III is a greenhouse linked to Case-I and Case-II with the same structure.

If the greenhouse air temperature drops below the set point, additional heat energy must be supplied to increase the greenhouse temperature by ΔT to the set point, as is given by

$$Q_e = k_b A_b \Delta T \quad (14)$$

where is the overall heat transfer coefficient of the south-facing roof, $W \cdot m^{-2} \cdot k^{-1}$; and A_b is the area of the south-facing roof of the greenhouse, m^2. The value of was calculated from the thermal resistance of the insulating blanket (2-cm PE foam in Case-I and Case-II; 2-cm recycled needled felt in Case-III), and convective heat transfer coefficients were derived from the interior and exterior surfaces of the south-facing roof (Chai et al., 2007; He, Fu, Yin, Ding, & Li, 2015).

Additional heat energy must be collected during the day and released at night. The required solar collector area and storage volume were calculated using the tank temperature model for three nights of three cases based on the lowest greenhouse air temperatures measured during the experimental period. The results for different desired air temperatures for three cases are presented in Table 5 (a-c), which shows that for every one-degree increase in the air temperature set point, an additional 2 m^2 of solar collectors is required. The solar collectors currently cover an area of 46 m^2 in Case-I but no area in Case-II or Case-III. To realize a set-point of 12℃, approximately 14 m^2 of additional solar collector area is needed for Case-I, approximately 60 m^2 is needed for Case-II and approximately 62 m^2 is needed for Case-III. The north-facing wall is 54% covered by the AHS in Case-I; thus, there is sufficient space to add more solar collectors. The additional storage tank volume required is 0.1 m^3 for every one-degree increase in the air temperature set-point, which is insignificant. This prediction suggests that the current storage tank (8 m^3) used in Case-I is unnecessarily oversized. An optimal configuration of AHS systems must be determined based on economic considerations. Future studies should carry out cost-benefit analyses (i.e., weighing additional installation costs against potential yield increases resulting from higher night temperatures) to determine the optimal configuration of an AHS system.

Table 5 (a-c)　The average air temperature in the CSG, ; the average outdoor air temperature, ; solar collector area and storage tank volume by simulated need for different greenhouse air temperature set points during the three coldest nights of the experimental period.

a　Case-I

Date			T=12℃			T=13℃			T=14℃			T=15℃		
			Q_e	A	V	Q_e	A	V	Q_e	A	V	Q_e	A	V
	℃	℃	W	m^2	m^3	W	m^2	m^3	W	m^2	m^3	W	m^2	m^3
20-21 Dec., 2014	10.8	−8.3	615	52	4.6	1128	54	4.7	1640	56	4.8	2152	58	4.9
26-27 Dec., 2014	11.2	−7.8	256	60	4.8	769	62	4.9	1281	64	5.0	1794	66	5.1
3-4 Jan., 2015	11.1	−8.0	359	58	4.7	871	60	4.8	1384	62	4.9	1896	64	5.0

b　Case-II

Date			T=12℃			T=13℃			T=14℃			T=15℃		
			Q_e	A	V	Q_e	A	V	Q_e	A	V	Q_e	A	V
	℃	℃	W	m^2	m^3	W	m^2	m^3	W	m^2	m^3	W	m^2	m^3
21-22 Dec., 2014	6.4	−8.3	4681	46	4.4	5517	48	4.5	6353	50	4.6	7189	52	4.7
27-28 Dec., 2014	7.5	−4.5	3761	50	4.6	4597	54	4.8	5433	58	5.0	6269	62	5.2
6-7 Jan., 2015	7.4	−6.1	3845	60	4.9	4681	62	5.0	5517	64	5.1	6353	66	5.2

c　Case-III

Date			T=12℃			T=13℃			T=14℃			T=15℃		
			Q_c	A	V	Q_c	A	V	Q_c	A	V	Q_c	A	V
	℃	℃	W	m^2	m^3	W	m^2	m^3	W	m^2	m^3	W	m^2	m^3
17-18 Dec., 2013	9.5	−11.8	2782	50	4.4	3895	52	4.5	5008	54	4.6	6121	56	4.7
7-8 Jan., 2014	8.8	−5.9	3561	62	5.1	4674	66	5.3	5787	70	5.5	6900	74	5.7
8-9 Jan., 2014	8.6	−12.3	3784	52	4.5	4897	54	4.6	6010	56	4.7	7122	58	4.8

12　Conclusions

An AHS system was used to store solar energy during the daytime and to heat a CSG at night during the winter season. Experimental results show that the AHS system increased greenhouse night air temperatures by an average of 3.7℃. The indoor-outdoor temperature difference in the AHS greenhouse ranged from 14.3℃ to 21℃, with an average of 18.3℃. The average COP was 3.5, which can be improved by proper design of the water flow system.

The AHS heating cost was 267.8% and 53.6% lower than EHS and GFH respectively, but was higher than CFH by 28.2%.

The effective collector absorptivity was measured at 0.79. The heat collecting efficiency of the solar collectors decreased with the temperature difference between the solar collectors and the surrounding air. The heat transfer coefficient between the solar collectors and air was 14.7 $W·m^{-2}·k^{-1}$ during the day and 7.1 $W·m^{-2}·K^{-1}$ at night.

The model predicted water temperatures in the storage tank with an accuracy of 0.4℃. The model was then used to determine the required solar collector area and storage tank volume for different desired greenhouse temperatures. For the 272-m^2 experimental greenhouse equipped with 46 m^2 of solar collecting plates, an additional 14 m^2 of solar collector plates is needed to maintain greenhouse night air temperatures above 12℃ with outdoor average air temperatures of less than −7.8℃. And for other two greenhouses with the same structure but without AHS, 60 m^2 and 62 m^2 of solar collector plates should be equipped to maintain greenhouse night air temperatures above 12℃ with outdoor average air temperatures between −4.6℃ and −12.3℃. For every one-degree increase in the air temperature set-point, roughly 2 m^2 of additional solar collectors is needed. The required increase in storage tank volume is insignificant (roughly 0.1 m^3 for every one-degree increase in the air temperature set-point).

13　Acknowledgements

We thank Professor G. P. A. Bot from Wageningen University, Professor In-bok Lee from Seoul National University, Dr. Shuhai Li for editing the manuscript. This work was sponsored through the National Natural Science Foundation of China (No. 51508560) and the China Scholarship Council State Scholarship Fund (201603250064).

References

Bai, Y., Chi, D., Wang, T., et al., 2006. Experimental research of heating by fire-pit and underground heating exchange system in a solar greenhouse. *Transactions of the CSAE*, *22*, 178-181.

Bakker, J. C., Bot, G. P. A., et al., 1995. Greenhouse climate control: An integrated approach. Netherlands: Wageningen Academic Publishers.

Bartzanas, T., Tchamitchian, M., & Kittas, C. 2005. Influence of the heating method on greenhouse microclimate and energy consumption. Biosystems Engineering, 91, 487-499.

Bot, G. P. A. 1983. Greenhouse Climate: From Physical Processes to a Dynamic Model (PhD Dissertation) Wageningen Agricultural University, Netherlands.

Bot, G. P. A. 2001. Developments in indoor sustainable plant production with emphasis on energy saving. Computers and Electronics in Agriculture, 30, 151-165.

Chai, L., Ma, C., & Ni, J. Q. 2012. Performance evaluation of ground source heat pump system for greenhouse heating in northern China. Biosystems Engineering, 111, 107-117.

Chai, L., Ma, C., Ji, X., et al., 2007. Present situation and the performance analysis of the energy-saving material in solar greenhouse of Beijing region. Journal of Agricultural Mechanization Research, 8, 17-21.

Chen, D. 1994. Advance of the research on the architecture and environment of the Chinese energy-saving sunlight greenhouse. Transactions of the CSAE, 10, 123-129.

Fang, H., Li, W., Lu, W., et al., 2013. An active heat storage-release system using circulating water to supplement heat in a Chinese solar greenhouse. ISHS Acta Horticulturae, 1037, 49-56.

Fang, H., Yang, Q., Zhang, Y., et al., 2015. Performance of a solar heat collection and release system for improving night temperature in a Chinese solar greenhouse. Applied Engineering in Agriculture, 31, 283-289.

Gao, Q., Liang, W., & Duan, A. 2003. Light characteristics and its changing laws in solar greenhouse. Transactions of the CSAE, 19, 200-204.

Goldberg, B., & Klein, W. H. 1980. A model for determining the spectral quality of daylight on a horizontal surface at any geographical location. Solar Energy, 24, 351-357.

Grafiadellis, M. 1986. Development of a passive solar system for heating greenhouses. Acta Horticulturae, 245–252.

Hassanien, R. H. E., Li, M., & Lin, W. D. 2016. Advanced applications of solar energy in agricultural greenhouses. Renewable and Sustainable Energy Reviews, 54, 989-1001.

Hazami, M., Kooli, S., Lazaar, M., et al., 2005. Performance of a solar storage collector. Desalination, 183, 167-172.

He, F., Fu, J., Yin, Y., et al., 2015. Solar greenhouse environment test and analysis in winter of Beijing. Northern Horticulture, 20, 51-53.

Hui, F., Qichang, Y., & Ji, S. 2008. Application of ground-source heat pump and floor heating system to greenhouse heating in winter. Transactions of the Chinese Society of Agricultural Engineering, 24, 145-149.

Incropera, F. P., Autores, V., Dewitt, D. P., et al., 2007. Fundamentals of heat and mass transfer: With introduction to mass and heat transfer (6th ed.). New York: John Wiley & Sons.

Iordanau, G. 2009. Flat-plate solar collectors for water heating with improved heat transfer for application in climatic conditions of the Mediterranean Region (PhD Dissertation). Durham University, UK.

JiaDianJiaJu 2013. Retrieved from http://www.moa.gov.cn/zwllm/zwdt/201301/t20130131_3213056.htm

JiaDianJiaJu 2016. Retrieved from http://guoshu.jinnong.cn/news/2016/8/19/201681910354564906.shtml

Jinquan, P. 1996. Development of sunlight greenhouse in China. Journal of Shihezi Agricultural Academe University, 3, 1-4.

Kumari, N., Tiwari, G. N., & Sodha, M. S. 2006. Modelling of a greenhouse with integrated solar collector for thermal heating. International Journal of Ambient Energy, 27, 125-136.

Li, J., Bai, Q., & Zhang, Y. 2010. Analysis on measurement of heat absorption and release of wall and ground in solar greenhouse. Transactions of the Chinese Society of Agricultural Engineering, 26, 231-236.

Liang, H., Fang, H., Yang, Q., et al., 2013. Performance testing on warming effect of heat storage-release curtain of back wall in Chinese solar greenhouse. Transactions of the Chinese Society of Agricultural Engineering, 29, 187-193.

Liu, S., Zhang, J., Zhang, B., et al., 2003. Experimental study of solar thermal storage for increasing the earth temperature of greenhouse. Acta Energy Solaris Sinica, 24, 461-465.

Ma, C., & Miao, X. 2005. Agricultural biological environment engineering (1st ed.). Beijing: China Agriculture Press.

Ma, Z., Zou, P., & Lu, Y. 2015. Heating, ventilation and air conditioning (3rd ed.). Beijing: China Building Industry Press.

Mavrogianopoulos, G. N., & Kyritsis, S. 1993. Analysis and performance of a greenhouse with water filled passive solar sleeves. Agricultural and Forest Meteorology, 65, 47-61.

Öztürk, H. H. 2005. Experimental evaluation of energy and exergy efficiency of a seasonal latent heat storage system for greenhouse heating. Energy Conversion and Management, 46, 1523-1542.

Öztürk, H. H., & Başçetinçelik, A. 2003. Energy and exergy efficiency of a packed-bed heat storage unit for greenhouse heating. Biosystems Engineering, 86, 231-245.

People' s Bank of China. 2014. The exchange rate between Chinese Yan and U.S dollar. http://www.pbc.gov.cn/zhengcehuobisi/125207/125217/125925/17105/index26.html

Santamouris, M., Argiriou, A., & Vallindras, M. 1994. Design and operation of a low energy consumption passive solar agricultural greenhouse. Solar Energy, 52, 371-378.

Sethi, V. P., & Sharma, S. K. 2007. Experimental and economic study of a greenhouse thermal control system using aquifer water. Energy Conversion and Management, 48, 306-319.

Sethi, V. P., & Sharma, S. K. 2008. Survey and evaluation of heating technologies for worldwide agricultural greenhouse applications. Solar Energy, 82, 832-859.

Sun, W., Yang, Q., Fang, H., et al., 2013. Application of heating system with active heat storagerelease and heat pump in solar greenhouse. Transactions of the Chinese Society of Agricultural Engineering, 29, 168-177.

Tong, G., Christopher, D. M., & Li, B. 2009. Numerical modelling of temperature variations in a Chinese solar greenhouse. Computers and Electronics in Agriculture, 68, 129-139.

Tong, G., Christopher, D. M., Li, T., et al., 2013. Passive solar energy utilization: A review of cross-section building parameter selection for Chinese solar greenhouses. Renewable and Sustainable Energy Reviews, 26, 540-548.

Tong, G., Wang, T., Bai, Y., et al., 2003. Heat transfer property of wall in solar greenhouse. Transactions of the Chinese Society of Agricultural Engineering, 19, 186-189.

Vogelezang, J. V. M. 1993. Bench heating for potplant cultivation: Analysis of effects of root and air temperature on growth, development and production (PhD Dissertation). Wageningen Agricultural University, Netherlands.

Wang, F. 2004. Effect of a solar water auxiliary-heating system on increasing the temperature in solar greenhouse (Master Dissertation). China Agricultural University, Beijing.

Wang, J., Li, S., Guo, S., et al., 2014. Simulation and optimization of solar greenhouses in northern Jiangsu Province of China. Energy and Buildings, 78, 143-152.

Wang, S. S., Ma, C. W., Chai, L. L., et al., 2007. Equipment in sunlight greenhouse for collecting heat and adjusting temperature. Journal of Agricultural Mechanization Research, 2, 130-133.

Zhang, F., Zhang, L., Liu, W., et al., 2009. Experimental study on temperature increasing in solar greenhouse with underground pebble bed thermal storage. Renewable Energy Resources, 27, 7-9.

Zhang, Y., Yang, Q., & Fang, H. 2012. Research on warming effect of water curtain system in Chinese solar greenhouse. Transactions of the Chinese Society of Agricultural Engineering, 28, 188-193.

不同转速下不同介质搅拌产热效果研究

郭宇[1,2]，杨康[1,2]，孙先鹏[1,2]，邱昕洋[3]，李建明[1,2]，胡晓辉[1,2]

（1. 西北农林科技大学园艺学院，杨凌　712100；2. 农业部西北设施园艺工程重点实验室，杨凌　712100；3. 西北农林科技大学机械与电子工程学院，杨凌　712100）

摘要：为了解决风能致热中的关键问题，寻求一种高效的搅拌致热介质，本试验设计并搭建了一套搅拌致热实验系统，采用六直叶圆盘涡轮作为搅拌致热叶片，测量不同介质在不同转速下的产热效果。通过对比水、石蜡油、淀粉悬浊液、沙子悬浊液、饱和NaCl溶液、9%$NaHCO_3$溶液以及石蜡油与水等比例混合作为搅拌致热介质在不同转速下的产热效果，得知相同转速下石蜡与水等比例混合是效果最好的搅拌致热介质。

关键词：风能致热；搅拌致热；搅拌介质

Study of Heating by Direct Stirring with Different Media at Different Speed

Guo Yu[1,2]，Yang Kang[1,2]，Sun Xianpeng[1,2]，Qiu Xinyang[3]，Li Jianming[1,2]，Hu Xiaohui[1,2]

（1. *College of Horticulture*，*Northwest A&F University*，*Yangling*　712100；2. *Key Laboratory of Horticulture Engineering in Northwest China*，*Ministry of Agriculture*，*Yangling*　712100；3. *College of Mechanical and Electric Engineering*，*Northwest A&F University*，*Yangling*　712100）

Abstract: In order to solve the key problems caused by wind-heating，to seek an efficient mixing heating medium，this experiment designed and built a stirring heating experiment system with six straight blade turbine blade as the mixing blade，measuring heat production effect of different media at different rotational speeds. Through the comparison of water，paraffin oil，starch suspension，sand suspension，saturated NaCl solution，9%$NaHCO_3$ solution and paraffin oil and water as the heat effect caused by the mixing ratio of heat medium at different speeds，that same ratio and water mixed paraffin ratio is the best mixing heating medium.

Key words: Wind-heating；Stir heating；Stirring medium

1　引言

设施农业作为一种高效现代化农业模式，近几年在我国得到广泛的推广和快速发展。这种规模化、现代化的农业生产模式已成为中国农业发展的趋势。但在我国北方地区，由于冬季光照时长短、光强小、温度过低，温室内的温度根本不能满足作物正常生长的要求。为了实现茄果类等喜温作物在温室内的正常生产，必须采取增温措施。目前温室冬季

基金项目：国家大宗蔬菜产业技术体系（CARS-23-C-05）

通讯作者：胡晓辉（1977—），博士、教授、博士生导师，主要从事设施农业理论与生产技术方面的研究。E-mail：hxh1977@163.com

供暖主要采用热风炉、电热风机、燃油燃煤锅炉等方式[1]，而这些采暖方式都存在能源效率低、环境污染大、运行成本高的缺陷。加温耗能已成为温室冬季运行的主要障碍。降低能耗，是目前提高温室冬季生产效益的最直接手段[2]。

我国北方冬季气温较低，但其风能资源储量非常丰富。风能是一种清洁的可再生能源，研究、开发和应用风能具有明显的现实意义[3]。利用其丰富的风能资源转化为热能为冬季温室补温，降低冬季温室增温能耗，节约运行成本。此外风热转化设备简单、能量转化效率高、投资低、使用和维护方便，因此开发风能辅助补温装置，对缓解我国能源压力、减轻环境污染、提高温室生产效益、增加农民收入具有重要的现实意义。风力致热是近几十年来兴起的一种高效率的风能利用新技术，日本、美国、荷兰等[4]农业技术发达国家，研制的风力致热装置利用风能对水等工质加热，为畜禽舍、温室、住宅、浴室和一些低温工艺等采暖、供热已初见成效，进入示范试验阶段。国内，中国农业大学、西安交通大学、沈阳工业大学、上海电力学院等[5]高校学者针对不同的致热方式开展了多层次的试验研究，获得了宝贵的成果。

目前风能致热方式主要有搅拌液体致热、油压阻尼孔致热、固体摩擦致热、电涡流致热和压缩空气致热等[6]，其中搅拌液体致热转化效率高，致热装置结构简单、工作可靠。本文通过改变搅拌致热装置中的搅拌介质，探究不用介质对搅拌致热效果的影响。

2 风能搅拌致热试验

2.1 搅拌致热装置与原理

风能搅拌致热的原理是利用风轮转动通过传动机构带动搅拌器转子高速旋转，转子与定子上都有叶片，并且都有一定的间隙，当转子搅拌桶内液体时，会产生涡流运动，运动的液体冲击转子、定子和容器壁面，不停地相互摩擦；液体自身分子的不规则运动碰撞也会相互摩擦，从而将吸收到的动能转换为热能，致使介质的温度逐渐升高。

本试验设计的风能致热装置系统由变频器、三相异步电动机、扭矩传感器、减速器、致热器、热电偶及数据采集器等组成（图1）。试验利用三相异步电动机代替自然风驱动的风力机，通过变频器控制电动机输出不同转速，致热器在不同转速下连续运行；致热器外部填充保温材料，以保证致热器严格保温；用扭矩传感器以及扭矩仪记录搅拌不同介质的功率；在致热器内多点设置热电偶，通过数据采集器实时记录致热器内工质的温度变化情况。

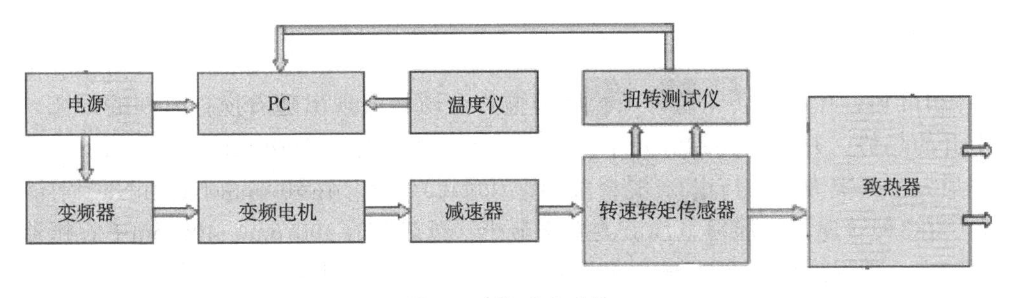

图1　致热试验系统

2.2 技术参数

搅拌桶是搅拌致热系统的主要装置，其由圆柱形桶体、固定在桶壁上的定子和安装在桶盖上的转子、桶内的搅拌介质（图2）。

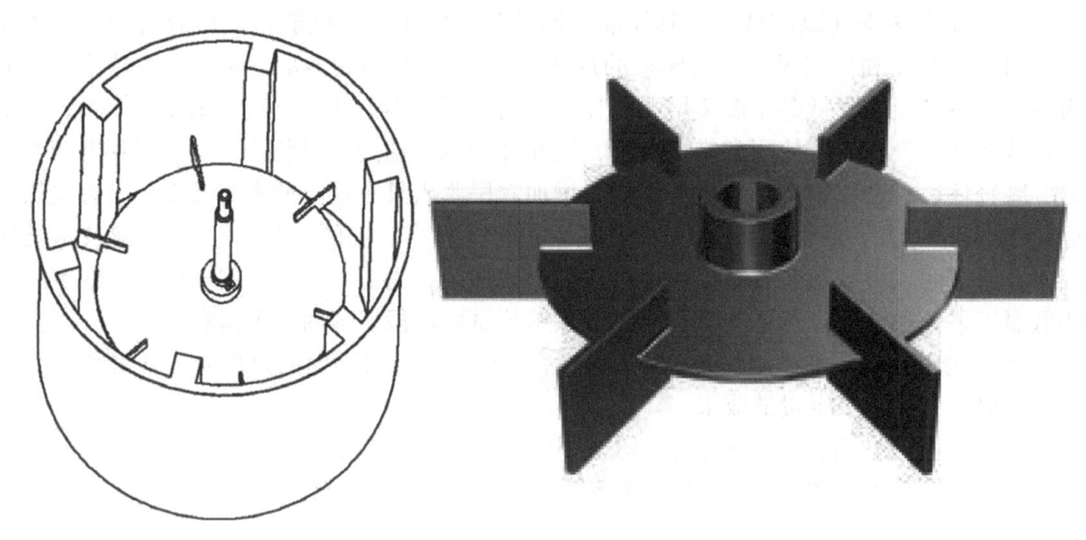

图 2 搅拌桶结构示意图　　图 3 六直叶圆盘示意图

致热器内阻流板的设计，防止了“圆柱状回转区”的出现，增大致热器内的扰动，使致热器内湍流加强，致热效果更加明显。本文设计的致热器搅拌桶内径为425mm，桶高为450mm；阻流板的长度为300mm，宽度为5mm，数量为4片，对称安装在搅拌桶内壁。实验设计的转子叶片为六直叶圆盘涡轮叶片，圆盘直径260mm，每片叶片长60mm，宽50mm，厚6mm，叶片形状如图3所示。

2.3 试验过程

试验主要测量搅拌功率及致热器内液体的温度，以此来计算该系统的各项参数。在环境温度为10℃的条件下连续搅拌介质3h，搅拌介质选用水、石蜡油、淀粉悬浊液、沙子悬浊液、饱和NaCl溶液、9%$NaHCO_3$溶液以及石蜡油与水等比例混合，致热器内液体体积为20L，通过变频器控制电动机轴的转速分别为300rpm、400rpm和500rpm。致热器内设有热电偶，每分钟记录一次致热器内温度，$NaHCO_3$同时通过扭矩传感器测得每种介质的搅拌功率和转速。

3 结果与分析

由图4可知，在不同转速条件下，所有搅拌介质均表现出随着搅拌时间的延长，温度不断上升的趋势，并且这种趋势几乎均为直线上升。其中在300rpm、400rpm下，石蜡油的斜率最大，其次为石蜡与水1:1混合液和饱和NaCl溶液，在500rpm下饱和NaCl溶液的斜率最大，在3种转速下，淀粉悬浊液的斜率最小。这表明在300rpm、400rpm下石蜡油的温升速率最高、效果最好，在500rpm下饱和NaCl溶液的温升速率最高、效果最好，在3种转速下淀粉悬浊液的温升速率均为最低、效果最差。

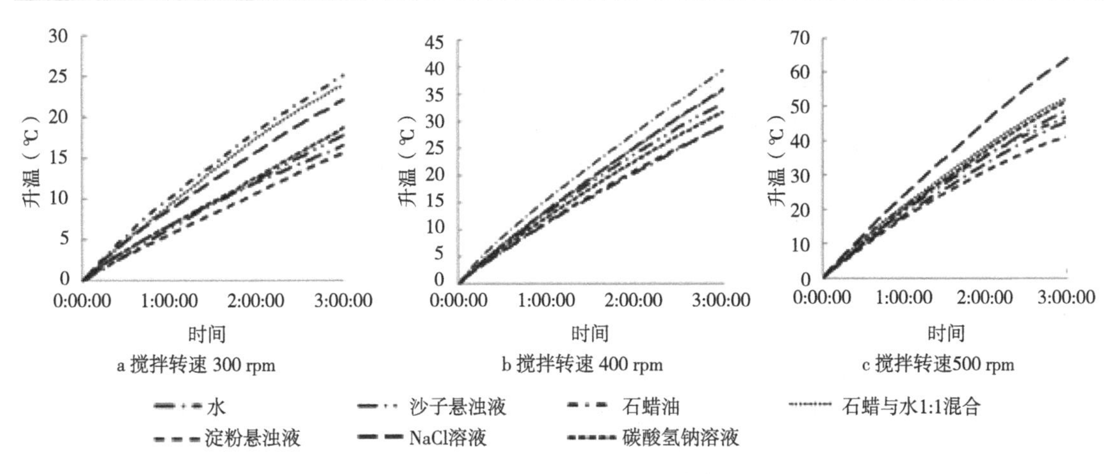

图 4　不同转速下不同介质搅拌温升变化曲线

为了能进一步反映各种搅拌介质的优劣，引入致热效率计算，其表达式为：

$$\eta=\frac{\sum \Delta T\times m_{介质}\times C_{介质}}{\sum_{i=1}^{n}P_i\times\Delta t}\times 100\%$$

式（1）中，η表示效率；ΔT表示段时间内的温升，℃；m表示介质质量，kg；C表示介质比热容，J/（kg·℃）；P表示搅拌功率，W；Δt表示对应做功时间，s。

通过公式（1）计算出每种介质在不同转速下的致热效率（表1）。

表 1　不同介质在3种转速下的致热效率

介质	300rpm	400rpm	500rpm
水	63%	53%	50%
淀粉悬浊液	59%	53%	52%
沙子悬浊液	56%	53%	49%
饱和NaCl溶液	57%	51%	51%
石蜡油	85%	63%	42%
9%$NaHCO_3$溶液	51%	50%	47%
石蜡与水等比例混合	91%	69%	68%

由表1可以看出，石蜡与水等比例混合搅拌致热效率在不同转速下均为最高；石蜡油在低转速下搅拌致热效率较高，高转速下效率较低；水的搅拌效率与淀粉悬浊液、沙子悬浊液、饱和NaCl溶液以及9%$NaHCO_3$溶液几乎一致。

4　讨论与结论

通过试验可知，采用不同液体作为搅拌介质是可以改变搅拌致热效果的。其中石蜡与水等比例混合搅拌致热效率在不同转速下均为最高，并且在不同转速下的温升也较高，是本试验中效果最好的搅拌介质。造成这种现象的原因可能是水和油混合后由于分子间不亲和和密度不同会发生分层，加外力搅拌时会破坏分子间原有的这种状态，使得分子间的碰撞加剧，从而提高了致热效率。而在水中添加各种物质配成的溶液或悬浊液虽然可以提高水的致热温度，但是在致热效率上并没有太大的影响。这可能是添加的物质改变了水的密度以及比热容，从而造成温度上的差异，而这些溶于水的物质配成的溶液在搅拌时由于与水亲和并不会增加分子碰撞，从而对搅拌致热效率没有产生太大影响。但以上猜想还需要大量的试验来进一步证明。

参考文献

[1] 孙先鹏，邹志荣，郭康权，等. 可再生能源在我国设施农业中的应用［J］. 北方园艺. 2012（11）：46-50.

[2] 方慧，杨其长，孙骥. 地源热泵—地板散热系统在温室冬季供暖中的应用［J］. 农业工程学报. 2008（12）：145-149.

[3] 赵建柱，毛恩荣，董生，等. 风能利用与可持续发展［J］. 农机化研究. 2004（06）：40-42.

[4] 王元凯. 国外风力致热现状［J］. 太阳能. 1987（3）：11-13.

[5] 马林. 液压式风能致热系统的研究［D］. 南京理工大学，2015.

[6] 刘洋，胡以怀. 搅拌式风力致热装置的参数设计［J］. 太阳能学报. 2014（10）：1 977-1 980.

内置空气—卵石槽提高日光温室地温的研究

刘名旺[1, 2]，李子栋[1, 2]，曹晏飞[1, 21*]

（1. 西北农林科技大学园艺学院，杨凌　712100；2. 农业部西北设施园艺工程重点实验室，杨凌　712100）

摘要：为提高日光温室越冬栽培时土壤耕层温度，探索日光温室边际区域的界限和边际环境特点。试验在温室南墙内侧设置空气—卵石槽，及无空气—卵石槽（对照，CK）两种处理。通过监测各处理温室内距南墙不同距离、距地表深度20cm处的土壤温度，对日光温室南侧加设空气—卵石槽的保温效果进行分析评价。温室南墙内侧填埋卵石，具有良好的蓄热保温性，可使土壤温度提高1.0～2.4℃、边际界点南移至少60cm，扩大了日光温室冬季种植的土壤温度稳定区域。研究结果可为减少低温灾害对温室越冬茬作物生产的影响提供依据，在越冬茬作物栽培中具有推广意义。

关键词：日光温室；卵石；土壤温度；边际效应

Study on Increasing Soil Temperature in Solar Greenhouse with Air-pebble Groove

Liu Mingwang，Li Zidong，Cao Yanfei

（1 *Department of Horticulture*，*Northwest A & F University*，*Yangling*　712100；2. *Key Laboratory of Protected Horticultural Engineering in Northwest China*，*Ministry of Agriculture*，*Yangling*　712100）

Abstract：In order to improve the temperature of arable soil，and explore the boundary and characteristics of marginal zone in the winter-used solar greenhouse. An air-pebble groove was set inside the south wall of the greenhouse，while air-pebble groove treatment was considered as the control（CK）. The temperatures of soil at depth of 20 cm- from the surface of the inside ground at different distances away from the south wall in the greenhouse was investigated to analyze the effect of heat preservation of the air-pebble groove which added to the south wall. The air-pebble groove presented a good thermal insulation performance. The soil temperature can be raised 1.0～2.4℃ on average，and the boundary point can be moved at least 60 cm southwards，which expanding the stable soil temperature area in the greenhouse. The results can provide the basis for reducing the impact of meteorological disasters on the winter cropping of greenhouses，which is of great significance in the winter cropping cultivation.

Key words：Solar greenhouse；Pebble；Soil temperature；Edge effect

---

［基金项目］宁夏回族自治区重点研发计划重大项目（2016BZ0901）；现代农业产业技术体系建设专项（CARS-25）；博士科研启动基金（2452015274）

［作者简介］刘名旺（1996—），男，河南驻马店人，在读本科生，主要从事设施园艺研究，E-mail：18821714581@163.com

［通讯作者］曹晏飞（1986—），男，湖南娄底人，讲师，博士，主要从事设施农业结构与环境方向的研究，E-mail：caoyanfei@nwsuaf.edu.cn

近年来日光温室由于其良好的保温蓄热性能，发展类型较多，在我国北方地区得到了大面积的推广，深受广大农民的喜爱。目前，日光温室冬季越冬茬作物栽培仍然以传统土壤栽培方式为主。日光温室室内一般无加温设备，冬季极端低温环境会延迟了作物的发育进程、降低了作物品质[1]，其中关键影响因素是地温[2]。根区土壤温度降低对作物根系生长、水分和养分的吸收、土壤微生物的生存等均有影响，从而影响作物地上部分的生长发育[3]。

作物生长越冬期间，温室等设施结构周围存在着边际效应，针对温室边际区域土壤温度的界限以及环境特点，研究学者开展了一系列的研究[4-6]。其中温室南侧是土壤横向传热的主要途径，经常导致温室南侧部位土壤温度较低，影响作物的光合特性等[5]。为了提高日光温室室内土壤温度，研究人员采用覆盖地膜、设置防寒沟[7]以及墙体内置泡沫板[8]等技术。

卵石是一种廉价、良好的蓄热材料，研究人员尝试将这种卵石材料应用到温室中[9-10]。但有关在温室南墙内侧填埋卵石材料对温室土壤温度和边际效应的影响的研究尚未见报道。本试验在南墙内侧填埋一定深度的卵石，研究其对设施内边际土壤耕层温度的影响，旨在分析空气—卵石槽保温性能，以期为设施越冬茬栽培提供新的防寒技术和日光温室优化设计提供依据。

1 材料和方法

1.1 供试温室及空气—卵石蓄热系统

1.1.1 供试温室

试验安排在陕西省杨陵示范区西北农林科技大学北校区园艺场内进行，选取1号日光温室进行数据监测。该温室为砖砌后墙钢骨架结构，棚膜为无滴膜，外覆盖保温被，通常08:00～09:00揭帘，16：00～17:00盖帘。温室长50m，跨度8.0m，脊高3.5m，后墙宽1.0m，高2.2m，沿长度方向分为五个隔间，试验选取第三隔间为试验区域，选取第五隔间为对照区域，隔间长度为8.0m。

1.1.2 空气—卵石蓄热系统

在温室试验隔间的南侧下挖长6m，宽500mm，深500mm的沟槽，槽底部和南侧铺设有用于隔热的毛毡，槽内填埋直径1～10cm的卵石，中部铺设直径为90mm的PVC管道通风，沿管道长度方向每隔30cm钻直径为2cm的小孔用于排风。进风口位于卵石槽西侧，距离室内地面高度1.0m处垂直放置风机。

风机功率为22W，转速1 600r/min，风量127m^3/h。9：00～16：00期间，当室内空气温度＞25℃时开启风机。

1.2 测点布置及监测方法

1.2.1 测点布置

（1）土壤中温度测点设置（图1）：A-1至A-7表示试验区域距地表面20cm处的土壤温度，均布置于试验温室隔间东西方向的中央位置，A-1距离温室南侧基础700mm处，A-1到A-6各个点之间距离200mm，A-7布置于温室中央；对照隔间土壤温度测点与试验隔间测点位置相同，标号T-1到T-7。

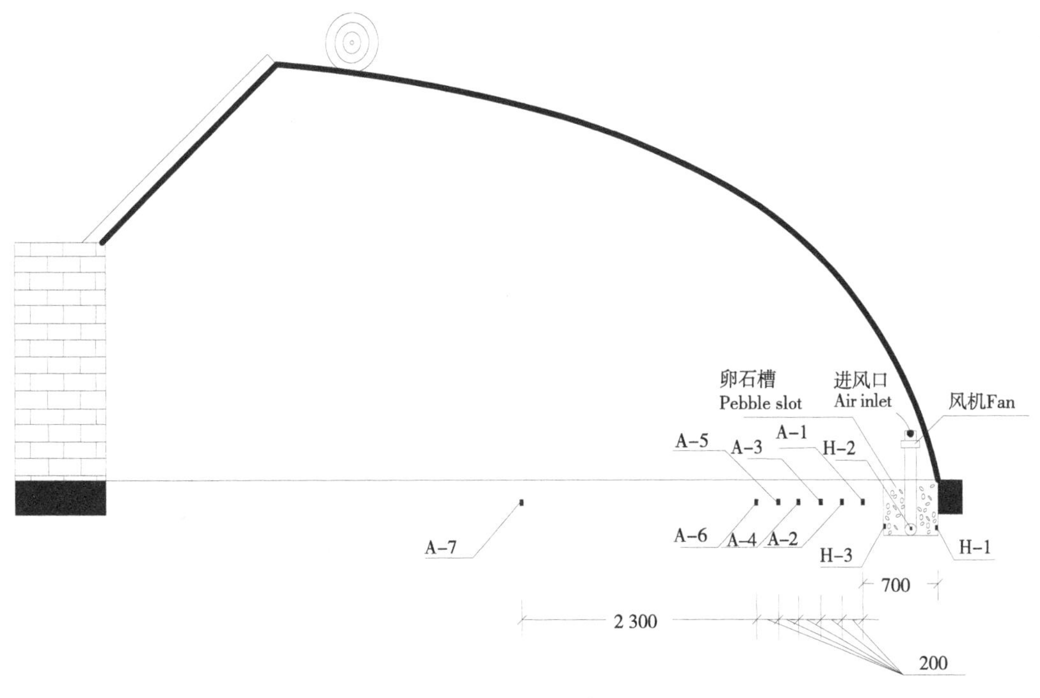

图 1　试验温度测点布置

Figure 1　Test temperature measuring point layout

（2）卵石槽中温度测点的设置（图1）：H-1至H-3表示试验温室南侧空气—卵石槽内距离地表面40cm处的温度测点，布置于试验温室隔间东西方向的中央位置，其中，H-3为卵石槽靠近温室内土壤的槽壁温度测点，H-2为管道内温度测点，H-1为卵石槽靠近温室南侧基础的槽壁温度测点。

1.2.2　数据监测

H-1至H-3温度测点采用HOBO UX120-014M四通道热电偶温度记录仪（测量范围：−20～70℃，精度：±0.21℃）监测，A-1至A-7和T-1至T-7温度测点采用T型热电偶测量，利用安装在安捷伦34 970A数据采集器中的34 901A模块进行数据记录。数据采集时间为2016年12月至2017年2月，数据采集间隔为10min。温室外侧设有一座气象站，可采集太阳辐射、空气温度、以及距地表面20cm处的土壤温度等数据。

1.3　温室边际效应界点确定方法

本试验主要考虑温室南侧卵石槽对土壤温度变化的影响，故忽略其他因素影响。假设同一深度土壤温度受温室内温度影响均匀，且温室中部土壤温度不受室外变化影响。Walker[11]研究表明，根温变化1℃就能引起植物生长的明显变化。若温室边际土壤温度与中部土壤温度的平均差值大于1℃时，该部分的土壤温度会影响作物的生长发育。为此，选用1℃作为确定温室边际效应界点的阈值。

2 结果与分析

2.1 温室土壤温度变化分析

图2显示出2017年1-16日—2017年1-21日温室内A-1/T-1至A-7/T-7等7个点的平均值随时间的变化曲线以及室外相同深度的土壤温度变化。其中，2017年1月16日、20日为多云，17、18日为阴天，19日、21日为晴天。可见，试验隔间土壤温度与对照隔间土壤温度变化趋势基本相同，室外距地表20cm处的土壤温度基本在4℃左右，且试验隔间土壤温度均高于对照隔间土壤温度，显著高于室外相同深度的土壤温度。经计算得，试验隔间土壤温度与对照隔间土壤温度平均温差为1.84℃。Walker[11]研究表明，根温变化1℃就能引起植物生长的明显变化。可见，相比室外土壤温度，在对照温室有较高土壤温度的基础上，试验隔间能进一步为植物提供更适宜的土壤温度。

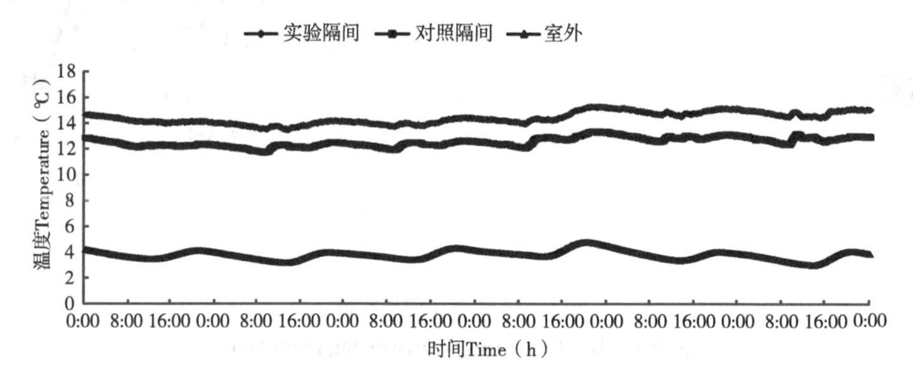

图 2 试验期间不同天气状况下土壤温度变化

Figure 2 Soil temperature changes under different weather conditions during the experiment

考虑到1月份是1年中最冷月，同时为了比较分析土壤温度测点在典型天气条件下的变化差异，为此选用试验隔间和对照隔间中距离南侧墙体最近测点（A-1/T-1）和最远测点（A-7/T-7）在2017-1-21（晴天）和2017年1月16日（阴天）的土壤温度数据进行分析（图3和图4）。

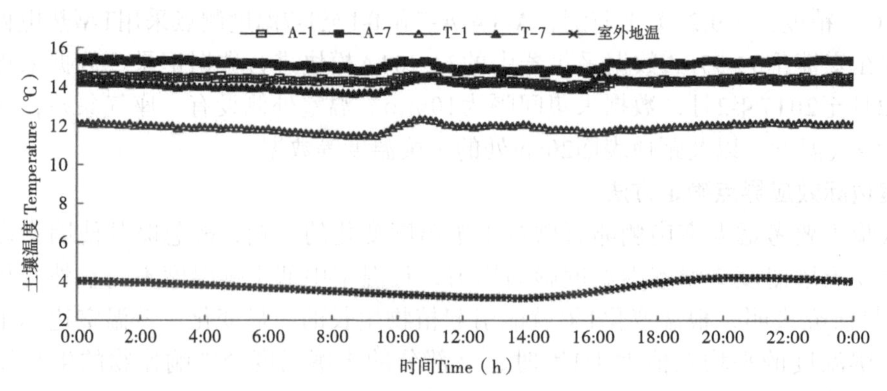

图 3 晴天条件下不同温室隔间土壤温度变化

Figure 3 Soil temperature changes in different greenhouse sections under sunny conditions

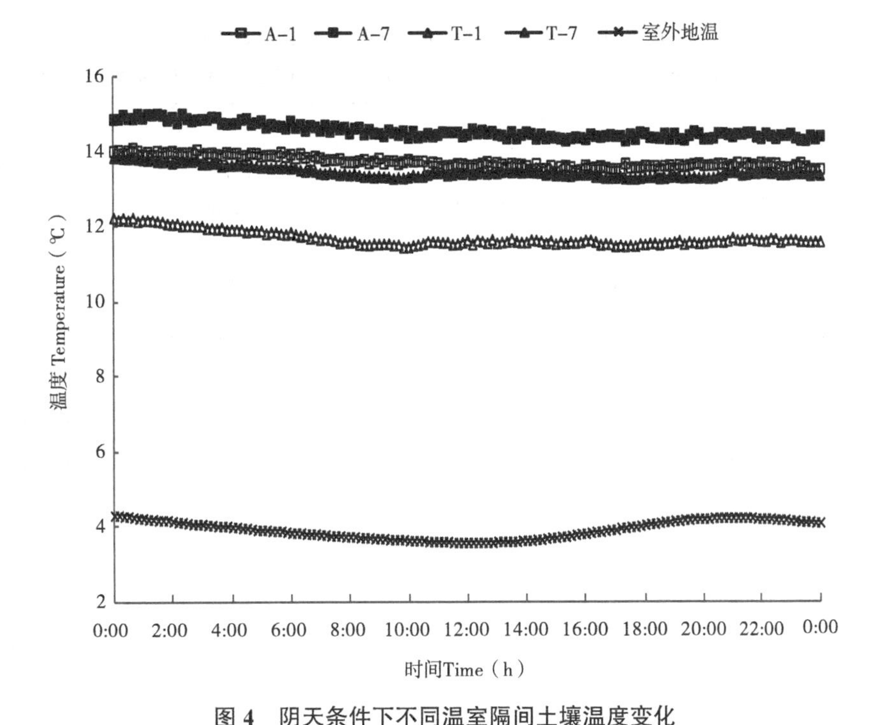

图 4　阴天条件下不同温室隔间土壤温度变化

Figure 4　Soil temperature changes in different greenhouse compartments under cloudy conditions

由图3可知，在晴天条件下，试验隔间与对照隔间的土壤温度均呈现下降—上升—下降—上升的变化趋势，但不同测点的变化幅度相差不大，T-1和T-7测点的变化幅度最大，均约为0.9℃。与对照隔间土壤温度相比，试验隔间中相同位置的土壤温度均有所提高，土壤温度平均提高1.0 ~ 2.4℃。试验隔间与对照隔间距离南侧墙体最远测点的土壤温度均要高于最近测点的土壤温度，其中试验隔间A-7与A-1测点的平均温差为0.8℃，对照隔间T-7与T-1测点的平均温差为2.2℃。

由图3可知，在阴天条件下，试验隔间与对照隔间的土壤温度均呈现缓慢下降的变化趋势，变化幅度均在1.0℃以内，且相同位置的土壤测点温度的变化幅度相比，试验隔间比对照隔间小。与对照隔间土壤温度相比，试验隔间中相同位置的土壤温度均有所提高，土壤温度平均提高1.1 ~ 2.0℃。试验隔间与对照隔间距离南侧墙体最远测点的土壤温度均要高于最近测点的土壤温度，其中试验隔间A-7与A-1测点的平均温差为0.8℃，对照隔间T-7与T-1测点的平均温差为1.8℃。说明在晴天、阴天条件下试验隔间中填埋卵石槽的处理有助于减少了外界土壤环境对室内土壤温度的影响，从而提高土壤温度。而且阴天条件下试验隔间的土壤温度也可保持在13℃以上，避免了冻害的发生，在不影响作物正常生长的条件下还可有效提高根系活力。

2.2　温室土壤边际效应界点变化

表1显示出2017年1月16日—2017年1月21日温室内各测点的白天（6：00 ~ 18：00）和夜晚（18：00 ~ 6：00）的平均土壤温度。

表 1　试验期间不同天气状况下各点土壤温度变化

Table 1　Each point of soil temperature changes under different weather conditions during the experiment

处理区域 Processing area	标号 Label	1.16		1.17		1.18		1.19		1.20		1.21	
		白天 Day time	夜晚 Night time	白天 Day time	夜晚 Night time	白天 Day time	夜晚 Night time	白天 Day time	夜晚 Night time	白天 Day time	夜晚 Night time	白天 Day time	夜晚 Night time
试验隔间 Test compartment	A-1	13.7	13.8	13.3	13.5	13.4	13.5	13.6	13.9	14.2	14.4	14.2	14.4
	A-2	14.1	14.2	13.6	13.9	13.8	14.0	14.1	14.5	14.6	14.9	14.5	14.8
	A-3	14.1	14.3	13.7	14.0	13.9	14.2	14.3	14.8	14.7	15.1	14.6	15.0
	A-4	14.3	14.5	13.8	14.2	14.1	14.4	14.5	15.0	15.0	15.4	14.8	15.2
	A-5	14.4	14.6	13.9	14.3	14.3	14.6	14.7	15.2	15.1	15.5	15.0	15.4
	A-6	14.6	14.8	14.1	14.5	14.5	14.8	14.9	15.4	15.3	15.7	15.1	15.5
	A-7	14.5	14.7	14.0	14.3	14.3	14.5	14.7	15.1	15.2	15.4	15.0	15.3
对照隔间 Control compartment	T-1	11.6	11.8	11.3	11.5	11.5	11.6	11.8	12.0	12.0	12.2	11.8	12.0
	T-2	11.9	12.2	11.7	11.9	11.9	12.1	12.3	12.6	12.4	12.7	12.3	12.5
	T-3	12.1	12.4	11.9	12.2	12.1	12.3	12.5	12.8	12.7	13.0	12.6	12.7
	T-4	12.3	12.5	12.1	12.3	12.3	12.5	12.8	13.0	13.0	13.2	12.9	13.0
	T-5	12.4	12.7	12.3	12.5	12.5	12.7	12.9	13.1	13.2	13.4	13.1	13.2
	T-6	12.6	12.8	12.5	12.7	12.7	12.9	13.1	13.3	13.3	13.5	13.3	13.4
	T-7	13.4	13.6	13.2	13.4	13.5	13.6	13.7	13.9	14.0	14.2	14.0	14.2

由表1可知，每天各点的土壤温度变化趋势相同，无论白天还是夜晚与对照隔间土壤温度相比，试验隔间中相同位置的各点土壤温度均有所提高。在白天，A-1比T-1平均高2.1℃，A-2比T-2平均高2.0℃，A-3比T-3平均高1.9℃，A-4比T-4平均高1.9℃，A-5比T-5平均高1.8℃，A-6比T-6平均高1.9℃，A-7比T-7平均高0.9℃；在夜晚，A-1比T-1平均高2.1℃，A-2比T-2平均高2.1℃，A-3比T-3平均高2.1℃，A-4比T-4平均高2.0℃，A-5比T-5平均高2.0℃，A-6比T-6平均高2.0℃，A-7比T-7平均高1.1℃。每天各点夜晚的土壤温度高于白天，其中在试验隔间，测点夜间土壤温度平均高于白天0.3℃；在对照隔间测点夜间土壤温度平均高于白天0.2℃。

选用2017年1月21日（晴天）和2017年1月16日（阴天）的温度数据进行分析（表2和表3）。参考孙治强等[5]的时间段划分方法，将一天内的土壤温度数据根据变化情况分成4时段，分别为0：00～6：00、6：00～12：00、12：00～18：00、18：00～24：00，计算出每一段时间内试验隔间与对照隔间中不同土壤温度测点与中部土壤温度测点的平均绝对差值。

表 2 晴天条件下不同土壤温度测点与中部土壤温度测点的平均绝对差值

Table 2 Average absolute difference between temperatures at different soil temperatures and middle soil temperatures under sunny conditions

时间 Time	试验隔间 Test compartment						对照隔间 Control compartment					
	A-1	A-2	A-3	A-4	A-5	A-6	T-1	T-2	T-3	T-4	T-5	T-6
0：00～6：00	0.8	0.5	0.3	0.1	0.1	0.0	2.1	1.6	1.3	1.0	0.8	0.7
6：00～12：00	0.7	0.4	0.4	0.2	0.1	0.0	2.2	1.8	1.4	1.1	0.9	0.7
12：00～18：00	0.8	0.4	0.3	0.3	0.1	0.1	2.3	1.8	1.6	1.2	1.1	0.9
18：00～24：00	0.9	0.4	0.4	0.2	0.2	0.0	2.2	1.7	1.5	1.2	1.0	0.8

表 3 阴天条件下不同土壤温度测点与中部土壤温度测点的平均绝对差值

Table 3 Average absolute difference of measuring points between different soil temperatures and central soil temperatures under cloudy conditions

时间 Time	试验隔间 Test compartment						对照隔间 Control compartment					
	A-1	A-2	A-3	A-4	A-5	A-6	T-1	T-2	T-3	T-4	T-5	T-6
0：00～6：00	0.9	0.5	0.4	0.2	0.1	0.1	1.7	1.4	1.2	1.0	0.9	0.8
6：00～12：00	0.8	0.5	0.4	0.2	0.1	0.0	1.8	1.6	1.3	1.1	1.0	0.9
12：00～18：00	0.8	0.5	0.4	0.2	0.1	0.1	1.8	1.5	1.3	1.1	1.0	0.8
18：00～24：00	0.8	0.4	0.3	0.2	0.1	0.1	1.8	1.4	1.2	1.0	0.9	0.7

根据上述1.3节所提出的边际效应界点确定方法，以1℃作为确定温室边际效应界点的阈值。由表1、表2可知，晴天试验隔间A-1测点与隔间中部A-7测点的平均绝对差值低于1℃，说明试验隔间南侧土壤边际效应界点在距温室南墙70cm之内，而对照隔间T-4测点与隔间中部T-7测点的平均绝对差值在1℃左右，说明试验隔间南侧土壤边际效应界点在距温室南墙130～150cm；阴天下试验隔间和对照隔间的土壤边际效应界点分布与晴天相似，试验隔间内与中部A-7测点的平均绝对差值等于1℃的界点在A-1南侧，说明试验隔间南侧土壤边际效应界点在距温室南墙70cm之内，而对照隔间中与隔间中部T-7测点的平均绝对差值为1℃的界点在T-4和T-5，说明试验隔间南侧土壤边际效应界点在距温室南墙130～150cm。综上可知，无论晴天还是阴天，温室试验隔间中南端空气—卵石蓄热系统的使用一定程度上改变了温室内的土壤边际效应界点，使边际界点南移，土壤边际效应界点南移程度最小为60cm，至少增大了温室跨度7.5%的土地利用率。

2.3 温室地下空气—卵石槽温度变化分析

图5示出2017年1月16日至2017年1月21日卵石槽内H-1、H-2和H-3等3个测点随时间变化曲线。其中，2017年1月16日、20日为多云，17日、18日为阴天，19日、21日为晴天。

由图5可知，每天14：30时左右H-2都会出现一次高峰，且19日峰值最大，21日峰值次之，说明晴天白天有热空气通过风机进入了地下卵石槽中进行蓄热。H-1和H-3温度变化趋势相同，且变化相对较小，其中H-3>H-1。

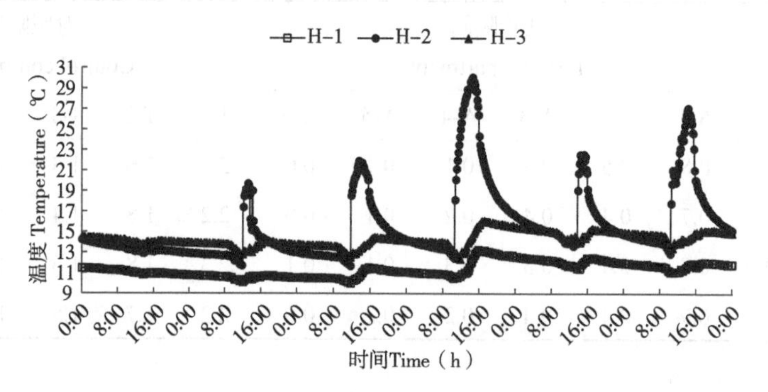

图5　试验期间不同天气状况下空气—卵石槽内不同测点的温度变化

Figure 5　Temperature variation at different measuring points in air-pebble grooves under different weather conditions during the experiment

同样以2017年1月21日（晴天）和2017年1月16日（阴天）的数据为例，分析空气—卵石槽中不同测点的温度变化情况（图6和图7）。

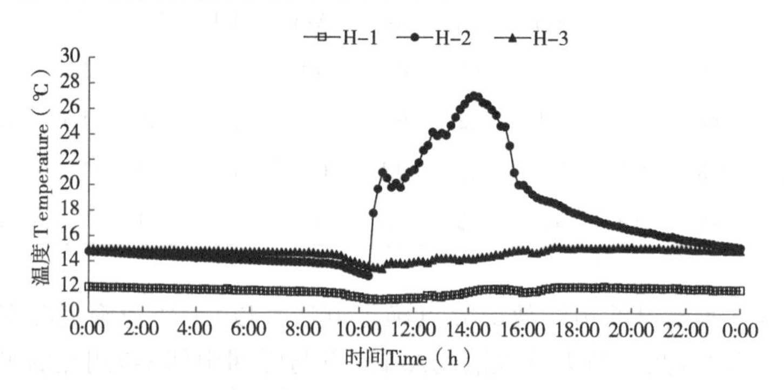

图6　晴天条件下空气—卵石槽内不同测点的温度变化

Figure 6　Temperature changes in different measuring points in air—pebble trough under sunny conditions

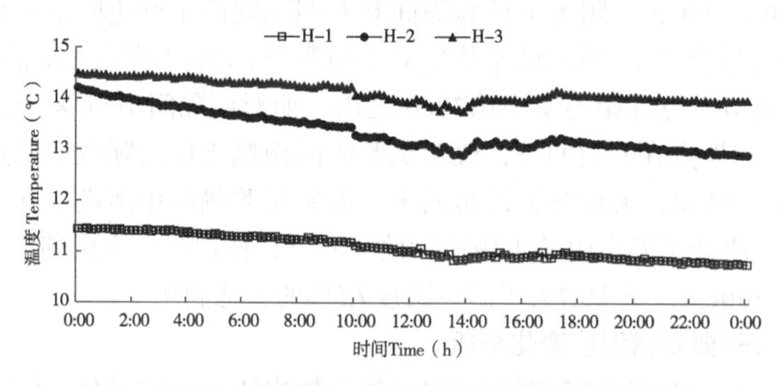

图7　阴天条件下空气—卵石槽内不同测点的温度变化

Figure 7　Temperature changes in different measuring points in air—pebble trough under cloudy conditions

由图6、图7可知，晴天条件下，受室内热空气的影响，H-2测点在风机开启后，温度有一个明显的上升，全天温度变化幅度最大，为14.2℃，H-1、H-3测点温度变化幅度分别为1.1℃、1.8℃，风机开启期间，H-2测点温度明显大于H-1和H-3测点温度，管道内温度上升明显，而管道两侧的测点温度也随之缓慢上升，说明晴天白天有热空气通过风机进入了地下卵石槽中进行蓄热。而在阴天，H-3测点的温度最大，H-2测点的温度次之，H-1测点的温度最小，说明热量由温室卵石槽内侧向外侧传递。其中，H-1测点全天的温度下降幅度小于1℃，根据李明等[12]提出的温波方法，该区域属于热稳定区域，说明卵石槽在阴天向外传递的热量少，有较好的保温性。

3 结论

（1）温室南墙内侧填埋卵石可以阻挡热量外流，具有良好的蓄热保温性，可平均提高土壤温度1.0～2.4℃。

（2）日光温室南侧空气—卵石槽系统的使用较大程度上改变了温室土壤的边际效应界点，促使边际界点至少南移60cm，扩大了日光温室冬季种植的土壤温度稳定区域。

（3）日光温室南墙内置的空气—卵石槽是一个有效的蓄放热体，白天蓄热，夜间放热。

参考文献

［1］刘玉凤，李天来，焦晓赤. 短期夜间亚低温及恢复对番茄光合作用和蔗糖代谢的影响［J］. 园艺学报，2011，38（4）：683-691.

［2］Tahir I S A，Nakata N，Yamaguehi T，et al. Influence of high shoot and root-zone temperatures on growth of three wheat genotypes during early vegetati ve stages［J］. Journal of Agronomy and Crop Science，2008，194（2）：141–151.

［3］傅国海，杨其长，刘文科. LED补光和根区加温对日光温室起垄内嵌式基质栽培甜椒生长及产量的影响［J］. 中国生态农业学报，2017，25（2）：230-238.

［4］王思倩，张志录，侯伟娜，等. 下沉式日光温室南侧边际区域土壤温度变化特征［J］. 农业工程学报，2012，28（8）：235-240.

［5］孙治强，孙丽，王谦，等. 日光温室土壤温度环境边际效应［J］. 农业工程学报，2009，25（5）：150-155.

［6］张海鑫，塔娜，康宏源. 日光温室边际土壤温度试验及模拟研究［J］. 北方园艺，2017（5）：41-48.

［7］白义奎，刘文合，柴宇，等. 防寒沟对日光温室横向地温的影响［J］. 沈阳农业大学学报，2004，35（z1）：595-597.

［8］刘旭，侯伟娜，张涛，等. 日光温室南墙内置泡沫板的保温效果［J］. 中国农业气象，2014，35（1）：26-32.

［9］张峰，张林华，刘文波，等. 带地下卵石床蓄热装置的日光温室增温实验研究［J］. 可再生能源，2009，27（6）：7-9.

［10］张洁，邹志荣，张勇，等. 新型砾石蓄热墙体日光温室性能初探［J］. 北方园艺，2016（2）：46-50.

［11］Walker J M. One degree increments in soil temperature affect maize seedling behavior［J］. Pro. Soc. Soil Sci. Amer.，1969，33：729-736.

［12］李明，周长吉，魏晓明. 日光温室墙体蓄热层厚度确定方法［J］. 农业工程学报，2015，31（2）：177-183.

黑龙江省保温墙体日光温室冬季热环境分析

李明，王平智，宋卫堂，赵淑梅，马承伟，田东坤
［中国农业大学水利与土木工程学院，农业部设施农业工程重点（综合）实验室，北京　100083］

摘要：为了验证日光温室是否可在加强土壤蓄热的基础上取消墙体蓄热功能，利用保温墙体替代传统保温蓄热墙体，对黑龙江省漠河保温墙体日光温室在冬季的光热环境进行了分析，对该温室在冬季的光、热环境进行了测试分析。结果表明：当夜间室外最低气温为-34.7℃时，日光温室内最低气温为7.0℃，平均温度为（10.3±2.5）℃，室内外平均温差为（38.4±2.9）℃，最大温差可达42.2℃，室内气温可满足耐寒作物的生长需求；日光温室保温被和后屋面具有较高的热阻，可有效阻止室内热量向室外散失；墙体蓄热性能较差，在夜间主要承担保温功能；另外，草莓栽培槽外表面夜间温度高于室内气温，可向室内放热，有利于温室维持较高的室内气温。因此，该类型日光温室具有良好的保温功能，配合A字型草莓栽培架，可在高纬度地区的严寒气候条件下维持较高的室内气温，满足作物生长需求。该结果还进一步完善了上述假设，即在加强温室保温性能的基础上，仅依靠土壤和A字型栽培架基质的蓄热量即可维持夜间室内气温，使用保温后墙替代传统保温蓄热后墙可行，未来可据此改革日光温室设计理论。
关键词：日光温室；保温后墙；蓄热性能；保温性能

Analysis of Thermal Environment inChinese Solar Greenhouse with Heat Insulation Wall Located in Heilongjiang Province During Winter

Li Ming，Wang Pingzhi，Song Weitang，Zhao Shumei，Ma Chenwei
（1. *Key Laboratory of Agricultural Engineering in Structure and Environment*，*Ministry of Agriculture*，*College of Water Resources and Civil Engineering*，*China Agricultural University*，*Beijing*　100083，*China*）

Abstract：Chinese solar greenhouse（here after referred to as "solar greenhouse"）is the most popular horticulture facility in Northern China. It can meet the temperature requirement of thermophilic crops without or with little extra heating，even when the lowest outdoor temperature is −28℃. The traditional north wall of the solar greenhouse could absorb the heat from solar radiation and hot air during day time，and release heat during night when the indoor air temperature was low. However，this kind of wall has the problems of low building efficiency and high building cost. Thus，it is not suitable for the development of solar greenhouse in future. To solve the problems，it is assumed that the new type wall build with heat insulation material

收稿日期：2017年12月15日，修订日期：2018年
基金项目：现代农业产业技术体系建设专项（CARS-23-06B）
作者简介：李明（1983年—），男，博士，讲师，从事设施园艺工程研究。中国农业大学水利与土木工程学院，100083。E-mail：lim_abe@cau.edu.cn
通信作者：宋卫堂（1968年—），男，教授，主要从事设施工程与设备研究。中国农业大学水利与土木工程学院，100083。E-mail：songchali@cau.edu.cn

can be applied, if the heat storage performance of the soil can be enhanced. The purpose of this paper was to verification this assumption with a solar greenhouse with heat insulation wall located in Mohe Conty, Heilongjiang Province (122.7° E, 53.48°). The north wall and the back roof of the solar greenhouse were constructed with 15cm thick polystyrene board and 31.5cm thick hollow polystyrene board respectively. The lower edge of the back roof was extended to the ground and formed interspace behind the north wall. The hollow glass was used as the transparent cover material of the solar greenhouse. In the greenhouse, 50cm thick soil was laid on the 10cm thick polystyrene board. During the test period, the solar greenhouse was used for growing celery and straw berry, which were grown on the soil and the A-type cultivation frames, which was composed with pipes filled with substrate. When the lowest outdoor air temperature was −34.7℃, the average and lowest indoor air temperature was (10.3 ± 2.5) ℃ and 7.0℃, respectively. The average and larges temperature differences between indoor and outdoor air were (38.4 ± 2.9) ℃ and 42.2℃, respectively. The results indicated that the solar greenhouse can be used for growing the chimonophilous crop under severe cold climate. It is also found that the thermal resistances of the heat preservation quilt and back roof was very high, so that they can preventing the heat in the solar greenhouse from losing to the outside. during the most time of the night, the inner surface of north wall was 0.2℃ lower than the indoor air temperature. The result indicated that the north wall cannot release heat into the solar greenhouse, but stop the heat in the solar greenhouse from losing to the outside. on the contrary, the exterior surface of the culture pipes was 1.9 ~ 2.2℃ higher than the indoor air temperature during the nighttime. On the other hand, the heat flux that released by the culture pipes of A-type cultivation frames was close to that lost from the front roof. Considering that the total surface area of the culture pipes was 290 m^2, the A-type cultivation frames were helpful for maintaining high indoor air temperature. In conclusion, the solar greenhouse has high levels of light transmittance and heat insulation ability. With the application of A-type cultivation frames, it can meet the reequipment of indoor crops to the environment under severe cold conditions. Thus, the results complete the assumption that mentioned above. It is possible to build the north wall with the heat insulation material when the following conditions were satisfied: (1) both soil and substrate were used to store heat, (2) the insulation performance of the solar greenhouse was improved.

Key words: Solar greenhouse; Heat insulation wall; Heat storage performance; Heat insulation performance

1 引言

日光温室东西北三面环墙、南面透光，是具有中国特色的一种园艺设施，可在我国北方冬季夜间不加温或少量加温的条件下维持较高的室内气温，保证室内蔬菜正常生长，具有明显的经济、社会和生态效益，应用非常广泛[1]。截至2014年，我国日光温室面积101.3万hm^2，占设施园艺总面积的24.6%，成为解决我国北方冬季蔬菜短缺，调整农业产业结构，促进农民增收的重要抓手[2]。此外，随着“一带一路”战略的推进，日光温室技术也是我国实施现代农业走出去的重要内容。

日光温室属于一种生产型土木结构，主要由北墙、前屋面、保温被和后屋面等构成[3]。温室的光照和温度水平，与温室尺寸、北墙结构、保温被材料等因素密切相关[3-5]。其中，日光温室北墙兼具保温和蓄热功能，可在日间吸收并储蓄来自太阳辐射和空气的热量，然后在夜间向室内放热[6-10]。为了实现其蓄热功能，北墙的建造需要使用大量的高密度、高比热容的材料，如夯土、黏土砖、碎石块等。常见的墙体类型有厚土墙式、黏土砖复合墙等。其中，厚土墙厚度较大，土地利用效率较低，还存在着耐久度较式、对耕地层破坏严重等问题。黏土砖复合墙虽然厚度小，土地利用效率较高，但存在着施工效率较低、建造成本过高等问题，也不利于日光温室产业的可持续发展[11]。

针对上诉问题，李明等提出一种新的日光温室设计理论，即在增加土壤蓄热量的基础上，取消墙体的蓄热功能，仅使用轻质保温材料建造日光温室后墙，提高施工效率和质量、降低建造成本[12]。但上述设想目前还没有得到实际验证。本文的目的是针对黑龙江省漠河县的保温后墙日光温室进行测试，对上述设想进行验证，并对影响该类日光温室冬季夜间室内气温的其他因素进行验证，为创新日光温室设计理论、优化建造工艺提供借鉴和参考。

2 试验温室与测试方法

2.1 试验温室

试验温室位于黑龙江省漠河县（122.7° E，53.48° N）。该温室坐北朝南，东西长度为60m，南北净跨7.9m，脊高5.4m，北墙高2.8m，后屋面在水平地面投影2.8m。室内地面下沉60cm；在距室内地面600mm深度处，先铺设100mm厚聚苯板，然后再回填500mm厚耕土。北墙为30cm厚聚苯板，温室内表面上抹1cm厚水泥砂浆。温室前屋面为钢筋焊接桁架结构，采用双层中空玻璃（8mm玻璃+12mm空气夹层+8mm玻璃）作为透明覆盖材料；安装防水保温被。后屋面为双层中空聚苯板（15cm聚苯板+1.5cm空气夹层+15cm聚苯板），从屋脊处延展到地面，并与温室北墙形成一个储藏室。储藏室地面铺设混凝土板，并在混凝土板下方建造了一个宽1.5m、深1.6m的地下室。试验温室结构如图1所示。

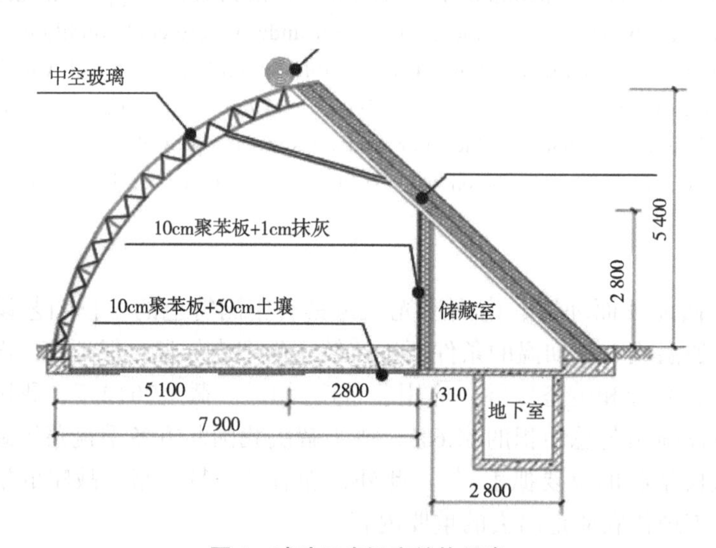

图 1　试验日光温室结构示意

Figure 1　Structure of tested Chinese solar greenhouse

试验温室在测试期间栽培有芹菜和草莓。芹菜直接栽培在温室土壤中，面积约36m^2；草莓在A字型栽培架进行基质栽培，每个栽培架装有7个栽培槽，4层布置，其中顶层安装1个栽培槽，其余3层对称安装，每层2个栽培槽。每个栽培槽长6.0m，内部填充当地的草炭土作为栽培基质，约0.024m³。温室内共放置22个栽培架。试验温室保温被揭开和关闭的时间分别为9:20和15:20。当白天室内温度较高时，开启山墙处的通风扇进行通风。

2.2 测点布置

日光温室室内外气温、北墙内表面温度、土壤地下10cm深度处温度、栽培槽外表面温度以及中空玻璃内墙外表面温度，使用铂电阻传感器测量（测量范围：−100～350℃，精度：±0.5℃）。栽培槽外表面温度测点，分别选择位于顶层、中间层和底层的栽培槽进行测量。室内外太阳辐射光合有效光量子通量密度（Photosynthetic Photon Flux Density，PPFD）采用光合有效辐射传感器测量（Hobo Co.，S-LIA-M003，测量范围：400～700μm；精度：±5%，适用温度范围：−40～70℃）。室内气温和太阳辐射PPFD测定设置，在温室中心、距离室内地面1.0m高的位置。上述传感器采用无纸记录仪采集和储存，数据记录间隔为10min。

试验测试时间为2016年12月27日至2017年1月1日，测试期间未遇到阴天，选择室外气温最低的1天（2016年12月31日09:20至2017年1月1日09:20）中所采集的数据进行分析。

3 结果与分析

3.1 温室光照及室内外气温变化

室内外太阳光的PPFD如图2所示。经测试，2016年12月31日室外PPFD的最大值为453μmol·m^{-2}·s^{-1}，出现在11:20。室内PPFD的变化规律与室外一致，其最大值为311μmol·m^{-2}·s^{-1}。芹菜和草莓的光照补偿点分别为110μmol·m^{-2}·s^{-1}[17]和10～70μmol·m^{-2}·s^{-1}[18]。试验温室全天室内PPFD超过110μmol·m^{-2}·s^{-1}的时间为4.3h，可基本满足室内芹菜和草莓生长对光照的需求。另外，日光温室在保温被揭开期间（12月31日9:20—15:20）的PPFD透过率为（54±10）%，其最大值和最小值分别为77%和48%，其中正午时分的透过率高于其他时间段。

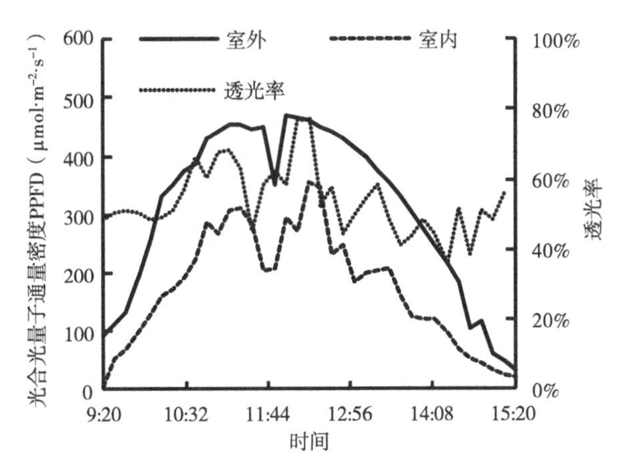

图2 2016年12月31日9:20—15:20室内外太阳光的光合有效光量子通量密度（PPFD）

Figure 2 Indoor and outdoor photosynthetic photon flux density (PPFD) of solar irradiation during the period from9:20—15:20 inDec. 29，2016

在上述气象条件下，室内外气温的变化如图3所示。在日光温室保温被揭开期间（12月31日9:20—15:20），室外气温从-26.2℃逐步上升，在13:40达到最大值-7.4℃，然后开始下降。室内气温的变化趋势与室外气温相似。清晨保温被揭开前室内气温为6.8℃，保温被揭开之后即开始缓慢上升，同样在13:40达到最大值，22.4℃，然后开始下降。该期间内室内外温差为（30.6 ± 1.2）℃。

在日光温室保温被关闭期间（12月31日15:20—1月1日09:20），室外气温随时间逐渐下降，在1月1日07:20达到最低值，-34.7℃。该期间室外气温平均值为（-28.1 ± 5.3）℃，室内气温变化与室外气温相似。当保温被关闭时，室内气温为18.0℃，然后随时间不断下降，在次日保温被揭开时刻（9:20）达到最低值7.0℃。该期间室内气温平均值为（10.3 ± 2.5）℃，室内外平均温差（38.4 ± 2.9）℃，最大温差出现在次日7:20，为42.2℃。

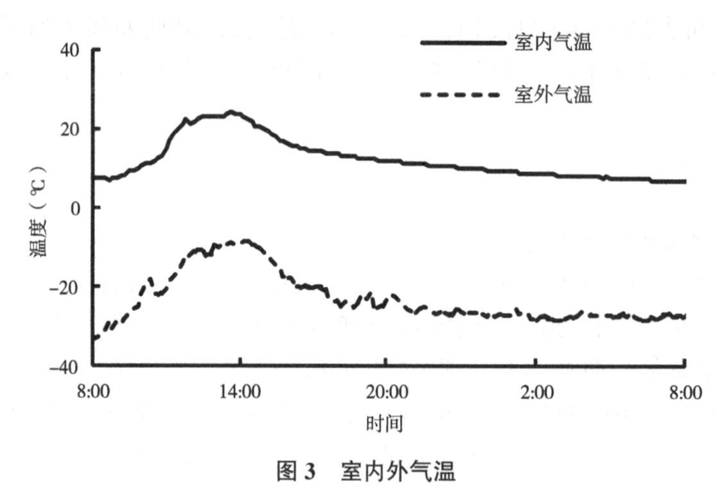

图3　室内外气温

Figure 3　Temperatures of indoor and outdoor air

3.2　北墙内表面、土壤及栽培槽温度变化

北墙内表面和地下10cm处土壤温度如图4所示。上述温度变化情况均是先升高然后逐渐下降。在保温被揭开期间，北墙内表面温度最大值和最小值分别为34.9℃和6.3℃，分别出现在12月31日13:20和1月1日9:20。在保温被揭开期间（9:40—15:20），北墙内表面温度均高于室内气温（图3），两者温度之差为（8.5 ± 3.5）℃。北墙在上述期间一直向室内放热，表明北墙通过吸收并储蓄太阳辐射能来提高自身温度。在保温被关闭时刻（15:20），北墙内表面温度为23.5℃，较室内气温高5.5℃。此后，北墙内表面温度逐逐渐下降，在次日保温被揭开时刻（9:20）达到最小值，6.8℃。该期间北墙内表面温度超过室内气温的时间仅维持了3h。此后，北墙内表面温度与室内气温基本一致。上述结果表明：北墙蓄热性能较差，在夜间主要承担了保温功能，即防止室内空气热量通过北墙向室外方向散失。

地下10cm处土壤温度为（10.5 ± 0.2）℃，最大值和最小值分别为10.8和9.7℃，变化幅度较小。

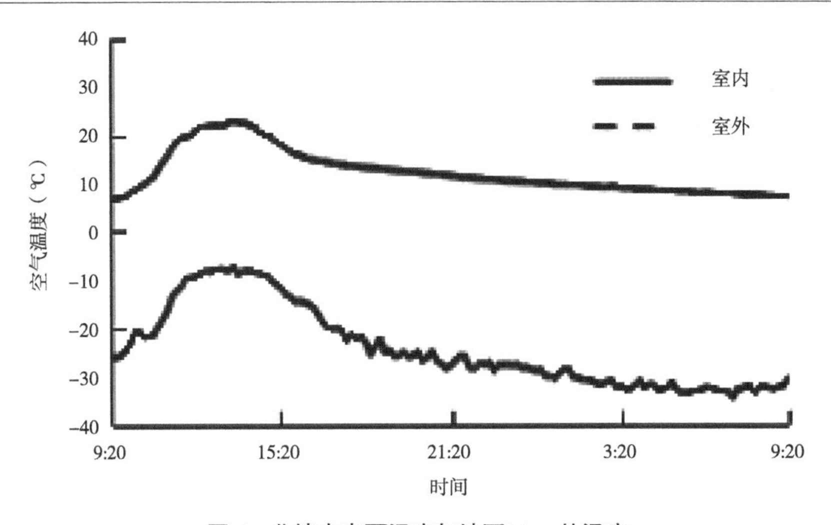

图 4　北墙内表面温度与地下10cm处温度

Figure 4　Inner surface temperature of north-wall and soil temperature at depth of 10cm

图5所示为草莓栽培架各层栽培槽外表面温度的变化情况，可以看出：各层栽培槽外表面温度随时间变化均是先升高而后降低；顶层、中层和底层栽培槽的外表面温度最大值分别为19.1℃、18℃和16.1℃，均在15:20达到，较室内气温发生了1h40min的延迟，然后缓慢下降。在10:00—15:10期间，栽培槽外表面温度均低于室内气温（图3），可见，栽培槽可以从室内热空气中吸收并储蓄热量于栽培基质中。在保温被关闭之后（15:20—），栽培槽的外表面温度始终高于室内气温，并且二者之差逐渐增大，然后又逐渐减小。在该期间，顶层、中层和底层栽培槽外表面温度与室内气温的差值分别为（1.8 ± 0.8）℃、（2.3 ± 0.2）℃和（1.9 ± 0.2）℃。其中，顶层栽培槽表面温度的下降幅度达到10.8℃，分别较中层和底层栽培槽外表面温度的下降幅度高1.6℃和3.2℃。

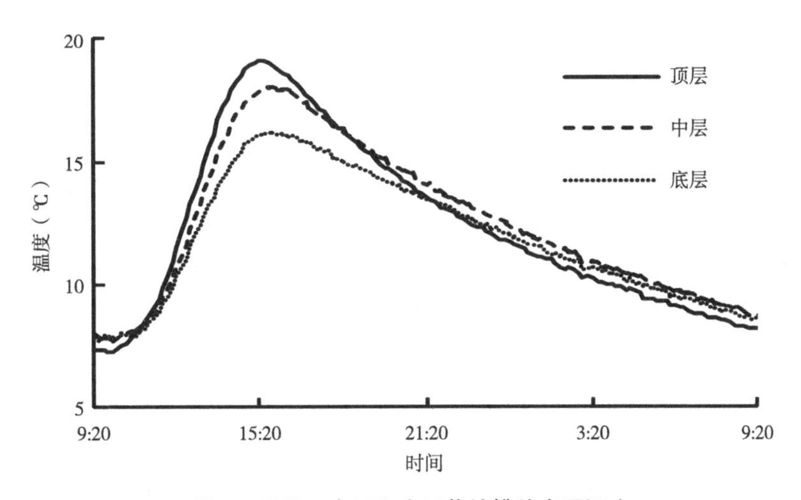

图 5　顶层、中层和底层栽培槽外表面温度

Figure 5　External surface temperature of top，middle and bottom layer cultivation tubes

3.3 前屋面温度变化

日光温室前屋面透明覆盖材料内外表面温度随时间的变化情况如图6所示。在保温被揭开期间（9:20—15:20），中空玻璃内表面与室内空气之间的温度差异很小（见图3）。在保温被关闭后（15:20—），中空玻璃内表面温度较室内气温低（2.3 ± 0.6）℃，这表明夜间室内空气通过前屋面向室外散失热量。

在保温被揭开之后（9:02—），中空玻璃外表面温度下降了6.6℃，然后开始回升，直到到达最高点后开始下降。而在保温被关闭后（15:20—），中空玻璃外表面温度是先上升了3.8℃，然后逐步下降。中空玻璃外表面温度全天低于内表面，在保温被揭开期间（9:20—15:20），中空玻璃外表面温度较内表面温度低（3.7 ± 0.5）℃。

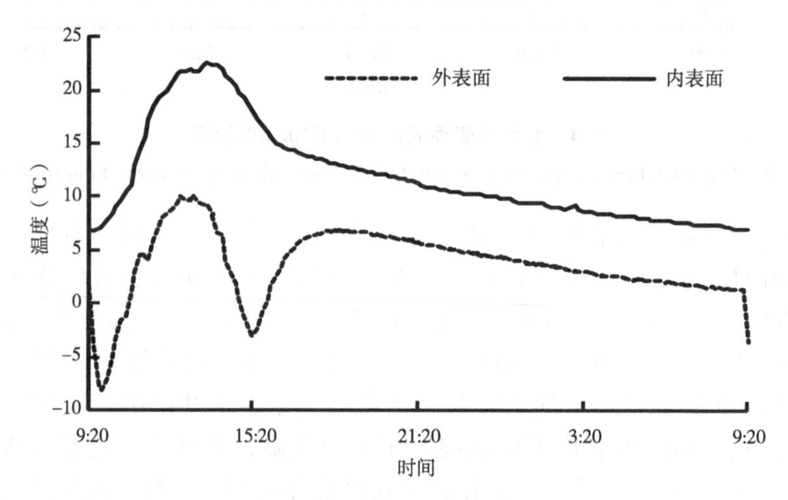

图 6　中空玻璃内外表面温度

Figure 6　Inner and external surface temperatures of hollow glass

4　讨论

4.1 温室保温性能

根据日光温室热平衡分析，前屋面、后屋面和温室结构缝隙是引起日光温室室内热量损失的主要部位。其中，前屋面面积大，覆盖材料传热系数较大，夜间日光温室通过前屋面损失的热量可达温室总失热量的63%[13]。

前屋面的散热量除了受室内外温差影响外，还要受其覆盖材料热阻的影响。根据传热学原理，日光温室夜间通过前屋面的热流密度可根据下式计算：

$$q_f = \frac{t_i - t_e}{1/\alpha_i + R_{sheet} + R_{glass} + 1/\alpha_e} \tag{1}$$

$$q_f = \alpha_i (t_i - t_{i,fr}) \tag{2}$$

式中：q_f为日光温室夜间通过前屋面流向室外的热流量，W/m^2；t_i、t_e和$t_{i,fr}$分别为室内气温、室外气温和中空玻璃内表面温度，℃；R_{sheet}和R_{glass}分别是保温被和中空玻璃的热阻，（m^2·K）/W；α_i和α_e分别为内外表面的对流换热系数，W/（m^2·℃）。

式（2）可进一步改写为：

$$R_{sheet}=\frac{t_i-t_e}{q_f}-R_{glass}-1/\alpha_i-1/\alpha_e \quad (3)$$

测试温室的覆盖材料为双层中空玻璃（8mm玻璃+12mm空气+8mm玻璃），其中8mm玻璃的导热系数为0.76W/（m·K），热阻为0.1（m^2·K）/W，空气间层的热阻为0.14（m^2·K）/W。则R_{glass}为0.34（m^2·K）/W[14-15]。另外，根据测试结果计算，夜间前屋面热流密度为（17.9±0.2）W/m^2，总热阻为2.0（m^2·K）/W，保温被热阻达到了1.76（m^2·K）/W，在目前现有保温被中处于较高水平。因此，测试日光温室的前屋面散热量较小，保温性能良好[16-17]。

在测试中发现，中空玻璃外表面温度在保温被揭开之后首先下降，然后快速升高。而在保温被关闭之后，中空玻璃外表面温度则先升高，然后开始逐步下降。这是由于保温被揭开之后，中空玻璃直接与室外冷空气相接触，导致外表面失热过多而引起的。而在保温被关闭之后，中空玻璃外表面不再与室外冷空气直接接触，外表面向室外散失的热量得到有效抑制，使得由内表面传导的热量在外表面蓄积，并将外表面温度抬升，以达到新的热平衡。

测试日光温室的后屋面为“15cm聚苯板+1.5cm空气间层+15cm聚苯板”。由于聚苯板为保温材料，蓄热性能较差，主要承担防止室内热量向室外散失的功能。因此，后屋面热阻是影响后屋面热流量密度的主要因素。通过后屋面的热流密度可根据下式计算。

$$q_B=\frac{t_i-t_e}{1/\alpha_i+R_B+1/\alpha_e} \quad (4)$$

$$q_b=\alpha_i(t_i-t_{i,b}) \quad (5)$$

式中：q_b为日光温室夜间通过后屋面流向室外的热流量，W/m^2；$t_{i□b}$为后屋面内表面温度，℃；R_b是后屋面热阻；α_i和α_e分别为内外表面的对流换热系数，W/（m^2·℃）。

根据计算，测试日光温室的后屋面热阻为7.47（m^2·K）/W，其中空气夹层热阻为0.145（m^2·K）/W，15cm聚苯板热阻为3.66（m^2·K）/W[14，，16]。测试期间，日光温室夜间通过后屋面流向室外的热流密度为（5.2±0.2）W/m^2。在此基础上，将R_b增加或减少1（m^2·K）/W，则q_b相应减少0.6W/m^2或增加0.8W/m^2，变化范围很小。因此可以认为，通过后屋面的热流量较小，后屋面的保温性能较好。

由于试验温室所在地区室外气温较低，甚至在白天都低于0℃，温室结构缝隙处的冷凝水很容易结冰。在测试过程中就发现，温室的缝隙处出现结冰现象。冷凝水结冰可以有效密封缝隙，将温室的通风换气量降低到最低程度。虽然日光温室依然会有部分热量通过缝隙的冰层向室外流失，但缝隙处的冰层可避免室外热空气直接通过缝隙流向室外，有利于减少室内空气的热量损失，维持较高的室内气温。

4.2 温室蓄热性能

根据热平衡分析，北墙和土壤是温室夜间最主要的放热部位[13]。一般的日光温室中，晴天北墙的放热强度高于土壤，但由于土壤面积较大，所以北墙的总放热量小于土壤[12]。一般北墙通过升高或降低自身热量来储蓄或释放热量。根据相关测试分析，不同结构北墙可储蓄热量的范围在0.3～0.5m[18-22]。而测试温室的北墙结构为“1cm混凝土砂浆+30cm聚苯板”，聚苯板的蓄热作用很弱，主要依靠来1cm厚的砂浆层蓄积热量，可用于储热的墙体区域较小，储热能力较差。根据测试结果，北墙对热量的蓄积效果较差，放热时间短、强度弱，在夜间大多数时间内北墙内表面温度略低于室内气温，主要承担保温功能。另一方面，北墙热阻低于后屋面，但夜间北墙内表面温度与室内气温之差几乎为零，远低于后屋面与室内气温之差。这可能是由于北墙后方的菜窖以及延展的后屋面进一步将北墙与室外冷空气隔绝，使得北墙的热流量密度显著降低。

土壤放热量与土壤温度密切相关。但是测试温室地下10cm处的土壤温度较低，仅在夜间23:00—次日10:00高于室内气温，且最大温差仅3.1℃。考虑到土壤具有较高的比热容和密度，其内部传热为非稳态型式，还需要进一步测量土壤表面温度才能正确评价土壤在夜间向室内的放热性能。

日光温室土壤可以吸收来自室内地面太阳辐照度和空气温度热量，将其温度抬升。土壤表面温度变化幅度最大。但在温度波向土壤深层传递的时候，土壤温度的变化幅度随土壤深度而逐渐减小[23]。因此，测试温室-10cm处土壤温度变化范围较小。另一方面土壤还会向地下温度较低的位置传导热量；导致土壤温度下降。漠河地区冬季-40～-80cm处土壤温度低于0℃，容易导致土壤向下传递热量，影响土壤温度。在地下设置隔热苯板有助于减少土壤向下传递热量，维持较高的室内气温。刘生财等研究发现：在黑龙江，日光温室设置3cm厚的聚苯板就能很好的阻止地中热传导[24]。测试温室用于隔热的聚苯板厚度为10cm，超过上述研究中所使用聚苯板厚度，可以有效防止地中热传导，但还需要进一步测试分析确定测试温室地中热传导对土壤夜间放热性能的影响。

测试发现，栽培槽外表面温度在日间大部分时间低于室内气温，在夜间的大部分时间都高于室内气温，说明栽培槽基质可在日间吸收热量，在夜间向室内放热。而且其夜间放热强度可达16.6～20.2W/m^2，与日光温室前屋面的热流量密度相当。这可能是由于栽培过程中基质含水率较高，其热容和导热系数较干燥状态有所改善，基质储热性能提升，能够在日间储存大量的热量，并在夜间室内气温较低的时候向室内放热。另外，温室内栽培槽表面积和基质表面积分别为290m^{2}和185m^2，其总面积与前屋面面积相当。假设基质表面温度与栽培槽温度相当基质表面，则栽培基质在夜间向室内的放热量与前屋面相当。因此，栽培槽基质亦是该温室重要的热量来源之一，对维持室内气温具有非常重要的意义。但还需要进一步研究才能精准确定栽培槽基质夜间放热量对室内气温的贡献程度。

5 结论

本文针对黑龙江省漠河保温后墙日光温室的光、热环境进行了测试，通过分析温室的透光率、围护结构、土壤、立体栽培槽等的保温蓄热性能，得出以下结论。

（1）测试期间日光温室太阳光PPFD的透过率为（54±10）%，室内PPFD超过110μmol·m^{-2}·s^{-1}的时间为4.3h，可满足室内芹菜和草莓的生长需求。

（2）当夜间室外最低气温为-34.7℃时，日光温室室内最低气温为7.0℃，平均温度为（10.3±2.5）℃，可满足喜凉作物的生长需求；同时，室内外平均温差（38.4±2.9）℃，最大温差达到了42.2℃。

（3）日光温室保温被和后屋面保温性能较好，可有效防止室内热量散失；北墙蓄热性能较差，主要承担保温功能；草莓A字型立体栽培槽的基质可在夜间向室内释放大量的热能，有利于温室维持较高的室内气温。证明北墙蓄热可由土壤蓄热和A字型栽培架的基质蓄热来替代，从而实现降低日光温室建造成本的目的。

综上所述，黑龙江省漠河县保温后墙日光温室主要依靠围护结构良好的保温性能以及A字型立体栽培系统的蓄热性能来维持较高的室内气温。该结果表明可在采取其他简单可行的蓄热或保温方法的基础上取消后墙蓄热功能，为创新日光温室设计理论奠定了基础。

参考文献

［1］ Wang Junwei，Li Shuhai，Guo Shirong，et al. Simulation and optimization of solar greenhouses in Northern Jiangsu Province of China［J］. Energy & Buildings，2014，78（4）：143-152.

［2］ 魏晓明. 日光温室的设计与建造技术［J］. 农民科技培训，2015（10）：42-44.

［3］ 陈青云. 日光温室的实践与理论［J］. 上海交通大学学报（农业科学版），2008，26（5）：343-447.

［4］ 张勇，邹志荣，李建明. 倾转屋面日光温室的采光及蓄热性能试验［J］. 农业工程学报，2014，30（1）：129-137.

［5］ 张勇，邹志荣. 日光温室主动采光机理与透光率优化试验［J］. 农业工程学报，2017，33（11）：178-186.

［6］ Zhang Xin，Wang Hongli，Zou Zhirong，et al. CFD and weighted entropy based simulation and optimisation of Ch.

［7］ inese Solar Greenhouse temperature distribution［J］. Biosystems Engineering，2016，142：12-26.

［8］ 赵丽玲，樊东隆，杨爱华，等. 冬季辽阳型与白银型日光温室的温、湿度特性比较［J］. 北方园艺，2014（15）：40-43.

［9］ 徐凡，马承伟，曲梅，等. 华北五省区日光温室微气候环境调查与评价［J］. 中国农业气象，2014，35（1）：17-25.

［10］ 史宇亮，王秀峰，魏珉，等. 日光温室不同厚度土墙体蓄放热特性研究［J］. 农业机械学报，2017，48（11）：359-367.

［11］ 凌浩恕，陈超，陈紫光，等. 日光温室带竖向空气通道的太阳能相变蓄热墙体体系［J］. 农业机械学报，2015，46（3）：336-343.

［12］ 李明，魏晓明，齐飞，等. 日光温室墙体研究进展［J］. 新疆农业科学，2014，51（6）：1162-1170，1176.

［13］ 李明，周长吉，周涛，等. 日光温室土墙传热特性及轻简化路径的理论分析［J］. 农业工程学报，2016，32（3）：175-181.

［14］ 赵宗. 辽沈Ⅳ型大跨度节能日光温室围护结构蓄散热规律试验研究［D］. 沈阳：沈阳农业大学，2006.

［15］ 马承伟. 农业生物环境工程［M］. 北京：中国农业出版社，2005.

［16］ 中华人民共和国国家标准　住房和城乡建设部. GB50716-2016，民用建筑热工设计规范［S］.中国标准出版社.

［17］ 刘晨霞，马承伟，王平智，等. 日光温室保温被传热的理论解析及验证［J］. 农业工程学报，2015，31（2）：170-176.

［18］ 王平智，马承伟，赵淑梅. “西北非耕地温室结构与建造技术”项目成果汇报（10）：温室覆盖材料保温性能测试台改进及保温被的选型［J］. 农业工程技术：温室园艺，2014（10）：26-26.

［19］ 管勇，陈超，凌浩恕，等. 日光温室三重结构相变蓄热墙体传热特性分析［J］. 农业工程学报，2013，29（21）：166-173.

［20］ 彭东玲，杨其长，魏灵玲，等. 日光温室土质墙体内热流测试与分析［J］. 中国农业气象，2014，35（2）：168-173.

［21］ 温祥珍，李亚灵. 日光温室砖混结构墙体内冬春季温度状况［J］. 山西农业大学学报：自然科学版，2009，29（6）：525-528.

［22］ 李明，周长吉，魏晓明. 日光温室墙体蓄热层厚度确定方法［J］. 农业工程学报，2015，31（2）：177-183.

［23］ 王铁良，李晶晶，李波，等. 不同灌溉方式对日光温室土壤温度的影响［J］. 北方园艺，2009（2）：147-149.

［24］ 刘生财，胡兆平，于锡宏. 地中隔热材料对温室土壤温度的影响［J］. 北方园艺，2005（3）：20-21.

夏季遮阳对大棚甜椒生长与蒸腾作用的影响

梁浩[1,2]，陈春秀[1,2]，刘明池[1,2]，季延海[1,2]，王宝驹[1,2]，武占会[1,2*]

（1. 北京市农林科学院蔬菜研究中心，北京 100097；2. 农业部华北都市农业重点实验室，北京 100097）

摘要：利用遮阳网对塑料大棚进行顶部外覆盖，是夏季棚室降温的主要途径之一。本文分析了遮阳覆盖对棚室小气候和作物长势产量的影响，进而通过改进的Penman-Monteith模型，预测了遮阳与不遮阳条件下，大棚内蒸腾速率。结果表明，遮阳处理使棚内光合有效辐射强度平均降低了54.4%，温度降低4.7℃，甜椒长势产量提高1.54kg/m^2；同时，计算得出大棚外遮阳可使蒸腾速率降低52.4%，有利于甜椒的生长发育。

关键词：遮阳网；温室小气候；Penman-Monteith模型；甜椒

Effect of Screen-shading on Transpiration and Growth of Sweet Pepper in Plastic Tunnel

Liang Hao[1,2]，Chen Chunxiu[1,2]，Liu Mingchi[1,2]，Ji Yanhai[1,2]，Wang Baoju，Wu Zhanhui[1,2*]

（1. *Vegetable Research Center of Beijing Academy of Agricultural and Forestry Sciences*，*Beijing* 100097；
2. *Key Laboratory of Urban Agriculture of North China*，*Ministry of Agriculture*，*Beijing* 100097）

Abstract：Screen-shading is one of the main ways for greenhouse summer cooling. This paper analyzed the effect of greenhouse screen-shading on transpiration and crop development，with the improved Penman-Monteith model，transpiration of the pepper canopy was predicted. The results show that with the net shading，PPFD was decreased by 54.4%，the temperature was reduced by 4.7℃，the yield of sweet pepper was increased by 1.54kg/m^2；meanwhile，transpiration rate decreased by 52.4% with screen net covering.

Key words：Screen shading；Greenhouse microclimate；Penman-Monteith model；Sweet pepper

北京西北部地区，地处华北平原与内蒙古高原的连接地带，地理、气候特征独特，夏季光照资源充足，干燥凉爽，较为适合从事果菜类越夏设施生产。夏季设施生产需解决棚内高温、强光对蔬菜生长发育的影响[1]。一方面，高温环境会抑制植物光合作用叶片萎蔫，褐变[2]，影响持续坐果能力；另一方面，果菜对光照较为敏感，当光照强度达到并超过光饱和点后，发生病毒病的风险大大增加[3,4]，尤其在坐果期阳光直射果面极易产生日灼，影响产品产量和品质。解决这一问题直接且有效的方式是采用遮阳覆盖，通过减

通讯作者：武占会，男，北京市农林科学院蔬菜研究中心，副主任/研究员。主要从事无土栽培与植物工厂研究

项目支持：北京市科技重大专项课题（D171100002017002），北京市农林科学院院创新能力建设课题（CZZX001-03），国家特色蔬菜产业技术体系资助项目（CARS-24-B-02）

少部分太阳辐射直射到温室内，降低光照直射和太阳辐射能量在棚内的过分积累[5]，从而缓解了高温、强光对设施蔬菜的威胁。

同时，温光环境的变化还会直接影响植物蒸腾作用和水分需求规律。作物需水量主要受作物本身的生长发育状况和环境条件等因素的影响，这些因素相互联系，较为复杂，准确预测作物需水量是一个重要的科学问题[6-8]。以往研究中较多的利用称重法[9]或仪器测量法[10]直接获得，这些方法准确性较高，但对于实际生产过程中的预测指导的时效性不足。Boulard等[11-12]和Wang等[13]利用室内能量平衡和Penman-Monteith方法（以下简称P-M方程）推导出基于温室外气象数据的温室作物蒸腾量计算模型是目前预测作物蒸发蒸腾量的有效途径。

本文利用棚内气象数据，以P-M方程为基础，系统分析遮阳对棚内环境因子，甜椒冠层温湿度，蒸腾特性的影响，并建立了基于温室内气象数据的大棚甜椒需水量估算模型，为设施遮阳降温、节水措施的改进和推广提供理论依据和数据支持。

1 材料与方法

1.1 试验场地与设计

实验在北京市延庆区公司基地进行，选择了两栋结构、走向相同的单跨温室，两座温室间距6m，东西走向，钢架结构，单层膜覆盖，总高3.5m，跨度8m，长50m，单棚面积约400m^2。其中一栋用透光70%的黑色遮阳网进行覆盖，另一栋无遮阳网覆盖作为对照处理；两座棚内均种植甜椒，品种为长剑，高畦种植，株行距40cm × 50cm。

1.2 测定项目与方法

环境参数测试：采用美国LI-COR公司生产的光量子仪（LI-1400）测试温室内外光合有效辐射强度；采用清华同方温湿度自动记录仪（RHLOG）测试室内外温湿度；于2016年7月3—9日（坐果期），选取有代表性的植株5株，每60min采集一次环境数据。

每处理随机取5株幼苗用直尺测定株高，游标卡尺测茎粗，电子天平称量果实重量。采用浙江托普仪器公司生产的叶面积指数仪测定积叶面积指数。

1.3 大棚甜椒需水量模型

根据Penman-Monteith改进模型，在温室环境条件下，作物蒸腾模型可表述为[14]：

$$\lambda E = A \cdot R_s + B \cdot VPD$$

其中，λ为水的汽化潜热，MJ·kg^{-1}，E为作物冠层蒸腾量，mm/d，R_s（W·m^{-2}）为太阳辐射强度，VPD（kPa）为饱和蒸汽压差，A为辐射系数，B为空气动力学系数，A，B系数可通过叶面积指数来确定，经验公式如下：

$$A = \alpha\left[1 - \exp\left(-k\ LAI\right)\right]$$

$$B = \beta\ LAI$$

其中，k是消光系数（甜椒可取0.64），LAI为叶面积指数。

$$ET = A\left[1 - \exp\left(-\alpha\ LAI\right)R_s + B \cdot LAI \cdot VPD\right]$$

1.4 数据处理

采用Microsoft Excel 2010软件进行计算分析与作图；采用SPSS 13.0软件的Duncan极差法在$P<0.05$的显著水平下对测试数据进行显著性分析。

2 结果与分析

2.1 遮阳网对温室小气候环境的影响

由图1可知，覆盖遮阳网对温室内光环境影响较大，7d的连续测试中，2d为多云天气，其他均为晴天。无遮阳处理与遮阳处理晴天棚内最大光合有效辐射强度最大值平均分别为970.9μmol·m^{-2}·s^{-1}和548.9μmol·m^{-2}·s^{-1}。

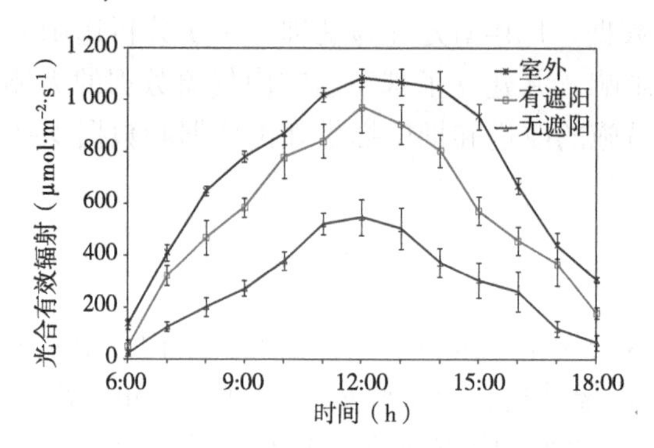

图 1　温室内外光照环境对比

Figure 1　Solar radiation inside and ambient the greenhouse

透光率，即相对辐射强度，是室内光合有效辐射强度与室外光合有效辐射强度之比，在午间（10:00～14:00）无遮阳网处理的温室的平均光合有效辐射透过率为84.5%，而覆盖遮阳网处理温室的平均光合有效辐射透过率为45.6%。

如图2所示，与无遮阳网覆盖温室相比，遮阳温室白天（8:00～17:00）平均气温降低了4.7℃，两棚夜间温度无显著差异；遮阳温室的平均相对湿度表现为白天较高，夜间较低，但这种差异并不显著（图3）。

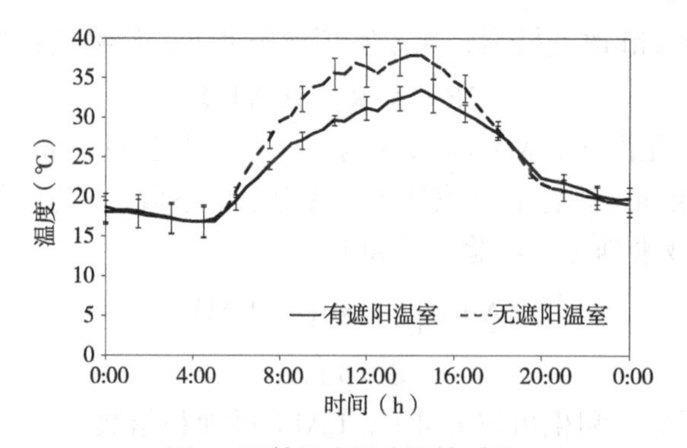

图 2　两栋温室温度环境对比

Figure 2　Temperature performance of the two greenhouses

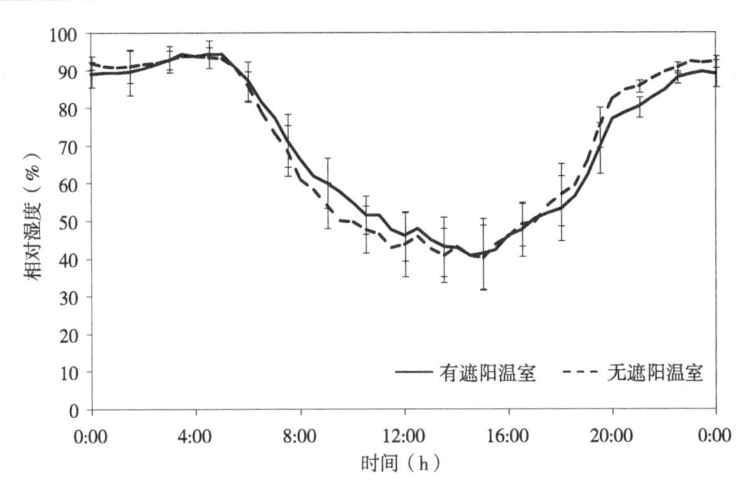

图 3　两栋温室相对湿度对比

Figure 3　Relative humidity of the two greenhouses

2.2　遮阳对甜椒蒸腾作用影响

图4是根据P-M模型计算的7d蒸腾速率的平均值。无遮阳网覆盖温室的蒸腾速率是遮阳温室的2.1倍，棚室内更多的水分以汽化潜热方式散发到空气中，同时由于无遮阳棚室内温度和饱和蒸气压更高，所以并没有引起相对湿度的升高。

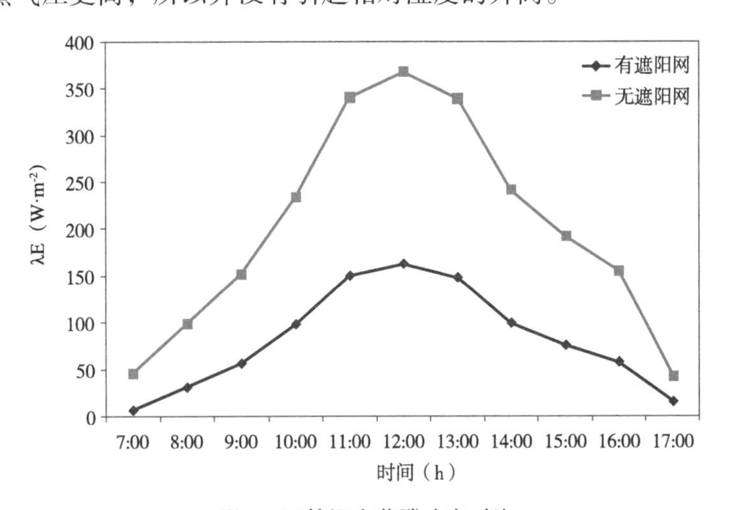

图 4　两栋温室蒸腾速率对比

Figure 4　trans-evaporation of the two greenhouses

然而午间光照过强，对气孔导度产生影响，不利于作物的光合作用，可能对植物生长发育产生影响[15]。

2.3　遮阳对甜椒生理参数的影响

由表1可以看出，有遮阳处理大棚甜椒的株高、茎粗，叶面积指数和产量指标均显著高于无遮阳大棚。遮阴作用下，叶片蒸腾压力显著下降，促进了作物的营养转化和光能利用效率，甜椒产量增加1.54kg/m^2。

表 1　两种处理下甜椒采摘期的生理及产量指标

Table 1　Growth and yield indexes of the sweet peppers under two treatments

类别	株高（mm）	茎粗（mm）	叶面积指数	产量（kg/m^2）
有遮阳处理	1 419a	23.3a	3.88a	5.46a
无遮阳处理	878b	19.6b	3.09b	3.92b

3　结果与讨论

遮阳网处理能够降低进入到温室内部的太阳辐射热量，能够有效抑制设施内部温度高温，遮阳率达到54.4%，有效降低温室白天温度4.7℃。同时，遮阴可以降低作物对水分的需求，蒸腾速率降低47.6%，同时从甜椒的表现来看，遮阳网处理下的甜椒株高、茎粗、叶面积指数及单位面积产量均显著高于无遮阳的处理，有利于甜椒的生长发育。

参考文献

［1］　张振贤，王培伦，刘世琦，等. 蔬菜生理［M］.北京：中国农业科技出版社，1993.

［2］　潘宝贵，袁希汉. 高温胁迫对不同辣椒品种光合作用的影响［J］. 江苏农业学报，2006，22（2）：137-140.

［3］　钱妙芬，潘永. 塑料遮阳网大棚小气候观测与分析［J］. 成都信息工程学院学报，2001（1）：32-33.

［4］　汪波，刘建，李波，等. 夏季遮阳网覆盖对塑料薄膜大棚小气候的影响［J］. 江苏农业科学，2015，43（10）：479-483.

［5］　Tanny，J.，Cohen，S. and Teitel，M. 2003. Screenhouse microclimate and ventilation：An experimental study. Biosyst. Engin. 84：331-341.

［6］　Möller，M.，Assouline，S. 2007. Effects of a shading screen on microclimate and crop water requirements. Irrig. Sci. 25：171-181.

［7］　Desmarais，G. 1996. Thermal Characteristics of Screenhouse Configurations in a West-African Tropical Climate. PhD Thesis. McGill University，Quebec，172-177.

［8］　贾志银，巩振辉，许红娟，等. 高温胁迫对辣椒幼苗生长及生理性状的影响［J］. 北方园艺，2010（12）：5-8.

［9］　周晓慧，吴阳清，胡艺春，等. 勋章菊不同品种蒸腾速率比较研究. 北方园艺. 2014（01）：62-65.

［10］郭阿君，岳桦，王志英. 9种室内植物蒸腾降温作用的研究. 北方园艺. 2007（10）：141-142.

［11］Boulard T，Baille A. Modelling of air exchange rate in a greenhouse equipped with continuous roof vents. Journal of Agricultural Engineering Research，1995，61：37-48.

［12］Boulard T，Wang S. Greenhouse crop transpiration simulation from external climate conditions. Agricultural and Forest Meteorology，2000，100：25-34.

［13］Wang S，Boulard T. Predicting the microclimate in a naturally ventilated plastic house in a Mediterranean climate. Journal of Agricultural Engineering Research，2000，75：27-38.

［14］Baille，M.，Baille，A. and Delmon，D. 1994. Microclimate and transpiration of greenhouse rose crops. Agric. For. Meteorol. 1994，71：83-97.

［15］任旭琴，周强，刘浩如. 涂料和遮阳网覆盖对大棚环境和甜椒光合特性的影响. 东北农业大学学报，2012，43（11）：87-91.

温室多层立体栽培下增加人工补光对生菜品质的影响

刘庆鑫，杨其长，魏灵玲，魏强

（中国农业科学院环境与可持续发展研究所，北京　100086）

摘要：试验采用意大利奶油生菜为研究对象，研究温室多层立体栽培条件下不同栽培层的光照强度情况，以及对中、下两层增加人工光源进行补光对生菜生长形态、产量、硝酸盐、蛋白质、糖、叶绿素含量的影响。试验栽培架为三层立体栽培架，上层为全自然光，中、下两层为自然光加人工补光。结果表明：栽培架中层平均总太阳辐射强度是上层的25.15%～55.42%，下层平均总太阳辐射强度是上层的22.66%～37.46%。不同层间温度差异在5.6℃范围内。增加人工补光后生菜的形态有明显的变化，生菜叶片更加宽大密实，干鲜重有明显提高。同时补光后生菜的硝酸盐含量有明显的降低，蛋白质、糖、叶绿素和胡萝卜素含量有了明显的升高。这表明在温室多层立体栽培条件下，适当的增加人工补光，对生菜的品质、产量有明显的提高。

关键词：立体栽培；太阳辐射强度；人工补光；品质

Influence on the Quality of Lettuce Increased Greenhouse Under Artificial Light Stereo Multilayer Institute of Environmental and Sustainable Development

Liu Qingxin，Yang Qichang，Wei Lingling，Wei Qiang

（*Chinese Academy of Agricultural Sciences*，*Beijing*　100086）

Abstract：The research object use the Italy butter lettuce. Three-dimensional cultivation conditions of greenhouse cultivation layer multilayer under different light intensity，and the middle and lower two layers increased artificial light for light effect on the morphology，yield，nitrate，protein，sugar and chlorophyll content of lettuce. The cultivation frame for three layer stereo cultivation frame upper for natural light，and the two layer is the natural light and artificial light. The results showed that the average solar radiation intensity in the middle layer of the cultivation frame was 25.15%～55.42% on the upper level，and the average total solar radiation intensity in the lower layer was 22.66%～37.46% in the upper layer. The temperature difference between different layers is within the range of 5.6℃. The increase of artificial lighting after lettuce showed significant morphological changes，lettuce leaves more spacious dense，fresh and dry weight increased significantly. At the same time fill after the nitrate content significantly decreased，protein，sugar and chlorophyll content of Russell and carotenoid have obviously increased. This shows that the three-dimensional cultivation conditions in greenhouse under the proper increase of multilayer，artificial lighting，improve the quality and yield of lettuce.

Key words：Stereo cultivation；Solar radiation；Artificial lighting；Quality

---

作者简介：刘庆鑫　联系方式：lqx13121259299@icloud.com

项目资助：科技部科技伙伴计划资助（KY201702008）

生菜的营养价值丰富、市场需求量大，传统露底或日光温室栽培的生菜无论从数量上还是品质上都不能满足人们日益增长的需求[1-2]，它的工厂化栽培迫在眉睫[3]。植物工厂作为一种蔬菜作物栽培的新型种植技术，具有环境智能可控、水肥一体化管理和周年生产的优势，能够全年稳定种植生产洁净无污染的蔬菜，近年来得到了快速发展。然而，植物工厂需要采用人工环境及人工光源需要消耗大量的能源，其中人工光源的电能消耗是植物工厂的主要能源消耗，约占总能耗的80%[4]，由于植物工厂前期的巨大投资、后期巨额的维护费用以及在使用过程中能源的消耗，造成了蔬菜产品成本较高，限制其走向更广阔的市场[5]。

为了获得温室内生菜栽培的高产，近年来发展了以立柱式为主的立体栽培技术[6]。由于多层栽培架的结构特点，存在层架间光、温分布不均匀的问题，尤其是光环境的不均匀，很大程度上影响了不同层生菜的产量和品质。对于立体栽培条件下的光温分布对作物的影响目前国内已经有了一些研究，但是供试作物主要是草莓和水稻秧苗[7-8]，对于生菜在立体栽培条件下的进行人工补光对其产量及品质的应用少有报道。

本试验通过对立体栽培条件下中、下层的生菜进行定时、定量的补光后，通过与完全自然光条件下生菜的产量、形态、可溶性蛋白、可溶性糖以及硝酸盐等含量的变化，为以后生菜的多层立体栽培进行补光提供参考。

1 材料与方法

1.1 材料

试验品种为意大利奶油生菜（产地荷兰）。试验使用的三层栽培架：长3.72m，宽0.6m，高1.2m，上下栽培面的间隔0.6m，补光间距0.42m。上层UP，中层M，下层D，每层栽培架可种植96棵生菜。

1.2 试验设计

试验地点在中国农业科学院农业环境与可持续发展研究所楼顶的玻璃温室（自然光植物工厂）内，试验时间选择在2017年9—10月。试验于9月1日进行浸种催芽，采用水培方式在植物工厂内进行育苗，待小苗长至四叶一心时，移栽到温室（自然光植物工厂）内。温室（自然光植物工厂）内的栽培架编号为A、B。A栽培架为对照组不进行任何补光处理，三层均为太阳光。B栽培架，在9:00—13:00对中、下两层生菜进行补光处理，上层为太阳光不做补光处理，补光强度为110μmol/m^2·s。试验期间对试验组以及对照组进行统一的营养液循环管理，每天营养液循环3h，确保根系营养液养分以及氧气含量的供给充足。

1.3 测定项目及方法

在栽培架的不同层布点放置TR-7wf对温湿度进行温湿度环境监测，每5min记录一个数据；SPN1太阳辐射传感器对三层栽培面上的全天太阳的总辐射、散射辐射、直射辐射进行数据采集，设置每5min对记录数据1次，用CR-1000对数据进行采集。

试验于2017年10月14日进行取样，分别称量地上、地下部分的鲜重，以及生菜生长叶长、叶面积形态指标，采用烘干法称量生菜的干物质重量。用SPAD叶绿素仪（SPAD-

502，Konica Minolta，日本）测定生菜叶片SPAD的含量。采用苯酚法测定可溶性糖、考马斯亮蓝G-250法测定可溶性蛋白、分光光度计测定硝态氮的含量。

1.4 统计与分析

采用Microsoft Excel 2010对测定的数据进行绘图和做表，用IBM SPSS Statistics软件对数据进行差异分析$P<0.05$为差异显著。

2 结果和分析

2.1 自然光植物工厂内不同层栽培面太阳辐射和温湿度日变化情况

在9—10月试验期间，9月28日（晴）10月3日（阴）生菜冠层太阳辐射和温湿度变化如图1至图6所示。图1是晴天时太阳辐射随时间日变化情况，日平均最高的总太阳辐射强度上、中、下层分别为608w/m^2、368w/m^2、256w/m^2，图2是阴天时太阳辐射随时间日变化情况，日平均最高总太阳辐射强度上、中、下层分别为325w/m^2、189w/m^2、123w/m^2。由图可以看出中层平均总太阳辐射强度是上层的25.15%～55.42%，下层平均总太阳辐射强度是上层的22.66%～37.46%。在生菜的生长期间内不同栽培层间的太阳辐射强度有较大的差异。

图3是晴天时湿度随时间日变化情况，上层的湿度变化在90.7%～46.8%，中层湿度变化在91.7%～66.1%，下层湿度变化在88.9%～65.2%，最大湿度差为43.9%。图4是阴天时湿度随时间日变化情况，上层的湿度变化在74.4%～40.5%，中层湿度变化67.7%～30.2%，下层湿度变化在66.9%～30.7%，最大湿度差为37.5%。图5是晴天时温度随时间日变化情况，图6是阴天时温度随时间日变化情况。上层与中层日最高温度差为3.5℃，上层与下层日最高温度差为5.6℃。日平均最高温湿度在每天10:00—15:00，上层与中下两层温湿度有明显差异。因此，可以看出生菜的形态、品质的差异主要是由于不同层栽培架上的光照以及温湿度耦合结果所导致的，其中引起温湿度的主要原因是太阳辐射强度的不同。

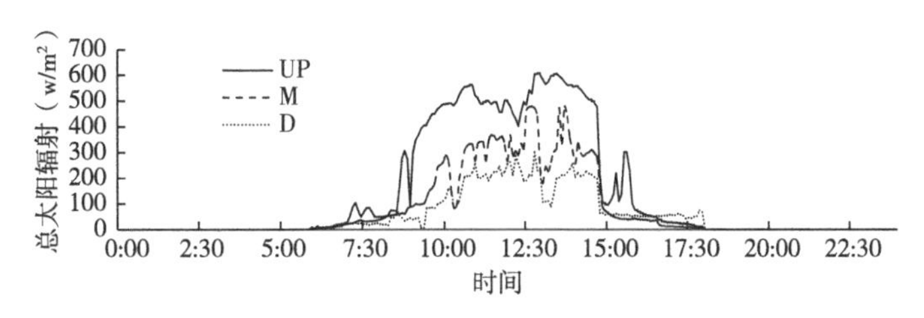

图 1 晴天时太阳辐射时间变化情况

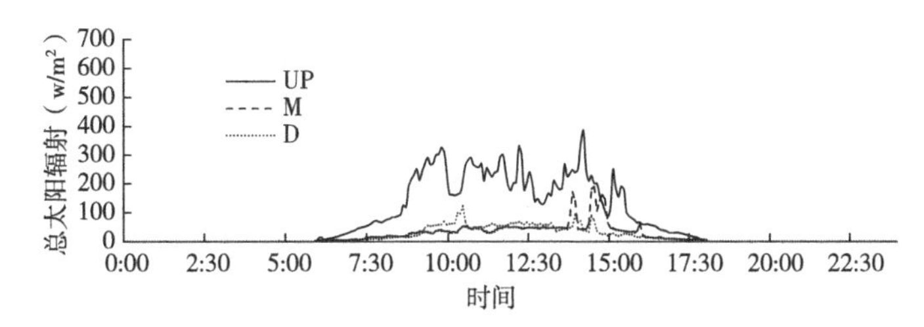

图 2 阴天时太阳辐射时间变化情况

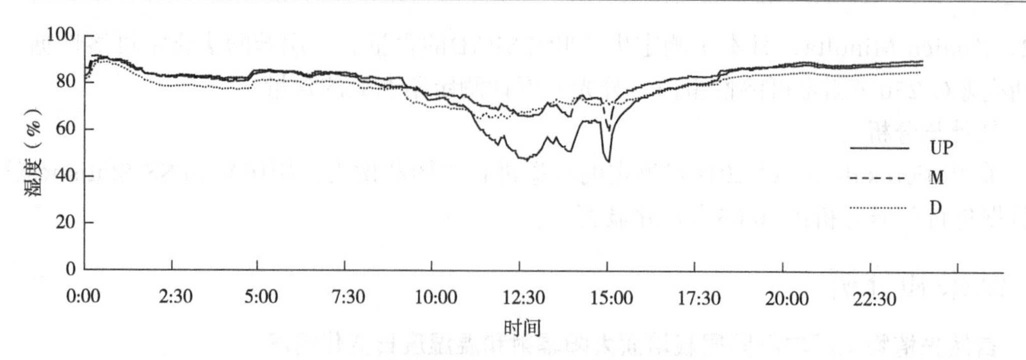

图 3　晴天时湿度随时间日变化情况

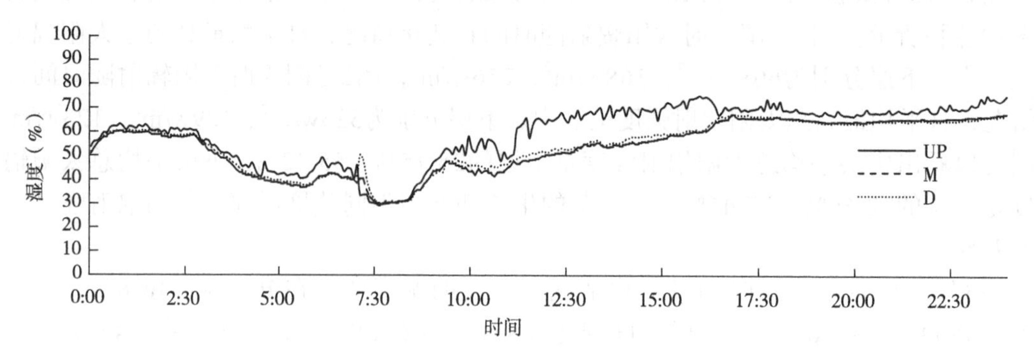

图 4　阴天时湿度随时间日变化情况

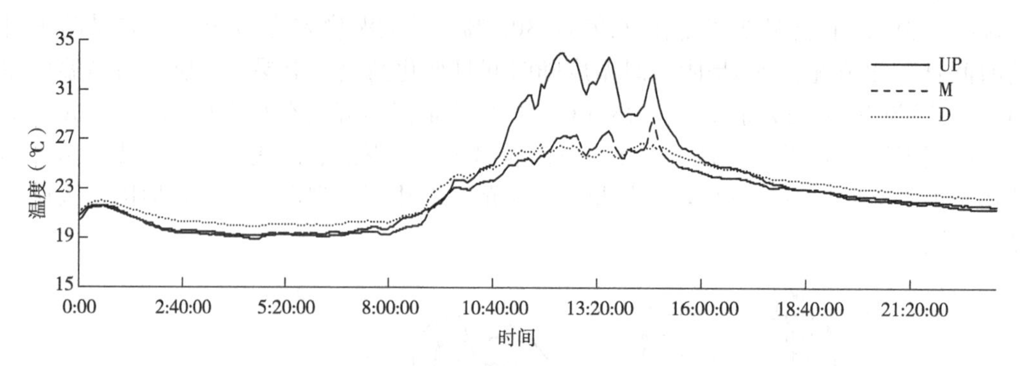

图 5　晴天时温度随时间变化情况

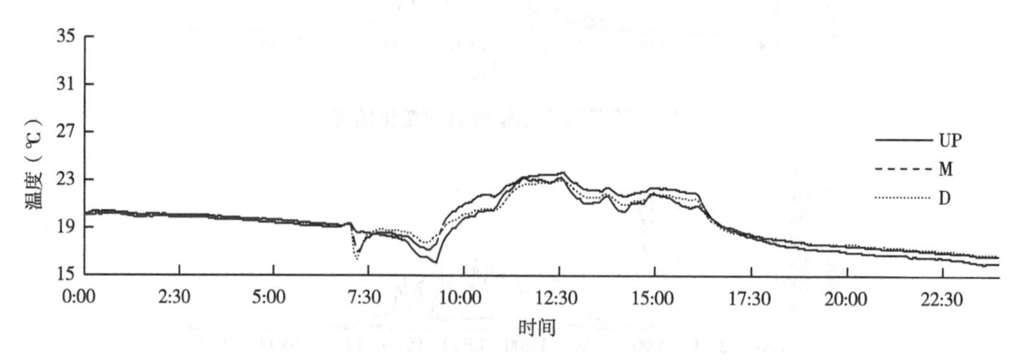

图 6　阴天时温度随时间变化情况

2.2 增加补光对不同层架上生菜生长形态的影响

（1）自然光下不同栽培层对生菜生长形态的影响由于中层和下层处于弱光条件下，所以生菜的叶片形状上发散，根系更加的细长。由图7和图8可以看出A-UP处理下的生菜叶片紧凑，且叶片颜色较深，根系粗壮；A-M处理下的生菜叶片相对较发散，颜色也较A-UP处理的生菜更浅一些，根系细长；A-D处理下的生菜叶片更加发散，叶脉也更加细长，叶片颜色更加浅，根系细长。由破坏性测量不同处理下的生菜总叶面积可以得到，A-UP处理下的生菜总叶面积是A-M处理下的29.05%、是A-D处理下的66.45%。

图7　A-UP、A-M和A-D处理下的生菜叶片生长情况

图8　A-UP、A-M和A-D处理下的生菜根系生长情况

（2）补光对中、下层生菜生长形态的影响图9是中层和下层补光与不补光的对照图片，可以看出补光后生菜的叶片不再向四周发散、叶面积增大、生菜整体更加密实，叶面积增大，B-M补光处理后总叶面积增加了31.45%，B-D补光处理后总叶面积增加了27.47%。

图9　补光对中、下层生菜生长形态的影响

2.3　增加补光对不同层架上生菜生理的影响

2.3.1　补光处理对生菜产量、形态的影响

由于不同层光照的差别，上层栽培架不受遮挡的影响，因此光照强度要强于中下层，导致在此处理下的生菜鲜重、干重有较大的提高。由表1可以得到，不同层栽培条件下，生菜的鲜重、干重都有很大的差别，差异性显著。在纯自然光条件下A-M处理较A-UP处理下的生菜鲜重减少了37.23%、干重减少了43.18%，A-D处理下较A-UP处理下的生菜鲜重减少了49.39%、干重减少了57.07%。纯自然光栽培下上层A-UP处理下的SPAD较A-M、A-D处理下有显著性差异，这与不同层受太阳光照情况有着直接的影响。

增加人工补光后，B-M处理较A-M处理生菜的鲜重增加了27%、干重增加了19.8%。补光后SPAD有所增加，但是差异并不显著，这可能是由于即使进行了人工补光，但是在一天中的一段时间内，生菜的生长依然处于弱光条件之下（表1）。

表1　不同处理对生菜产量、形态的影响

处理	叶鲜重（g）	根鲜重（g）	根冠比（鲜）	叶干重（g）	根干重（g）	根冠比（干）	SPAD
A-UP	54.32^{d}	6.86^{d}	0.126	2.703^{d}	0.437^{e}	0.162	23.52^{b}
A-M	35.16^{b}	3.24^{ab}	0.092	1.595^{b}	0.189^{b}	0.118	18.18^{a}
A-D	28.12^{a}	2.84^{a}	0.101	1.198^{a}	0.150^{a}	0.125	18.22^{a}
B-UP	60.56^{e}	7.78^{e}	0.128	2.983^{e}	0.468^{e}	0.157	27.04^{c}
B-M	47.76^{c}	4.84^{c}	0.101	2.205^{c}	0.288^{c}	0.131	20.3^{a}
B-D	35.06^{b}	3.52^{ab}	0.100	1.639^{b}	0.204^{b}	0.319	19.12^{a}

2.3.2　补光处理对生菜品质的影响

由表2可以得出，补光后生菜的亚硝酸盐含量较纯自然光条件下有了明显的减低，叶绿素含量有所提高，但是差异并不显著，可溶性糖提高了8.3%～6.9%、可溶性蛋白含量提高了8.5%。这表明人工补光处理有助于生菜的品质的提高。

表 2　不同处理对生菜品质的影响

处理	叶绿素a（mg/g）	叶绿素b（mg/g）	胡萝卜素（mg/g）	亚硝酸盐（mg/g）	可溶性糖（%）	可溶性蛋白（mg/g）
A-UP	0.39^{c}	0.12^{c}	0.099^{d}	12.13^{ab}	1.52^{bcd}	60.11^{ab}
A-M	0.34^{ab}	0.11^{ab}	0.081^{abc}	13.32^{b}	1.33^{ab}	57.44^{ab}
A-D	0.35^{ab}	0.11^{ab}	0.082^{abc}	17.74^{b}	1.21^{a}	54.13^{a}
B-UP	0.41^{c}	0.12^{c}	0.104^{d}	11.69^{ab}	1.75^{c}	68.95^{c}
B-M	0.38^{ab}	0.11^{c}	0.089^{abcd}	11.89^{ab}	1.43a	62.85^{bc}
B-D	0.34^{ab}	0.10^{ab}	0.079^{ab}	11.88^{ab}	1.32^{ab}	59.13^{ab}

3　结果与讨论

试验表明，由于多层立体栽培架的挡光影响，中层平均总太阳辐射强度是上层的25.15%～55.42%，下层平均总太阳辐射强度是上层的22.66%～37.46%，生菜的产量受此影响，中层较上层的生菜鲜重减少了37.23%、干重减少了43.18%，下层较上层的生菜鲜重减少了49.39%、干重减少了57.07%。品质上补光处理后生菜的亚硝酸盐含量有所降低，可溶性糖、可溶性蛋白含量、叶绿素含量有明显提高。

综上所述，在多层立体栽培条件下，对中、下两层的生菜进行每天4h，补光强度为110umol/m^{2}·s的补光，对生菜的产量和品质都有一定程度上的提高。这大大提高了温室内的土地利用效率，成倍的提高了生菜的产量。此外，试验工程中的补光时段以及补光的强度还不是最佳的方案，这由于每天的太阳光变化是不可控制的，对于全天不同栽培面上的光照分布情况还需要进一步研究，对于不同层人工补光的方案还需要进一步试验研究。

参考文献

[1]　魏灵玲，杨其长，刘水丽. LED在植物工厂中的研究现状与应用前景［J］.中国农学通报，2007，23（11）：408-410.

[2]　郭世荣. 无土栽培学［M］.北京：中国农业出版社，2003.

[3]　闻婧，杨其长，魏灵玲，等.不同红蓝LED对生菜形态与生理品质的影响［A］.寿光国际设施园艺高层学术论坛. 2015.

[4]　OhyamaK，KozaiT，YoshinagaK. 2006. Electricenergy，water and carbon dioxide utilization efficiencies of a closed-type transplant production system［A］.200：28-32.

[5]　李志鹏.植物工厂光斑可调式LED节能光源研制及应用研究.硕博学位论文［D］. 北京：中国农业科学院，2016.

[6]　李止正，龚颂福.立柱式无图栽培系统及其在生菜栽培上的应用［J］.应用于环境生物学报，2002，8（2）：142-147.

[7]　陈宗玲，刘鹏，张斌，等.立体栽培草莓的光温效应及其光合的影响［J］.中国农业大学学报，2011，16（1）：42-48.

[8]　马旭，林超辉，齐龙，等.不同光质与光照对水稻温室立体育秧秧苗素质的影响［J］。农业工程学报，2015，31（11）：228-235.

[9]　Lee H J，Titus J S. Relation ship between it rate reductive activity and level of soluble car bohyd rate sunder prolonged Darkness in MM.106 apple leaves［J］.Journal of Horticultural Sciences，1993，68（4）：589-596.

营养液作为贮热体系的热传导规律分析

马宇婧，温祥珍，杜莉雯，李亚灵

（山西农业大学园艺学院，太谷　030801）

摘要：试验测定了多孔定植板条件下叶菜生产系统内营养液温度变化，认为营养液作为贮热体系是具有实际意义的。在体系营养液深度为21.5cm条件下，不同层次日较差变化符合对数关系：y=−2.619ln（x）+ 4.2152。根据各深度日较差的变化幅度，将营养液划分为热交换层（>3℃）、热缓冲层（1～3℃）、热稳定层（0～1℃），分别位于液面表层0～5cm、5～10cm、10～15cm。秋季营养液0cm、5cm、10cm、15cm最高温分别出现在14:00、16:00、17:40、20:00，并随着营养液深度的增加而快速降低。上述结果表明能量在营养液中是以传导方式为主进行的。

关键词：营养液；叶菜生产系统；贮热体系；热传导

Analysis of Heat Conduction Law of Nutrient Solution as Heat Storage System

Ma yujing，Wen xiangzhen，Du liwen，Li yaling

（*Shanxi Agricultural University*，*College of horticulture*，*Taigu*　030801）

Abstract：The experiment measured the temperature change of nutrient solution in the leaf-producing system under the condition of porous colonies，and considered the nutrient solution as the heat-storage system is of practical significance. In the system of nutrient solution depth of 21.5cm conditions，the different levels of daily variation in line with the logarithmic relationship: y = −2.619ln（x）+ 4.2152. The nutrient solution was divided into heat exchange layer（>3℃），thermal buffer layer（1 ～3℃），heat stability layer（0 ～ 1℃），which located in the liquid surface layer 0～5 cm，5～10 cm，10～15 cm，respectively. The highest temperature of 0 cm，5 cm，10 cm and 15 cm of nutrient solution in autumn appeared at 14:00，16:00，17:40 and 20:00，respectively，and decreased rapidly with the increase of nutrient solution depth. The above results show that the energy in the nutrient solution is heat transfer by conduction.

Key words：Nutrient solution；Leaf vegetable production system；Heat storage system；Heat conduction

基金项目：国家自然科学基金重点项目“温室环境作物生长模型与环境优化调控”（编号：61233006），山西省煤基重点科技攻关项目“设施蔬菜高效固碳技术研究与示范”（编号：FT201402-05）。

已申请国家发明专利“叶菜类蔬菜智能控制生产系统”（专利申请号：201710908963.5）

作者简介：马宇婧（1995-），女（汉），山西孝义人，硕士，研究方向：设施园艺

通讯作者：温祥珍（1960-），男（汉），山西原平人，教授，博士生导师，研究方向：设施园艺，Tel: 13935439298；E-mail: 330821473@qq.com

1 引言

叶菜生产系统将日光温室的结构特征和水培系统结合起来，采用现代化技术和低成本材料，通过蓄存水体，将营养液作为蓄放热的载体来调节系统内环境。根据大海调温原理[1]，将营养液作为一种新的缓热介质，代替传统的土墙、砖墙及土壤储热，既可为叶菜提供生长营养，又可调节温度，具有一定的热稳定性，是叶菜生产系统中最重要的一部分，其热传导规律尚不明确。

目前有关热传导规律的研究主要集中在温室墙体方面，如杨艳红等[2-3]将墙体划分为热交换层、热缓冲层和热稳定层，分别位于墙体从内向外的0～15cm、15～25cm和25cm以后；史宇亮[4]发现后墙体内侧约0.7m的厚度为昼夜间蓄放热的关键厚度；易东海[5]在墙体内找到了变温层与恒温层的界限，并提出该界限在不同季节不同；马承伟等[6]提出了以墙体夜间放热量作为评价指标的日光温室墙体保温蓄热性能评价的方法。也有温室土壤方面的研究，何芬等[7]对日光温室土壤热传导过程进行定量分析，采用有限差分法构建了日光温室非稳态二维地温模拟模型。

掌握营养液的热传导变化规律，对叶菜生产系统的蓄热保温性能分析评价有重要意义，可为今后的系列试验提供依据。为此，试验测定了多孔定植板条件下不同深度的营养液温度。

2 材料与方法

试验采用的叶菜生产系统是由山西农业大学设施农业工程研究所设计，为对称拱圆形状，东西侧高0.4m，中间高1m，宽5.4m，南北长6m，剖面图如图1所示。其中，上部采光部分骨架由圆钢焊接组装而成，采用2/3固定阳光板和1/3活动PO膜，可通过卷帘的方式控制塑料薄膜覆盖面积进行通风。下部贮液池由角钢焊接成长方体，角钢内嵌入4cm厚加密聚苯板作为隔热保温材料，减少内外能量传递。贮液池内铺防渗膜，注入7m^{3}营养液，深度约21.5cm。营养液表面铺满厚5cm、162孔的泡沫定植板，试验期间不进行作物栽培，夜间顶部用铝箔被进行保温。

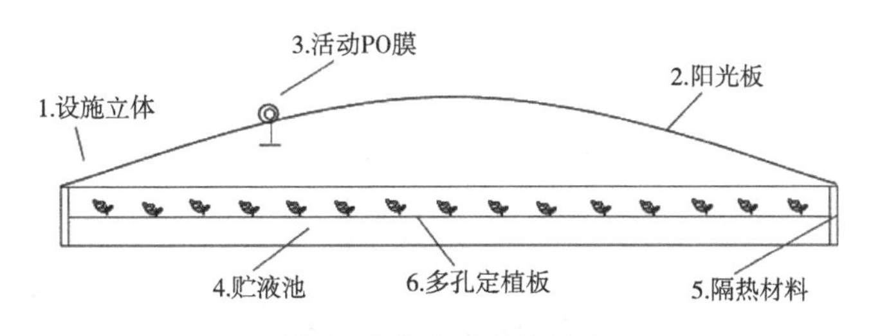

图1 叶菜生产系统剖面

为了了解热量在营养液中的传递规律，我们对营养液进行分层，从上到下每隔5cm设置一个测试位点，即测定营养液表面0cm、液面以下5cm、10cm、15cm共4个层次液温，并在贮液池内选取3个位置进行重复（1.东北角，2.中间，3.西南角）。测试区域不同深度测试位点截面见图2。所有温度数据通过SH-16路温度巡检仪测定，传感器为镍铬—镍硅

（K型）热电偶，测温范围为-50～300℃。设置每10min自动记录一组温度数据。试验于2017年11月2日至2017年11月26日进行（共25d）。数据采用WPS软件进行处理分析。

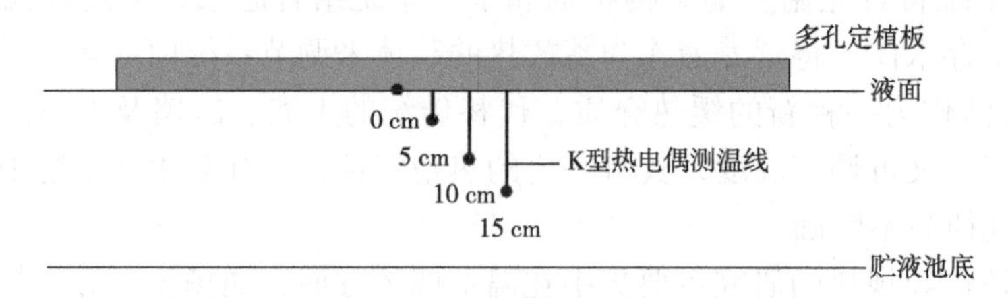

图2　测试区域营养液不同深度测试位点截面

3　结果与分析

3.1　营养液不同位点的温度日变化

试验测定了3个不同位点营养液不同深度的温度，如图3所示，依次代表0cm、5cm、10cm、15cm处液温。由图可看出，白天蓄热期间营养液温度位点东北角＞中间＞西南角，夜晚放热期间中间温度较东北角和西南角高。由于西南角白天接受太阳辐射迟，升温较东北角和中间滞后。受边际效应的影响，夜间东北角和西南角较中间温度低。3个测试位点温度虽有差异，但远不及深度带来的差异，不影响营养液整体传导规律，故使用3个位点的平均值来进行下一步分析。

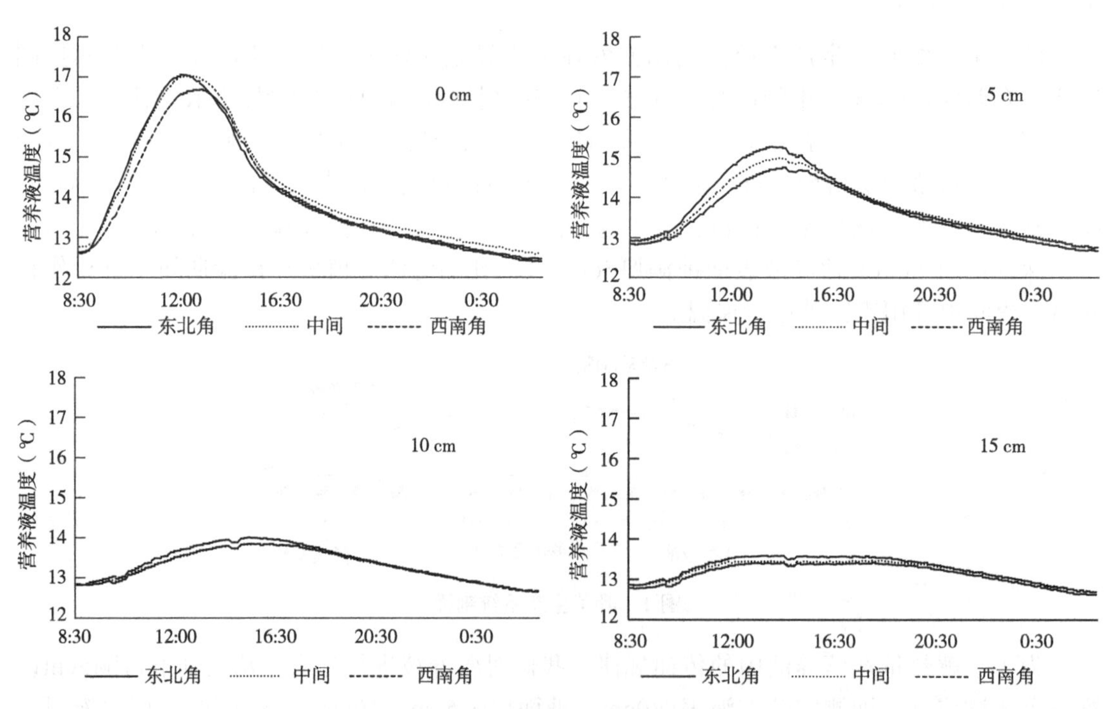

图3　营养液不同位点的温度日变化

注：图中每个数据点为2017年到11月2日到共25d的平均值。

3.2 营养液不同深度的温度日变化

在2017年11月2日至26日测定了叶菜生产系统中营养液不同深度的温度变化，结果如图4所示。从中看到：不同深度营养液温度变化显著不同，表层变幅最大，说明能量交换最多，越往深变幅越小，能量交换越少。各层平均温度分别是14.1℃、13.7℃、13.4℃、13.3℃，变幅依次为4.1℃、2.0℃、0.9℃、0.5℃，变幅越来越小，营养液温度越来越稳定。表层蓄积热量后逐层传递，0cm、5cm、10cm及15cm处分别在14:00、16:00、17:40、20:00达到最高温，分别为16.9℃、15.0℃、13.9℃、13.5℃，深层较浅层滞后。表1列出营养液各深度最高温、最低温及日较差的变异系数，营养液越深，温度变化越小，温度逐渐稳定。

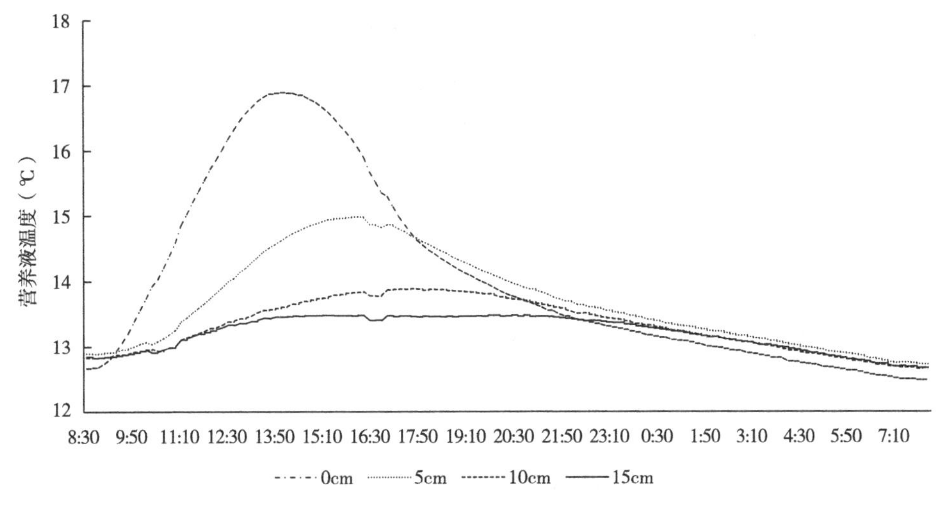

图4　营养液不同深度的温度日变化

注：图中每个数据点为3个位点2017年11月2—26日共25d的平均值。

表1　营养液各深度温度情况

类别		平均值	最大值	最小值	极差	标准差	变异系数
平均温	0cm	14.1	17.1	10.8	6.2	2.15	0.152 6
	5cm	13.7	16.4	10.6	5.8	2.04	0.148 3
	10cm	13.4	15.9	10.3	5.6	1.97	0.147 5
	15cm	13.3	15.6	10.3	5.4	1.93	0.145 4
最高温	0cm	17.0	20.7	13.2	7.5	2.54	0.149 9
	5cm	15.0	18.1	11.6	6.5	2.19	0.146 1
	10cm	13.9	16.6	10.8	5.7	2.03	0.145 4
	15cm	13.6	15.9	10.6	5.3	1.95	0.143 3
最低温	0cm	12.6	15.0	9.5	5.6	1.96	0.155 8
	5cm	12.8	15.3	9.8	5.5	1.92	0.149 6
	10cm	12.8	15.1	9.7	5.4	1.92	0.150 1
	15cm	12.8	15.2	9.8	5.4	1.92	0.150 1

（续表）

类别		平均值	最大值	最小值	极差	标准差	变异系数
日较差	0cm	4.4	5.8	1.7	4.1	0.98	0.225 7
	5cm	2.2	2.9	1.0	2.0	0.51	0.234 2
	10cm	1.2	1.5	0.6	0.9	0.22	0.186 9
	15cm	0.8	1.1	0.6	0.5	0.12	0.142 6

注：表中数据根据2017年11月2日—11月26日每天平均温、最高温、最低温、日较差求出各自的平均值、最大值、最小值、极值以及标准差和变异系数。其中标准差通过STDEV函数计算而得，变异系数CV=标准差/平均值×100%[8]。

3.3 不同深度营养液昼夜温差的频度及区域划分

温度的变化反映着能量的交换，试验对各深度的营养液日较差进行了频度分析，结果如表2所示。由表可见，在11月2到26日共25d的观测中，液面0cm、液面以下5cm、10cm、15cm处的日较差分别集中在3～6℃、1～3℃、0～2℃、0～1℃，各有23d、24d、25d、23d，分别占观测总天数的92%、96%、100%、92%。进一步从能量交换量看（图5），0～5cm、5～10cm、10～15cm各占比为56.69%、32.35%、10.96%，即0～10cm达到近90%，是主要的能量交换部位。

综合表2和图5得出营养液日较差快速下降的部位位于液面表层0～5cm，变化幅度为3～6℃；日较差变化平缓部位位于5～10cm，变化幅度为2～3℃；日较差变化稳定部位位于10～15cm，变化幅度在0～1℃。据此，将营养液划分为热交换层（>3℃）、热缓冲层（1～3℃）、热稳定层（0～1℃）。

表2　不同深度营养液昼夜温差的频度（d）

营养液深度/cm	不同的温度区范围					
	0～1℃	1～2℃	2～3℃	3～4℃	4～5℃	5～6℃
0cm	0	1	1	5	11	7
5cm	2	7	17	0	0	0
10cm	5	20	0	0	0	0
15cm	23	2	0	0	0	0

注：图中数据来自各测点2017年11月2—26日的日较差

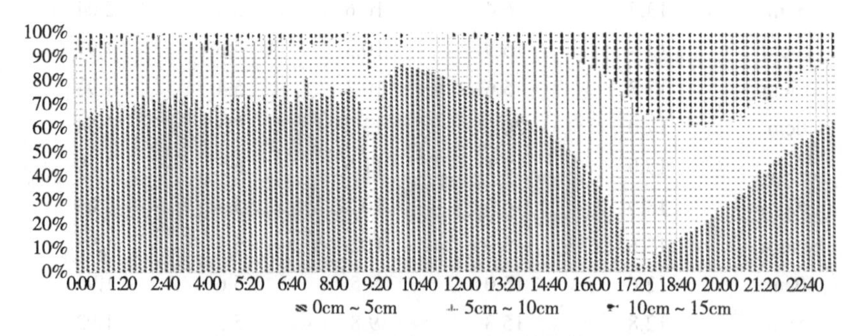

图5　一天中不同液层能量交换比率变化

注：图中每个数据点通过每时刻不同液层$|T_{浅}-T_{深}|/(|T_0-T_5|+|T_5-T_{10}|+|T_{10}-T_{15}|)\times100\%$计算而得。

3.4 营养液日较差变化趋势

为了解营养液内部热传递变化规律，根据试验期间每日各深度的日较差，以营养液深度为横坐标，日较差大小为纵坐标绘制得图6。无论天气好坏，25d的趋势线均表现出营养液温度随深度增加变化稳定，日较差渐小，呈对数趋势递减。25d每日各深度日较差平均值绘制的趋势线方程为y=−2.619ln（x）+4.215 2，由此关系式可计算出当日较差y=0℃时x=5，即液面以下20cm处日较差为0℃。

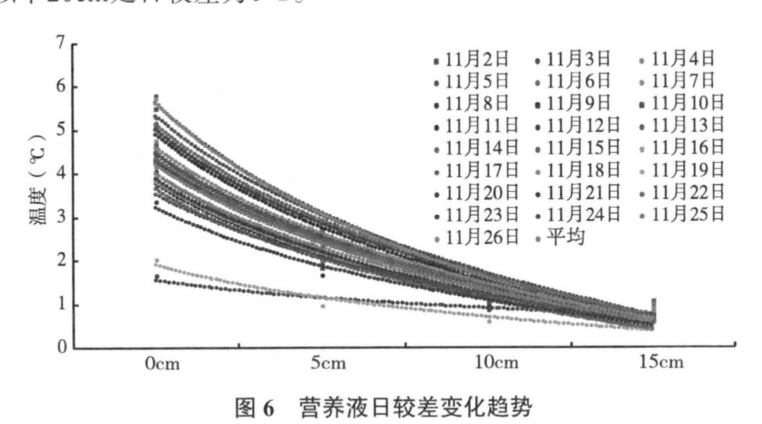

图6　营养液日较差变化趋势

注：图中每个数据点日较差为2017年11月2—26日每日最高温与最低温的差。

3.5 系统热交换的速度变化

通过叶菜生产系统营养液各层热量传输过程中变化的温度，计算出各层变化的热量，从而得出系统热量交换速率，绘制出图7有关要项的变化情况，图7中正值表示热量损失速率，负值表示热量蓄存速率。由图看出营养液0～5cm白天热损失速率最高可达772kcal/cm，出现在13:20；5～10cm在15:20热损失速率最高为382kcal/cm；10～15cm在17:50热损失速率最高为138kcal/cm。5～10cm和10～15cm在20:00后热损失速率降到100kcal/cm以下，于0:00后趋于稳定。

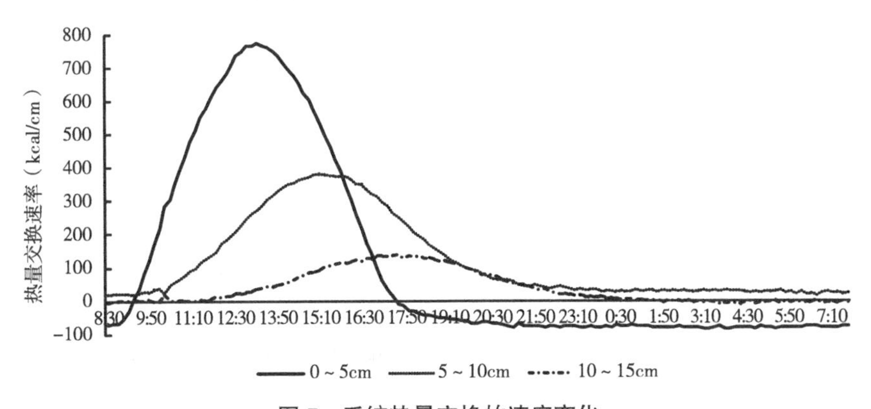

图7　系统热量交换的速度变化

注：图中每个数据点通过（$T_{浅}$-$T_{深}$）×1 620kcal/5cm计算而得（1m^{3}即10^{6}g水升高1℃需10^{3}kcal能量，5cm深营养液体积为6×5.4×0.05=1.62m^3）。

4 讨论与结论

从热量传递规律上来讲，营养液类似于日光温室墙体的变化规律，能量在营养液中主要是以传导方式进行的。朱开玲等[9]认为，烤烟漂浮育苗设施系统热量分布为基质：水：空气：水泥结构=11.4:62.3:1:20.1，证实了水在漂浮育苗设施系统中起重要作用。本研究通过对多孔定植板条件下叶菜生产系统内营养液不同深度温度的测定分析，认为营养液作为贮热体系是具有实际意义的。试验得出不同测试位点对营养液温度有一定影响，但不影响整体传导规律。不同深度营养液温度变化显著不同，表层变幅最大，说明能量交换最多，越往深变幅越小，温度变化越来越稳定。各层液温分别在14:00、16:00、17:40、20:00出现峰值，并随着营养液深度的增加而快速降低，变异系数减小，深层较浅层滞后。营养液作为缓冲介质，在能量交换上可发挥一定作用。

在体系营养液深度为21.5cm条件下，不同层次日较差变化符合对数关系：$y=-2.619\ln(x)+4.2152$。0～5cm、5～10cm、10～15cm各层能量交换量各占总交换量的56.69%、32.35%、10.96%。根据各深度日较差的变化幅度，将营养液划分为热交换层（>3℃）、热缓冲层（1～3℃）、热稳定层（0～1℃），分别位于液面表层0～5cm、5～10cm、10～15cm。营养液0～5cm、5～10cm、10～15cm白天热损失速率最高可达772kcal/cm、382kcal/cm、138kcal/cm，分别出现在13:20、15:20、17:50。

事实上，由于本实验的叶菜生产系统体积较小，空气所占容积小，气温变化1℃仅需6.33kcal，而1m^3营养液变化1℃则需1 000kcal，可见营养液中能储存大量能量，在环境调控以及蓄热上能发挥作用。

上述结果表明营养液作为贮热介质具有明显的热传导规律，能量在营养液中是以传导方式为主进行的，与常规意义上的水通过上下交换传递热量有所不同。

参考文献

[1] 张宇雷，吴凡，管崇武，等. 环渤海地区热能资源分布及海水养殖水体调温模式研究［J］. 江苏农业科学，2014，42（6）：229-231.

[2] 温祥珍，李亚灵. 日光温室砖混结构墙体内冬春季温度状况［J］. 山西农业大学学报（自然科学版），2009，29（6）：525-528.

[3] 杨艳红，李亚灵，马宇婧，等. 日光温室北侧墙体内部冬春季的温度日较差变化分析［J］. 山西农业大学学报（自然科学版），2017，37（8）：594-599.

[4] 史宇亮. 日光温室不同厚度土墙蓄放热特性研究［D］. 泰安：山东农业大学，2017.

[5] 易东海. 日光温室土质墙体温度特性研究［D］. 郑州：河南农业大学，2011.

[6] 马承伟，卜云龙，籍秀红，等. 日光温室墙体夜间放热量计算与保温蓄热性评价方法的研究［J］.上海交通大学学报（农业科学版），2008（5）：411-415.

[7] 何芬，马承伟，周长吉，等. 基于有限差分法的日光温室地温二维模拟［J］.农业机械学报，2013，44（4）：228-232.

[8] 明道绪. 田间试验与统计分析［M］. 北京：科学出版社，2005.

[9] 朱开玲，段淑辉，李良勇，等. 烤烟漂浮育苗设施系统的温热特性及其对烟苗生长的影响［J］.湖南农业科学，2015（8）：122-126.

日光温室蔬菜物联网智能预警预测专家决策系统的开发与应用

贺超兴[1*]，陈芳[2]，周进[2]，刘娜[2]

（1. 中国农业科学院蔬菜花卉研究所，北京 100081；
2. 新疆生产建设兵团第六师农业科学研究所，五家渠，831300）

摘　要：本文基于对日光温室温湿度和蔬菜病害发生关系的研究,建立了基于温室温湿度环境因子的蔬菜物联网智能预警预测专家决策系统，通过实际应用，根据预警信息早期采取农药等预防措施，进行了实际应用检验，结果表明：基于物联网的环境模型具有较好的病害预警功能，可以用于日光温室的病害早期防控，为我国日光温室病害预警控制和预测提供了智能化管理平台。

关键词：温湿度；病害预警；早期防控；专家决策系统；日光温室

The Development and Application of the Vegetable Intelligent Disease early Warning and Prediction Expert Decision System in Solar Greenhouse

He Chaoxing[1]，Chen Fang[2]，Zhou Jin[2]，Liu Na[2]

（1. *Institute of vegetables and flowers, Chinese Academy of Agricultural Sciences, Beijing* 100081；2. *Xinjiang production and construction corps sixth division agricultural science research institute, Wujiaqu* 831300）

Abstrac： Based on study of the relationship between solar greenhouse humidity and temperature change and the vegetable disease occur ratio, the vegetables intelligence expert decision system based on the solar greenhouse temperature and humidity factors was established to early warning and forecast the vegetable disease. After practical application, the results showed the model of environment based on Internet has disease early warning function, which can be used to the early prevention and disease control. the software in solar greenhouse may provide a intelligent management platform for disease warning and early control.

Key words： Temperature and humidity；Disease warning；Early prevention；Expert decision system；Solar greenhouse

设施农业的发展为蔬菜周年高产高效生产带来了契机，它通过应用先进的科学技术、

基金项目：新疆生产建设兵团科技攻关与成果转化计划项目（2015AC022），国家特色蔬菜产业技术体系岗位专家项目（CARS-24-B-04）和农业部园艺作物生物学与种质创制重点实验室资助项目

*通讯作者：贺超兴，研究员，研究方向为 设施蔬菜栽培与产地环境综合治理，E-mail: hechaoxing@126.com

连续的生产方式和高效的智能化管理模式，以一定的生产设施为载体，通过控制生产环境，为作物提供相对适宜的温度、湿度、光照、水肥和气候环境等条件，从而可以高效、均衡地生产各种蔬菜、水果、花卉和药用植物等。设施农业包含设施栽培、各类温室、大棚及地膜覆盖，还包括所有进行农业生产的保护设施。这种高产高效，最具活力的生产模式已经成为各个国家研究的热点。我国温室管理以管理人员的经验决策为主，管理不当使得温室生产存在着许多问题：在低温高湿条件下易发生病害而过度施用化学农药；过量施肥使温室土壤板结；滥施农药化肥，造成成本上升，环境污染，产品的数量和质量达不到预期的目标，所以通过科学、规范的技术施肥减少化肥农药施用是当前设施农业发展的重要问题[1]。

目前国内外设施蔬菜包括番茄、黄瓜、甜椒等栽培面积很大，市场广阔，但由于很多种植者缺乏相应的温室管理知识和技术，基本是凭经验凭感觉来管理，没有在恰当的时候采取合理的措施，造成投入和产出不成比例，就会导致经济损失和不良后果。温室病害的预警预防是日光温室管理中非常重要的技术环节，病害严重发生不仅威胁着蔬菜的生产，影响果实品质，还对蔬菜安全构成影响[2, 3]。因此早期预警病害发生，对症下药及时防治或及时预报早期预防，才可最大限度减少损失。

1 设计思路

为应对优质蔬菜需求量的不断增长趋势，实现在农业物联网的基础上对日光温室蔬菜生产的智能化管理与预警预测，并对生产和销售提供决策支持指导来设计软件。当用户通过系统人机交互界面提供温室生产管理中需要决策的问题后，系统开始搜集数据信息，并根据知识库中的知识来判断和识别问题，并给出相应的结论。通过农业物联网技术可以实现日光温室生产管理过程中对作物、土壤、环境从宏观到微观的实时精准监测，定期获取作物生长发育状态、病虫害、水肥状况、生产管理过程，以及相应生态环境的实时信息，形成温室作物实时大数据库，再结合专家知识经验库，为温室栽培科学管理、病虫害及时预测预防和蔬菜种植决策调控提供技术支持和数据支撑，不仅达到合理使用农业资源、降低生产成本、改善生态环境、提高农产品产量和品质，提高农户经济收益的目的，还能有效地普及生产知识、传播实用的先进技术、提高生产者的科学素质等，具有重大的社会价值和现实意义[4]。

基于Web的结合温室环境监控系统和日常操作管理记录功能的专家系统，能够让农业生产企业管理者对整个农场或多个生产基地进行全面、直接的了解，方便管理，节约成本。温室管理专家系统还有一个明显的好处就是它的非时空限制性，它可以任意时间拷贝任意系统中任何计算机中的数据，可以作为一个智能程序永久保存，并能准确、高效、周密、迅速而不疲倦地工作，因而可以很好地解决温室数量与专家人数增长不对等的矛盾。

1.1 技术特点

基于温室周年环境数据库及BP神经网络算法结合病害发生条件可以对病害进行预警，并可基于蔬菜生长发育期的生理天数根据采收期预测播种期或根据苗龄和定植期预测开花期及采收期，从而达到对生产进行预警预测的目的[5, 6]。遵循人工智能的技术原理，专家决策系统还具有数据和知识半自动获取、知识库求精、确定性推理、不确定性推理、模型自动解析等功能，实现对特定农业领域问题的定性推理和定量决策，具有高度智能化的特点。整个系统分成三大部分：Web服务器程序、后台管理程序和数据库。

1.2 软件的设计思想

软件设计的目的是开发基于网络化监测和自动化预警与控制管理相结合的智能化专家决策系统平台，以提高温室蔬菜科学管理水平，增加温室生产的经济效益。软件应用到生产实践当中，可为农业科技人员和农业生产者提供环境管理知识、病害早期预警和病虫害识别与防治知识、蔬菜发育时期预测和产量产期预测信息，辅助完成基于环境监测的日常水肥管理控制与光温湿环境管理数据记录等功能。

本软件通过与农用通等物联网环境数据平台链接，对温室环境进行监测和提出病害预警或防控建议，并为专家系统提供数据库，如温室环境和设施蔬菜发育数据库、病虫害诊断知识数据库、温室管理知识数据库，温室日常操作管理参数设置及记录数据库。在指导生产实践方面为用户提供相应的生产预测管理基础知识，通过具有实际应用价值的发育阶段预测功能，便于生产者制订生产计划；通过对病虫害的发生时间进行预测，提供相应的防治方法，有利于及时防治病虫害。

农业物联网技术广泛应用于农业生产、流通等环节，促进了农业生产经营从粗放式、经验性管理到精细化、科学化管理的快速转换、提高农产品的产量与品质、降低农业生产成本、保护农业环境，是加快推进现代农业建设的必然选择，对促进农业增效、农民增收、农村稳定具有重大推动作用。

2 系统构成

2.1 系统设计基础

日光温室蔬菜物联网智能预警预测专家决策系统以农业物联网为设计基础，物联网的主流架构体系始终贯彻IaaS（架构即服务）、SaaS（软件即服务）、PaaS（平台即服务）的设计思想，在设计理念和应用效果上呈现“可视化、泛在化、智能化、个性化、一体化”的特点，采用三层架构：感知层、传输层和应用层。

2.2 系统体系结构

日光温室蔬菜物联网智能预警预测专家决策系统是基于“浏览器/Web服务器/数据库系统”三层分布式计算结构体系的网上查询、诊断、学习等综合性管理系统，普通用户可利用浏览器登录访问。完成温室蔬菜生产中的病虫害预测、诊断及管理模式化等功能。同时，领域专家也可通过登录维护页面对知识库进行更新和维护，实现知识的更新升级功能。其系统体系结构如图1所示。

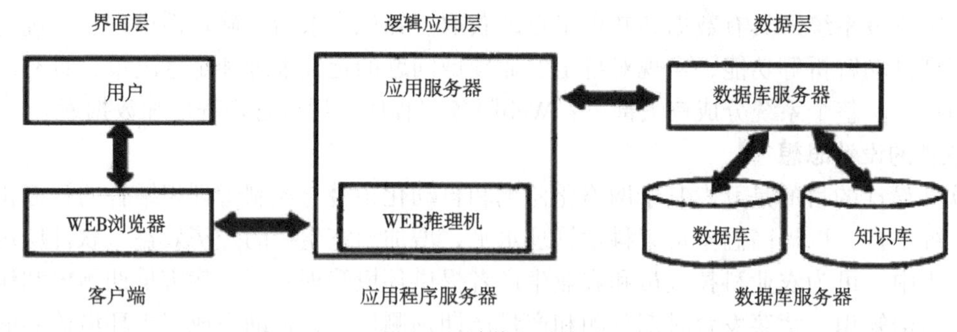

图1　系统体系结构

3　系统模块功能

日光温室蔬菜物联网智能预警预测专家决策系统功能模块整体分布如下（图2）。

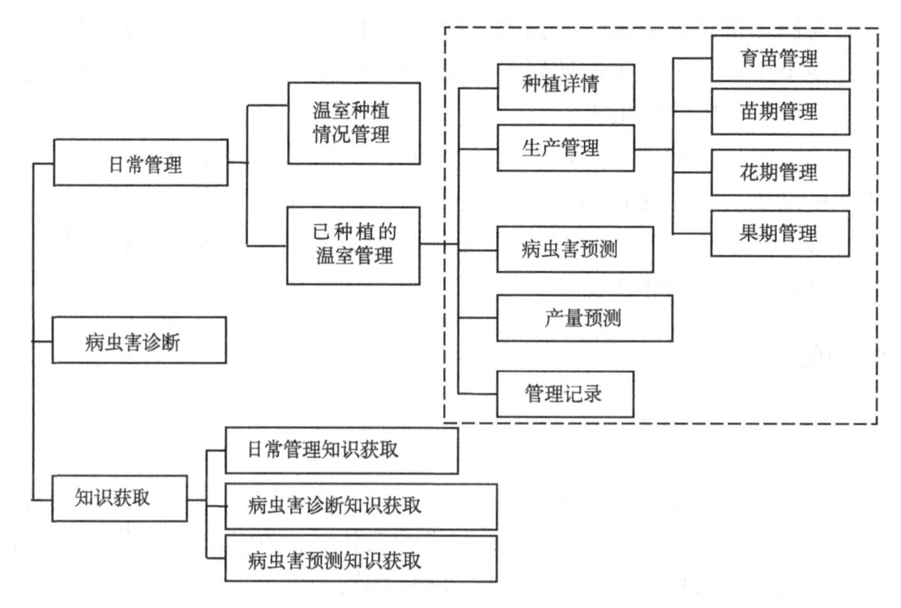

图2　系统整体布局

3.1　系统登录

系统登录界面，通过用户名和密码达到验证登录者身份的目的，保证了系统安全性。

3.2　系统首页

系统首页主要包含系统的功能模块导航和系统研制单位简介。

3.3　温室管理

系统的温室管理部分主要包含“温室选择”和“温室维护”两大模块。

（1）温室选择。主要是对不同温室信息的选择、温室介绍包含作物种类、所处地区、温室结构参数的详细信息等。

（2）温室维护。主要是对温室植物维护管理信息的记录，包含温室名称、所处地区、种植作物、温室面积、株距、行距、联系人、联系电话、创建时间、描述、操作等。

3.4 育苗管理

系统的育苗管理部分是对育苗信息的记录，包含“育苗管理”和“育苗知识”两大模块。

（1）育苗管理。主要是育苗不同时期所需环境要素信息的记录及图片信息。

（2）育苗知识。主要包含育苗方面的视频资料方便用户查看学习。

3.5 栽培管理

系统的栽培管理部分是对作物栽培信息的记录，包含“田间档案管理”“生长管理”“绿色食品”和“栽培知识”等四大模块。

（1）田间档案管理。主要是对田间信息的记录，包含温室名称、工作人员、管理条目、操作日期、操作记录等。

（2）生长管理。主要是对作物生长时期所需环境要素信息的记录和相关生长期图像。

（3）绿色食品。主要是对相关绿色食品信息的介绍，包含绿色食品基本知识、绿色食品农药使用准则、无公害绿色与有机食品有何区别、无公害蔬菜生产技术保障体系、有机食品的国际标准等。

（4）栽培知识。主要包含作物栽培知识的相关视频资料。

3.6 病虫害诊断

系统的病虫害诊断部分是对作物病虫害信息的记录，包含“病害概述”“虫害概述”、病害诊断”和“虫害诊断”等四大模块。

（1）病害概述。主要是对各种病害信息的记录，包含病情原理、病害症状、防治方法等。

（2）虫害概述。主要是对各种虫害信息的记录。包含虫害种类、作物症状，防治方法等。

（3）病害诊断。主要是对蔬菜不同部位感染的病害进行智能诊断，以此作为蔬菜感染何种病害的防治依据。

（4）虫害诊断。主要是对蔬菜不同部位感染虫害后的结果阐述，以此作为蔬菜感染何种虫害的诊断依据。

3.7 预警预测

系统的预警预测部分主要包含“产量预测”“产期预测”和“病虫害预测”三大模块。

（1）产量预测。根据不同蔬菜单株座果情况结合定植时输入的栽培参数，包括栽培密度、座果数和单果重的数据进行预测，其中单果重受栽培蔬菜及品种的影响，座果数易受环境影响，栽培密度则由定植密度包括定植的行株距确定。

（2）产期预测。根据不同蔬菜开花至采收的生理发育天数结合温室周年数据库的日光温室环境参数确定的不同月份的生长天数来进行预测，受季节和品种果实大小等的影响，将相关参数输入后即可根据生长环境进行产期预测。

（3）病害预测。根据温室周年环境数据库的温度环境参数及高湿持续时间结合设施蔬菜主要病害发病的环境参数需求特点来确定不同月份的发病指数进行初步预警预测，然后随着蔬菜栽培结合更新的环境数据进行病害预测修正，另外还以根据田间档案记录的往年历史发病情况在病害高发时期的早期对病害发生进行预警预报，同时系统还可以提供病害预防措施和药物防病技术，从而通过早期防病实现防病于未然的目标。例如白粉病在发生主要是在新疆夏季栽培时，高温高湿环境下发生的[7, 8]，所以可基于环境数据进行病害预警预测。

3.8 知识库管理

系统的知识库管理部分主要包含“知识库管理”和“知识库更新”两部分，知识库更新是新增病害的补充，如图3所示。

图 3　知识库更新

3.9 系统设置

系统设置部分主要包含“用户管理”“角色管理”“菜单管理”和“权限分配”几大模块。①“用户管理”主要是对登录此系统用户的管理，包括用户的详细信息，及添加删除用户等。②“角色管理”是对登录专家决策系统不同角色的管理，包括“专家学者”“系统用户”和“系统管理员”等。③“菜单管理”是对整个专家决策系统各级目录菜单的管理，可增加删除等。④“权限分配”主要是对系统模块不同权限进行划分。

3.10 系统维护功能

系统维护功能包括环境数据库维护、温室环境设备管理、知识库维护以及网站维护等。温室环境数据采集系统在运行过程中会产生大量数据，用户需要定期使用系统提供的工具对这些数据进行整理或删除。同时，本系统具有可扩展性，用户能够根据系统的提示添加新的蔬菜品种或者设施类型，添加新的生长模型以及病虫害数据。

4 预警防治效果

2017年夏季的6月，在新疆五家渠农六师农科所日光温室应用专家系统，当环境数据库连续病害预警3天后，温室环境适合白粉病发生且无改变温室环境措施情况下，通过对监测温室部分及时施用50%醚菌酯药剂结合硫磺粉熏蒸对温室黄瓜白粉病进行了药剂预防，对照未做处理，每周防病处理1次，2周后不同处理的发病情况和防治效果有明显差异，如下表所示。

表　日光温室黄瓜白粉病预警预防效果比较

处理	7d病指	14d病指	防效（%）
处理1	2.2	6.1	−93.9
处理2	1.1	2.2	−97.8
CK	66.29	100	

处理1：50%醚菌酯WG 3000倍液；处理2：硫磺熏蒸器；CK：对照

由上表可以看出，在环境适宜的病害高发期，通过环境预警后采取措施对病害进行早期防治，可以很好地降低病害发生率，取得事半功倍的防效，其中处理1的效果不如处理2，表明采用硫磺粉熏蒸处理，在不增加空气湿度的情况下对病菌进行杀灭有更好的防病效果，因此在日光温室黄瓜白粉病的早期预防中应用硫磺熏蒸的效果好于药液防治。可见采取预警预防可以达到减少防治次数，提高防效的目的。

5　系统应用总结

基于温湿度周年数据库和环境监测建立的日光温室蔬菜物联网智能预警预测专家决策系统可以辅助温室管理人员对日光温室进行合理有效的管理，充分发挥专家系统不受时间和空间限制的优势，可解决我国设施蔬菜栽培管理专家与温室增长数量不对等的矛盾。本设计总结如下。

（1）利用物联网温室环境监控系统平台，构建了基于实时环境数据、温室周年数据库和基于设施蔬菜生长发育生理发育天数模型的温室专家决策系统，充分发挥了生长发育模型和专家系统各自的优势。应用蔬菜生长发育模型完成蔬菜生长期的定量计算，应用专家系统完成数据处理及解释功能。系统为品种参数、生长发育模型、环境数据库、病虫害数据库等设计了统一的接口，可以根据需要添加新的模型和病虫害样本，具有较大的灵活性。

（2）建立了基于规则推理的日常管理和病害预警诊断子系统。通过构造日常管理决策树，获取每一个生长阶段温室环境目标值，水分管理、肥料管理、植株管理知识，并以规则形式存储，推理方法采用产生式推理。病害预警是根据病害发生的环境需求特点根据温室环境周年数据库的描述初步判断为病害高发期，之后再通过田间档案历史病害发生情况加权打分，计算病害发生的概率，根据发生率情况进行预警，并提出预防有关病害的技术措施。

（3）应用蔬菜主要生长发育期的生理天数结合不同月份的生理天数值预测产期，明确果实发育期蔬菜果实生长速率与温度之间的关系。通过基于周年数据库的办法来初步获取采收期数据，再结合日光温室实际数据进行修正实现产期的精准预测，从而实现从任何一个生育期预测其他生育期的起始日期。通过引入实时温度数据更正机制，修正预测偏差，提高了预测精准度。

参考文献

[1] 乔晓军，余礼根，张云鹤，等. 设施蔬菜病虫害绿色防控系统研制与初步应用 [J]. 农业工程技术，2017，37(31)：29-31.

[2] 赵洁 宋建华.老菜区黄瓜白粉病流行原因及无公害防治技术 [J]. 长江蔬菜，2016(17)：58-59.

[3] 王晓蓉，吕雄杰，贾宝红.基于物联网技术的日光温室黄瓜白粉病预警系统研究.农学学报，2016，6(8)：50-53.

[4] 武向良，高聚林，赵于东，等. 农业专家系统研究进展及发展方向 [J].农机化研究，2008(1)：235-238.

[5] 王冀川，马富裕. 基于生理发育时间的加工番茄生育期模拟模型，应用生态学报 2008，19(7)：1 544-1 550

[6] 马丽丽，纪伟，贺超兴，等. 番茄专家系统环境数据库在病害预测中的应用 [J].农机化研究，2008(6)：161-163.

[7] 白晓丽，吴凤芝.黄瓜白粉病的传播途径和防治方法.现代化农业，2012(10)：68-69.

[8] 王萍.黄瓜白粉病的发生与防治.农业科技与信息，2015(13)：30-31.

日光温室装配式墙体结构创新与热工性能初探

邹志荣，鲍恩财，申婷婷，张勇，曹晏飞

（西北农林科技大学园艺学院，农业部西北设施园艺工程重点实验室，杨凌 712100）

摘要： 日光温室是具有中国特色的高效节能型温室设施，在中国设施园艺的发展过程中起到了重要的作用。该文在前期研究基础上，采用不同施工工艺建造装配式主、被动蓄热墙体，对传统主动蓄热墙体日光温室（G1）、回填装配式主动蓄热墙体日光温室（G2）、模块装配式主动蓄热墙体日光温室（G3）、现浇混凝土装配式墙体日光温室（G4）进行冬季室内环境测试，分析其热工性能。试验结果表明，连续晴天条件下，G1、G2、G3、G4的夜间平均气温分别为15.2℃、16.0℃、17.3℃、15.9℃；连续阴天条件下，4座温室的夜间平均气温分别为11.3℃、12.9℃、13.0℃、12.5℃。对典型天气的蓄放热量进行分析，典型晴天G1、G2、G3、G4的蓄热量分别比放热量多155.83、299.52、309.84、198.56MJ；典型阴天G1、G2、G3、G4的蓄热量分别比放热量少52.34、29.39、3.48、60.20MJ，故G3的热量累计最多，室内气温最高。相同种植面积的4座温室建造成本表现为：G1>G2>G4>G3。因此，装配式日光温室主动蓄热墙体的技术方案可行，且成本较低，在适宜日光温室发展的地区具有一定的推广价值。

关键词： 装配式；施工工艺；墙体；主动蓄热；日光温室

Structural Innovation and Preliminary Study on Thermal Performance of Assembled Wall in Chinese Solar Greenhouse

Zou Zhirong，Bao Encai，Shen Tingting，Zhang Yong，Cao Yanfei

（*College of Horticulture*，*Northwest A&F University*，*the Agriculture Ministry Key Laboratory of Protected Horticultural Engineering in Northwest*，*Yangling* 712100）

Abstract: Solar greenhouse is the energy efficient greenhouse with Chinese characteristics and plays an important role in the development of Chinese protected horticulture. Based on the previous research，the paper constructed the assembled active and passive heat storage wall by different construction techniques. We tested the indoor environment of traditional active storage wall in Chinese solar greenhouse（G1），backfill-assembled active storage wall in Chinese solar greenhouse（G2），module-assembled active storage wall in Chinese solar greenhouse（G3），Cast-in-place concrete wall solar greenhouse（G4）in winter，and analyzed the thermal performance. The results showed that under continuous sunny conditions，the night average temperature of G1，G2，G3 and G4 were 15.2℃，16.0℃，17.3℃，15.9℃ respectively; Under continuous cloudy conditions，the night average temperature of G1，G2，G3 and G4 were 11.3℃，12.9℃，13.0℃，12.5℃ respectively. The heat

基金项目：宁夏回族自治区重点研发计划重大项目（2016BZ0901）；陕西省科技统筹创新工程项目（2016KTCL02-02）

作者简介：邹志荣，男，陕西延安人，教授，博士，博士生导师，主要从事设施园艺方面的研究。Email：zouzhirong2005@163.com storage amount of G1、G2、G3 and G4 in typical sunny day were 155.83、299.52、309.84、198.56 MJ more than the heat release amount of them respectively; The heat storage amount of G1、G2、G3 and G4 in typical sunny day were 52.34、29.39、3.48、60.20 MJ less than the heat release amount of them respectively. The construction cost of the four greenhouses with the same planting area is as follows: G1> G2> G4> G3. Therefore, the technical scheme of the active solar thermal storage wall in an assembled solar greenhouse is feasible and low cost, and has certain promotion value in a region suitable for the development of solar greenhouse.

Key words: Assembled; Construction technology; Wall; Active heat storage; Chinese solar greenhouse

1 引言

日光温室是具有中国特色的高效节能型园艺设施，具有完全的自主知识产权，在中国设施园艺的发展过程中起到了重要的作用，为提高城乡居民的生活水平、稳定社会做出了历史性贡献[1]。日光温室的墙体是其与其他类型设施的最大区别所在，主要起承重、蓄热、保温的作用。目前对日光温室的墙体材质与结构的研究较多，在材质方面，前人研究发现任何单一材料墙体的保温性能均低于多层异质复合墙体，且异质复合墙体还具有厚度薄、节省材料的特点[2]；在结构方面，诸多学者研究了墙体的适宜厚度、开发了墙体高效蓄热结构[3-5]。

日光温室的墙体结构除了保温、蓄热的作用，还需兼顾其现代化发展的趋势和低成本的施工建造与生产应用[6]。近年来，日光温室墙体的装配式建设逐渐成为行业热点，闫俊月等[7]报道了一种以预制混凝土板夹芯填土作为储热层、以聚苯板作为保温层的轻简化装配式后墙；张洁等[8]设计了一种装配式砾石蓄热墙体日光温室；张义等[9]研究了一种轻简装配式日光温室，该温室由基于水媒介质蓄热的主动蓄放热系统来实现蓄放热功能，由预制装配式复合墙体来实现保温隔热功能。通过简化施工流程、减少施工用材、降低人工投入来提高施工效率、降低成本。本课题组研究的传统主动蓄热墙体具有较好的蓄热效果[4-5]，但也存在施工工艺复杂、人工用量大、建造速度慢的问题，制约了该类日光温室的推广应用，为此，本文根据日光温室主动蓄热墙体结构特点，采用不同施工工艺对主、被动蓄热墙体进行装配式建造，进行冬季室内环境测试，分析其热工性能，从而为日光温室装配式墙体结构的进一步发展提供有益参考。

2 材料与方法

2.1 试验温室

4座供试温室均位于陕西省杨凌示范区旭荣农业基地（34°16′ N，108°06′ E），建成于2017年10月，测试期间室内种植作物均为番茄（于2017-11-05定植），采用基质袋培，灌溉方式为滴灌。夜间采用保温被覆盖，9:00开启，17:00关闭。试验期间晴天的12:00～14:00打开通风口。

试验温室结构如图1所示。传统主动蓄热墙体日光温室（G1），南北跨度为10m，东西长32m，方位南偏东5°，脊高5.0m，后墙高3.6m。后墙结构为100mm聚苯板+120mm粘土砖墙+960mm相变固化土+120mm黏砖墙（从外向内），相变固化土由当地黄土添加8%

掺量（质量比）的相变固化剂搅拌均匀制成，相变固化剂配方见文献[10]。温室采用卡槽骨架，间距1m，后屋面采用100mm聚苯板，前屋面为直屋面，覆盖PO膜。

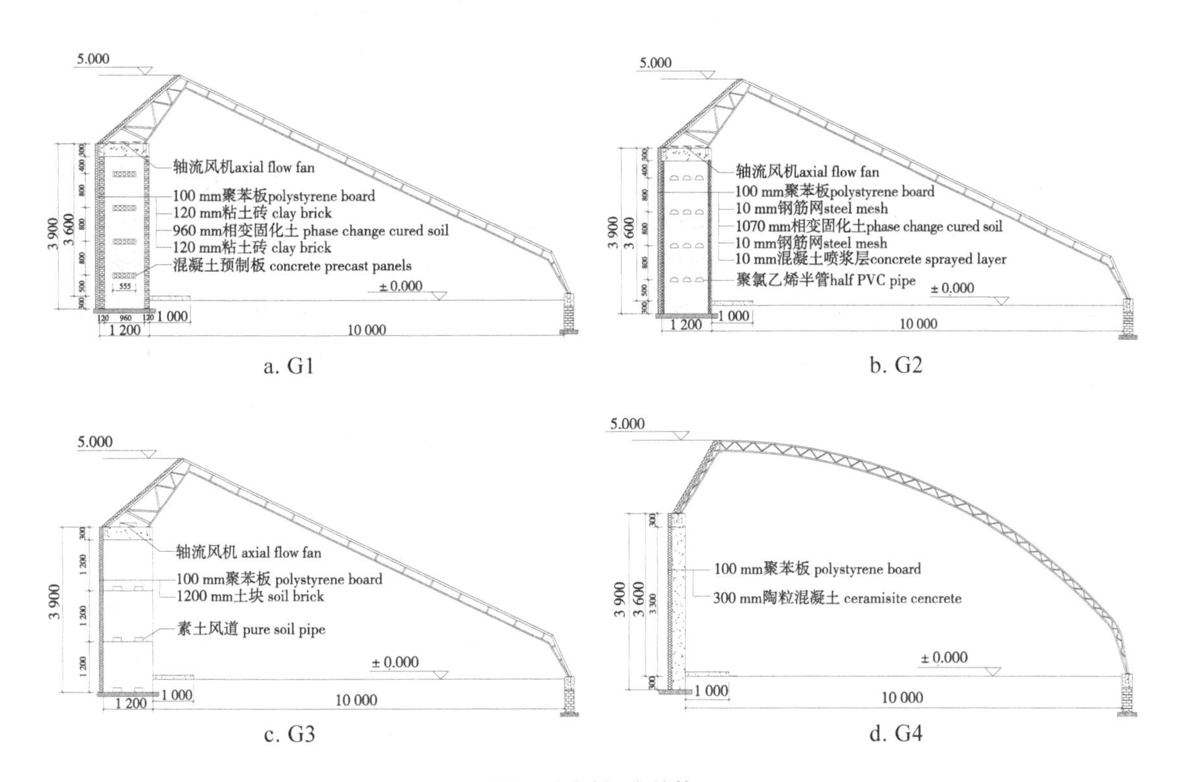

图 1　测试温室结构

Figure 1　Schematic diagram of tested solar greenhouses

回填装配式主动蓄热墙体日光温室（G2），后墙结构为100mm聚苯板+10mm钢筋网+1 070mm相变固化土+10mm钢筋网+10mm混凝土喷浆涂层（从外向内）。其他参数与G1一致。

模块装配式主动蓄热墙体日光温室（G3），后墙结构为100mm聚苯板+1 200mm素土模块墙（从外向内），单个素土模块尺寸为1 200mm×1 200mm×1 200mm，由当地黄土添加2%掺量（体积比）的麦草秸秆搅拌均匀后通过速土成型机（陕西杨凌旭荣农业科技有限公司生产）压实而成，制作方法见文献[11]。其他参数与G1一致。

现浇陶粒混凝土装配式墙体日光温室（G4），后墙结构为100mm聚苯板+300mm现浇混凝土墙（从外向内），前屋面为弧形屋面，其他参数与G1一致。

2.2　施工工艺

4座温室墙体的主要区别在于墙体蓄热部分和主动蓄热循环系统的施工工艺不同。G1墙体的建造主要采用人工砌筑围护砖墙和拉结砖墙、人工回填并夯实相变固化土；G2墙体的建造主要采用人工砌筑拉结砖墙、人工焊接围护钢筋网、人工回填并夯实相变固化土、机械喷浆；G3墙体由机械压制素土模块后全部采用叉车搬运堆砌；G4墙体先支模，随后现场机械搅拌均匀后浇筑陶粒混凝土，强度达到要求后再拆除模板。G1的主动蓄热循环系统采用人工砌筑竖向砖墙风道、横向风道采用混凝土预制板人工安装；G2的主动蓄热循环系统采用人工安装竖向圆管风道和横向半管风道；G3的主动蓄热循环系统在素土模块制作过程中将风道压制成型，墙体堆砌过程中自然对接而成。

2.3 试验方法

4座供试温室内部各布置2个空气温湿度测点（HOBO UX100-011温湿度记录仪，美国onset公司生产，精度：温度±0.2℃、相对湿度±2.5%）、2个光照测点（HOBO UA002-64型光照强度记录仪，美国onset公司生产，精度：±10lx），分别布置在温室长度方向3等分截面处，跨度方向中部，所有测点均位于地面以上1.5m高度处。室外环境数据测点布置在距G1正西方10m处的空旷场地，水平高度均与室内测点一致。后墙顶部主动蓄热循环系统的进、出风口处各布置1个空气温湿度测点；墙体表面热流密度采用HFP01SC热流传感器（荷兰Hukseflux公司生产，精度：±3%），布置于后墙中部距离地面1.5m高处，连接到34 970A数据自动采集仪（美国Agilent公司生产）。试验数据采集时间为2017-11-01至2018-01-31，所有记录数据的时间间隔均为30min。

2.4 数据处理

本文试验数据采用Excel 2007进行数据分析及图表的制作。选择连续晴天（2017-12-30 9:00至2018-1-2 9:00）、连续阴天（2018-1-13 9:00至2018-1-16 9:00）、典型晴天（2017-12-31 9:00~次日9:00）、典型阴天（2018-1-14 9:00~次日9:00）的数据进行光照强度、热工性能的分析。

根据主动蓄热循环系统进、出口空气温湿度值结合空气风速和管径等参数，计算得到系统运行过程中的传热量，按下式计算：

$$Q_{act}=\sum v_{\tau}\cdot A\cdot(\frac{1}{V_{in}}-\frac{1}{V_{out}})\cdot\Delta H_{\tau}\cdot\Delta t_{\tau}/10^{3} \tag{1}$$

式中：Q_{act}为主动蓄热循环系统的传热量，MJ；$v\tau$为τ时段风道内空气流速，m/s；A为风道截面面积，m^2，G1、G2、G3的当量截面面积分别为0.152、0.235、0.120m^2；V_{in}、V_{out}分别为τ时段进、出口空气的比容，$m^3\cdot kg^{-1}$，按文献[12]计算；$\Delta H\tau$为τ时段空气焓差，$\Delta H\tau=H_{in}-H_{out}$，$H_{in}$、$H_{out}$分别为τ时段进、出口空气的焓值，$kJ\cdot kg^{-1}$，按文献［12］计算；$\Delta t\tau$为测试期间记录数据的时间间隔，即1 800s。

主动蓄热墙体除了主动蓄热循环系统运行时为主—被动联合传热外均为被动传热，按下式计算：

$$Q_{pas}=\sum q_{\tau}\cdot S\cdot\Delta t_{\tau}/10^{6} \tag{2}$$

式中：Q_{pas}为后墙被动传热量，MJ；$q\tau$为τ时段墙体表面热流密度，W/m^2；S为主动蓄热后墙表面积，m^2，即后墙长度与高度之积，取值115.2。

3 结果与分析

3.1 温室内外光照强度对比分析

日光温室室内光照强度直接影响到主动蓄热后墙的传热，对供试温室光照强度进行分析，是后续分析的前提和基础。图2显示了典型晴天供试温室室内外光照强度的日变化。由图2可知，4座供试温室室内的光照强度曲线变化趋势与室外基本一致。G1、G2、G3、G4白天平均光照强度分别为72 192lx、71 638lx、71 351lx、71 955lx，不存在显著性差异（$P<0.05$），室外为110 165lx。因此，在同一天气条件下，4座供试温室的平均光照强度差距较小，室内温度环境的差异不是由光照强度的差异造成的。

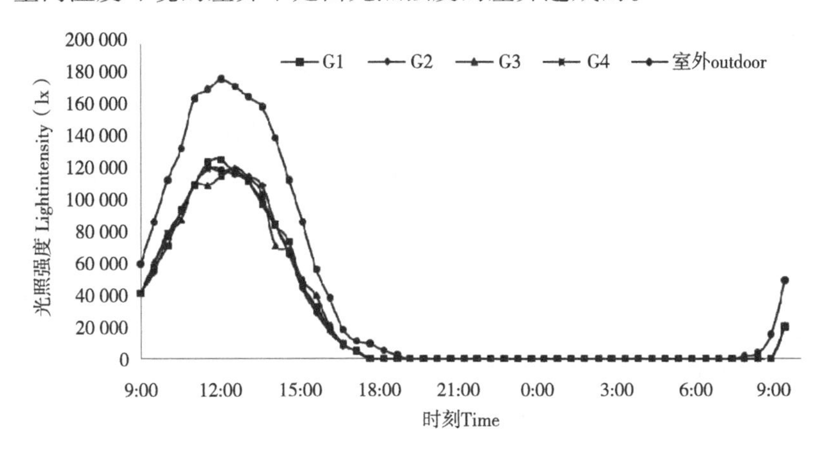

图2　不同温室室内外光照强度的变化

Figure 3　Variations of indoor and outdoor light intensity in different solar greenhouses

3.2　温室内外气温对比分析

3.2.1　连续晴天条件下温室内外气温对比分析

图3a和图3b显示了连续晴天温室室内外气温变化。由图3可知，连续晴天条件下，G1、G2、G3、G4、室外的平均气温分别为19.9℃、20.1℃、20.8℃、19.2℃、–1.1℃，夜间平均气温分别为15.2℃、16.0℃、17.3℃、15.9℃、–3.5℃，夜间平均最低气温分别为13.4℃、14.7℃、15.6℃、14.5℃、–6.3℃。

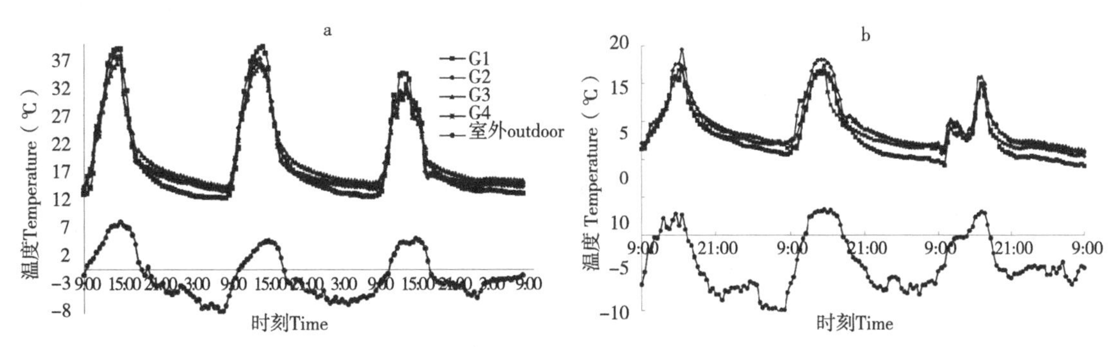

图3　不同温室室内外气温变化

Figure 3　Variations of indoor and outdoor air temperatures in different solar greenhouses

3.2.2　连续阴天条件下温室内外气温对比分析

低温及寡照是影响日光温室安全生产的主要因素，而冬季连续阴天又是这两种限制因素的主要表现形式，因此，有必要对冬季连续阴天条件下日光温室的保温性能进行比较分析。本文规定日照时数≤2h连续3d及以上时为连续阴天条件的统计标准。图3b显示了连续阴天温室室内外气温变化。由图可知，连续阴天条件下，G1、G2、G3、G4、室外的平均气温分别为12.9℃、14.4℃、14.3℃、13.6℃、-3.7℃，夜间平均气温分别为11.3℃、12.9℃、13.0℃、12.4℃、-5.9℃，夜间平均最低气温分别为9.8℃、11.5℃、11.7℃、10.9℃、-8.2℃。

3.3 后墙传热性能分析

对典型晴天、典型阴天条件下后墙主、被动蓄热量与放热量的计算结果如表1所示。

表1 典型天气条件下4座温室后墙的传热量

Table 1 Heat transfer amount of four solar greenhouses under typical weather conditions MJ

传热量Heat transfer amount	晴天Sunny day								阴天Cloudy day							
	蓄热阶段 Heat storage process				放热阶段 Heat release process				蓄热阶段 Heat storage process				放热阶段 Heat release process			
	G1	G2	G3	G4	G1	G2	G3	G4	G1	G2	G3	G4	G1	G2	G3	G4
Q_{act}	16.05	19.92	10.24	—	2.24	1.95	0.28	—	—	—	—	—	0.97	0.35	0.04	—
Q_{pas}	280.58	412.06	415.56	408.11	138.57	130.51	115.68	209.55	12.21	17.87	36.83	19.44	63.57	46.91	40.27	79.63

由表1可知，典型晴天蓄热量均大于放热量，G1、G2、G3、G4的蓄热量分别比放热量多155.83、299.52、309.84、198.56MJ，这部分热量在墙体内部蓄积，用于提高墙体内部蓄热土壤温度，故G3墙体内部的温度最高，宏观表现为G3的室内气温最高；典型阴天蓄热量均小于放热量，G1、G2、G3、G4的蓄热量分别比放热量少52.34、29.39、3.48、60.20MJ，这部分热量来源于晴天墙体内部蓄积的热量，G4在晴阴天的放热量均为最多，这是因为其墙体最薄，热稳定性最差。

3.4 经济性分析

表2为4座温室的建筑成本比较，G3的每平方米造价最低，分别较G1、G2、G4降低71.2元、162.1元、69.1元，价格差异主要体现在墙体和主动蓄热循环系统的建设上。从性能及造价角度来说，模块装配式主动蓄热墙体的的性价比最高。

表2 不同温室的建筑成本

Table 2 Costs of different greenhouses yuan/m^2

名称Name	G1	G2	G3	G4
温室钢骨架Steel frame	72.6	72.6	72.6	69.5
温室基础Foundation	42.5	42.5	42.5	34.0
温室墙体Wall	246.5	193.5	110.2	210.6
温室覆盖、保温材料及控制系统Plastic film and thermal insulation quilt with control system	43.4	43.4	43.4	45.5
主动蓄热循环系统active heat storage cycle system	40.1	23.6	18.5	—
其他费用Others	16.0	14.3	11.8	8.5
合计 Total	461.1	389.9	299.0	368.1

4 结论

针对原有日光温室主动蓄热墙体结构施工工艺复杂的问题，本文采用不同施工工艺建造主动蓄热墙体，对传统主动蓄热墙体日光温室（G1）、回填装配式主动蓄热墙体日光温室（G2）、模块装配式主动蓄热墙体日光温室（G3）、现浇陶粒混凝土装配式墙体日光温室（G4）的热工性能进行了测试分析，得出以下结论：

（1）连续晴天条件下，G1、G2、G3、G4的夜间（17:00～次日9:00）平均气温分别为15.2℃、16.0℃、17.3℃、15.9℃，夜间平均最低气温分别为13.4℃、14.7℃、15.6℃、14.5℃；连续阴天条件下，4座温室的夜间平均气温分别为11.3℃、12.9℃、13.0℃、12.4℃，夜间平均最低气温分别为9.8℃、11.5℃、11.7℃、10.9℃。

（2）典型晴天G1、G2、G3、G4的蓄热量分别比放热量多155.83、299.52、309.84、198.56MJ，G3的累积热量最多，墙体内部与室内气温最高；典型阴天G1、G2、G3、G4的蓄热量分别比放热量少52.34、29.39、3.48、60.20MJ，G4因墙体最薄，热稳定性最差。故G3墙体的整体保温蓄热能力最好。

（3）G1、G2、G3、G4每平方米造价分别为461.1元、389.9元、299.0元、368.1元，G3施工工艺最简、施工用材最少、施工造价最低，性价比最高。

综上，装配式墙体结构的保温蓄热效果较传统主动蓄热墙体好、造价较低，其中模块装配式主动蓄热墙体的性价比最高，在适宜日光温室发展的地区具有一定的推广价值。

参考文献

［1］李天来.我国日光温室产业发展现状与前景［J］.沈阳农业大学学报，2005，36（2）：131-138.

［2］白义奎，王铁良，李天来，等.缀铝箔聚苯板空心墙体保温性能理论研究［J］.农业工程学报，2003，19（3）：190-195.

［3］杨建军，邹志荣，张智，等.西北地区日光温室土墙厚度及其保温性的优化［J］.农业工程学报，2009，25（8）：180-185.

［4］张勇，高文波，邹志荣.主动蓄热后墙日光温室传热CFD 模拟及性能试验［J］.农业工程学报，2015，31（5）：203-211.

［5］鲍恩财，朱超，曹晏飞，等.固化沙蓄热后墙日光温室热工性能试验［J］.农业工程学报，2017，33（9）：187-194.

［6］李天来，邹志荣，马承伟，等.节能日光温室设计建造规程［M］.北京：中国农业出版社，2017.

［7］闫俊月，李明，张秋生，等."西北非耕地温室结构与建造技术"项目成果汇报（4）轻简化装配式后墙［J］.农业工程技术·温室园艺，2014（4）：62-63.

［8］张洁，邹志荣，张勇，等.新型砾石蓄热墙体日光温室性能初探.［J］北方园艺，2016，（2）：46-50.

［9］张义，方慧，周波，等.轻简装配式主动蓄能型日光温室［J］.农业工程技术·温室园艺，2015，（25）：36-38.

［10］鲍恩财，邹志荣，张勇.日光温室墙体用相变固化土性能测试及固化机理［J］.农业工程学报，2017，33（16）：203-210.

［11］邹志荣，鲍恩财，申婷婷，等.模块化组装式日光温室结构设计与实践［J］.农业工程技术·温室园艺，2017，（31）：50-55.

［12］马承伟，苗香雯.农业生物环境工程［M］.北京：中国农业出版社，2005.

湿热—高压静电场混合引发恢复洋葱种子潜在活力的生物学机理研究

赵颖雷[1]，胡鸣鹤[2]，高照[2]，陈晓雪[2]，郏惠彪[3]，潘学勤[1, 3]，黄丹枫[1, *]

（1. 上海交通大学农业与生物学院，上海，200240；2. 上海交通大学机械与动力工程学院，上海，200240；3. 上海惠和种业有限公司，上海）

摘要：洋葱（*Allium cepa*）的种子即便是贮藏在最理想的环境下也会逐渐丢失其活力。但某些因素导致的失活是可逆的，且可以通过引发技术恢复活力。高压静电场引发可瞬间提升种子活力并加快萌发，但效果的有效期较短。湿热（水）引发因无任何化学试剂的清洁引发流程而被最广泛的应用，但引发时间较长，且需要引发环境的湿度精确控制。此文将湿热引发与高压静电场混合引发相结合形成湿热—高压静电场混合引发技术（Hydro-Electrostatic Hybrid Priming，HEHP），并尝试使用这种方式恢复洋葱种子的活力。研究发现，与两种单一引发方式相比，混合引发能够恢复洋葱种子更多的活力，且有效期更长。圆二色光谱法与电子顺磁共振揭示了该技术的机理：高压静电场改变了种子中超氧化物歧化酶的空间构象从而提高了酶的活性，使种子中的自由基被更高效的清除。扫描电镜与透射电镜的观察结果显示，混合引发能将种子胚细胞器与细胞膜更快更完整地修复。混合引发能够不借助任何化学试剂而缩短湿热引发的时间，延长高压静电场引发效果的有效期，并提升高引发效果。

关键词：电场；水；混合；种子引发

Biological Mechanisms of a Novel Hydro-Electro Hybrid Priming Recovers Potential Vigor of Onion Seeds

Zhao Yinglei[1]，Hu Minghe[2]，Gao Zhao[2]，Chen Xiaoxue[2]，
Jia Huibiao[3]，Dan Xueqin[1, 3]，HuangDanfeng[1, *]

（1. *Department of Plant Science*，*School of Agriculture & Biology*，*Shanghai Jiao Tong University*，*Dongchuan Road* 800，*Shanghai* 200240；2. *School of Mechanical Engineering*，*Shanghai Jiao Tong University*，*Dongchuan Road* 800，*Shanghai* 200240）

Abstract：Onion（*Allium cepa*）seeds gradually lose vigor during storage，even under optimal conditions. Some of the loss in vigor is reversible and can be recovered by priming. Electrostatic field priming can instantly increase seed vigor and accelerate germination，though the effects are short-lived. Hydro-priming is most widely used

---

通讯作者: 黄丹枫，教授. Tel.: +86 21 34206943；fax: +86 21 34206943. E-mail: hdf@sjtu.edu.cn

Corresponding author: Prof. Danfeng Huang. Tel.: +86 21 34206943；fax: +86 21 34206943. E-mail: hdf@sjtu.edu.cn

项目资助:上海市种业发展项目（2017）4-2：Integration and Application of Vegetable Seeds Priming and Industrial Seedling System；上海市农委（2016）1-11：嘉定白蒜脱毒、纯化与扩繁技术的应用于示范；上海交通大学大学生创新实践项目（IPP15053）。

as it does not use chemicals and is a clean process，but requires a long priming time and precise seed moisture control. Here，we investigated the potential of combined hydro-electrostatic hybrid priming on the recovery of vigor in onion seeds. Hydro-electrostatic hybrid priming lead to improved seed vigor recovery. Circular dichroism spectroscopy and electron paramagnetic resonance revealed exposure to an electrostatic field enhanced SOD activity in seeds by changing the secondary structure of the enzyme and led to more efficient free radical scavenging; scanning and transmission microscopy revealed these effects promoted more complete and rapid healing of embryo cell organelles and plasma membranes during the incubation stage of hydro-priming. Hydro-electrostatic hybrid priming has the advantages of greater recovery of potential vigor and a shorter processing time than single hydro-priming，leads to longer lasting effects than single electro-priming，and does not require any chemicals.

Key words：Electric field；Hydro；Hybrid；Priming；Seed

1 介绍

种子会随贮藏时间的增加而逐渐老化失活。引起种子老化的机理有很多种，有些老化甚至是可逆的，如细胞膜轻微恶化或酶的失活。这类老化可通过一些物理或化学处理进行恢复。种子引发作为一种最常用的种子活力恢复技术被大量运用于种子生产。传统湿热引发通常需要5天以上的引发舱湿度精确控制，这增加了种子间含水量的差异与有害微生物的数量。不同作物品种对应不同的引发参数与介质，这种工艺与流程控制的复杂性使引发技术具有较高的技术应用门槛。

电对植物生长影响的研究开始于1746。在随后的研究中发现，高压静电场可瞬间提升种子活力并加快萌发，但效果有效期较短，如电场提高向日葵与大麦种子萌发力的的效果会在20天左右出现明显消退。

本文中，湿热引发与高压静电场处理被结合成湿热—高压静电场混合引发。研究发现，在加入电场处理后，湿热引发恢复洋葱种子活力的效率与效果能够被进一步提升。混合引发的机理通过研究进行分析与解释，并与两种单一引发方式的机理做对比。

2 材料与方法

2.1 实验材料

“帝黄”洋葱种子（上海惠和种业），产于2014年，储藏于5℃、50%相对湿度环境。

2.2 实验方法

引发处理在上海交通大农业与生物学院实验室内进行，步骤与流程如表1所示，每个处理含50粒种子，重复3次。浸种过程为种子在22℃蒸馏水中浸泡5h；高压静电场发生器电源型号为BM-201（产于中国江苏），电场正负极为两块10cm × 10cm × 1mm的纯铜板，两极板间隔1cm；高压静电场照射过程为使用10kv/cm的场强处理种子40s，电场方向与地球大气电场方向相同；湿度维持过程为将种子置于22℃、98%相对湿度的人工气候箱（QHX-300BSH-Ⅲ，产于中国上海）中24、48与96h（黑暗环境）；回干过程为使用鼓风干燥箱30℃下干燥48h。

表1 处理编号与流程

Table 1 Priming treatments and processing protocols

编号	引发处理流程			
	Step 1：浸种	Step 2：高压静电场照射	Step 3：湿度维持	Step 4：回干
CK	×	×	×	×
EF	×	√	×	×
Hyd24	√	×	24h	√
Hyd48	√	×	48h	√
Hyd96	√	×	96h	√
HEHP24	√	√	24h	√
HEHP48	√	√	48h	√
HEHP96	√	√	96h	√

注：HEHP，湿热—高压静电场混合引发；Hyd，水引发；EF，纯高压静电场引发。

2.3 种子萌发测定

各处理的洋葱种子播于湿润的泥炭后（泥炭与水，*w/w*=3:1）（peat and water，*w/w*=3:1）置于人工气候箱（QHX-300BSH-Ⅲ，产于中国上海），设定22℃与80%相对湿度；人工光源使用TLD 30W/865日光灯（PHILIPS，泰国产），设定光强为16 000lx；每个处理含3个样本，重复3次。

播种后每48h统计一次萌发的种子数，4d时的萌发率定义为出芽势（GP，%），6d时的萌发率定义为最终芽率（GR，%），6d时统计配根长度（S，mm）与鲜重（FW，g）。

平均出芽时间（MET）计算公式：

$$\bar{T}=\frac{\sum Dt\times n}{\sum n} \tag{1}$$

注：*Dt*为播种后的天数，*n*是第*Dt*日新萌发的种子数[1]．

种子发芽指数（GI）计算公式（Association of Official Seed Analysts，1983）：

$$\mathrm{GI}=\sum\frac{\mathrm{Gt}}{\mathrm{Dt}} \tag{2}$$

注：*Gt*是第*Dt*日新萌发的种子数[2]．

种子活力指数（VI）计算公式（International Seed Testing Association，1985）：

$$\mathrm{VI}=\mathrm{GI}\times \mathrm{S} \tag{3}$$

2.4 电导率测定

每个处理50粒种子（约0.19g）浸于40mL、25℃的中5h。因Hyd与HEHP在引发过程中已有5h浸种，部分导电物质已被浸出，因此EF与CK再额外浸种5h，测定种子浸提液电导率，值设为a2。种子与浸提液一起入沸水浴1h，测定种子浸提液电导率，值设为a3。每个处理含3个样本，重复3次。蒸馏水的电导率值设为a1。

相对电导率（REC）计算公式（Association of Official Seed Analysts，1983）：

$$REC(\%) = \frac{a2 - a1}{a3 - a1} \times 100 \quad (4)$$

2.5 MDA的测定

每个处理50粒种子（约0.19g）播于湿润的泥炭，并在6d时回收。使用MDA试剂盒（MAK08550；Sigma-Aldrich）并按说明书要求测定。每个处理含3个样本，重复3次。

2.6 总SOD活性的测定

每个处理50粒种子（约0.19g），使用SOD试剂盒（19160-1KT-F；Sigma-Aldrich）并按说明书要求测定。每个处理含3个样本，重复3次。

2.7 种子活力的主成分分析

使用SPSS 23.0对7个萌发指标（发芽势、发芽率、配根长、鲜重、平均萌发时间）与3个生理指标（SOD、MDA、EC）进行主成分分析（PCA）以综合评价种子活力。

2.8 SOD的纯化

取方法材料与方法2.6中的上清液2mL，按参考文献中方法操作[3]。每个处理含3个样本，重复3次。

2.9 SOD酶蛋白结构分析

取方法材料与方法2.8中的提取物。使用CD光谱仪（J-815，Jasco，日本产）按参考文献中方法操作[4]。每个样品3次重复。

2.10 SOD同工酶种类的鉴定

取方法材料与方法2.8中的提取物，按参考文献中方法操作[5]。SDS-PAGE电泳使用不连续凝胶系统，5%浓缩胶（附录，表1a）与10%分离胶（附录，表1b），使用考马斯亮蓝染色。电泳仪为Bio-Rad Mini-Protean II（美国产）。每个处理1次重复。

2.11 自由基含量的测定

每个处理50粒种子（约0.19g）。EPR设备为EMX-8（Bruker BioSpin Corp.，德国产）。按参考文献中方法操作[6]。实验在华东理工大学分析测试中心进行。

2.12 扫描与透射电镜镜检

取所有的处理中2d萌发、4d萌发与6d未萌发的种子完整的胚。使用日立E-1045（日本产）按参考文献中方法喷金元素[7]。所用扫描电镜为MIRA3-LHM&XFlash6|60（EDS）（TESCAN&BRUKER，捷克产），共观察了72个胚组织。按参考文献中方法包埋[8]。使用莱卡UC6-FC6（德国产）冷冻切片机切片。所用透射电镜为Tecnai™ G2 Spirit BioTWIN（美国产），共观察了72个胚组织。扫描电镜与透射电镜观察均在上海交通大学分析与测试中心进行。

2.13 混合引发效果有效期测试

各处理的种子存放于5℃与50%相对湿度环境中60d，按上述主成分分析方法评价种子活力。

2.14 数据分析

采用单因素ANOVA进行多重比较，采用Duncan差异显著性分析（$p<0.05$），使用SPSS Statistics 23软件进行统计分析。

3 结果与分析

3.1 三种引发均不同程度提升洋葱种子萌发力

如表2中所示，与对照相比，纯电场引发（EF）、湿热引发（Hyd）与HEHP混合引发均显著优化了与萌发相关的各个指标。随引发处理时间的延长，Hyd与HEHP的各处理逐渐缩短平均出芽时间，增加其他所有萌发相关指标，但随着引发时间的增加，4d时的发芽势、发芽率与平均出芽时间的差异逐渐变得不显著。HEHP96的发芽率、胚根长、鲜重、发芽指数与活力指数值最大，平均出芽时间最短。Hyd96第4d时的发芽势值最大，但与各HEHP处理的差异不显著。

表2 不同引发方式对洋葱种子萌发的影响

Table 2 Effects of different priming protocols on onion seed germination

处理编号	发芽势（day 2，%）	发芽势（day 4，%）	发芽率（day 6，%）	胚根长（mm）	鲜重（g）	平均出芽时间（d）	发芽指数	活力指数
CK	12.8 ± 1.1e	63.6 ± 7.3c	71.2 ± 5.9d	21.01 ± 2.2f	0.34 ± 0.02e	3.85 ± 0.17a	10.18 ± 0.75f	212.87 ± 14.86f
EF	40.0 ± 4.9c	76.8 ± 4.1ab	82.0 ± 6.0ab	30.26 ± 1.0c	0.49 ± 0.02c	3.14 ± 0.25cd	15.03 ± 0.26cd	454.85 ± 15.83c
Hyd24	27.6 ± 11.0d	70.0 ± 5.1bc	76.8 ± 2.3c	23.08 ± 0.7e	0.37 ± 0.02e	3.46 ± 0.22b	12.77 ± 1.27e	294.52 ± 28.74e
Hyd48	37.2 ± 11.4c	75.6 ± 8.2ab	79.2 ± 4.8bc	26.31 ± 0.4d	0.42 ± 0.02d	3.16 ± 0.30c	14.40 ± 1.85d	378.69 ± 47.70d
Hyd96	44.0 ± 7.1bc	81.2 ± 8.8a	83.6 ± 1.7ab	30.94 ± 0.7c	0.49 ± 0.02bc	3.01 ± 0.20cd	15.85 ± 0.86bc	490.26 ± 26.12c
HEHP24	43.6 ± 5.0bc	80.8 ± 5.0a	83.2 ± 2.7ab	30.26 ± 1.0c	0.49 ± 0.02bc	3.01 ± 0.14cd	15.75 ± 0.46bc	476.47 ± 14.23c
HEHP48	50.4 ± 7.9ab	80.0 ± 2.4a	85.2 ± 1.1a	33.39 ± 0.7b	0.53 ± 0.03ab	2.94 ± 0.15cd	16.73 ± 0.98ab	559.03 ± 40.47b
HEHP96	55.2 ± 4.8a	80.4 ± 3.3a	86.0 ± 2.4a	35.18 ± 1.5a	0.56 ± 0.07a	2.84 ± 0.17d	17.42 ± 0.55a	612.38 ± 21.94a

注：差异显著性水平$p<0.05$.

3.2 湿热引发与高压静电场照射相结合可有效降低种子电导率

种子浸提液的电导率（EC）与种子细胞膜通透性呈正比，EC值越高说明细胞膜破裂程度越高，而引发可降低EC值。对照的种子的EC值最高。Hyd96、HEHP48与HEHP96的EC值较对照种子显著降低。在Hyd与HEHP的各处理中，EC值随引发时间的延长而逐渐降低，于HEHP96处获得最低值（图1B）。

3.3 高压静电场照射提高了湿热引发抗脂质过氧化的效果

所有引发处理的种子（EF、Hyd和HEHP）的MDA含量在引发处理结束后与对照的差异不显著，和说明引发不会清除已经产生的MDA。而在播种6d后，对照处理的MDA增量最大，达到0.384 ± 0.024mg/g。而处理EF、Hyd96、HEHP24、HEHP48与HEHP96的增量则显著小于对照（图1C）。Hyd与HEHP各处理对MDA增量的抑制随引发时间的延长而逐渐增强，以HEHP96在6d内的增量最小，6d时含量为0.216 ± 0.024mg/g。

3.4 高压静电场照射进一步提高SOD活性

与对照相比，EF、Hyd96与所有HEHP处理在引发完后SOD的活性显著提升，但处理Hyd24与Hyd48则不显著。HEHP96的SOD活性最高，但与HEHP24、HEHP48的差异不显著。所有HEHP处理的SOD活性均随引发时间的延长而提高（图1A）。

3.5　PCA分析结果：混合引发可获得更好的综合效果并缩短引发时间

进行主成分分析（PCA）对7个萌发指标（发芽势、发芽率、配根长、鲜重、平均萌发时间）与3个生理指标（SOD、MDA、EC）进行了综合评价。结果显示，第一主成分解释了总变异的79.484%，初始特征值7.948，且是唯一数值大于1.000的成分（附录，表2）。因此第一主成分可以代表原始的10个指标。计算变量矩阵（Appendix Table 3）根据如下公式：

$$CV=\frac{CM}{\sqrt{TIE}} \tag{5}$$

CM：Component Matrix；CV：Compute Variable.

输出的洋葱种子萌发的PCA模型为：

$$Z=0.282X_1+0.299X_2+0.336X_3+0.327X_4-0.314X_5+0.341X_6+0.347X_7+0.321X_8-0.314X_9-0.273X_{10} \tag{6}$$

最初的10个生长相关指标被标准化后代入公式（6），获得了PCA输出的Z值（Table 3）。与对照相比，所有引发处理的种子拥有更高的Z值，代表了种子更好的萌发与更高的种子活力。对照处理的Z值最低为-5.101。而在相同的引发时间条件下，HEHP能比Hyd恢复出更高的种子活力。HEHP24的Z值升至高于Hyd48，接近Hyd96。HEHP48的Z值则全面超越Hyd96，说明电场照射处理可缩短湿热引发的时间。

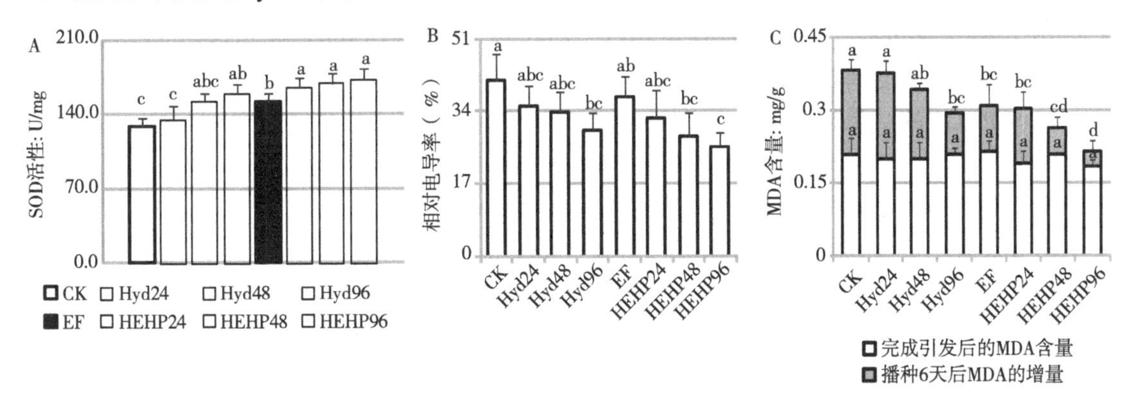

图1　电场引发、湿热引发与混合引发对洋葱种子生理的影响

Figure 1　Effect of electro-priming（EF）, hydro-priming（Hyd）and hydro-electro hybrid priming（HEHP）on onion seed.（A）所有引发处理完成后的种子SOD活性，（B）所有引发处理完成后的种子相对电导率，与（C）播种6天内种子MDA含量的变化。

表3　不同引发处理对洋葱种子萌发的PCA分析Z值

Table 3　The PCA result of the effects of different priming protocols on onion seed germination

排位	处理编号	Z值
1	HEHP96	3.645
2	HEHP48	2.532
3	Hyd96	1.386

（续表）

排位	处理编号	Z值
4	HEHP24	1.208
5	EF	0.182
6	Hyd48	−0.879
7	Hyd24	−2.974
8	CK	−5.101

3.6　电场照射处理改变SOD二级结构组成

使用圆二色光谱法分析了SOD的结构组成。结果显示，所有样本均在191nm与197～240nm波长附近获得正负吸收峰，为典型的β-折叠构象[9]。经过电场照射后的处理（EF与HEHP24）在这两处吸收峰处的光谱分别表现为2～3.5与1～1.5nm的红移，对比没有电场照射的处理（CK与Hyd）吸收峰值显著升高（图2A-D）。

如图2E所示，对比未经电场照射的处理（CK、Hyd24、Hyd48与Hyd96），经过电场照射的处理（EF、HEHP24、HEHP48与HEHP96）的β-折叠含量显著增加，无规则卷曲含量显著降低（图3）。HEHP48与HEHP96处理较EF具有更高的种子含水量与更长的引发时间，从而表现出更高的β-折叠与无规则卷曲含量。Spearman相关性分析的结果显示SOD活性与β-折叠含量呈正相关（r=−0.651；P=0.001），与无规则卷曲含量呈负相关。α-螺旋（r=−0.106；P=0.622）与β-转角（r=0.071；P=0.741）含量与SOD活性的相关性不显著。

3.7　洋葱种子中仅含有Cu/Zn-SOD

SDS-PAGE的结果显示，从所有处理中提取纯化的SOD均在35kDa处显示明显条带，说明洋葱种子中的SOD同工酶类型仅为Cu/Zn-SOD。

3.8　高压静电场照射提高了湿热引发对自由基的清除效果

EPR光谱正负吸收峰的总面积与自由基含量呈正相关。如图4与表4所示，对照处理的吸收峰总面积最大（30.62±2.51a），说明了其自由基含量最高。引发处理EF并没有显著降低吸收峰总面积。处理Hyd与HEHP的吸收峰总面积随引发时间的延长而降低，且在相同引发时间内，HEHP比Hyd具有更低的值，以HEHP96的值最低（14.06±0.63）。

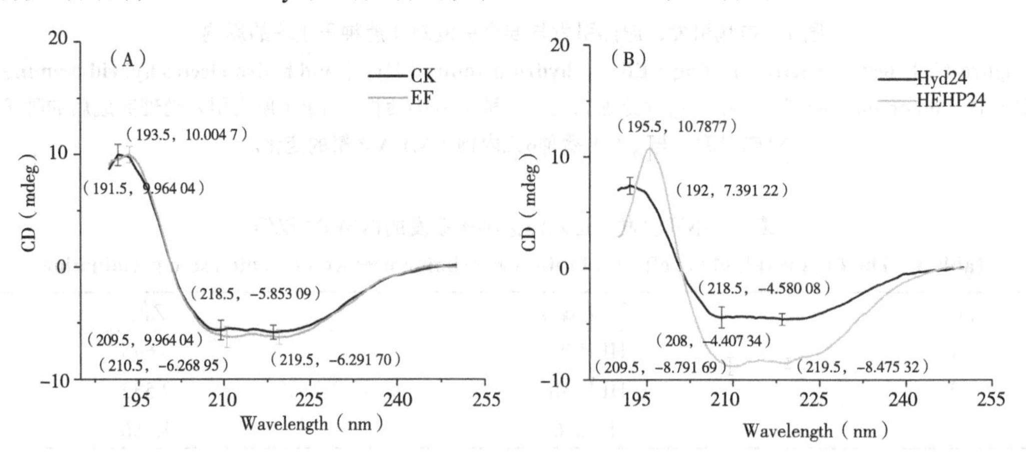

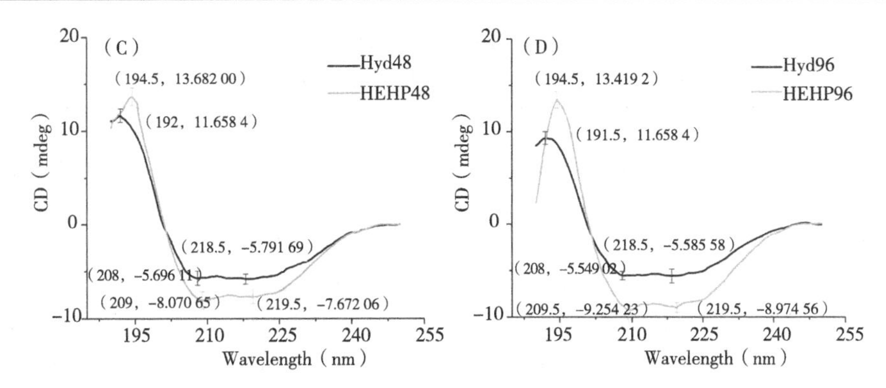

图 2　所有处理中的SOD圆二色光谱分析。有电场照射与无电场照射处理的光谱图的两两对比：（A）CK与EF；（B）Hyd24与HEHP24；（C）Hyd48与HEHP48；（D）Hyd96与HEHP96。（E）各处理的SOD结构含量对比

Figure 2　CD spectra analysis of SOD extracted from onion seeds after different priming protocols. Comparison of the CD spectra of（A）CK and EF；（B）Hyd24 and HEHP24；（C）Hyd48 and HEHP48；and（D）Hyd96 and HEHP96.（E）Effects of priming on the secondary structure composition of SOD.

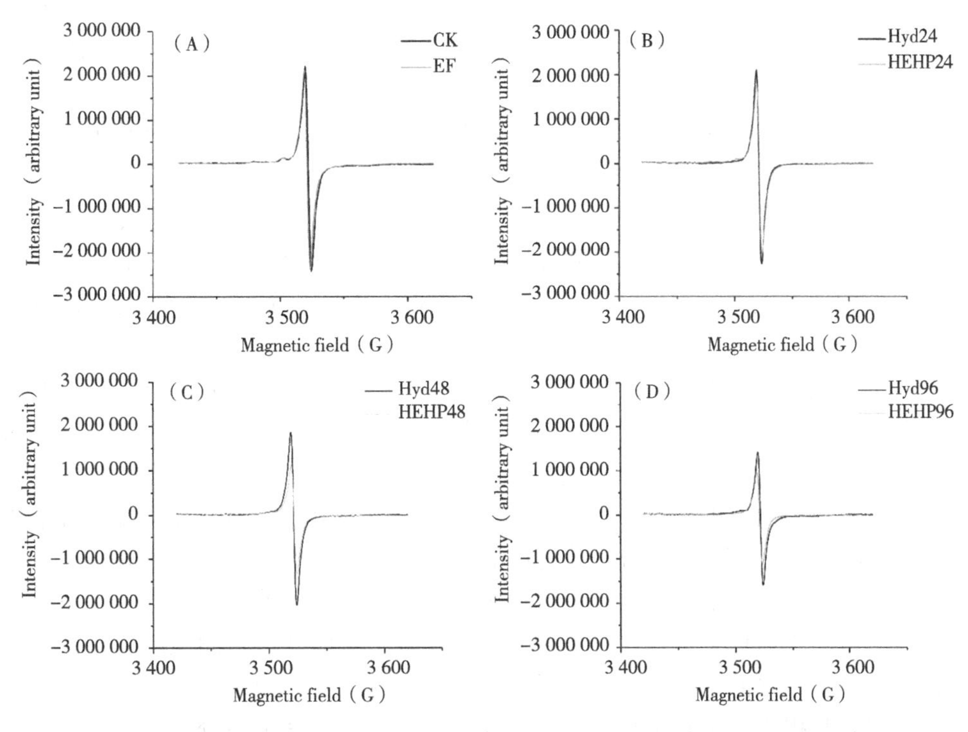

图 3　电场引发、湿热引发与混合引发对种子中自由基含量的影响。EPR光谱图之间两两对比：（A）CK与EF；（B）Hyd24与HEHP24；（C）Hyd48与HEHP48；（D）Hyd96与HEHP96。吸收峰面积与自由基含量呈正比

Figure 3　Effect of electro-priming（EF），hydro-priming（Hyd）and hydro-electro hybrid priming（HEHP）on the free radical content of onion seeds. EPR spectra for control（CK）and EF（A），Hyd24 and HEHP24（B），Hyd48 and HEHP48（C），and Hyd96 and HEHP96（D）. Integrated peak area is proportional to free radical content

表 4　EPR吸收峰面积

Table 4　Integrated EPR absorption peak areas

处理编号	综合面积（$\times 10^6$）		
	正峰面积	负峰面积	总面积
CK	15.02 ± 1.66a	15.60 ± 1.24a	30.62 ± 2.51a
EF	13.91 ± 3.06a	14.81 ± 1.73ab	28.72 ± 4.72ab
Hyd24	12.35 ± 2.58ab	13.39 ± 1.25ab	25.74 ± 1.97b
HEHP24	12.36 ± 1.11ab	13.08 ± 2.29b	25.44 ± 3.37b
Hyd48	12.23 ± 1.74ab	13.23 ± 1.08ab	25.46 ± 2.67b
HEHP48	8.47 ± 0.69c	9.16 ± 0.52cd	17.63 ± 1.21cd
Hyd96	9.27 ± 0.48bc	10.53 ± 0.66c	19.80 ± 0.85c
HEHP96	6.7 ± 1.18c	7.36 ± 0.55d	14.06 ± 0.63d

3.9　高压静电场照射增强湿热引发对胚组织的修复与细胞膜的完整性

扫描电镜的观察结果显示，播种6d仍未萌发的种子，其胚表面存在严重的恶化与穿孔现象，胚表面颜色深浅不一，细胞壁模糊不清（图5A1与A2）。这种状态的种子在任何引发处理前就已丧失大部分生命力，无法正常发芽或完全不能发芽。严重的恶化甚至可造成胚内部组织的暴露（图5B1）。

但是，细胞组织的自我修复机制会在湿热引发阶段或播种后的预萌发阶段，对细胞器与细胞膜的恶化、穿孔等各种损伤进行修复。从图5C1中可看到，损伤的边缘正在逐渐愈合，最终留下一条清晰的愈合痕迹（图5B2）并在周边部位呈现出光滑完整的表面（图5C2）。在各处理中，已愈合或即将愈合的痕迹常见于2d内萌发的种子胚表面，而愈合的过程常见于4d内萌发的种子胚表面。引发操作就是将更多比例的种子恢复到图5B与图5C的状态（表 2）。

透射电镜的观察结果显示，播种6d仍未萌发的种子，其胚细胞膜与细胞壁变窄、畸形，脂质组织发生降解（图6A1与A2）。这样的种子通常含有大量的自由基并以丧失大部分生命活力，无法正常发芽或完全不能发芽。但是，播种后或引发过程预萌发阶段中的抗氧化机制可实现自由基的清除，并部分修复受损的脂质组织（图6B2）或减轻播种后的组织降解（图6B1）。但脂质体仍旧排列松散、无序，大小不一，甚至可观察到核膜的受损（图6B1）。这种现象普遍存在于播后4d萌发的种子胚上。若自由基已被大量的清除，则观察不到降解的脂质体或细胞膜（图6C1）。此时脂质体外形圆润饱满，排列紧密、整齐，细胞质与细胞壁之间的细胞膜厚度均一，界限清晰（图6C2），常见于播后2d萌发的种子胚上。与对照相比，所有湿热引发与混合引发处理均显著减少处于图6A1与A2状态的种子数（表2）。

3.10　纯电场引发与时间较短的混合引发的有效期较短

各处理经过5℃与50%相对湿度环境下贮藏60d后再播种，使用相同的主成分分析获得的Z值如表5中所示。HEHP24与EF的Z值排位发生变化分别从原来的第4与第5位下降到第5与第7位。

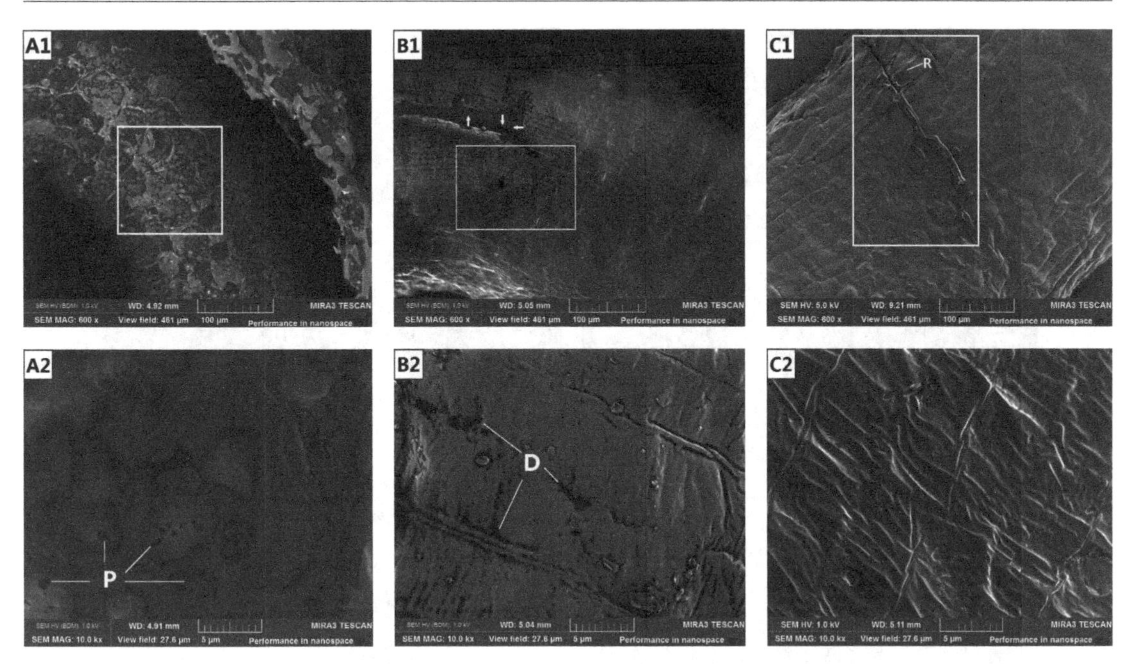

图 4　洋葱种子的扫描电镜镜检结果。（A1）播后6d未萌发的胚表面，600x放大，未得到恢复的恶化如白色框中所示，此样本取自一个对照处理；（A2）对A1中白框区域的10 000x放大；（B1）播后4d萌发的胚表面，600x放大，恢复中的恶化如白色框中所示，受损部分沿箭头方向恢复，此样本取自一个Hyd48处理；（B2）对B1中白框区域的10 000x放大，恶化的恢复痕迹如白线所示；（C1）播后2d萌发的胚表面，600x放大，损伤愈合痕迹清晰可见，该痕迹随后会完全愈合并逐渐消退，此样本取自一个HEHEP96处理；（C2）对C1中白框区域以外部分的10 000x放大，胚表面已完全恢复，无任何恶化迹象；P，细胞穿孔；D，恢复进行中的恶化；R，恢复后残留的痕迹

Figure 4　SEM examination of onion seeds embryos.（A1）Surface of the embryo of a seed that did not germinate by day 6 after sowing，600x SEM. Unrecovered deteriorations are indicated by the white square. One image of a CK seed is shown in A1.（A2）Area of the white square in A1 viewed at 10 000x.（B1）Surface of the embryo of a seed that germinated on day 4 after sowing，600x SEM. Deterioration with recovery in progress is evident in the white square；wound healing progresses in the direction of the arrow. One image of a Hyd48 seed is shown in B1.（B2）Area of the white square in B1 viewed at 10 000x. Trails of deterioration with recovery in progress are marked by white lines.（C1）Surface of the embryo of a seed that germinated on day 2 after sowing，600x SEM. Distinct healed wounds are evident in the white square，indicative of good recovery. Gradually fading，indistinct trails of recovery are indicated by white lines. One image of a HEHP96 seed is shown in C1.（C2）Area of the white square in C1 viewed at 10 000x. The embryo surface was recovered or un-deteriorated. The surface of embryo was well-recovered or un-deteriorated. P，perforations；D，deterioration with recovery in progress；R，indistinct trails of recovery.

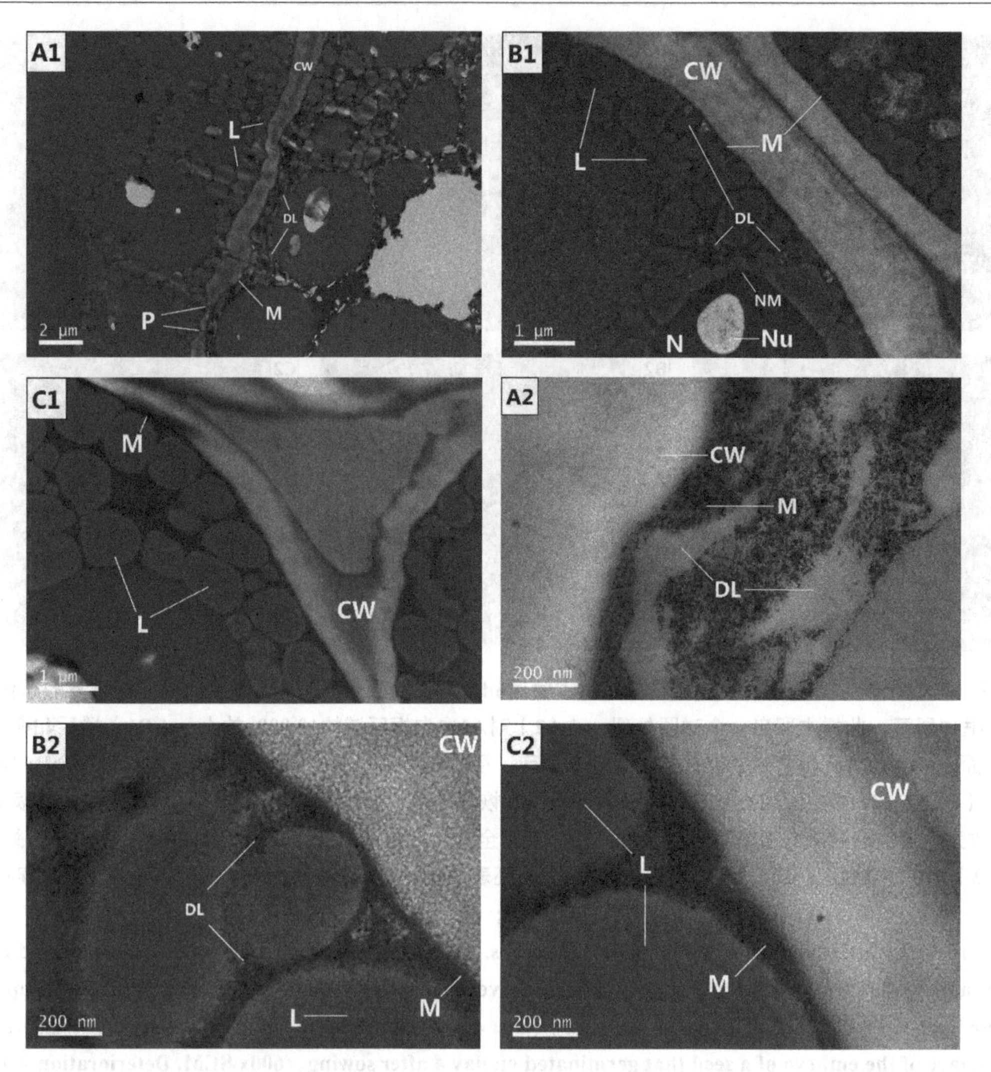

图 5 洋葱种子的透射电镜镜检结果。（A1）播后6d未萌发的胚细胞，4 500x放大，此样本取自一个对照处理；（A2）对A1的87 000x放大，观察到溶解的脂质体；（B1）播后4d萌发的胚细胞，4 500x放大，此样本取自一个Hyd48处理；（B2）对B1的87 000x放大，观察到轻微或部分溶解的脂质体；（C1）播后2d萌发的胚细胞，4 500x放大，600x放大，此样本取自一个HEHP96处理；（C2）对C1的87 000x放大，观察到无脂质体溶解；DL降解的脂质体；M，细胞膜；L，脂质体；CW，细胞壁；N，细胞核；Nu，核仁；NM，核膜恶化

Figure 5 TEM examination of onion seed embryo cells.（A1）Cells of the embryo of a seed that did not germinate by day 6 after sowing，4 500x. One image of a CK seed is shown in D1.（A2）High-power view of decomposed and dissolved liposomes in A1（87 000x）.（B1）Cells of the embryo of a seed that germinated on day 4，4 500x. One image of a Hyd48 seed is shown in B1.（B2）High-power view of mildly decomposed and dissolved liposomes in B1（87 000x）.（C1）Cells of the embryo of a seed that germinated on day 2，4 500x. One image of a HEHP96 seed is shown in C1.（C2）High-power view revealed a lack of decomposition or dissolution of any lipoid tissue（87 000x）. DL，decomposition and dissolution of liposome；M，cell membrane；L，liposome；CW，cell wall；N，nucleus；Nu，Nucleoli；NM，nucleus membrane deterioration.

表 5　贮藏60d后个处理种子的萌发表现与PCA分析结果

Table 5　Comparison of the effects of electro-priming（EF）, hydro-priming（Hyd）and hydro-electro hybrid priming（HEHP）on indexes of germination and PCA Z-values of onion seeds after 60 days storage

排位	处理编号	发芽势（第4天，%）		发芽率（%）		胚根长（mm）		鲜重（g）		平均出芽时间（d）		种子发芽指数		种子活力指数		Z值
1	HEHP96	80.4 ± 1.2	0	84.7 ± 2.2	↓1.3	35.04 ± 1.1	↓0.14	0.55 ± 0.02	↓0.01	2.79 ± 0.16	↑0.32	16.24 ± 0.15	↓1.18	569.09 ± 24.99	↓43.29	4.091
2	HEHP48	77.9 ± 2.0	↓2.1	83.7 ± 1.7	↓1.5	31.52 ± 1.7	↓1.87	0.49 ± 0.03	↓0.04	3.17 ± 0.18	↑0.33	15.47 ± 0.25	↓1.26	487.39 ± 17.99	↓71.64	2.653
3	Hyd96	79.1 ± 3.1	↓2.1	82.8 ± 0.4	↓0.8	30.10 ± 1.5	↓0.84	0.47 ± 0.01	↓0.02	3.49 ± 0.16	↑0.02	14.13 ± 0.29	↓1.72	425.05 ± 14.93	↓65.21	1.853
4	Hyd48	74.5 ± 2.0	↓1.1	77.7 ± 1.3	↓1.5	26.60 ± 1.2	↑0.29	0.40 ± 0.01	↓0.02	3.61 ± 0.21	↑0.45	12.58 ± 0.32	↓1.82	334.87 ± 22.40	↓43.83	−0.002
5	HEHP24	72.4 ± 1.2	↓8.4	75.9 ± 1.1	↓6.2	23.30 ± 1.5	↓6.46	0.35 ± 0.02	↓0.11	3.68 ± 0.24	↑0.67	11.75 ± 0.27	↓4.00	273.85 ± 10.23	↓202.62	−1.051
6	Hyd24	68.7 ± 2.0	↓1.3	74.0 ± 0.6	↓2.8	22.33 ± 1.3	↓0.75	0.34 ± 0.02	↓0.03	3.78 ± 0.24	↑0.32	11.28 ± 0.22	↓1.49	251.93 ± 10.44	↓42.59	−1.689
7	EF	63.3 ± 1.2	↓13.5	71.7 ± 2.5	↓10.3	22.92 ± 1.4	↓7.34	0.34 ± 0.03	↓0.15	3.85 ± 0.20	↑0.71	10.03 ± 0.18	↓5.00	229.68 ± 7.68	↓225.17	−2.379
8	CK	59.0 ± 1.2	↓4.6	67.6 ± 1.5	↓3.6	20.65 ± 2.7	↓0.36	0.31 ± 0.01	↓0.03	4.07 ± 0.16	↑0.22	9.68 ± 0.31	↓0.50	199.67 ± 2.69	↓13.20	−3.477

Note：KMO（Kaiser-Meyer-Olkin）= 0.725.

4 讨论

种子引发技术通常可通过调节引发液渗透压、温度、湿度或添加植物激素实现对引发效果的优化。本文介绍并研究了一种简便的混合引发方式，该种方式使用电场提升湿热引发对种子活力恢复的效果。

评价种子萌发的传统指标都只能间接反应引发效果，一些与萌发相关的，可间接反应引发效果的生理指标如SOD活性、MDA含量与EC值均未包含在这些传统指标中。由于有些传统指标的差异不显著，因此利用主成分分析整合7个传统发芽指标与3个种子生理指标进行综合评价，来有效的区分与衡量引发效果。

超氧自由基与羟基自由基是植物组织中主要的活性氧类型。SOD直接催化分解超氧自由基（$2O_2\cdot$-+2H+→H_2O_2+O_2）。羟基自由基仅产生于SOD（或Fe^{2+}与Fe^{3+}）与H_20_2反应的过程中（$\cdot O_2$-+H_2O_2→2OH·-+O_2），产生需要水环境且其寿命极短（约10^{-9}s）。在干种子中，羟基自由基不会大量产生，而是以H_2O_2的形式存在。本文中ERP所使用的样本为干洋葱种子研磨后的粉末，在此样品中自由基主要以超氧自由基形式积累。EPR吸收峰的总面积与自由基含量呈正比。虽然超氧自由基与羟基自由基的EPR反应中心场强均处于3 500 ± 100G处，但本文获得的EPR信号主要属于超氧自由基并且反应了洋葱种子中自由基的清除水平。

本文中的所有引发处理均提升SOD活性，且提升程度随湿热引发时间延长而加大。通过对细胞脂质组织与细胞膜穿孔的扫描电镜与透射电镜观察，我们发现高活性的SOD加快清除自由基以抗击脂质过氧化，但这个过程必须以种子吸涨为前提，即必须有湿热引发过程或在播种后。纯电场引发只提升萌发过程中SOD活性与抑制MDA的产生，引发过程中并没有修复细胞膜或清除自由基。这一观点在EPR、MDA含量与EC值的测定中得到了证实。湿热引发（Hyd）与混合引发（HEHP）均在引发过程中提升了膜完整性（更低的EC值）并抑制了自由基的产生（更小的EPR峰面积）。纯电场应发后的种子的细胞膜完整性与自由基含量与对照种子是一样的。虽然Hyd与HEHP均能抑制MDA的产生并降低EC值，但仍然需要引发时间大于48h才能使效果显著。

自由基和抗氧化酶系统在种子吸涨后被同时激活。如果抗氧化酶的活力不足以及时清除自由基，则脂类物质的降解加速。同理，只有以吸水为前提的引发（Hyd与HEHP）才能激活自由基清除机制，阻碍细胞继续恶化，修复受损的种胚。我们注意到，只有在生理上得到修复的种子，其引发效果才能保持较长久。当引发处理完成并即刻播种后，EF与HEHP24处理的PCA-Z值均高于Hyd24与Hyd48，甚至接近Hyd96。但在理想条件下贮藏60d后，EF与HEHP24这两个处理的播后的PCA-Z值均显著下降。这说明了这两个处理之前的高得分很大程度上是依赖于电场照射产生的即时SOD活力提升效果，这进一步证明了纯电场应发过程不清除自由基或修复细胞膜，而HEHP24（或Hyd24）则是由于湿热引发的时间不够而不能产生明显的修复和清除效果。但HEHP48与HEHP96的引发时间则足够实现充分的修复与清除，这也解释了混合引发能够延长纯电场引发的效果的机理。

研究认为，外电场照射可以改变蛋白质二级结构中a-螺旋、β-转角、β-折叠与无规则卷曲的含量，从而增强其功能活性。SOD结构与活性可以被特定的电场强度和电场照射时间改变。本文发现，与CK和Hyd对比，EF与HEHP在播种后的初始SOD活性较高。我们认为是电场改变了SOD的结构组成，减少了无规则卷曲含量，提高了β-折叠的含量，从而强化了酶活性。Xu用特定的电场强度抑制了SOD的活性，并发现β-折叠与β-转角的含量降低，而α-螺旋与无规则卷曲的含量上升，这与本文中的结果相似[10]。此外，在高种子含水量条件下，SOD的结构变得更容易改变，如用EF与HEHP48和HEHP96作比较。在将纯电场引发与对照进行对比，与将混合引发与湿热引发进行对比后发现，二级结构含量发生了显著变化。如果电场照射是刺激了SOD相关基因的过表达并造成SOD含量的增加，而不是改变了酶的二级结构，则在测得SOD活性提高的同时各处理间二级结构组成的差异应为不显著的。同时，目前为止未见有关电场照射处理能够改变SOD相关基因表达量的报道出现。

植物组织中仅存在3种SOD同工酶（Mn-SOD、Fe-SOD与Cu/Zn-SOD）。虽然文中所用的试剂盒仅能测总SOD的活性，但SDS-PAGE的鉴定结果表明洋葱种子中的SOD类型仅为Cu/Zn-SOD。该类型的SOD活性中心的构象主要为β-折叠，而Mn-与Fe-SOD则为α-螺旋。这一结果也与电场照射显著提高SOD的β-折叠含量和总酶性的结果相吻合。

高活力的种子往往具有形态完整、功能正常的胚细胞与细胞器，低活力或老化的种子则表现出胚细胞受损与恶化的迹象。在使用扫描电镜与透射电镜观察的过程中发现，对照和所有引发处理均存在三种类型的胚：（Ⅰ）严重恶化并已丧失大部分生物功能而无法萌发；（Ⅱ）存在恶化与损伤迹象但仍具有生物学功能，经过引发后被部分修复，并在播后4d萌发；（Ⅲ）仅有轻微或没有恶化与损伤迹象，经过引发后被完全修复，在播后2d快速萌发。本文认为，只有仍具有生物学功能的胚才能通过引发修复恶化与损伤。但是，具有此类胚的种子即便不经过引发，在播种后也会经历从类型Ⅱ到Ⅲ的自我修复过程，而湿热引发中的种子湿度保持过程就是将这种修复提至播种前并更高效更充分地完成，并且让更高比例的种子恢复到类型Ⅱ与Ⅲ阶段，电场照射则进一步强化了湿热引发的这一过程。但是，在引发结束后，我们无法判断哪粒种子正处于哪个阶段。虽然同样可以通过扫描电镜与透射电镜观察干种子胚的损伤程度，但我们无法判断哪种程度的恶化与损伤会延缓或阻碍种子萌发。因此，本文选择将种子播种，通过观察不同天数萌发与未萌发的种子的胚的恶化与受损程度，并根据表2中不同天数的萌发量来判断引发修复的效果。

个别种子的自由基含量较高，SOD活性较低且受损严重，但是仍然具有生理活性。这些种子即时经过96h的引发也无法有效修复，但能够在播种6d以后的某个时间萌发。而且，回干过程同样会对未充分修复的胚造成二次损伤。因此本文认为进一步延长引发时间是有必要的，并且可以进一步提升出芽率。

5 结论

基于对电场引发、湿热引发与水引发机理的对比，本文得出以下结论：湿热引发（图6c）提高种子含水量至一定水平并维持特定的时间，激活了种子内的SOD与自由基清除机制，为种子在播种前提供了一个额外的“预萌发”阶段以修复组织受损，以上过程取决于种子自身的生理活力。电场照射能够提高SOD的活性，因此混合引发（图6d）中的湿热引发的过程得以加速，并清除更多的自由基，使胚组织更快更完整的被修复。纯电场引发（图6a）通过电场照射增加SOD二级结构中β-折叠的含量以提高其即时活性，加速了种子播后自由基的清除与胚组织修复，从而缩短预萌发时间与提高发芽率。但与湿热引发和混合引发不同的是，纯电场引发并未在引发阶段进行任何自由基的清除或组织修复。经过混合引发的种子比经过湿热引发或纯电场引发的种子有更高的组织完整性。混合引发缩短了湿热引发的时间，延长了电场引发的有效期，综合引发效果更佳。

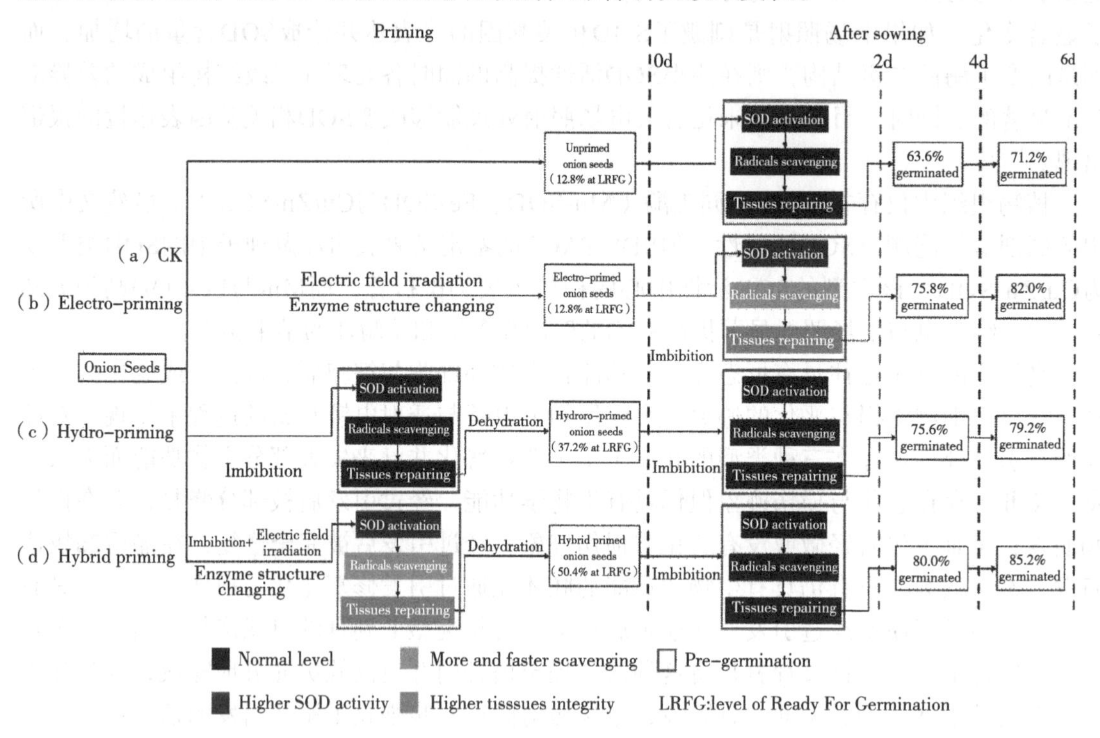

图6　纯电场引发、湿热引发与混合引发的机理对比。（a）对照；（b）纯电场引发；（c）湿热引发，以Hyd48为例；（d）混合引发，以HEHP48为例。

Figure 6　Comparison of the mechanisms and effects of electro-，hydro- and hybrid priming.（a）Control.（b）EF.（c）Hydro-priming，Hyd48 as example.（d）Hybrid priming，HEHP48 as example.

参考文献

［1］ Farooq，M.，Sma，B.，Rehman，H.，et al.，Seed priming enhances the performance of late sown wheat（triticum aestivum l.）by improving chilling tolerance. Journal of Agronomy & Crop Science，2008，194（1），55–60.

［2］ Dezfuli，P.M.，Sharifzadeh，F.，& Janmohammadi，M. Influence of priming techniques on seed germination behavior of maize inbred lines（zea mays l.）. Arpn Journal of Agricultural & Biological Science，2008，126（4），499-499.

［3］ Yu，X. Y.，Zhao，S. L.，Li，T.，et al.，Purification and some properties of superoxide dismutase from cactus. Fine Chemicals，2002，19（4），234-237.

［4］ Whitmore，L.，& Wallace，B.A. Protein secondary structure analyses from circular dichroism spectroscopy：methods and reference databases. Biopolymers，2008，89（5），392-400.

[5] Laemmli，U.K. Cleavage of structural proteins during the assembly of the head of bacteriophage t4. Nature，1970，227（5259），680-685.

[6] Naglreiter，C.，Reichenauer，T.G.，Goodman，B.A.，et al.，Free radical generation in pinus sylvestris and larix decidua seeds primed with polyethylene glycol or potassium salt solutions. Plant Physiol Biochem，2005，43（2），117-123.

[7] Clay，C. S.，& Peace，G. W. Ion beam sputtering：an improved method of metal coating sem samples and shadowing ctem samples.Journal of Microscopy，2011，123（1），25-34.

[8] Manabe，Y.，Sugimoto，T.，Kawasaki，T.，et al.，Nanometer-scale direct observation of the receptor for the leaf-movement factor in plant cell by a novel tem probe. Tetrahedron Letters，2007，48（8），1341-1344.

[9] Jr，W.C.J. Protein secondary structure and circular dichroism：a practical guide. Proteins Structure Function & Bioinformatics，1990，7（3），205-214.

[10] Xu，Q.，Wang，M.，& Yang，Z.H. Study of the effect of electric field on superoxide dismutase activity by spectroscopy. Advanced Materials Research，2011，236-238，2445-2448.

太阳能和生物质能开发利用初探

李建明，肖金鑫，张俊威
（西北农林科技大学园艺学院，杨凌　712100）

摘要： 新能源利用技术是解决现今现代农业中的能源消耗和环境污染问题的重要手段，主要体现在太阳能、地热能、生物质能的利用。本文研究了农业废弃物的发酵产热利用和地下深层蓄热装置的应用效果分析，研究表明农业废弃物发酵可以产生大量热量，可用于温室加温，具有时效长、热量大的优点，其中试验处理中以S3和S5两种处理的性能最好，整个酿热槽在典型晴天和雪天分别向外界释放$7.5X10^7$J和$1.1X10^8$J的热量。地下深层蓄热装置模型试验中，验证了该方法可行，具有实践意义。在温室应用中，典型晴天条件下地下卵石—空气热交换蓄热装置可蓄热195MJ，平均蓄热速率19.5MJ/h，放热140.4MJ，平均放热速率10.03MJ/h；地下土壤—空气热交换蓄热装置可蓄热106.3MJ，平均蓄热速率8.86MJ/h，放热82.7MJ，平均放热速，6.89MJ/h；地下水—空气热交换蓄热装置蓄热77.9MJ，平均蓄热速率8.66MJ/h，放热60.3MJ，平均放热速率4.02MJ/h。白天平均温度分别比对照低2.3℃、3.3℃、1.1℃，夜间平均温度分别比对照高2.5、2.3、1.0℃。

关键词： 太阳能；生物质能；地下深层蓄热

Study on the Development and Utilization of Solar energy and Biomass Energy

Li Jianming，Xiao Jinxin，Zhang Junwei
（*College of Horticulture of Northwest A&F University*，*Yang Ling*　712100）

Abstract: The technology of new energy utilization is an important way to solve the problem of energy consumption and environmental pollution in modern agriculture，which is mainly reflected in the utilization of solar energy，geothermal energy and biomass energy. This paper studies the application of agricultural waste fermentation heat utilization and underground heat storage device of the analysis，the research shows that agricultural waste fermentation can produce a large quantity of heat，can be used for greenhouse heating，in the test process，the performance of S3 and S5 was the best.，the heating groove in the a typical sunny day and snow respectively release to the outside world 7.5×10^7J and 1.1×10^8J of heat. Deep underground storage device model test，verified the feasibility of this method has practical significance，in the greenhouse application，under typical sunny conditions，the underground pebble-air heat exchange heat storage device can reheat 195MJ，the average heat storage rate is 19.5MJ/h，and the heat release is 140.4MJ，and the average exothermic rate is 10.03MJ/h.

基金项目：国家大宗蔬菜产业技术体系（CARS-23-C05）和陕西省重点研发项目（2017ZDXM-NY-003）

作者简介：李建明（1966-），男，陕西洛川人，教授，博士生导师，主要从事设施园艺工程及生理研究。杨凌　西北农林科技大学园艺学院，农业部西北设施园艺工程重点实验室，712100。Email：lijianming66@163.com

Underground soil - air heat exchange heat storage device can heat up 106.3MJ，average heat storage rate 8.86MJ/h, heat release 82.7MJ，average heat release rate，6.89MJ/h. Ground water - air heat exchange heat storage device accumulates 77.9MJ，the average heat storage rate is 8.66MJ/h，and the heat release rate is 60.3MJ，and the average exothermic rate is 4.02MJ/h. Day average temperature is lower than controls respectively 2.3℃，3.3℃，1.1℃，night average temperature is respectively higher than control 2.5℃，2.3℃，1.0℃.

Key words：Solar energy；Biomass energy；Deep underground heat storage

1 引言

我国是能源消耗大国，截至2017年，全国能源消费总量约44.9亿吨标准煤，预测2018年全国能源消耗45.8亿吨[1]。其中煤炭仍然是能源供给的主要来源，化石燃料等传统能源的大量消耗造成了环境污染、危害人类身体[2]。太阳能[3]、潮汐能[4]、风能[5]、地热能[6]和生物质能[7]等可再生能源的开发和利用成为解决新时代能源紧张和环境污染的有效途径[8]。

可再生能源技术对于我国许多行业建设具有重要意义，我国是农业大国，现正在向现代农业转型。可再生能源在现代农业中主要体现在太阳能[9-11]、地热能[12，13]、生物质能[14，15]等方面。在我国现代农业的发展进程中，生物质能的利用主要体现出农业废弃物的资源利用化[16]，我国是农业废弃物的生产大国，这部分利用技术仍处于初级阶段[17]。太阳能和地热能主要体现在温室蓄热中，我国温室面积截至2013年已有187.4万hm^2[18]。但是温室也是高耗能产业[19]，现今大多温室仍是被动式蓄热，主要为被动接受太阳能，进行蓄热加温。主动式蓄热温室应用较少，现今主动式温室主要是利用地中热交换系统[20]、相变材料[21]，热泵技术[22，23]、浅层地热利用技术[12]。新能源利用在现代农业中的应用已是必然趋势，本团队所研究的农业废弃物发酵产热和地下深层蓄热装置就是对生物质能、太阳能的利用研究，以期对未来温室与新能源的结合提供理论支撑。

2 农业废弃物发酵的效果热量分析

我国农业废弃物主要包括农业生产、农产品加工、畜禽养殖业和农村居民生活排放的废弃物，这类废弃物。农业废弃物资源化利用是控制农业环境污染、实现农业可持续发展的有效途径。利用形式主要为固态废弃物燃料化、肥料化、基料化、饲料化、材料化[17]。农业废弃物是一个巨大潜力的资源库，现今国内农业废弃物主要为堆肥技术和沼气技术，资源利用技术方面仍有巨大的提升空间。设施农业是我国农业发展的新形式，现今得到了快速的发展，但生产过程中也产生了很多问题，其中冬季夜间温室内的低温冷害一直威胁着作物的生长发育和良好的经济效益。本团队所进行的农业废弃物发酵是为温室提供热源的一种新形式，主要探究农业废弃物发酵的效果热量分析。

2.1 材料与方法

整个试验在大跨度非对称大棚内进行，本试验的发酵材料为番茄秸秆、黄瓜秸秆、小麦秸秆、猪粪、牛粪、菇渣，在酿热槽内设置了5个不同堆肥发酵处理，分别是S1（小麦秸秆+猪粪+菇渣）、S2（黄瓜秸秆+菇渣+猪粪）、S3（番茄秸秆+猪粪+菇渣）、S4（菇渣+牛粪）、S5（菇渣+猪粪），每个处理按照C/N比为20:1进行配比，堆肥体积为

2.0m×1.0m×0.8m，加入发酵物总质量3%的EM菌剂，相对含水量调至60%左右。2015年11月2日加水进行发酵，表面用薄膜封住，各处理位置位于酿热槽长度方向上四等分点处，如图1所示。基质发酵温度用精创RH-4型号温度记录仪长期记录，放置在每个发酵处理的中间位置，酿热槽覆盖膜表面温度及热流密度采用北京世纪建通环境技术公司生产的温度热流计JTR01，安装在酿热槽覆盖膜表面中间处。

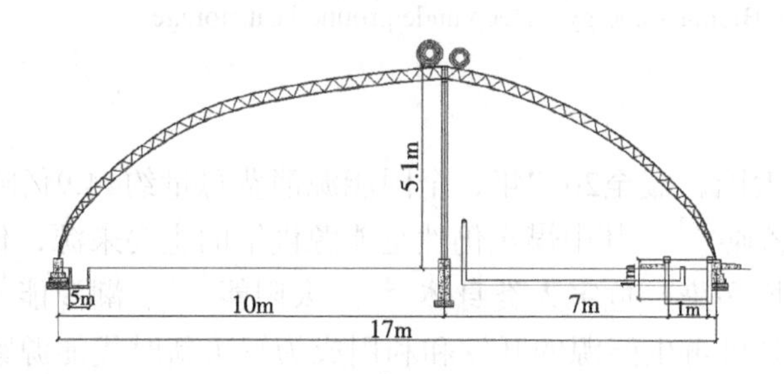

a　大跨度非对称大棚

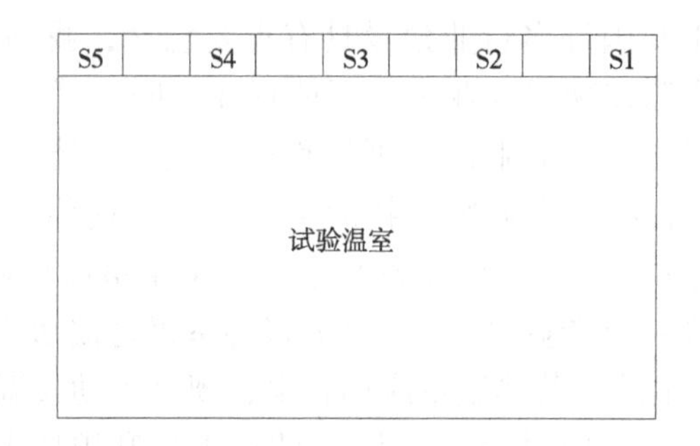

b　酿热槽处理示意图

图1　试验处理示意

2.2　结果与分析

图2是基质发酵的每日平均温度，从图2中可以看出，各处理发酵温度均呈现先上升后下降的趋势。S3处理温度升高速度最快，发酵温度在45℃以上的时间持续了32d，在五个处理中持续时间最长，其中最高温度达到51.9℃。S1、S2、S3、S4、S5处理发酵温度在40℃以上的持续时间分别为32d、25d、64d、28d、38d。在2月份，S5处理的发酵温度下降缓慢，发酵温度最高，其它处理发酵温度下降快。在11月初添加8m³的农业废弃物到酿热槽内进行发酵，11月24月左右酿热槽平均温度达到35℃以上，且高温持续时间在71d左右。S1、S2、S3、S4、S5处理的发酵积温分别为4 321.3℃、4 171.6℃、4 735.4℃、4 228.0℃、4 643.6℃，可以看出S3和S5处理的性能最好。整个酿热槽在典型晴天00:00～09:00和18:00～24:00时间段内，通过热流计测得酿热槽表面薄膜与大棚内空气的平均热流密度为46.2J/m·s^{-1}，可计算出在这段时间内，酿热槽向外界释放的总热量为7.5×10^7J，平均每小时释放热量为5.0×10^6J，典型雪天情况下，通过热流计测得酿热槽表面薄膜与大棚内空气的平均热流密度为68J/m·s^{-1}，可计算出在这段时间内，酿热槽通向外界释放的总热量为1.1×10^8J，平均每小时释放热量为7.3×10^6J。

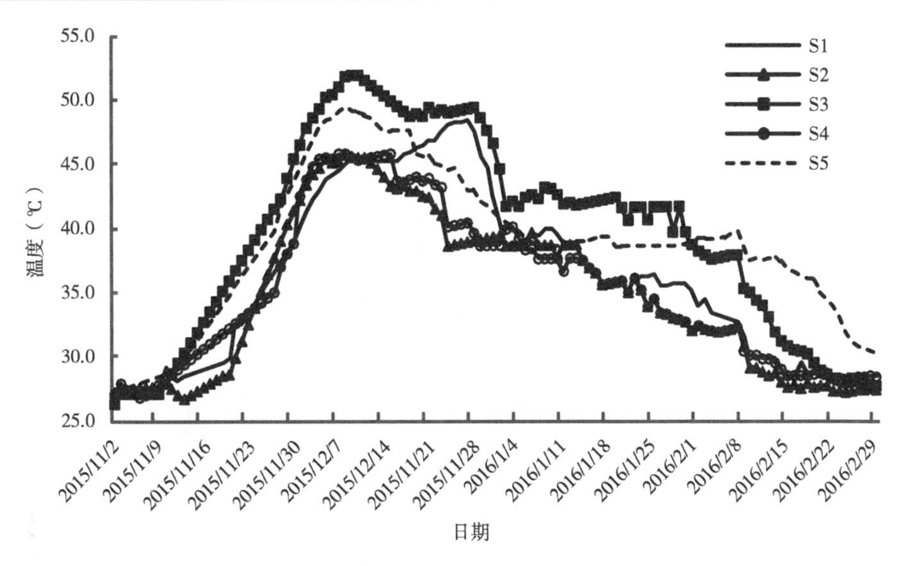

图 2　酿热槽不同处理基质发酵温度的变化

3　地下深层空气热交换蓄热装置效果分析

地下深层蓄热装置主要采用地中热交换系统的思想，地中热交换系统是一种有效利用太阳能作为热源，将温室内空气和地下的蓄热材料进行热交换的一种增温系统[24]。地中热交换系统的基本原理就是采用主动的蓄热方式将温室中的热量蓄积到地下的蓄热介质中，即可用于夏季的白天的降温也可以用于冬季夜间的加温。本团队在2017年6—8月进行了模型验证试验，主要探究地下蓄热装置的可行性、换热管道材质的选择和排布方式。2017年11月至2018年2月探究地下蓄热装置应用于温室中的效果。

3.1　地下深层蓄热装置模型试验

3.1.1　材料与方法

供试塑料大棚、蓄热装置结构参数及测点布置。试验用塑料大棚位于陕西杨凌西北农林科技大学南校区模型温室中的塑料大棚，长度10m，跨度5m，脊高2.5m。用塑料薄膜将大棚隔开分成三个隔间（图3）。卵石—空气热交换装置和水—空气热交换装置均埋深于地下1.1m处，尺寸均为1.1 × 1.1 × 1.1m，卵石和水的体积都为1m³。两个蓄热装置底部、顶部和四周均覆盖5cm厚泡沫板，水—空气热交换装置内另覆盖塑料薄膜，防止水渗漏（图4和图5）。

试验仪器使用哈尔滨五个电子仪器公司生产的多路温度记录仪PDE-R4，测量范围−40 ~ 80℃，准确度 ± 0.1℃，测量各装置进出风口温度以及内部温度。卵石—空气热交换装置进出风口、卵石中心位置表面下部0cm、10、20cm、30cm、40cm各一个温度测点，水—空气热交换装置进出风口和内部中心水表面下部40cm处各1个温度测点（图6）。

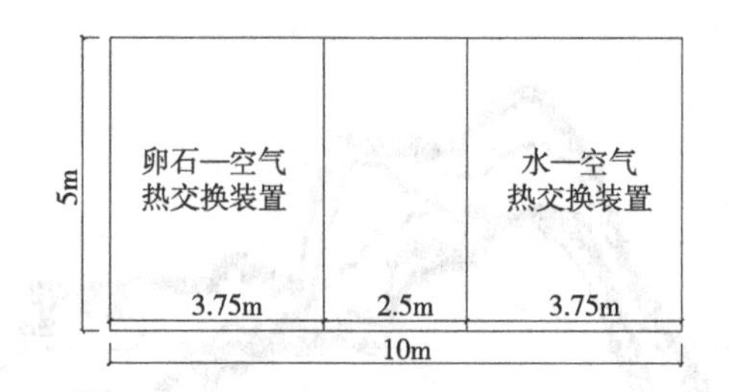

图 3　试验温室布置

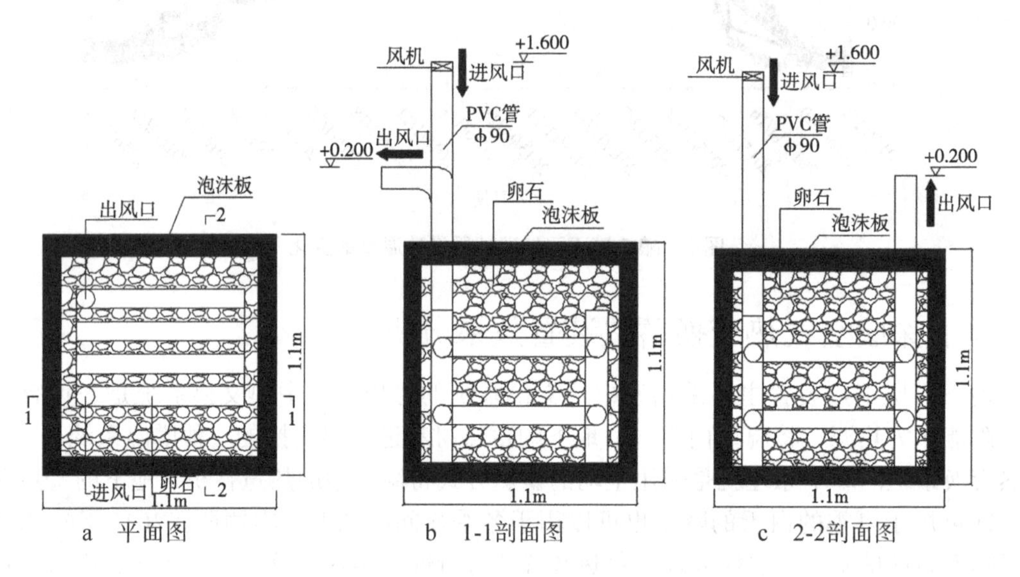

图 4　卵石—空气热交换装置详图

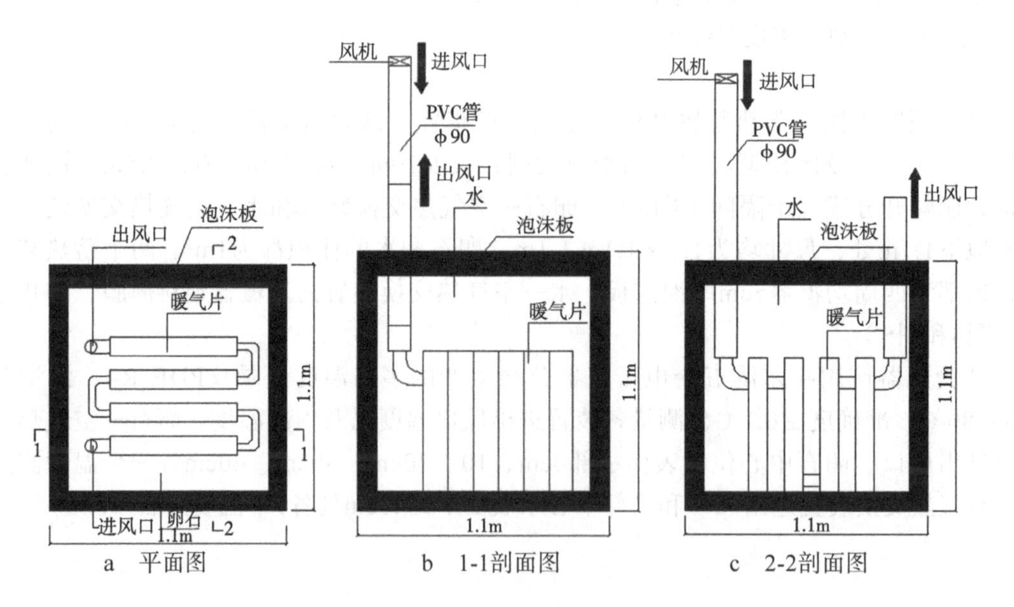

图 5　水—空气热交换装置详图

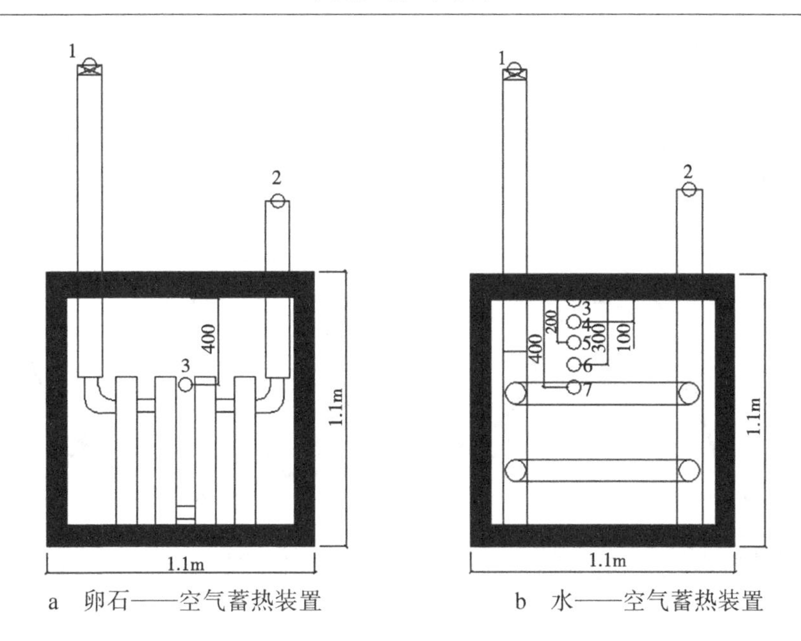

图 6　测点布置

3.1.2　结果与分析

（1）卵石—空气热交换装置。由图7可以看出进风口温度从早上9:00后开始高于出风口，一直持续到17:30，这段时间装置处于蓄热状态；最大温差5.2℃。从17:30到第二日早上8:00，出风口温度一直高于进风口，装置处于放热状态，最大温差为3.5℃。

卵石蓄热装置测量了0cm、10cm、20cm、30cm、40cm 5个深度的温度，由图8可以看出，因为40cm深度接近换热管道，所以该处温度变化幅度最大，30cm深度次之、其余深度变化基本一致，图8中可以看出卵石在9:00～19:00间蓄热，温度平均上升10℃，估算1m³卵石蓄热量为1.95×KJ；其余时间放热，温度平均下降9.4℃.说明放热1.83×KJ。

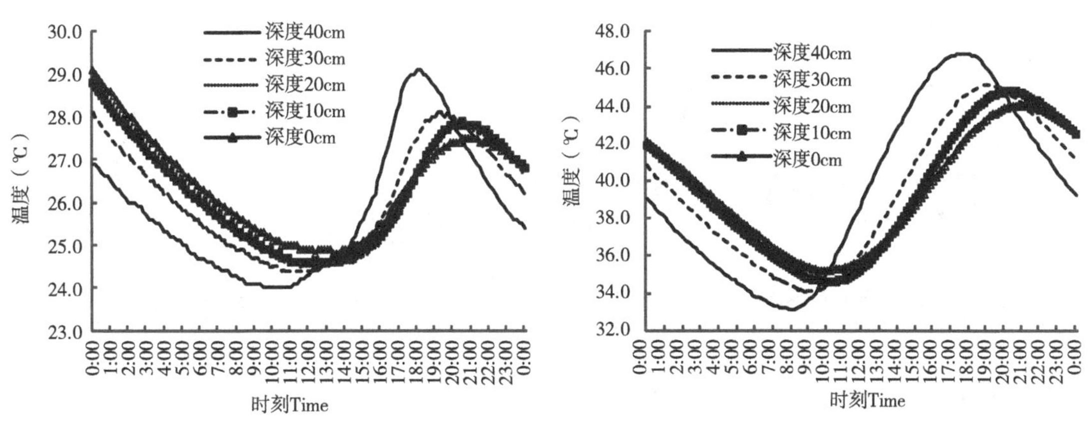

图 7　典型晴天装置进出风口温度　　**图 8　典型晴天卵石内部温度变化**

由图9得出阴雨天气时装置的进出风口温度随天气变化较为明显，白天当室外温度下降较快时，进风口温度会短暂的低于出风口，夜间出风口温度一直高于进风口温度，处于放热阶段，最大温差为2.2℃。

图10中可以看出卵石40cm深度处在11:00～18:00间蓄热，其余深度处在13:00～20:00蓄热，整体温度平均上升3.5℃，卵石蓄热量为6.83×KJ；其余时间放热，温度平均下降2.7℃，说明放热5.27×KJ。

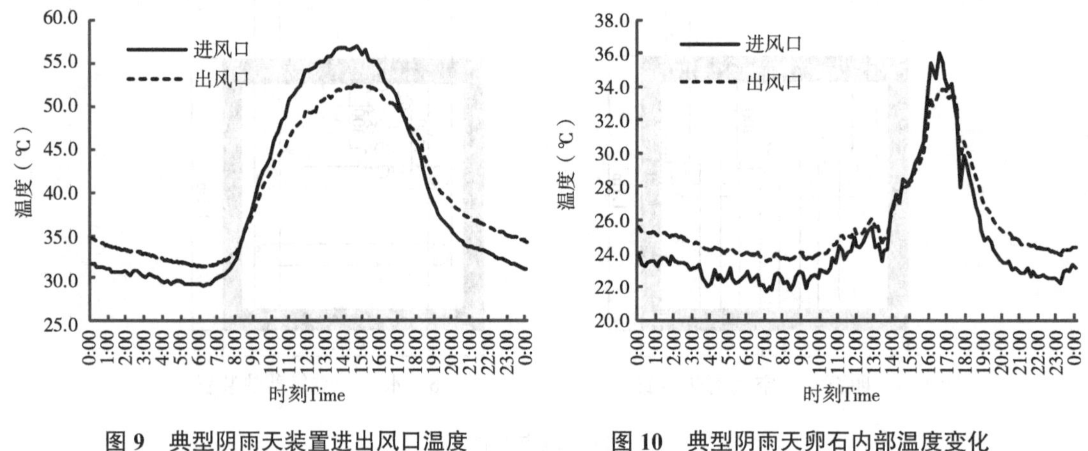

图9　典型阴雨天装置进出风口温度　　图10　典型阴雨天卵石内部温度变化

（2）水—空气热交换装置。由图11可以看出蓄热装置从8:10～20:30间进风口温度高于出风口温度，说明水在蓄热，由于水的比热容巨大，直到12:40温度才出现明显上升，一直持续到22:00，期间温度上升1.3℃，蓄热装置蓄热5.46×KJ，夜间温度下降1℃，放热4.2×KJ。

由图12可以看出阴雨天气时装置的进出风口温度随天气变化较为明显，夜间出风口温度一直高于进风口温度，处于放热阶段，水温基本处于下降趋势，整体下降1.7℃，全天放热7.14×KJ

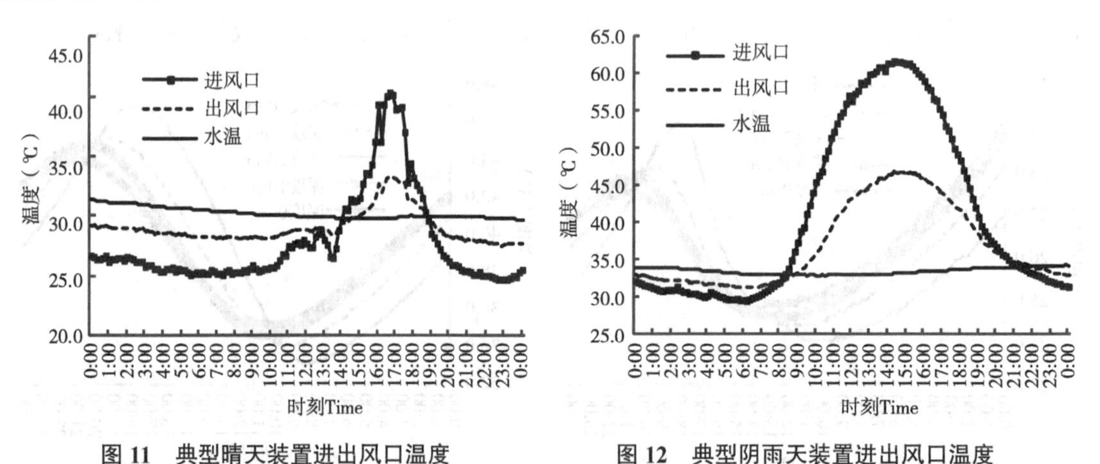

图11　典型晴天装置进出风口温度　　图12　典型阴雨天装置进出风口温度

3.1.3　结论

典型晴天卵石—空气热交换装置蓄热量为1.95×KJ；放热量为1.83×KJ。水—空气热交换装置蓄热量为5.46×KJ，放热量为4.2×KJ。典型阴雨天卵石—空气热交换装置蓄热量为

6.83 × KJ；放热量为5.27 × KJ。水—空气热交换装置放热量为7.14 × KJ。证明方案可行。

3.2 地下深层蓄热装置在温室中的应用效果分析

3.2.1 材料与方法

（1）供试温室、蓄热装置结构参数。试验在陕西杨凌揉谷农园内进行，试验大棚为17m非对称双层大棚，长度45m，跨度17m，外棚脊高6m，内棚脊高5.4m。外覆保温被，北屋面全天覆盖，南屋面按正常保温被管理方式揭起。试验共用4个大棚，3个分别放置蓄热设施，1个作为对照，内膜揭起，外膜全天封闭（图13）。

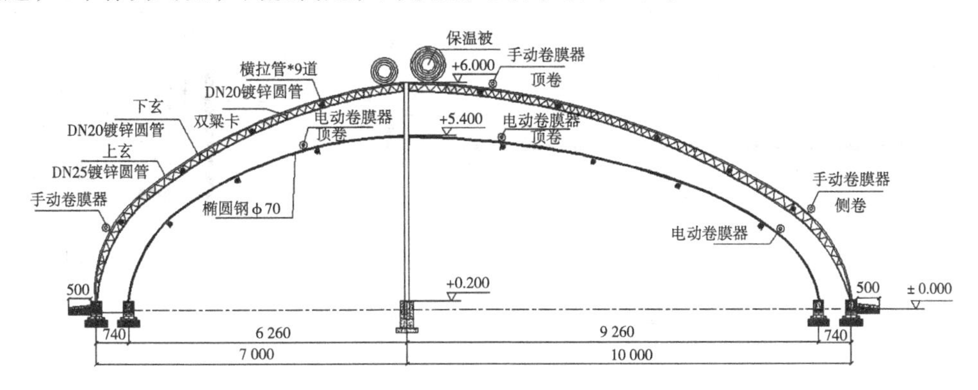

图 13　试验温室详图

试验共有3种蓄热装置，地下卵石—空气热交换蓄热装置：以卵石为蓄热介质，PVC管作为换热管，深埋于1.3m × 1.8m × 30m的深坑中，四周及底部覆盖10cm的泡沫板，上部覆盖塑料薄膜，埋管方式为两层敷设，中间1个进风口，两边各7.5m、15m 1个出风口（图14）；地下土壤—空气热交换蓄热装置与地下卵石—空气热交换蓄热装置建造一样，仅把蓄热介质变成土壤（图15）；地下水—空气热交换蓄热装置采用不锈钢波纹管作为换热器，盘在三个钢架上，每个钢架留出1个进风口和出风口，放置在直径1.2m，高1.6m的塑料桶中，桶外覆盖保温被。3个进风口通过塑料软管和PVC管连接风机（图16）。

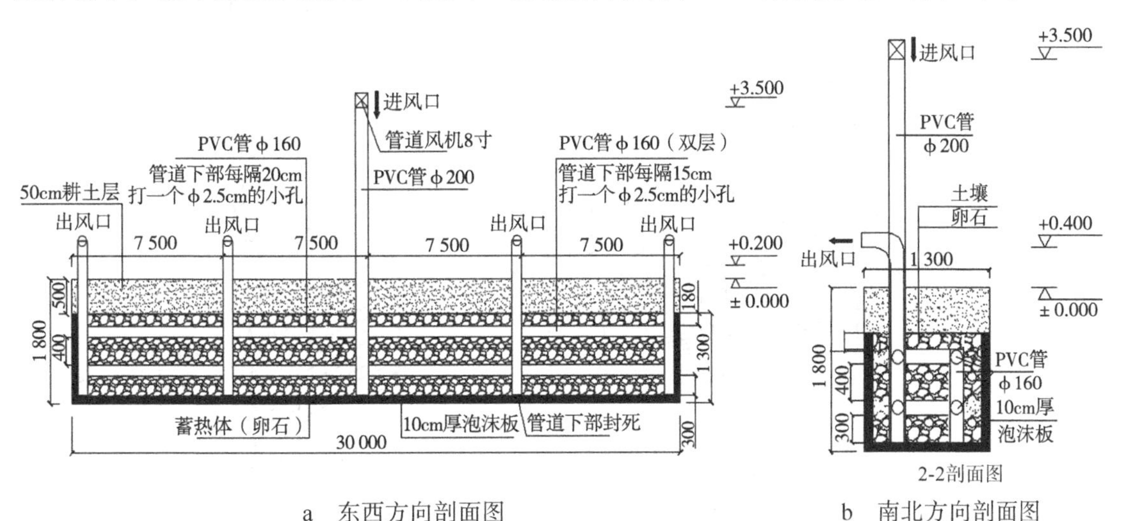

a　东西方向剖面图　　　　b　南北方向剖面图

图 14　卵石—空气热交换系统图

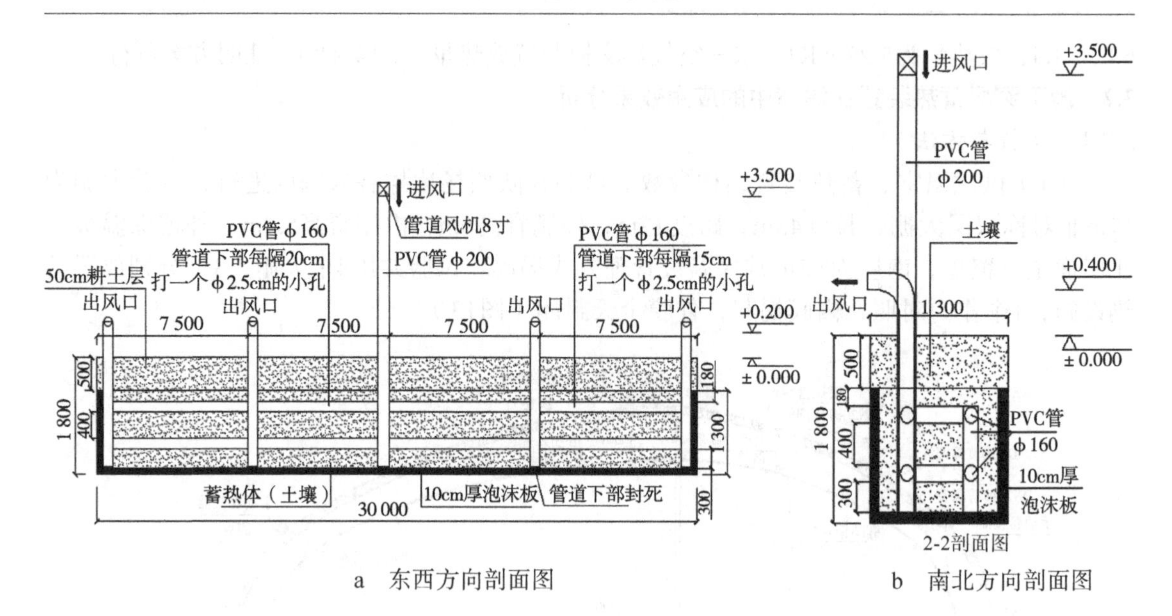

a　东西方向剖面图　　b　南北方向剖面图

图 15　土壤—空气热交换系统图

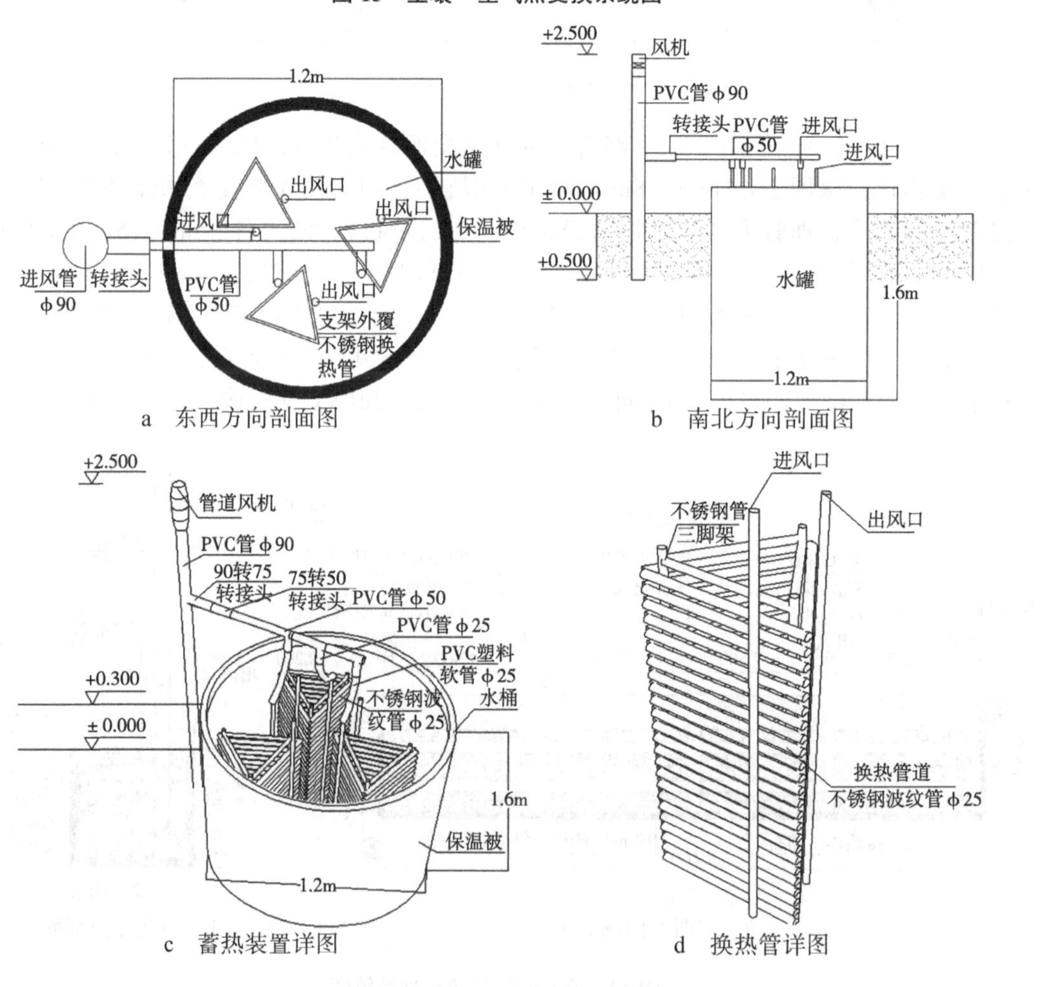

a　东西方向剖面图　　b　南北方向剖面图

c　蓄热装置详图　　d　换热管详图

图 16　土壤—空气热交换系统图

（2）试验仪器和测点布置。

试验采用哈尔滨物格公司的长期数据记录仪PDE-KI记录室内温湿度和光照强度，温室中心沿长度方向每1/4处布置1个测点，共3个，高度1.5m。采用热电偶测量蓄热装置温度测点：从西往东每隔7.5m在风口、90cm、130cm深度处布置温度测点共15个测点（图17），每10min记录1次数据，所有热电偶连接到国产的安柏多路数据记录仪AT4524。风机采用手持式泰仕风速仪AVM07测量蓄热装置进出风口风速。测得地下卵石—空气热交换装置和土壤—空气热交换装置进风口风速为7.62m/s，4个出风口速度均为2.85m/s；地下水—空气热交换装置进风口风速为2.00m/s。

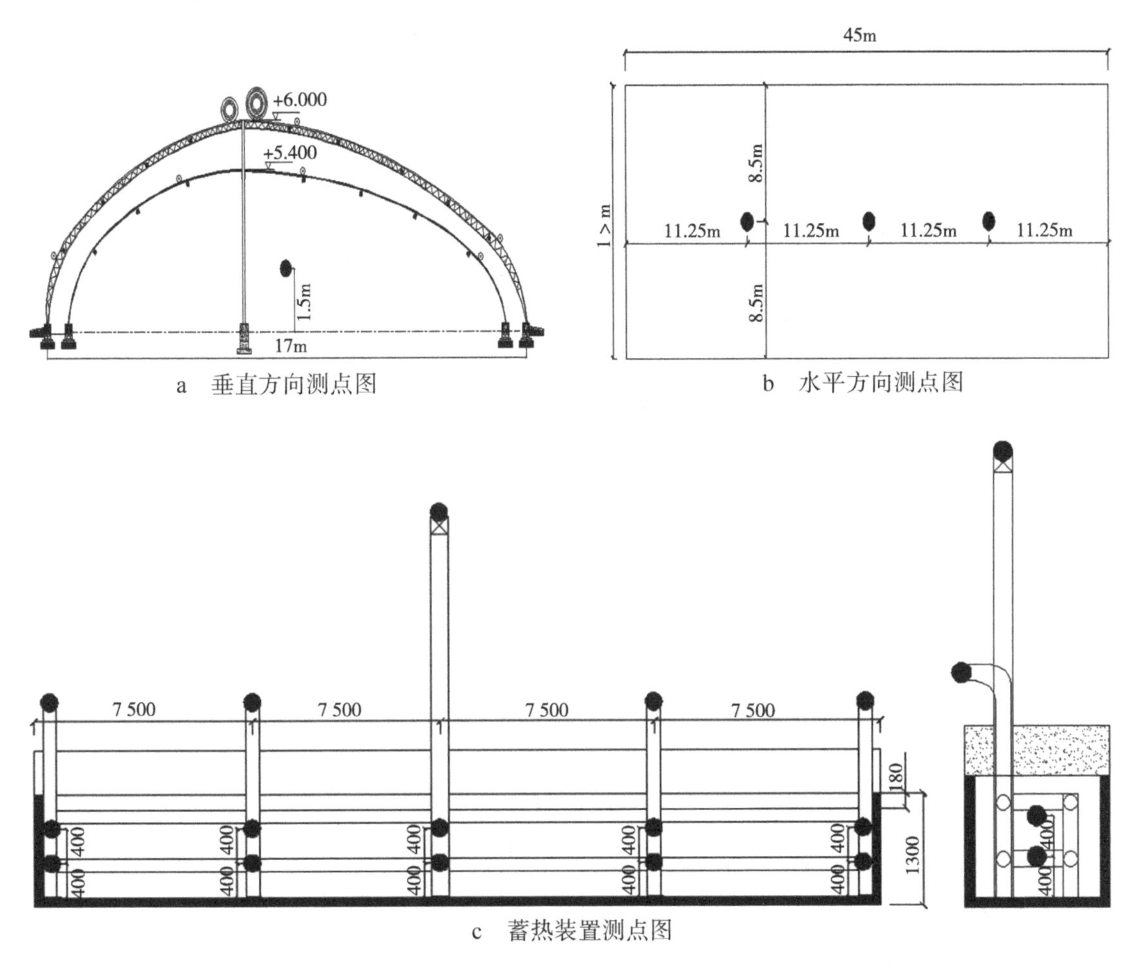

图 17　测点布置图

3.2.2　结果与分析

（1）地下卵石—空气热交换蓄热装置。由图18可以看出，在9:10～18:00间进风口温度高于出风口温度，其余时间出风口温度高于进风口温度，但进风口与地下卵石距离4m，输送过程存在热量损失，故判断蓄放热状态应从卵石内部分析。

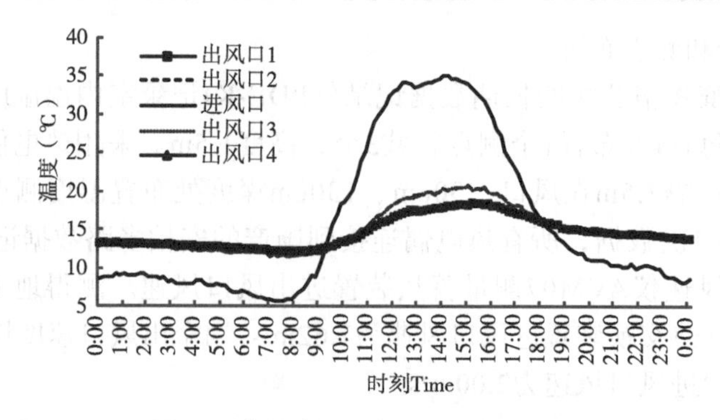

图 18　典型晴天进出分口温度变化图

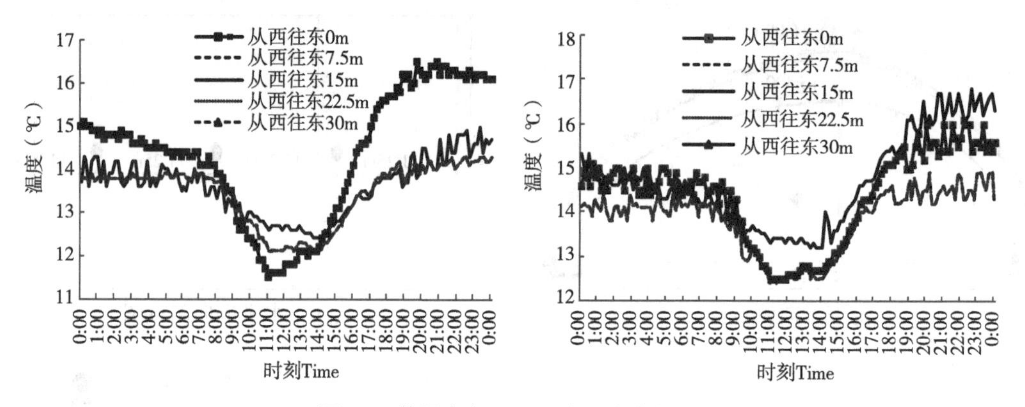

图 19　装置内部不同深度温度变化图

从图19可以看出内部卵石都是从12：00左右时刻开始出现局部蓄热，但升温平缓，14:00以后全部呈现升温并且温度上升迅速，说明该时刻后卵石整体开始蓄热，这说明进风口空气经管道输送导致热量损失，需进一步探究管道内部和卵石内部的热量变化情况。蓄热前后平均温差为2.5℃，说明典型晴天可蓄热195MJ，全天蓄热10h，平均蓄热速率19.5MJ/h，放热前后平均温差1.8℃，可放热140.4MJ，全天放热14h，平均放热速率，10.03MJ/h，

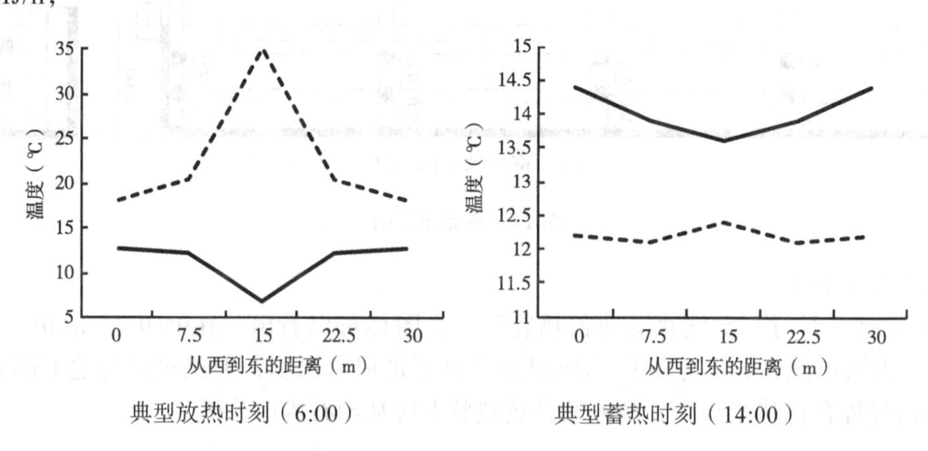

图 20　典型时刻不同位置蓄放热分析图

从图20中可以看出在典型蓄热时刻卵石内部靠近进风口的位置最高，说明中间部分最先开始蓄热。因为0 ~ 7.5m和22.5 ~ 30m之间换热管每隔15cm打孔，7.5 ~ 22.5m之间是每隔20cm打孔，0 ~ 7.5m和22.5 ~ 30m换热量多，故出现缓慢上升。放热时刻，两侧高于中间位置，说明放热时也是先从中间部分开始的。

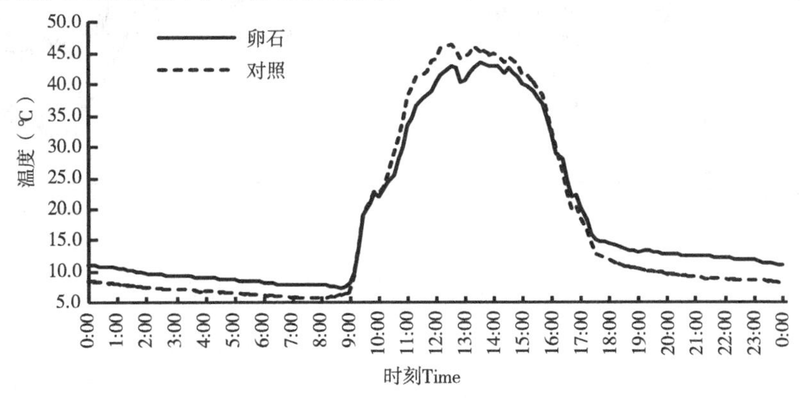

图 21　蓄热装置对温室气温的影响

由图21可以看出从9：00 ~ 16：10地下卵石蓄热温室气温低于对照温室，最大温差5.0℃，其余时间段高于对照温室，夜间可有效提高温度，最大温差为3.5℃。地下卵石—空气热交换装置应用于温室白天可使温度平均降低2.3℃，夜间可平均提升温度2.5℃，后半夜具有良好的放热能力，说明地下卵石—空气热交换蓄热装置可有效提升温室性能。

（2）地下土壤—空气热交换蓄热装置。由图22可以看出在9:20 ~ 17:30间进风口温度高于出风口温度，说明白天蓄热体在蓄热；其余时间出风口温度高于进风口温度，说明夜间在放热。但进风口与地下土壤距离4m，输送过程存在热量损失，故判断蓄放热状态应从土壤内部分析。

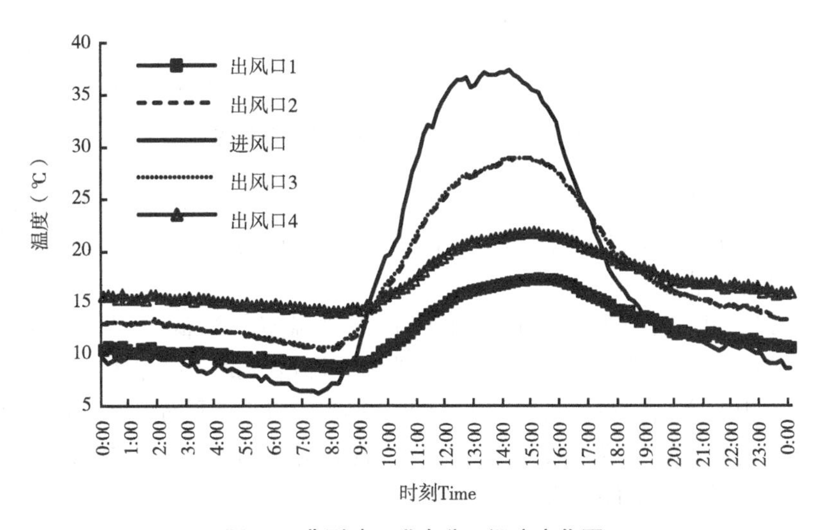

图 22　典型晴天进出分口温度变化图

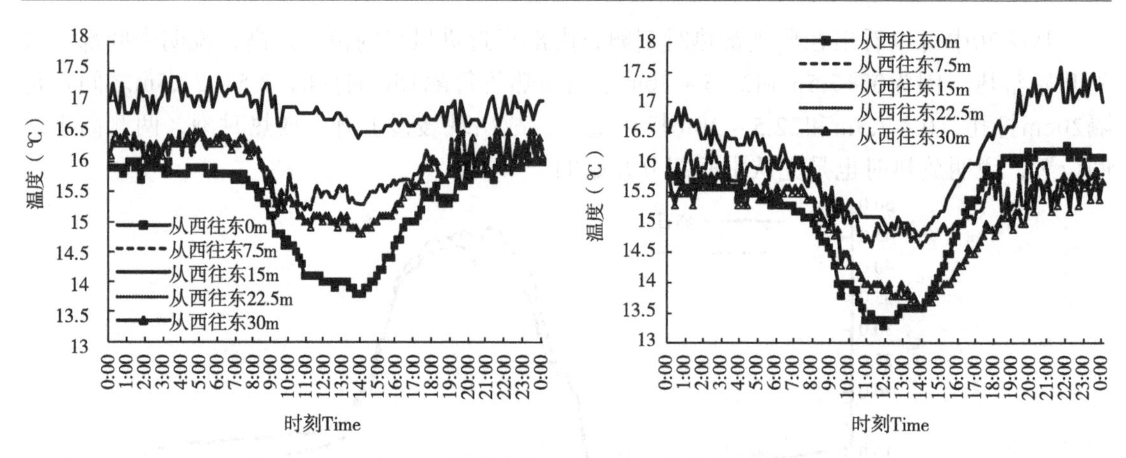

图 23　装置内部不同深度温度变化图

从图23中可以看出地表下90cm深度土壤夜间温度变化较为平缓，开始蓄热时间在14：00，而地表下130cm深度土壤夜间降温较为迅速，开始蓄热时间为12:30，这说明蓄热装置首先调动深层土壤进行蓄放热。蓄热前后平均温差为1.8℃，说明典型晴天可蓄放热106.3MJ，全天蓄热12h，平均蓄热速率8.86MJ/h，放热前后平均温差1.4℃，可放热82.7MJ，全天放热12h，平均放热速，6.89MJ/h。

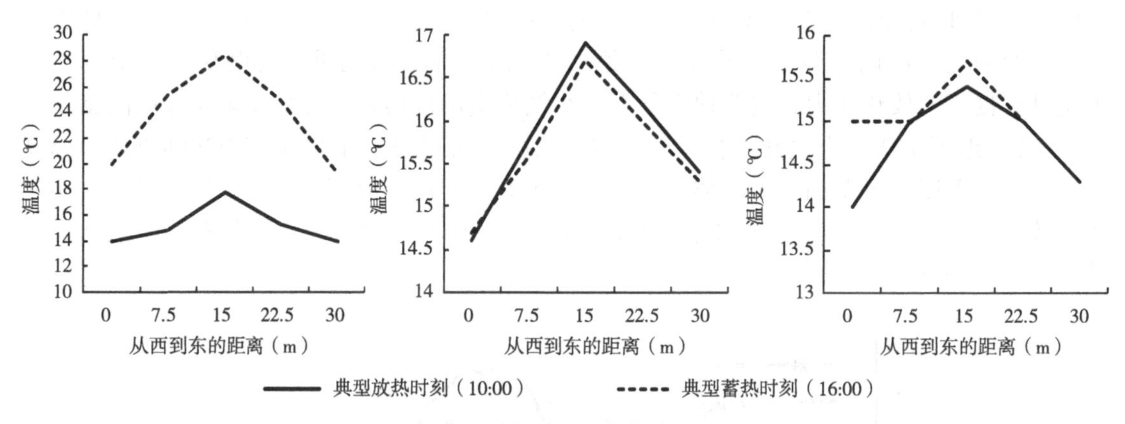

图 24　典型时刻不同位置蓄放热分析图

从图24中可以看出在与卵石—空气热交换装置存在明显差异。典型蓄热时刻和放热时刻均是中间位置高于两侧，说明土壤内部存在热量传导的过程，中间部分的土壤吸收热量较多，并向两侧传导热量。

由图25可以看出从8:50 ~ 16:10地下卵石蓄热温室气温低于对照温室，最大温差6.1℃，其余时间段高于对照温室，夜间可有效提高温度，最大温差为3.1℃。土壤—空气热交换装置应用于温室白天可使温度平均降低3.3℃，夜间可平均提升温度2.3℃，说明地下土壤—空气热交换蓄热装置可有效提升温室性能，并具有良好的储能性质，可以对抗连阴天和雨雪天等不利天气条件的影响。

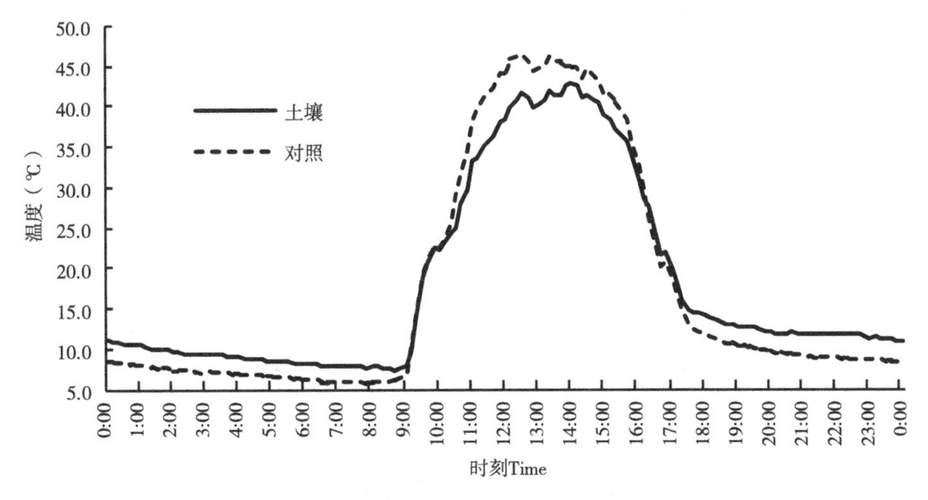

图 25　蓄热装置对温室气温的影响

（3）地下水—空气热交换蓄热装置。由图26可以看出，在9:50～17:30间进风口温度高于出风口温度，处于蓄热阶段；其余时间出风口温度高于进风口温度，处于放热阶段期间白天水温温度上升5.3℃，蓄热装置蓄热77.9MJ，蓄热9h，平均蓄热速率8.66MJ/h，夜间温度下降4.1℃，放热60.3MJ，放热15h，平均放热速率4.02MJ/h。

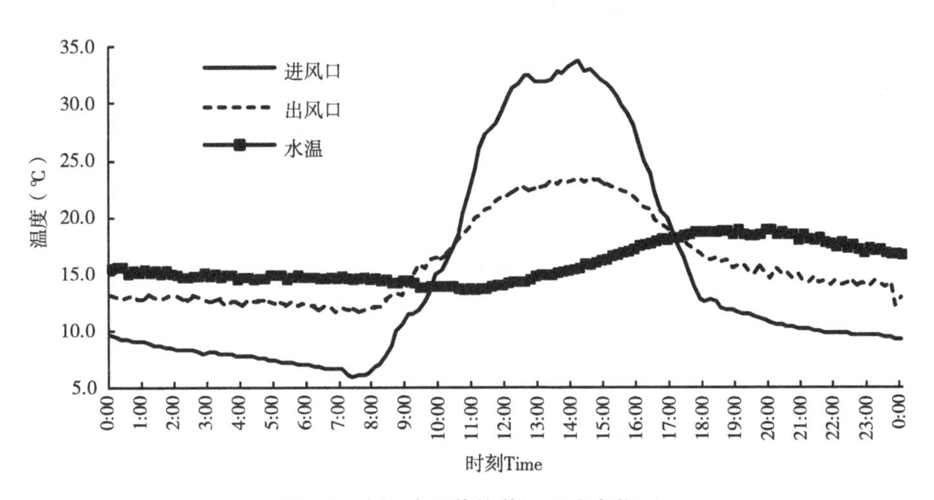

图 26　典型晴天蓄热装置温度变化图

由图27可以看出从09:00～16:10地下水蓄热温室气温低于对照温室，最大温差2.9℃，其余时间段高于对照温室，夜间可有效提高温度，最大温差为1.2℃。水—空气热交换装置应用于温室白天可使温度平均降低1.1℃，夜间可平均提升温度1.0℃，说明地下水—空气热交换蓄热装置可有效提升温室性能，并且装置稳定同时具有良好的储能性质，典型阴天与连阴天还需进一步研究。

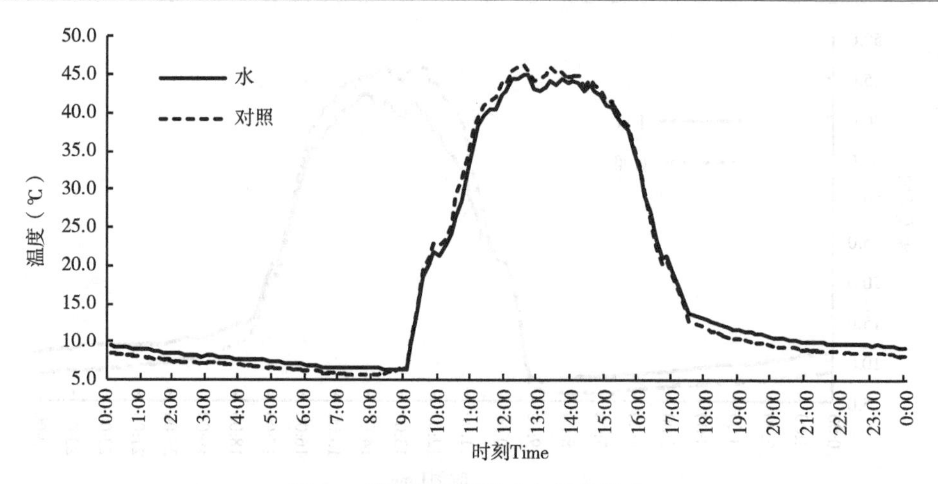

图27　蓄热装置对温室气温的影响

3.2.3　结论

（1）典型晴天地下卵石—空气热交换蓄热装置可蓄热195MJ，平均蓄热速率19.5MJ/h，放热140.4MJ平均放热速率10.03MJ/h，地下土壤—空气热交换蓄热装置可蓄热106.3MJ，平均蓄热速率8.86MJ/h，放热82.7MJ，平均放热速，6.89MJ/h，地下水—空气热交换蓄热装置蓄热77.9MJ，平均蓄热速率8.66MJ/h，放热60.3MJ，平均放热速率4.02MJ/h。白天平均温度分别比对照低2.3℃、3.3℃、1.1℃，夜间平均温度分别比对照高2.5℃、2.3℃、1.0℃。

（2）3种蓄热设施均能提高温室性能，其中地下卵石—空气热交换蓄热装置效果最优，地下土壤—空气热交换蓄热装置次之，地下水—空气热交换蓄热装置最末。但从成本和性价比考虑，地下水—空气热交换蓄热装置最好。

（3）试验后期还需要对地下卵石—空气热交换蓄热装置和地下土壤—空气热交换蓄热装置内部温度场、换热管道内热量变化规律、最佳埋管方式、最佳管径和管内流速进行详细分析，通过不同运行方式比较何种方式能达到最优的蓄热方案。

参考文献

［1］　肖宏伟. 2017年我国能源形势分析及2018年预测［J］. 科技促进发展. 2017（11）：902-908.

［2］　宋彦，金泰廙，Perera F. P. 化石燃料燃烧对儿童健康的多重威胁：空气污染和气候变化的影响［J］. 环境与职业医学. 2017（8）：751-757.

［3］　闫云飞，张智恩，张力，等. 太阳能利用技术及其应用［J］. 太阳能学报. 2012（S1）：47-56.

［4］　郑晨，于华明，宋彦，等. 瓯飞围海潮汐电站能量估算及环境影响研究［J］. 太阳能学报. 2017（7）：1 893-1 900.

［5］　闫占新，刘俊勇，许立雄，等. 风能等效转化利用模型及其效益研究［J］. 电力自动化设备. 2017（6）：48-54.

［6］　李德威，王焰新. 干热岩地热能研究与开发的若干重大问题［J］. 地球科学（中国地质大学学报）. 2015（11）：1 858-1 869.

［7］　吴创之，周肇秋，阴秀丽，等. 我国生物质能源发展现状与思考［J］. 农业机械学报. 2009，40（1）：91-99.

［8］　何凌云，吴梦，尹芳. 可再生能源投资总量和结构对碳排放的影响研究［J］. 中国地质大学学报（社会科学版）. 2017（1）：76-88.

［9］ 孙先鹏，邹志荣，赵康，等. 太阳能蓄热联合空气源热泵的温室加热试验［J］. 农业工程学报. 2015（22）：215-221.

［10］ 马承伟，姜宜琛，程杰宇，等. 日光温室钢管屋架管网水循环集放热系统的性能分析与试验［J］. 农业工程学报. 2016（21）：209-216.

［11］ 凌浩恕，陈超，陈紫光，等. 日光温室带竖向空气通道的太阳能相变蓄热墙体体系［J］. 农业机械学报. 2015（3）：336-343.

［12］ 王嘉维，王昭，杨俊伟，等. 苏南地区夏季浅层地热交换对大棚降温效果初探［J］. 浙江大学学报（农业与生命科学版）. 2017（4）：519-526.

［13］ 于威，王铁良，刘文合，等. 太阳能土壤加温系统在日光温室土壤加温中的应用效果研究［J］. 沈阳农业大学学报. 2010（2）：190-194.

［14］ 卞中华，王玉，胡晓辉，等. 外置式与内置式秸秆生物反应堆对番茄生长及光合性能的影响［J］. 应用生态学报. 2013（3）：753-758.

［15］ 杨圆圆，韩娟，王阳峰，等. 秸秆生物反应堆对日光温室微生态环境及草莓光合性能的影响［J］. 西北农业学报. 2014（8）：167-172.

［16］ 朱明，郭红宇，周新群. 现代农业产业工程体系建设方案研究［J］. 农业工程学报. 2010，26（1）：1-5.

［17］ 宋成军，张玉华，李冰峰. 农业废弃物资源化利用技术综合评价指标体系与方法［J］. 农业工程学报. 2011（11）：289-293.

［18］ 蒋卫杰，邓杰，余宏军. 设施园艺发展概况、存在问题与产业发展建议［J］. 中国农业科学. 2015（17）：3 515-3 523.

［19］ 刘文波. 日光温室地下蓄热技术的应用和研究［D］. 济南：山东建筑大学，2007.

［20］ 戴巧利，左然，李平，等. 主动式太阳能集热/土壤蓄热塑料大棚增温系统及效果［J］. 农业工程学报. 2009（7）：164-168.

［21］ 闫彦涛，邹志荣，李凯. 太阳能相变蓄热系统在温室加温中的应用［J］. 中国农业大学学报. 2016（5）：139-146.

［22］ 孙维拓，张义，杨其长，等. 温室主动蓄放热—热泵联合加温系统热力学分析［J］. 农业工程学报. 2014（14）：179-188.

［23］ 方慧，杨其长，张义. 基于热泵的日光温室浅层土壤水媒蓄放热装置试验［J］. 农业工程学报. 2012（20）：210-216.

［24］ 戴巧利. 主动式太阳能空气集—土壤蓄热温室增温系统的研究［D］. 镇江：江苏大学，2009.

温室茄子日参考蒸散量估算及评价研究

杨宜[1, 2]，陶虹蓉[1, 2]，李银坤[1, 3*]，郭文忠[1, 3]，李海平[2]，李灵芝[2]

[1. 北京农业智能装备技术研究中心，北京 100097；2. 山西农业大学园艺学院，太谷 030801；3. 农业部都市农业（华北）重点实验室，北京 100097]

摘要：评价修正前后Penman-Monteith（P-M）公式在温室内的适用性与精确性，以获得温室茄子各生育阶段日参考蒸散量的最佳估算方法。分别利用P-M公式与修正P-M公式估算了温室内秋茬茄子的逐日参考蒸散量，并以称重式蒸渗仪实测值为标准，对2种方法在各生育期的估算结果进行了验证。基于称重式蒸渗仪的温室内秋茬茄子的蒸散规律表现为苗期逐渐升高，在定植后41d（开花坐果期）达到峰值4.31mm后逐渐降低。P-M公式与修正P-M公式模拟的参考蒸散量与实测蒸散量变化规律基本一致，峰值均出现在开花坐果期，其中P-M公式估算值在58d（开花坐果期）达到峰值，与实测值相差2.26mm；修正P-M公式估算值在41d达到峰值，与实测值仅相差0.1mm。P-M公式估算值与实测值相比，整体表现为低估，估算值与实测值的平均偏差（*MBE*）均小于0；P-M公式在苗期估算值与实测值相关性较好，R^2为0.521 5（$P<0.01$），但二者之间差异较大，均方根误差（*RMSE*）和*MBE*分别为1.43mm·d^{-1}和−1.34mm·d^{-1}；在成熟期较接近实测值，*RMSE*和*MBE*分别为0.47mm·d^{-1}和−0.18mm·d^{-1}，但相关性不佳，R^2仅为0.221 9（$P<0.01$）。修正P-M公式在苗期估算值与实测值的相关性最好，R^2高达0.739 1（$P<0.01$），且二者之间差异较小，*RMSE*和*MBE*分别为0.44mm·d^{-1}和0.23mm·d^{-1}；其次为开花坐果期，估算值与实测值的相关系数R^2为0.507 4（$P<0.01$），*RMSE*和*MBE*分别为0.55mm·d^{-1}和0.26mm·d^{-1}。P-M公式在不同生育期的估算值与实测值的相关性较差且差异较大，不建议直接在温室中应用。修正P-M公式在各生育期的估算值与实测值较接近，在苗期的相关性最好，可用于估算温室内秋茬茄子各生育阶段的日参考蒸散量。

关键词：温室；称重式蒸渗仪；P-M公式；参考作物蒸散量（ET_0）

Estimation and Evaluation of Daily Evapotranspiration in Greenhouse Eggplant

Yang Yi[2], Tao Hongrong[2], Li Yinkun[1, 3*], Guo Wenzhong[1, 3], Li Haiping[2], Li Lingzhi[2]

[1. *Beijing Research Center of Intelligent Equipment for Agriculture*, *Beijing* 100097; 2. *College of Horticulture*, *Shanxi Agricultural University*, *Taigu*, *Shanxi* 030801; 3. *Key Laboratory of Urban agriculture* (*North China*), *Ministry of Agriculture*, *Beijing* 100097]

Abstract: The objective of this study was to evaluate the applicability and accuracy of the Penman-Monteith

基金项目：国家重点研发计划（2017YFD0201503）、国家自然科学基金（41501312）

作者简介：杨宜（1994—），女，河北张家口人，硕士研究生，研究方向为蔬菜栽培与生理。E-mail：Yangyi9401@163.com

通讯作者：李银坤（1982—），男，博士，高级工程师，研究方向为作物水肥高效利用。E-mail：lykunl218@163.com

（P-M）formula in greenhouses before and after correction，and to obtain the best estimation method of daily reference evapotranspiration at each growth stage of greenhouse eggplant. Daily reference evapotranspiration of autumn stubble Eggplant in greenhouse was estimated by using PM formula and modified PM formula respectively. The results of two methods at each growth period verification. The results showed that the law of evapotranspiration of autumn-stubble eggplants in greenhouse was gradually increased at seedling stage，and gradually decreased after reaching the peak value of 4.31 mm 41 days after flowering and fruit setting. The PM evapotranspiration and evapotranspiration simulated by PM formula and modified PM formula are basically the same，and the peak appears in the flowering and fruit setting period. The estimated value of PM formula reaches the peak at 58d（flowering and fruiting stage），which is 2.26 mm away from the measured value. Corrected PM formula estimates reached the peak in 41d，and the measured value difference of only 0.1 mm. Compared with the measured values，the PM formula shows an underestimation as a whole，and the average deviation（*MBE*）between the estimated and measured values is less than 0. PM formula has good correlation with the measured values at the seedling stage，R^2 is 0.521 5（$P<0.01$）. However，*RMSE* and *MBE* were 1.43 $mm \cdot d^{-1}$ and −1.34 $mm \cdot d^{-1}$，respectively. At maturity，the values of *RMSE* and *MBE* Respectively，0.47 $mm \cdot d^{-1}$ and −0.18$mm \cdot d^{-1}$，but the correlation was not good，R^2 was only 0.221 9（$P<0.01$）. The correlativity between estimated PM value and measured value was the best，R^2 was as high as 0.739 1（$P<0.01$），and the difference between them was small，*RMSE* and *MBE* were 0.44 $mm \cdot d^{-1}$ and 0.23 mm · The correlation coefficient R^2 between estimated and measured values was 0.507 4（$P<0.01$）. The *RMSE* and *MBE* values were 0.55 $mm \cdot d^{-1}$ and 0.26 $mm \cdot d^{-1}$ respectively. The correlation between P-M formulas and measured values at different growth stages is poor and different，which is not recommended for direct application in greenhouses. Corrected P-M formula in each growth period of the estimated value and the measured values are close，the correlation is best at the seedling stage，can be used to estimate the greenhouse autumn stubble eggplant during each stage of the reference evapotranspiration.

Key words：Greenhouse；Weighing Lysimeter；P-M Formula Reference；Evapotranspiration（ET_0）

1 引言

P-M公式是联合国粮农组织（FAO，Food and agriculture organization）提出的计算参考作物蒸散量（ET_0）的标准方法[1]，多应用于大田作物蒸散量的估算研究[2-3]。温室内环境郁闭，水热运移模式与大田有很大区别，研究表明在温室内直接应用P-M公式模拟ET_0有一定局限性，其估算值与实测值差异较大，而基于P-M的修正P-M公式在温室内应用效果更好。李振华等[4]通过对比修正前后P-M公式的逐时估算值，研究得出修正P-M公式在温室内稳定性更高，修正P-M公式在日尺度及月尺度下估算值变异系数均小于PM公式。陈新明等[5]研究得出，修正后P-M公式在温室内应用效果更好，修正前后P-M公式估算值之间差异较大，最大相对偏差为34.14%，绝对偏差达到1.38$mm \cdot d^{-1}$。称重式蒸渗仪（Lysimeter）作为一种获取实测蒸散量的技术手段，因其精度较高常被作为ET_0估算方法的评价标准[4-5]，而现有研究常以P-M公式估算值或水面蒸发量等为校核标准[6-7]，缺乏基于称重式蒸渗仪实测值的评价，且2种方法在各生育阶段的适用性尚不明确。因此，本研究针对修正前后P-M公式缺乏系统评价的问题，以秋茬茄子为试验作物，基于称重式蒸渗仪实测值验证修正前后P-M公式在各生育阶段的应用效果，旨在对温室灌溉预报及发展节水农业提供理论依据。

2 材料与方法

2.1 试验概况

试验于2017年8月8日在北京市农林科学院进行，供试玻璃温室长38m，宽11m，为南北走向。该地区位于东经116.29°，北纬39.94°，海拔56m，年平均总降水量为500～600mm，多年平均气温11.1℃，属于温带大陆性季风气候。土质为砂壤土，试验前0～20cm土壤理化性质为：土壤容重1.40g·cm^{-3}，田间体积持水量28.0%，有机质含量15.89g·kg^{-1}，全氮质量分数0.60g·kg^{-1}，速效钾含量0.15g·kg^{-1}。试验用称重式蒸渗仪的长宽高为：1m×0.6m×0.9m，每小时自动记录数据，分辨率为0.01mm水深。

供试茄子品种为“净茄黑宝”。定植时选取约3叶1心的茄子苗，移栽4株在称重式蒸渗仪上。保护行采用畦栽种植，畦宽0.75m，高0.1m。每畦栽种两行，株距为0.45m，行距0.4m。定植前底施商品有机肥30 000kg·hm^{-2}，定植后蒸渗仪灌定植水15mm，栽培畦灌定植水30mm，栽培畦为滴灌。温室内布置2个直径为20cm的蒸发皿，以每日18:00测定的水面蒸发量累积值乘以系数0.8作为灌水依据，灌水周期为7—10d。分别在定植后23d、42d和53d施用可溶性肥，每次15kgN/667m^2。

2.2 测定项目与方法

2.2.1 蒸散量

利用称重式蒸渗仪测定蒸散量，蒸渗仪每h记录一次土柱重量（分辨率0.01mm），蒸散量的计算原理如式（1）所示：

$$T \times A = (Wt-1-Wt)/\rho + I \tag{1}$$

式中，T为时间段内蒸散量，mm；A为蒸渗仪箱体表面积，0.6m^2；W_{t-1}、W_t为t−1时刻和t时刻蒸渗仪箱体内土、水的质量，g；ρ为水的密度，1g·cm^{-3}；I为时段内的灌水量，mm。

2.2.2 环境因子

利用布置在温室内的气象站对温度、湿度和太阳辐射等环境因子进行监测，采集间隔为1h；在温室中央放置2个直径为20cm的蒸发皿，取每日18:00记录的两个蒸发皿蒸发量平均值为温室水面蒸发量。

2.3 模型及评价指标

2.3.1 P-M公式

P-M公式是Monteith（1965）[8]在Penman（1948）[9]等人研究的基础上提出，结合了作物生理特征和空气动力学。1998年FAO发布第56号文件，推荐使用的P-M公式（FAO56），如式（1）所示：

$$ET_0 = \frac{0.408\Delta(R_n - G) + \gamma \frac{900}{T+273} u_2 (e_a - e_s)}{\lambda\left[\Delta + \gamma(1 + 0.34u_2)\right]} \tag{2}$$

ET_0为参考作物蒸散量（$mm \cdot d^{-1}$），λ为水汽化潜热，Δ为饱和水气压曲线斜率（$kPa \cdot ℃^{-1}$），e_a为饱和水汽压（kPa）e_s为实际水汽压（kPa），γ为干湿表常数（$kPa \cdot ℃^{-1}$），Rn为净辐射（$MJ \cdot m^{-2} \cdot d^{-1}$），$G$为土壤热通量（$MJ \cdot m^{-2} \cdot d^{-1}$），$G$=0，$T$为平均气温（℃），$u_2$为2m高度风速（$m \cdot s^{-1}$）。

由于温室中风速近似为零，将u_2=0代入式（2）后，温室内P-M公式如式（3）所示：

$$ET_0 = \frac{0.408\Delta(R_n - G)}{\lambda(\Delta + \gamma)} \quad (3)$$

2.3.2　修正P-M公式

由于式（3）中空气动力学项为零，但温室内仍存在热量输送，这与实际情况不符，因此式（3）能否直接应用于温室条件有待研究。陈新明等[5]以P-M公式为基础，引入冠层阻力R_a（70s/m）等因子，得到可在温室内应用的修正P-M公式，如式（4）所示，

$$ET_0 = \frac{(R_n - G) + \dfrac{\gamma 1694(e_a - e_s)}{T + 273}}{\lambda(\Delta + 1.64\gamma)} \quad (4)$$

2.3.3　模型评价方法与指标

以温室内蒸渗仪实测日值为标准，两种方法的估算值分别与实测值进行统计分析，并以均方根误差（*RMSE*）和平均偏差（*MBE*）为评价统计指标，评价模型与实测值差异程度。*RMSE*反映了两组数值间的平均偏离程度，反映了整体误差的情况；*RMSE*越小，两组数值越接近，模拟效果越好。*MBE*反应了模型估算的偏离程度，小于0为低估，大于0为高估；*MBE*越接近0，模拟效果越好。回归分析中的决定系数R^2代表了两组数值相关性的大小，一般来说R^2越接近1时相关性越好。具体公式分别为：

$$RMSE = \sqrt{\frac{\sum_{i=1}^{n}(P_i - O_i)^2}{n-1}} \quad (5)$$

$$MBE = \frac{\sum_{i=1}^{n}(P_i - O_i)}{n-1} \quad (6)$$

其中：Pi为模拟ET_0值（$mm \cdot d^{-1}$）；Oi为实测值（$mm \cdot d^{-1}$）；n为样本数；P和O分别为对应均值。

2.3.4　统计与方法

采用Microsoft2010进行试验数据的处理及相关图表制作，利用SPSS19.0进行统计分析。

3　结果与分析

3.1　估算值与实测值日变化分析

基于称重式蒸渗仪的温室内秋茬茄子的逐日蒸散量呈单峰曲线变化趋势，日蒸散量在苗期逐渐升高，开花坐果期（41d）达到需水高峰4.31mm·d^{-1}，在成熟期呈逐渐降低趋势。苗期、开花坐果期和成熟期的累积蒸散量分别为86.36mm、73.35mm和94.79mm。

与实测值相比，P-M公式的估算值变化幅度较小，在0.71～2.05mm·d^{-1}范围内波动，需水高峰期出现时间晚于实测值，在开花坐果期（58d）达到峰值2.05mm·d^{-1}，比实测值低估了2.26mm·d^{-1}。P-M公式在苗期的累积估算值相对接近于实测值，二者相差14.78mm。修正P-M公式估算值变化趋势与实测值基本一致，日蒸散量峰值出现时间与实测值相同，在开花坐果期（41d）达到峰值4.41mm·d^{-1}，仅与实测值相差0.1mm·d^{-1}；修正P-M公式在开花坐果期的累积估算值与实测值相比差异较小，二者仅相差6.22mm；其次为苗期，二者相差8.16mm（图1）。

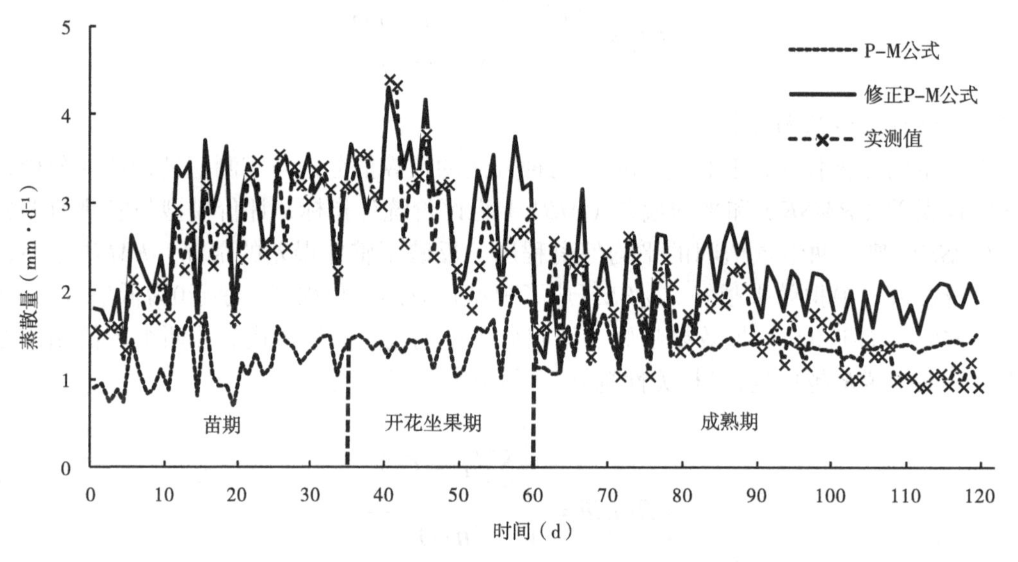

图1　不同方法估算值与实测值逐日动态变化

Figure 1　different methods of estimation and the measured daily dynamic changes

3.2　各生育期估算值与实测值相关性分析

图2为各生育阶段估算值与实测值的相关关系，除开花结果期的P-M公式估算值与实测值相关性不显著外（图2b）（$P>0.05$），其它生育阶段两种方法估算值与实测值均呈极显著正相关（$P<0.01$）。苗期P-M公式估算值与实测值的相关性较好，R^2为0.521 5（$P<0.01$）；在开花坐果期不均有显著相关性，R^2仅为0.007（$P>0.05$）。修正P-M公式估算值与实测值在苗期具有较好的相关性，R^2高达0.739 1（$P<0.01$），其次为开花坐果期，相关关系R^2为0.507 4（$P<0.01$）。

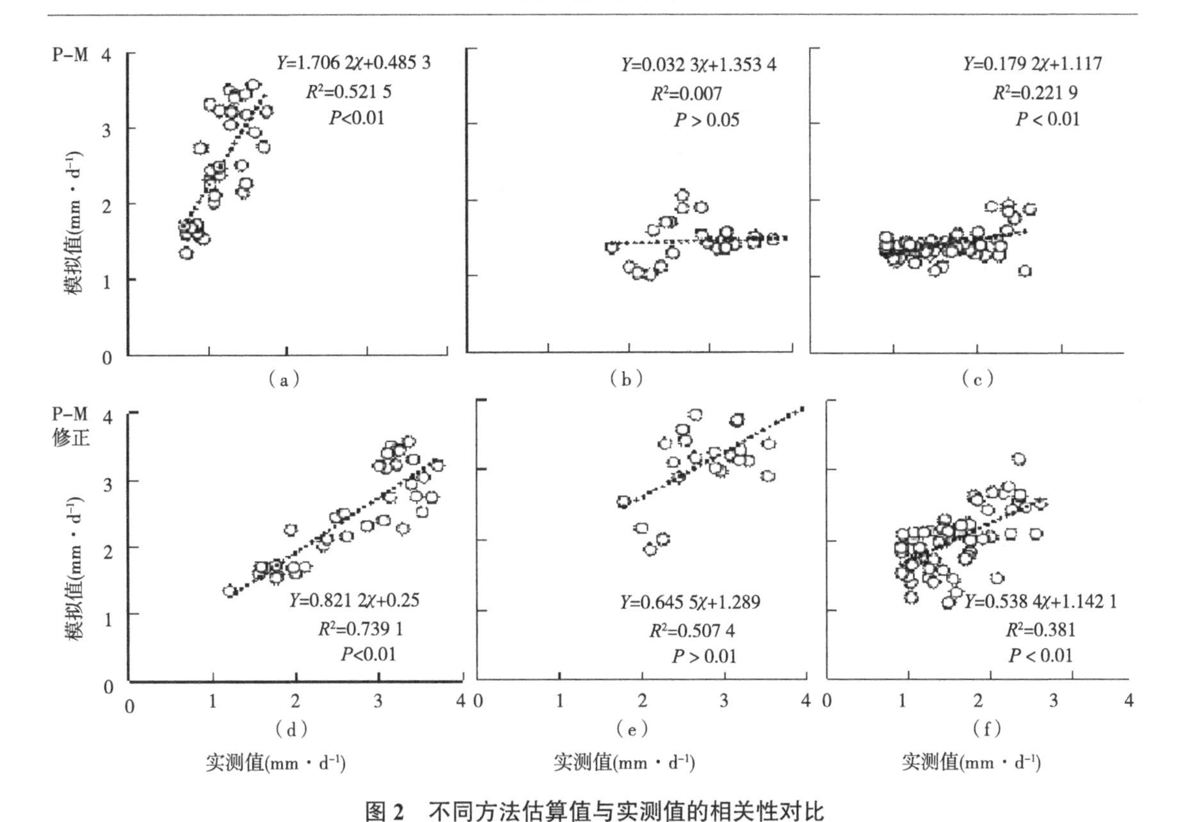

图 2　不同方法估算值与实测值的相关性对比

（a、b、c分别代表P-M公式在苗期、开花坐果期和成熟期与实测值的相关关系，d、e、f分别代表修正P-M公式在苗期、开花坐果期和成熟期与实测值的相关关系）

Figure 2　Comparison of different methods of estimation and the measured value of the correlation

（a，b and c represent the correlation between PM formula at the seedling stage，flowering and fruit setting and maturity，respectively，and d，e and f respectively represent the correlations between the corrected PM formula and the measured value at the seedling stage，flowering stage and maturity stage The correlation）

3.3　温室秋茬茄子估算值与实测值误差评估分析

下表为各生育期估算值与实测值的误差对比结果。P-M公式整体表现为低估，估算值与实测值的*MBE*均小于0。P-M公式估算值与实测值在苗期虽具有较好的相关性，R^2为0.521 5（$P<0.01$）（图2a），但估算值与实测值之间的差异较大，*RMSE*为1.43mm·d^{-1}，*MBE*为−1.34mm·d^{-1}。综合R^2、*RMSE*和*MBE*评价，各生育期P-M公式估算值与实测值的一致性排序表现为成熟期>苗期>开花坐果期。

各生育期修正P-M公式估算值较接近于实测值，*RMSE*均低于0.58mm·d^{-1}。修正P-M公式在苗期估算值与实测值的相关性较好，R^2高达0.739 1（$P<0.01$）（图2d），二者间差异也较小，*RMSE*为0.44mm·d^{-1}，*MBE*为0.23mm·d^{-1}；在成熟期修正P-M公式估算值与实测值相关性虽然相对低于其他生育期，R^2为0.381（$P<0.01$）（图2f），但二者之间的差异相对较小，*RMSE*为0.58mm·d^{-1}，*MBE*为0.41mm·d^{-1}。综合来看，各生育期修正P-M公式估

算值与实测值的一致性排序表现为苗期>开花坐果期>成熟期，如下表所示。

表 各生育期蒸散量实测值与两种方法估算值误差统计特征值

Table 1 Evapotranspiration measured during each growth period and two methods of estimation error statistical characteristics

生育期	P-M公式（$mm·d^{-1}$）		修正P-M公式（$mm·d^{-1}$）	
	RMSE	*MBE*	*RMSE*	*MBE*
苗期	1.43	−1.34	0.44	0.23
开花坐果期	1.67	−1.54	0.55	0.26
成熟期	0.47	−0.18	0.58	0.41
全生育期	1.12	−0.79	0.53	0.33

4 结论与讨论

作物蒸散量又称需水量，在农田水分消耗中占有很大比重，精确获取作物蒸散量是建立合理灌溉制度的前提，对于提高水分利用率，发展节水农业等具有重要意义。本试验基于称重式蒸渗仪实测值，对比研究了温室内修正前后P-M公式在温室秋茬茄子上的应用效果，结果表明P-M公式在各生育期的估算值与实测值相关性较差且差异较大，不建议在温室内直接使用。修正P-M公式在各生育期的估算值与实测值较接近，其中在苗期的相关性最好；修正P-M公式在温室条件下计算精度较高，应用效果更佳，可用于估算温室内秋茬茄子日参考蒸散量。

本试验分析得出P-M公式估算值与实测值在苗期虽具有较好的相关性，R^2为0.521 5（$P<0.01$），但二者之间的差异较大，*RMSE*为1.43mm·d^{-1}，*MBE*为−1.34mm·d^{-1}；修正P-M公式在苗期相关性最好，R^2高达0.739 1（$P<0.01$），二者间差异也较小，*RMSE*为0.44mm·d^{-1}，*MBE*为0.23mm·d^{-1}。王建等[6]对比了修正前后P-M公式估算值与水面蒸发量的相关性得出，P-M公式估算值与水面蒸发量相关性相对较低，R^2仅为0.643；修正P-M公式相关性较好，R^2为0.855，本试验与王建等[6]研究结论相似。造成P-M公式估算值与实际情况差异较大的原因，可能是温室内相对封闭，空气流速较慢，在计算时一般视风速为0，导致P-M公式仅有辐射项而忽略了土壤热通量G，造成了P-M公式缺少了空气动力项，进而影响了最终的估算结果；而修正P-M公式通过引入冠层阻力R_a与冠层高度h_c，可以有效避免空气动力项带来的误差，在温室内应用效果较好。另一方面，相对于其它生育期来说，修正P-M公式在成熟期的估算值与实测值差异虽较小，但相关性相对较低，R^2为0.381，可能是由于温室内秋茬茄子在成熟期的昼夜温差较大，日平均温度无法真实反映当天的热量平衡状况，可能会对估算结果造成一定影响。另外，本试验中净辐射R_n由气象站自动测定，成熟期开始后温度降低，温室内覆盖的保温被材料是否会对长波辐射净支出产生影响，且在不同典型天气（晴天与阴天）下，对修正前后P-M公式计算结果的精度是否有影响，有待于进一步研究。

参考文献

[1] 康绍忠，邵明安. 作物蒸发蒸腾量的计算方法研究［J］.中国科学院水利部西北水土保持研究所集刊（SPAC中水分运行与模拟研究专集），1991（1）：66-74.

[2] 刘晓英，李玉中，钟秀丽，等. 基于称重式蒸渗仪实测日值评价16种参考作物蒸散量（ET_0）模型［J］.中国农业气象，2017，38（5）：278-291.

[3] 曹金峰，李玉中，刘晓英，等. 四种参考作物蒸散量综合法的比较［J］.中国农业气象，2015，36（4）：428-436.

[4] 李振华，李毅，蔡焕杰，等. 温室逐时参考作物腾发量的估算［J］.灌溉排水学报，2013，32（4）：71-75.

[5] 陈新明，蔡焕杰，李红星，等. 温室大棚内作物蒸发蒸腾量计算［J］.应用生态学报，2007，（2）：317-321.［2017-09-21］.DOI：10.13287/j.1001-9332.2007.0053

[6] 王健，蔡焕杰，李红星，等. 日光温室作物蒸发蒸腾量的计算方法研究及其评价［J］.灌溉排水学报，2006（6）：11-14.

[7] 强小嫚，蔡焕杰，孙景生，等. 陕西关中地区ETO计算公式的适用性评价［J］.农业工程学报，2012，28（20）：121-127.［2017-09-21］.

[8] Monteith J L.Evaporation and the environment［J］.Symposium of the Society of Exploratory Biology，1965，19：205-234

[9] Penman H L. Natural evaporation from open water，bare soil and grass［J］.Proc of the Royal Soc（Series A），1948，193：120-145.

一种日光温室前屋面热贯流系数的估算方法

杨艳红[1]，李亚灵，温祥珍*

（山西农业大学　园艺学院，太谷　030801）

摘要： 针对目前日光温室前屋面覆盖材料种类繁多、性能各异，且热贯流系数易受工作环境和测试条件的影响这一问题。本试验根据能量平衡原理和日光温室夜间能量来源单一，气温变化对能量微量变化的敏感性确立了一种复合保温覆盖层热贯流系数的估算方法。试验温室复合保温覆盖层“塑料膜+保温被（3层杂棉+加强绒构成）”春季热贯流系数的变化范围在8～12KJ·m^{-2}·K^{-1}·h^{-1}之间，平均约为11.01KJ·m^{-2}·K^{-1}·h^{-1}。通过夜间不覆盖保温被来计算透明塑料薄膜的热贯流系数，进行验证试验，与现有文献所给的数值接近，表明这种估算温室前屋面热贯流系数的方法是可行的。这种方法可用于对不同外保温材料热贯流系数的测定与比较。

关键词： 日光温室；前屋面；热贯流系数；估算方法

An Estimation Method for the Thermal Flow Coefficient in the Front Roof of a Solar greenhouse

Yang Yanhong[1], Li Yaling, Wen Xiangzhen*

（*College of Horticulture*, *Shanxi Agricultural University*, *Taigu*　030801）

Abstract: In view of the various kinds and different performance of the cover materials in the front roof of a solar greenhouse, the thermal flow coefficient is easily affected by working environment and testing conditions. In this experiment, the thermal flow coefficient of composite insulation covering is estimated based on the principle of energy balance and energy source at night, and the sensitivity of air temperature to the energy change. The thermal flow coefficient of a composite thermal insulation covering “plastic film + thermal insulation quilt（3 layers of composite cotton + reinforced reinforcement）” of a solar greenhouse in the spring varies from 8～12KJ·m^{-2}·K^{-1}·h^{-1}, and the average value is about 11.01KJ·m^{-2}·K^{-1}·h^{-1}. A similar calculation method is used to verify the thermal flow coefficient of the transparent plastic film, which is close to the value given in the literature. It is proved that this method is feasible to estimate the thermal flow coefficient in the front roof of the solar greenhouse. This method can be used to determine and compare the thermal flow coefficient of different external thermal insulation materials.

Key words: A solar greenhouse; The front roof; The thermal flow coefficient; Estimation method

---

作者简介：杨艳红（1991—），女（汉），山西朔州人，研究生，硕士，研究方向：设施园艺

通讯作者：温祥珍（1960—），男（汉），教授，博士生导师，E-mail：wenxiangzhen2009@hotmail.com

基金项目：国家自然科学基金重点项目（61233006）；山西省煤基重点科技攻关项目（FT201402-05）

1 引言

前屋面是日光温室散热的主体，外保温是其节能的重要方面。随着科技的进步，外保温材料从最初的草帘、纸被、毛毡等发展到可实现自动卷放、轻简的保温被[1, 2]，但是目前生产厂家多，制作材料、工艺也有很大差异，导致其保温效果难以评价和比较，给用户带来选择的困惑，也会直接影响其温室的保温性，甚至是安全生产性能。热贯流系数被认为是衡量保温材料保温性能的重要指标[3]。张俊芳等[4, 5, 6]基于热箱法原理研制开发了园艺设施覆盖材料传热系数测试台，为温室覆盖材料传热测试工作的规范化、科学化、标准化奠定了基础，而目前关于温室覆盖材料热贯流系数的测定或计算方法不是很多。本文就是基于夜间日光温室热源单一，空气热容较小，对热量变化敏感这一实际，根据能量平衡原理，来分析前屋面覆盖材料的热贯流系数。

2 材料与方法

2.1 试验地概况

试验地位于山西省晋中市太谷县山西农业大学设施园艺示范中心（37° 42′ N，112° 55′ E），年平均气温9.9℃，无霜期176d，降水量462.9mm[7]。

2.2 试验温室概况

温室坐北朝南，东西延长，为砖墙结构，后屋面由“农膜+保温被”构成。农膜是覆盖2年的聚氯乙烯薄膜（河南珲春农膜），保温被由3层杂棉和加强绒构成，厚0.08m左右。温室结构及热性能参数分别如表1和表2[8, 9, 10]所示。

试验时间为2015年2月24日至5月5日，以3月26日为界，在这之前的温室夜间覆盖保温被，称之为试验前期，之后的夜间不覆盖保温被，称之为试验后期。

试验前期温室揭盖保温被的时间分别在9:00、16:00左右，不进行通风，室内种植红苋菜、油菜等；试验后期温室不进行保温被的揭放，白天以顶通风方式为主，附加前通风方式，夜间不进行通风，室内种植番茄，处于营养生长期。

表1 温室基本结构参数[8]

Table 1 The basic structural parameters of a greenhouse[8]

项目 Project	总长 Length	总跨度 Total Span	脊高 Ridge-Height	墙高 Wall Height	墙厚 Wall Thickness	后屋面面积Rear Roof Area	前屋面面积Front Roof Area	土地面积 Ground Area	墙面积 Wall Area	温室容积 Greenhouse Volume
单位unit	m	m	m	m	m	m^2	m^2	m^2	m^2	m^3
数量 amount	43.0	9.75	4.5	3.0	0.5	107.5	430.0	397.75	129.0	1 591.0

表 2　温室结构部分热性能参数[9, 10]

Table 2　Thermal performance of the structure parameters in a solar greenhouse[9, 10]

材料 Material	容重 Bulk Density	热导率 Thermal Conductivity	比热 Specific Heat	蓄热系数 Thermal Storage Coefficient
	$Kg \cdot m^{-3}$	$W \cdot m^{-1} \cdot K^{-1}$	$J \cdot kg^{-1} \cdot K^{-1}$	$W \cdot m^{-2} \cdot K^{-1}$
红砖墙Red Brick Wall	1 750	0.78	960	9.96
土壤Soil	1 500	0.84	1 400	10.36
空气Air	1.2	0.023	0.29	0.05

2.3　测点安排与测量仪器

2.3.1　墙体温度测量

为避免东西侧墙对北墙温度变化的影响，试验中的测点位于北墙中部距地面1.5m高处，从墙体内表面开始垂直于外侧，每5cm深放置一个传感器探头，分别为距墙体表面5cm、10cm……25cm、30cm。所用仪器为多点温度传感器，出自河北省邯郸市益盟电子有限公司，每30min自动记录一组温度数据，为瞬时数值。

2.3.2　地面温度测量

在温室地面中央，即距温室东西侧墙均为21.0m，距温室后墙4.5m，由浅及深地垂直布置测点，分别为距地面表面5cm、10cm……25cm、30cm。所用仪器同墙体温度测量仪器。

2.3.3　气温、湿度测量

在室内中央距地面1.5m高处，分别距前底脚3.0m、6.0m处，各悬挂1个欧宝；室外距前底脚1.0m处，高1.5m处悬挂1个欧宝。采用欧宝（HOBO）自动记录仪记录，可同时记录温度和湿度。室内气温和湿度的数值为两个测点的平均值。

2.4　试验方法

2.4.1　温室热平衡基本假设

为了简化考虑因素，在热平衡模型建立之前，先对日光温室热环境进行基本假设：①温室墙体各层（6层）、土壤各层（6层）传热视作均匀；②忽略墙体的垂直纵向传热和土壤的水平横向传热；③忽略夜间墙体、后屋面与前屋面上的水汽凝结所放出的热量；④由于室内作物的吸热、蒸发散热以及其他生理活动对环境的影响很小，可忽略不计；⑤相对于温室内、外环境传热的变化速率，保温被材料的热惰性较小，其中的传热过程视为稳态传热[11, 12]。在此基础上，日光温室夜间（覆盖保温被和不覆盖保温被时）的热平衡可用下述关系表述：

夜间覆盖保温被时，温室热平衡：$Q_{c1}+Q_{r}=Q_{w}+Q_{s}+Q_{a}$　　（1）

夜间不覆盖保温被时，温室热平衡：$Q_{c2}+Q_{r}=Q_{w}+Q_{s}+Q_{a}$　　（2）

其中：Q_{c1}代表温室夜间覆盖保温被时，复合保温覆盖层的热量变化，MJ；Q_{c2}代表温室夜间不覆盖保温被时，透明塑料薄膜的热量变化，MJ；Q_{r}代表温室后屋面的热量变化，MJ；Q_{w}代表温室北墙的热量变化，MJ；Q_{s}代表温室地面的热量变化，MJ；Q_{a}代表温室内空气质能的变化，MJ。

2.4.2 温室前、后屋面贯流传热量的计算

温室前屋面夜间覆盖保温被时，其下表面与固定透明覆盖材料相贴合，因此在研究过程中将二者视为一整体，称其为“复合保温覆盖层”，前、后屋面贯流传热量的计算表达式可表述为[13]：

复合保温覆盖层贯流传热量：$Q_{c+r} = A_{c+r} \cdot K_{c+r} \cdot (T_{in} - T_{out})$ （3）

透明前屋面贯流传热量：$Q_c = A_c \cdot K_c \cdot (T_{in} - T_{out})$ （4）

后屋面贯流传热量：$Q_r = A_r \cdot K_r \cdot (T_{in} - T_{out})$ （5）

其中：Q代表温室覆盖物的贯流传热量，MJ；A代表覆盖物的面积，m^2；K代表覆盖物的热贯流系数，$MJ \cdot m^{-2} \cdot K^{-1} \cdot h^{-1}$。下标c+r、c、r分别代表复合保温覆盖层、透明塑料膜、后屋面指数。T_{in}、T_{out}分别代表温室室内、室外气温，K。

数据分析：文中的数据处理主要采用Excell 2011软件进行分析。

3 结果与分析

3.1 试验期间不同时刻室内外温差的逐日变化

温室前屋面热贯流系数的计算求解与室内外温差密切相关，呈反比关系。由图1可知，试验期间，夜间不同时刻室内外温差变化没有明显的规律，但是受季节和试验处理的影响，2、3月份的室内外温差变化幅度明显高于4月份的。试验前期（2月24日至3月26日）的室内外温差偏大，变化范围在5～22℃，尤其是在3月3日至3月7日的室内外温差达20℃左右，这是因为这几天天气较晴朗，室外平均最大太阳辐射强度达450W·m^{-2}，而夜间的气温在白天能量较多的积累基础下不至于降得太低；试验后期（3月27日至5月5日）的室内外温差波动较稳定，逐日变化范围几乎在2～10℃。

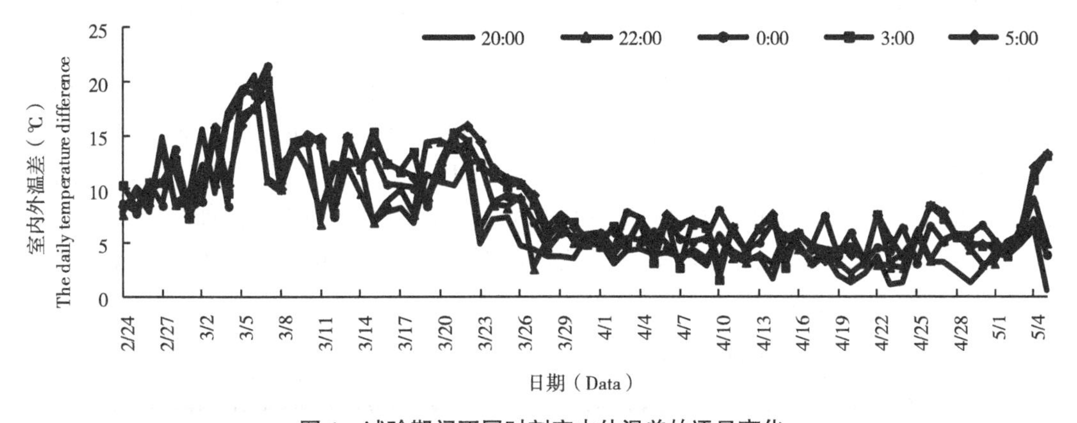

图 1 试验期间不同时刻室内外温差的逐日变化

Figure 1 The daily temperature difference variation between indoor and outdoor at different times during the test

注：图中数据点为5个不同时间点下室内外温差的逐日变化（2月24日—5月5日）。

Note: The data points are the daily temperature difference variation between the indoor and outdoor at 5 times in the figure（February 24th—May 5th）.

3.2 不同时刻温室墙体、地面的热量变化

杨艳红[8]在先前的试验中证实了墙体（土质或砖质）的热交换部位主要位于内表面到0.3m深处，热交换量是递减的，且土壤的耕作层也主要在0.3m深以上部位，为精确计算夜间它们对温室热量的贡献，将墙体、土壤均离散成6层（即每层厚度为0.05m），1d按0.5h划分成48个时段，每一块墙体或土壤的传热过程可认为是一维稳态导热，采用数值解法来计算求解，墙体或土壤的热流密度q_j和传热量q_k可用以下计算式[14]表达：

$$q_j = \rho_{1,2} \cdot C_{1,2} \cdot \frac{\left(t_{j+1} - t_j\right)}{\Delta t} \tag{6}$$

$$q_k = \sum_{j=1}^{n} q_j \cdot V \cdot \Delta t = \sum_{j=1}^{n} \rho_{1,2} \cdot C_{1,2} \cdot V \cdot \left(t_{j+1} - t_j\right) \qquad n=1，2，3\cdots\cdots47，48 \tag{7}$$

$$Q_{总} = \sum_{k=1} q_k \qquad k=1，2，3，4，5，6 \tag{8}$$

其中：q_j为墙体或土壤的热流密度，$MJ \cdot m^{-3} \cdot h^{-1}$；$_k$为某层墙体或土壤的热量变化，J；$_{总}$为墙体或土壤的总热量变化，J。$\rho_{1,2}$代表砖墙或土壤的容重，$kg \cdot m^{-3}$；$_{1,2}$代表砖墙或土壤的比热容，$KJ \cdot kg^{-1} \cdot K^{-1}$；代表砖墙或土壤的体积，$m^3$；代表时间间隔，；t代表温度，K；为各时刻的温度采集时间顺序（=1，为0:00时的采集值，=2，为0:30时的采集值，以此类推）；为墙体或土壤（k=6）的块数。

3.3 不同时刻温室内空气质能的变化[10]

$$Q_a = C_a \cdot m \cdot \left(T_{in} - T_{out}\right) = \rho_a \cdot C_a \cdot V \cdot \left(T_{in} - T_{out}\right) \tag{9}$$

其中：Q_a为温室内空气质能的变化，MJ；ρ_a代表空气的容重，$kg \cdot m^{-3}$；C_a代表空气的比热容，$KJ \cdot kg^{-1} \cdot K^{-1}$；代表温室体积，$m^3$。

夜间日光温室主要依靠维护结构储存的热量来维持室内气温，其放热量的多少对研究温室的保温性至关重要，通过上述公式计算出不同时刻各部位对温室的热贡献。图2-a、图2-b、图2-c、图2-d、图2-e分别代表温室墙体、地面、空气这三部分能量在20:00、22:00、0:00、3:00、5:00的逐日变化量，正负值是相对于温室的热贡献率而言的，负值代表其部分处于放热状态，正值代表其部分处于吸热状态。从图2可以看出，试验期间不同时刻日光温室墙体、地面和空气的能量变化趋势一致，均表现为空气的能量变化最小，几乎稳定在y=0MJ这条直线上，土壤总能量变化显著高于墙体，这是因为土壤的含水量高，其能量变化幅度也高于墙体。

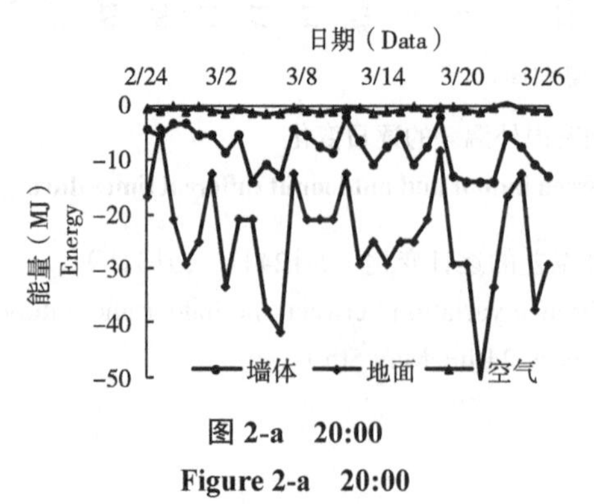

图 2-a　20:00

Figure 2-a　20:00

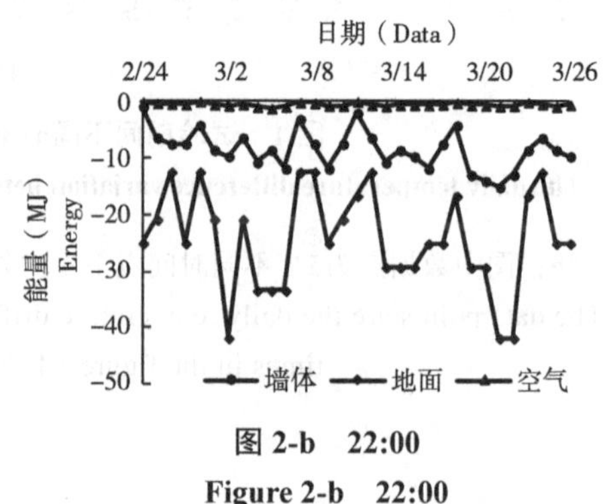

图 2-b　22:00

Figure 2-b　22:00

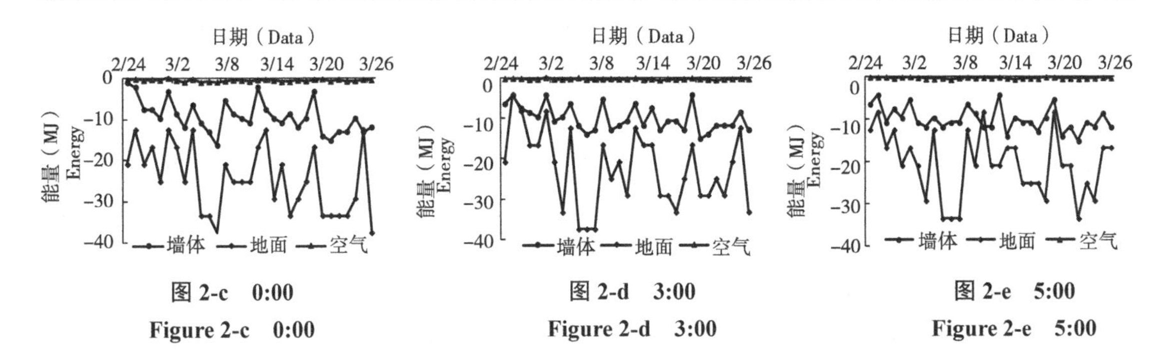

图 2-c　0:00

Figure 2-c　0:00

图 2-d　3:00

Figure 2-d　3:00

图 2-e　5:00

Figure 2-e　5:00

注：以上5个图中的数据点均代表不同时刻墙体、地面、空气的逐日能量变化。

Note：The data points represent the daily energy changes of north wall，ground and air at different times in the above 5 graphs.

3.4　不同时刻复合保温覆盖层热贯流系数的逐日变化

墙体、土壤向室内释放的热量通过屋面散出，室内温度的变化需要的能量很小，变化很敏感，其升降反映着能量的微小变化（图3）。在密闭的条件下，日光温室屋面在夜间的散热量可通过保温被复合体的热贯流系数估算出来。即联立方程式（1）、（3）、（6）、（7）、（8）和（9），可得到公式（10），为复合保温覆盖层热贯流系数的计算方法；同样在夜间不覆盖保温被的情况下，可以获得透明覆盖材料热贯流系数的计算方法，即联立方程式（2）、（4）、（5）、（6）、（7）、（8）和（9），可得到公式（11）。

$$K_{c+r} = \frac{Q_w + Q_s + Q_a}{A_{c+r} \cdot (T_{in} - T_{out})} \tag{10}$$

$$K_c = \frac{Q_w + Q_s + Q_a - Q_r}{A_c \cdot (T_{in} - T_{out})} \tag{11}$$

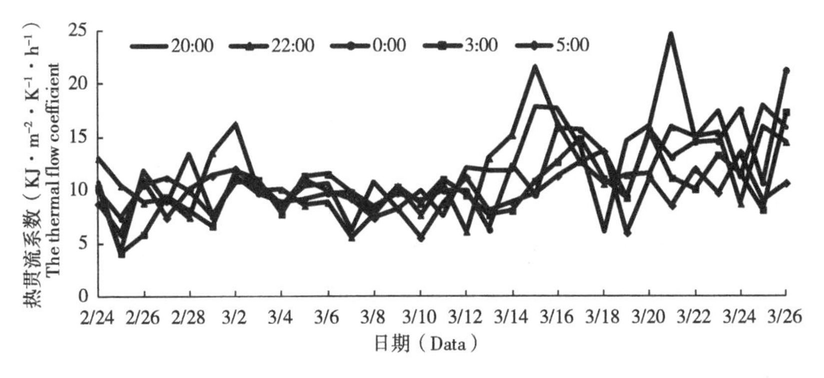

图 3　不同时刻复合保温覆盖层热贯流系数的逐日变化

Figure 3　The thermal flow coefficient of the composite insulating cover at different times

注：图中的数据点是根据公式（10）计算求得2月24日—3月26日（共31d）的热贯流系数逐日数值。

Note：The data points are the daily thermal flow coefficient calculated according to the formula（10）in the diagram.（February 24th—March 26th，total 31d）

用公式（10）计算出温室前屋面为复合保温覆盖层在不同时刻热贯流系数的逐日变化，见图4。由图4可知，5个不同时间点下热贯流系数的逐日变化规律不明显，变化范围在4.1～24.8KJ·m^{-2}·K^{-1}·h^{-1}之间，其中2月24日至3月12日的热贯流系数波动幅度较小，变化范围为4.1～16.0KJ·m^{-2}·K^{-1}·h^{-1}，平均为9.42KJ·m^{-2}·K^{-1}·h^{-1}；3月13日至3月26日的热贯流系数相对偏大，变化范围为6～24.8KJ·m^{-2}·K^{-1}·h^{-1}，平均为12.94KJ·m^{-2}·K^{-1}·h^{-1}，分析认为热贯流系数的这种变化受外界气象环境变化的影响较大。

3.5 前屋面热贯流系数的验证

根据前屋面热贯流系数的逐日分布，将其划分为不同的范围来分析不同时间点下出现数据的个数，进而来说明计算结果的准确性，再用类似的方法计算出透明塑料薄膜的热贯流系数以及在不同取值范围下的分布，来说明通过温室热平衡理论计算出的热贯流系数是可靠的，如表3所示。

表 3 前屋面热贯流系数在不同取值范围下的分布

Table 3 The distribution of thermal flow coefficient in front roof under different range

时刻 Time	复合保温覆盖层热贯流系数在不同取值范围下的分布/KJ·m^{-2}·K^{-1}·h^{-1} Distribution of thermal flow coefficient of composite insulating cover under different range				透明塑料薄膜热贯流系数在不同取值范围下的分布/KJ·m^{-2}·K^{-1}·h^{-1} Distribution of thermal flow coefficient of transparent plastic film under different range		
	4～8	8～12	12～16	16～25	8～15	15～35	35～45
20:00	5	14	6	6			
22:00	5	14	9	3	4	26	10
0:00	3	18	8	2	5	32	3
3:00	6	20	4	1	2	27	11
5:00	5	23	3	0			

注：复合保温覆盖层和透明塑料薄膜热贯流系数的总数据分别有30和40个。

Note：The total data of thermal flow coefficient of composite insulating covering layer and transparent plastic film are 30 and 40，respectively.

由表3可知，20:00、22:00、0:00、3:00、5:00时间点下，复合保温覆盖层的热贯流系数主要分布在8～12KJ·m^{-2}·K^{-1}·h^{-1}范围内，分别有14、14、18、20、23个数据，占总数据个数的45%、45%、58%、65%、74%；透明塑料薄膜的热贯流系数主要分布在15～35KJ·m^{-2}·K^{-1}·h^{-1}范围内，分别有26、32、27个数据，占总数据个数的65%、80%、68%，其他取值范围下的占比较小。

4 讨论与结论

4.1 讨论

日光温室前屋面热贯流系数与其屋面组成物质（材料、厚度）的导热率、对流传热率（室内外温差、温室密封性）和辐射传热率（材料）有关外，还受室外风速大小的影响[13]。虽然本文利用夜间温室传热理论计算所得的复合保温覆盖层热贯流系数比日本设施园艺协会产生的数据（复合材料保温被热贯流系数为7.56～8.64KJ·m^{-2}·K^{-1}·h^{-1}）稍高一点，但经分析认为与保温被已使用多年，密封性差、散热快、保温效果差有关；刘晨霞等[15, 16]分别分析了保温被表面的红外反射率、材料厚度、导热系数、室外风速大小对传热系数的影响，在不同条件下，传热系数的变化范围为1.8 ~ 11.88KJ·m^{-2}·K^{-1}·h^{-1}，这虽没有直接反映出热贯流系数的分布大小，但可以为热贯流系数的大小提供参考。通过对夜间塑料薄膜热贯流系数的计算，证实其数值与已知的热贯流系数23.01KJ·m^{-2}·K^{-1}·h^{-1}很接近[13]，进一步证明了该方法的可行性。

4.2 结论

本试验根据能量平衡原理和日光温室夜间能量来源单一，气温变化对能量微量变化的敏感性确立了一种复合保温覆盖层热贯流系数的估算方法。即在冬季不加热的密闭日光温室内，夜间的热量来源相对简单，通过测定室内、外气温和墙体、土壤各层次的温度，能够获得不同外覆盖材料下，日光温室前屋面的热贯流系数。

试验温室复合保温覆盖层“塑料膜+保温被（3层杂棉+加强绒构成）”春季热贯流系数的变化范围在8 ~ 12KJ·m^{-2}·K^{-1}·h^{-1}之间，平均约为11.01KJ·m^{-2}·K^{-1}·h^{-1}。通过夜间不覆盖保温被来计算透明塑料薄膜的热贯流系数，进行验证试验，与现有文献所给的数值接近，表明这种估算温室前屋面热贯流系数的方法是可行的。

这种方法可用于对不同外保温材料贯流散热系数的测定与比较。

参考文献

[1] 王云冰，邹志荣，杨建军，等. 高效保温材料在日光温室后屋面中的应用研究［J］. 西北农林科技大学学报（自然科学版），2010-1，38（1）：173-180.

[2] 张杰，温祥珍，王停停，等. 后屋顶结构材料对日光温室保温性的影响［J］. 新疆农业科学，2014，51（6）：1171-1176.

[3] 马承伟，张俊芳，覃密道，等. 基于覆盖层能量平衡法的园艺设施覆盖材料传热系数理论解析与验证［J］. 农业工程学报，2006-4，22（4）：1-5.

[4] 张俊芳，马承伟，覃密道，等. 温室覆盖材料传热系数测试台的研究开发［J］. 农业工程学报，2005-11，21（11）：141-145.

[5] Papadakis G，Briassoulis D，Mugnozza G S，et al. Radiometric and thermal properties and testing methods for greenhouse covering materials［J］. J Agric Engng Res，2000，77（1）：7-38.

[6] Feuilloley P，Issanchou G. Greenhouse covering materials measurement and modeling of thermal properties using the hot box method，and condensation effects［J］. J Agric Engng Res，1996，65：129-142.

[7] 百度百科——山西太谷. https：//baike.so.com/doc/6778615-6994713.html.

[8] 杨艳红，李亚灵，马宇婧，等. 日光温室北侧墙体内部冬春季的温度日较差变化分析［J］. 山西农业大学学报（自然科学版），2017，37（8）：594-599.

[9] 马承伟，卜云龙，籍秀红，等. 日光温室墙体夜间放热量计算与保温蓄热性评价方法的研究［J］. 上海交通大学学报（农业科学版），2008，26（5）：411-415.

[10] 李天来. 日光温室蔬菜栽培理论与实践［M］. 北京：中国农业出版社，2013.

[11] 陈青云，汪政富. 节能型日光温室热环境的动态模拟［J］. 中国农业大学学报，1996，1（1）：67-72.

[12] 徐克生，王琦，王述洋，等. 日光温室的热平衡计算［J］. 林业机械与木工设计，2004，7（32）：24-32.

[13] 李式军，郭世荣. 设施园艺学第二版［M］. 北京：中国农业出版社，2012.

[14] 史宇亮，王秀峰，魏珉，等. 日光温室土墙体温度变化及蓄热放热特点［J］. 农业工程学报，2016，32（22）：214-221.

[15] 刘晨霞，马承伟，王平智，等. 日光温室保温被保温性能影响因素的分析［J］. 农业工程学报，2015，31（20）：186-193.

[16] 刘晨霞，马承伟，王平智，等. 日光温室保温被传热的理论解析及验证［J］. 农业工程学报，2015，31（2）：170-176.

日光温室燃池辅助加温系统应用试验研究

于威，刘文合，徐占洋
（沈阳农业大学水利学院，沈阳 110866）

摘要：根据前期理论研究，在日光温室内土壤表面下，设计建造了高1.3m，半径1.8m的圆柱形燃池加温系统，用于东北地区日光温室越冬生产的辅助加温。试验测试了燃池周围空间3个维度方向的温度变化，并对试验结果进行了分析：土壤的最高温度出现在燃池顶板上表面，平均温度为38.28℃，随着与燃池距离的增加，温度逐渐降低，最低温度为13.2℃，无燃池区域土壤温度为8.14℃。结果表面燃池加温周围土壤温度场的分布情况，更加明确了圆柱型燃池对土壤加温的效果、加热范围，为圆柱形燃池在日光温室加温应用提供理论参考。
关键词：日光温室；燃池；加温；试验研究

Experimental Research on Fire Pit Warming System to Increase Temperature in Solar Greenhouse

Yu wei，Liu wenhe，Xu Zhanyang，
（*College of Water Conservancy*，*Shenyang Agricultural University*，*Shenyang* 110866）

Abstract：According to previous theoretical research，a cylindrical combustion pool heating system with high 1.3m and radius 1.8m has been designed and built under the soil surface in solar greenhouse，which is used for auxiliary heating of winter greenhouse in Northeast China. Test the temperature around the fire pit three spatial dimensions and direction，and analyses the test results: the highest temperature of the soil in burning pool on the roof surface，the average temperature of 38.28 degrees，with increasing distance of the fuel tank，the temperature gradually decreased，the lowest temperature is 13.2 DEG C，no burning temperature of soil pool area is 8.14 C. Results the distribution of soil temperature field around the surface burning pool was more clear，and the effect and the heating range of the cylindrical combustion pool on the soil heating were more clear，providing a theoretical reference for the application of cylindrical combustion pool in the greenhouse heating.
Key words：Solar Greenhouse；Fire Pit；Increase Temperature；Experimental research

燃池是利用阴燃原理，将燃料放入池中点燃后可自行发生反应，且持续时间长，阴燃过程稳定，放热均匀，可以有效地提高日光温室地温和气温的加温方式（王铁良，2002）。燃池运行过程中使用的燃料一般为农业废弃物，像秸秆、稻壳、锯末等，因此，利用燃池替换常规能源加温，在可再生能源利用方面具有十分重要的意义。

燃池最早是北方地区冬季采暖的一种方法，一般建造在室内地下，其构造主要由包括池体、顶部散热板、进出料口、通风道、烟囱和一些调温设施组成。能够低速缓慢地

释放能量（Ohlemiller T. J. and et al.，2002）。大连理工大学测试了燃池表面温度一般为35～40℃，燃池的热效率为15.4%。根据现有农村建筑集成燃池供暖时的问题，从造价和用户舒适度等方面提出了燃池供暖的优化方案，也验证了燃池供暖的可行性。日光温室的作物耕作于地面土壤，受地温影响很大，适宜的地温对农作物的种植及生长具有重要作用。燃池直接将热量传递给土壤，利用土壤蓄热和散热，同时也可以提高气温。

目前的燃池大多根据人为操作方便来设计，没有统一的规范，沈阳农业大学在日光温室内建造了长方形燃池，并对布设在土壤中的测点进行温度数据采集，得到长方形燃池对土壤加温后的温度变化规律。模拟了同体积条件下几种不同形状燃池的加温效果，得到了圆柱形燃池加温效果最优，因此，建造了圆柱形燃池并对其土壤温度分布规律及加温效果进行了试验研究。

1 试验方法

本试验在沈阳农业大学科学实验基地45号日光温室内进行。温室总长60m，净跨度8m，脊高3.5m，土壤宽度为7m，人行道宽度为1m。

新建燃池如图1所示，燃池的池体形状为圆形，体积130m^3，顶板面积10m^2，池深1.3m。日光温室外设有1.5m×1.5m的进出料口，深度为1.8m，燃池顶板上覆土厚400mm。

图1 日光温室燃池施工情景

Figure 1 The scene of construction of round fire-pit in solar greenhouse

试验时间为2013年1月9日至2014年2月28日，试验仪器为JTNT-C型地温采集仪和JTR11A型室内环境监测仪，对32个测点进行测量，数据采集时间间隔为10min，测试对象包括地温、室内的空气流速、温度以及太阳照度、辐射值等。以燃池中心地表点为坐标原点（x，y，z，=0，0，0），如图2所示，沿温室长度方向设为x轴，跨度方向设为y周，脊高方向设为z轴负方向，地表以下为Z轴正方向，各测点布置坐标如表1所示。

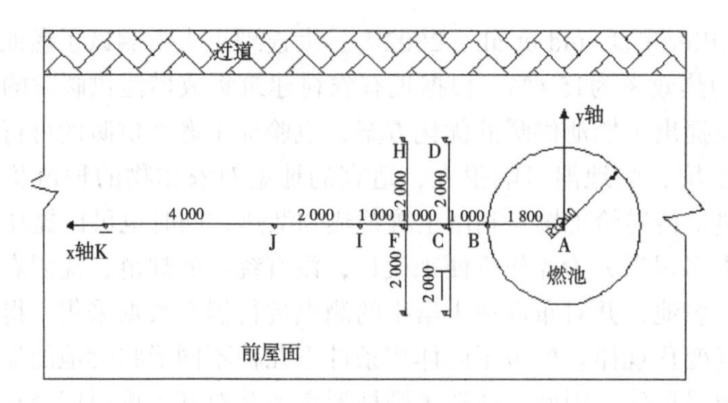

图 2　试验布置平面

Figure 2　Plane of the test

表 1　各测点坐标

Table 1　The coordinate of temperature measuring point

测点编号 Point number	坐标（m）coordinate		
	x	y	z
A1	0	0	0.05
A2	0	0	0.2
A3	0	0	0.4
B1	1.8	0	0.05
B2	1.8	0	0.2
B3	1.8	0	0.4
C1	2.8	0	0.05
C2	2.8	0	0.4
D	2.8	2	0.4
E	2.8	2	0.4
F	3.8	0	0.4
G	3.8	2	0.4
H	3.8	2	0.4
J	5.8	0	0.4
K	9.8	0	0.4

2　结果与分析

对池内秸秆燃烧的一个整周期数据进行采集，分析阴燃状态基本稳定是的数据，得到下面结果。

2.1　沿温室长度方向的加温变化规律

分析地面下0.4m深处土壤温度沿温室长度方向（x轴）的变化情况，取B-3、C-2、F、I、J以及试验区外的K点数据进行对比，温度波动小。距离燃池最近的B-3温度最高，平均值为37.71℃，随着温度测点与燃池距离的增大，土壤的温度逐渐降低。距燃池壁4m的J测点，土壤温度增加已不明显。距离燃池4m半径范围内，可提高地温5～20℃（图3）。

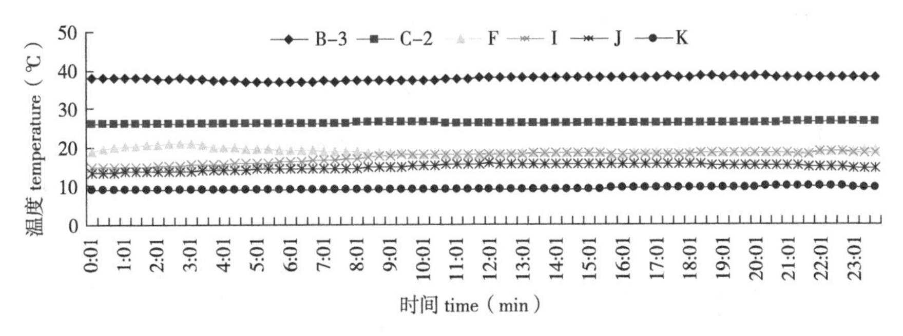

图 3　土层0.4m深土壤沿温室纵向温度变化规律

Figure 3　Changing regularity of soil temperature in length（−0.4m underground）

2.2　沿温室跨度方向的加温变化规律

分析地面下0.4m深处土壤沿温室跨度（y轴）的变化情况，取C-2、D、E和F、G、H测点数据进行分析，如图4所示。

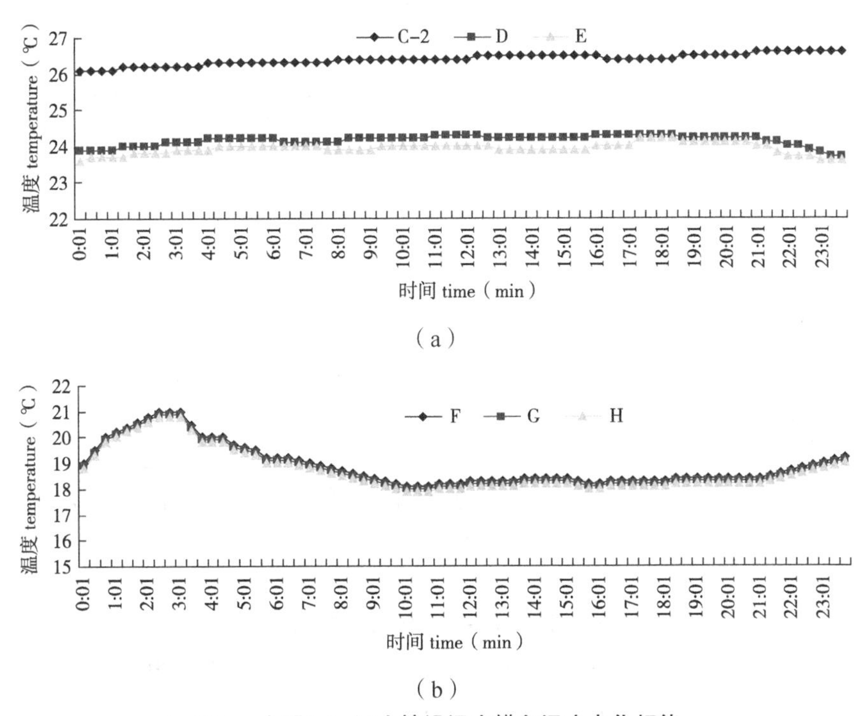

图 4　土层0.4m深土壤沿温室横向温度变化规律

Figure 4　Changing regularity of soil temperature in width（−0.4m underground）

距离燃池1m的测点C-2、D和E，平均温度分别为26.4℃、24℃和23.6℃。横向温度差异约2℃；距离燃池2m的跨度方向测点为F、G、H测点，平均温度分别为19.13℃、18.90℃和18.88℃，横向温差变化不大。随着与燃池距离的增大，跨度方向加温效果差异会减小。

2.3　沿土壤深度方向的加温变化规律

分析土壤深度方向测点A、B和C点，结果如图5所示。

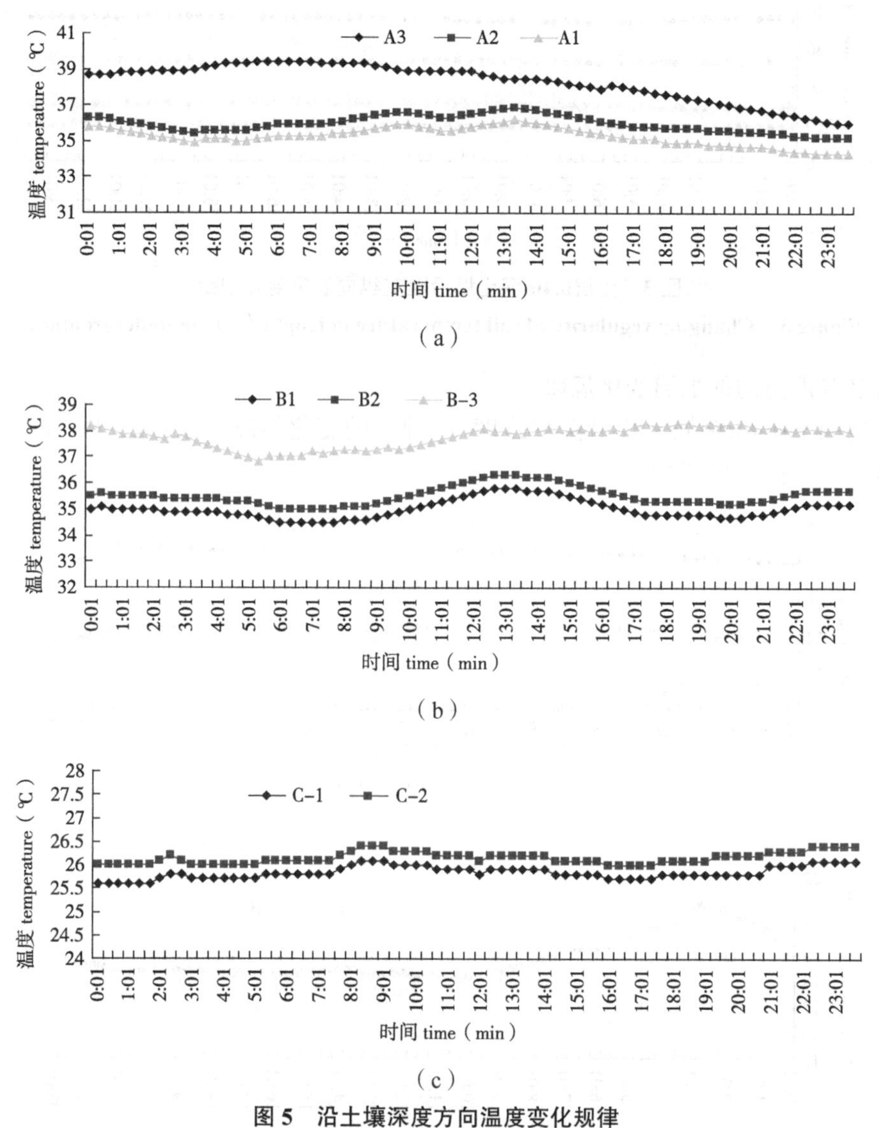

图 5　沿土壤深度方向温度变化规律

Figure 5　Changing regularity of soil temperature in depth

测点A-3、A-1点和A-2点的平均温度分别为38.2℃、35.2℃和36℃，测点B-3、B-1和B-2测点的平均温度分别为37.71℃、34.9℃和35.4℃，土壤温度差比A点相对小。测点C的两点个不同深度测点C1和C2的平均温度为26℃和26.4℃，温度差仅为0.4℃。

有温度曲线的形状可以看出，距燃池较近的土层随深度的变化温度变化较为明显，距离燃池较远位置，土层间基本不存在温度差。

3 结论与讨论

日光温室燃池辅助加温系统应用试验表明，燃池在冬季气温较低的情况下，能够有效提升土壤温度，加温效果明显。试验条件的燃池加温，加温有效半径大于6m。沿温室长度方向温度落差较大。燃池对日光温室的加温效果不仅作用于秸秆阴燃周期，燃烧熄灭后的2周内，仍可见余效。

燃池加热的温度很难控制。虽然目前提出了很多种燃池温度控制的办法，例如水管洒水、增加空气流量等，但在具体操作过程中很难控制，要做到温度随外界环境改变而自行控制就更加困难。因此，对燃池加温的控制问题还需要今后进一步的研究。

参考文献

［1］ 白义奎，王铁良，佟国红，等. 日光温室燃池加热系统传热机理与数学模型［J］. 农业机械学报，2006，37（9）：10.

［2］ 白义奎. 日光温室燃池——地中热交换系统研究［D］. 沈阳：沈阳农业大学，2007.

［3］ 李翔，陈滨，张雪研，等. 农村建筑集成燃池供暖优化方案与工程应用约束条件的关联性分析［D］. 大连：大连理工大学，2013.

［4］ 刘文合，许封.日光温室燃池形状和深度的模拟研究.北方园艺，2015，39（4）：44～49.

［5］ 路长，周建军，刘乃安，等. 阴燃材料受热升温过程分析［J］. 工程热物理学报，2006，27（2）：211-214.

［6］ 罗冰，何芳，高振强，等. 生物质燃烧和阴燃过程对比［J］. 山东理工大学学报，2013，27（1）：21-24.

［7］ 孙文策，解茂昭，郭晓平，等. 燃池内的阴燃过程的实验分析研究［J］. 工程热物理学报，2000，21（3）：393-396.

［8］ 田维治. 基于燃池利用供热方式的研究［D］. 大连：大连理工大学，2011.

［9］ 王铁良，白义奎，刘文合. 燃池在日光温室加热的应用试验［J］. 农业工程学报，2002，18（4）：98-100.

［10］ 张雪研，陈滨. 燃池供暖系统的设计建造方法研究［J］. 建筑科学，2014，30（2）：76-82.

温室全屋面外保温设计

展正朋
（中国农业科学院农业环境与可持续发展研究所，北京　100081）

摘要：温室日趋轻简化、大跨度、大空间发展，日光温室的大棚化、连栋化和塑料大棚的越冬周年性生产是两个重要的发展方向；为满足这一需求，必须解决其保温问题。其就温室外保温的现状和实际生产需求，设计了本套温室全屋面外保温设备。该套设备只用一个减速机提供动力进行联动，便可实现整个温室的保温被全棚面铺卷，既可避免卷起的保温被形成的固定阴影带而造成的遮光，将收起的保温被放置在天沟上又可以减轻温室拱架荷载。
关键词：温室；大棚；外保温；连栋；机械化

Greenhouse Thermal Insulation Overall Design

Zhan Zhengpeng
（*IEDA of CAAS*，*Beijing*　100081）

Abstract：Greenhouses are becoming lighter and more simplified with large-span and large-space development.In China，CSG and plastic sheds are two important directions of this development. In order to meet this demand，it is necessary to solve the problem of heat preservation problem. Faced with thermal insulation situation ，he designed this equipment which can meet the entire equipment' s power need only by one motor to achieve the insulation cover all over the shed，both can prevent the formation of the shadow by the roll of insulation，but can reduce the greenhouse' s arch load by stowing the insulation onto the gutter.
Key words：Greenhouse；Plastic sheds；External insulation；Multi-ridge；Mechanization

1　引言

日光温室作为我国北方特有的设施农业生产场所，因其在冬天不需要加温或极少加温就可以满足在北方寒冷地区的农业越冬生产，受到了广大农民的欢迎。但是，随着农民经济水平的提高和土地资源的日趋减少，日光温室单体空间小、环境调控能力弱、土地利用率低的缺点逐渐显现，极大限制了日光温室和北方设施农业的发展。日光温室轻简化、连栋化发展的需求日益强烈。同时，南方地区的塑料大棚常用于春提前和秋延后栽培，受其简单的结构和有限的环境调控能力制约无法进行越冬和越夏生产，随着土地租用费用的升高和人们对蔬菜消费的健康和丰富性需要，塑料大棚周年性节能生产需求也日趋迫切。日光温室和塑料大棚各有其优缺点，那么，是否可以将其优势互补——日光温室大棚化、连

作者简介：展正朋，安徽宿州人，主要从事设施农业工程方面研究。E-mail:1425314143@qq.com

栋化，塑料大棚温室化、节能化，即轻简化的装配式骨架结构，蓄热保温的节能性设施。而轻简化、连栋化、节能设计互为一体，既要走中国日光温室的节能特色路线，又要汲取塑料大棚的轻简结构，同时也应吸纳国外温室尤其是成熟的Venlo型温室标准化连栋设计的精华，走出一个中国特色新路子。

随着其他学科和工业水平的发展进步，众多科技工作者和爱思考的农民朋友进行了大量的实践设计，为其做出了不可磨灭的贡献。本文就其保温这一方面进行温室的外保温节能机械化设计，以求解决目前温室的保温问题并促进温室大型化连栋的发展。

2 材料与方法

2.1 材料

收集相关文献资料。

2.2 结果与分析

日光温室及塑料大棚的保温被卷放方式机械化水平参差不齐，就具体的某一卷放方式来说，适用的棚型也比较局限。虽然出现了一批改良型以及较为创新的卷放方式，但是依然没有很好的解决全棚面铺卷的倒坡形成的保温被在下坡卷收时铺卷不紧甚至溜坡的问题，也没能适应日光温室连栋化发展的趋势。因此，亟须研发出一种解决上述问题的保温被卷放设备。

3 解决方案

在此，提出一种温室保温被全屋面卷放设备：包括减速电机1、联动齿条2、轨道系统3、齿轮组4、保温被、卷被杆组5、牵引杆组6、卷被支架系统7。所述联动齿条2设置在温室的山墙两侧。所述轨道系统3设置在山墙轨道外侧，由拱架轨道31、槽轮32、滚轮33，牵引绳34组成。所述槽轮32固定在拱架轨道下方为牵引绳的运行节点，由L型支架321和深槽轮322组成。所述齿轮组4固定在卷被杆的两端。保温被放置在天沟上，其延温室长边一侧卷在卷被杆组5上，一侧固定在牵引杆组6上。所述卷被杆组有两根卷被杆分别放置在温室的两侧天沟上的卷被支架系统上，一侧卷被杆缠绕牵引绳带动牵引杆运动，另一侧卷被杆卷动保温被实现保温被的铺收。所述支架系统7由支座71和轴承72组成。该套设备只用一个减速机和齿条齿轮系统提供动力，便可实现整个温室的保温被全棚面铺卷，既可避免卷起的保温被形成的固定阴影带而造成的遮光，将收起的保温被放置在天沟上又可以减轻温室拱架荷载。

3.1 用于连栋温室时

（1）请参阅图1、图3、图4和图6，其特征在于：包括减速电机1、联动齿条2、轨道系统3、齿轮组4、保温被、卷被杆组5、牵引杆组6、卷被支架系统7。

（2）将保温被用卷被杆卷起放置在天沟支架上，另一侧固定在牵引杆上，由在固定在另一侧天沟上支架的卷被杆缠绕牵引绳带动运动实现铺盖；反之，逆向转动实现收起。所有动力通过电机和齿条实现动力的供给。

（3）所述联动齿条2设置在连栋温室的山墙两侧，齿条长度L与跨度d、拱架弧跨比α

满足以下关系式：$L=N\cdot d+\alpha+L_0$，N为连栋数，L_0为常数确保因安装误差造成的齿条短缺。

（4）所述轨道系统3设置在山墙轨道外侧，由拱架轨道31、槽轮32、滚轮33，牵引绳34组成。

（5）所述槽轮32固定在拱架轨道下方为牵引绳的运行节点，由L型支架321和深槽轮322组成。

（6）所述槽轮32优选的设置7个，均布在两侧山墙拱架的下方，作为牵引杆运动的变向节点，减小牵引绳的张力，增加保温被的拉铺流畅程度。其中两个优选的放置在近拱架天沟处，便以将保温被贴紧温室棚面。

（7）所述滚轮33设置在牵引杆两端，在轨道杆上限位滚动运行。

（8）所述齿轮组4固定在卷被杆的两端，通过卷被杆在支架和齿条上转动。

（9）保温被放置在天沟上，其延温室长边一侧卷在卷被杆组5上，一侧固定在牵引杆组6上。优选的，保温被在东西向温室上从北侧向南侧铺盖，在南北向温室上从东侧向西侧铺盖。

（10）所述保温被为促进连栋温室跨度的模数化发展，优选温室单跨为3.2m，保温被优选厚度为6mm及其以下。

（11）拱架弧长S与卷被杆直径D、保温被厚度a满足以下关系式：$S=\pi[nD+n(n+1)a]$，n为保温被在卷被杆上卷起层数；若单跨为3.2m，5mm厚保温被即可保证全覆盖，卷起后直径为15cm，放置在天沟上可以保证不额外遮阴。

（12）所述卷被杆组有两根卷被杆分别放置在单栋温室的两侧天沟上的卷被支架系统上，一侧卷被杆缠绕牵引绳带动牵引杆运动，另一侧卷被杆卷动保温被实现保温被的铺收。

（13）所述支架系统7由支座71和轴承72组成。支座高度稍高于保温被半径，优选配套模数化3.2m跨度温室，设置高度为150mm。

3.2 用于单栋温室时

（1）请参阅图3、图4、图5和图6，本发明提供一种单栋温室。大棚外保温铺卷设备，其特征在于：包括减速电机1、联动齿条2、轨道系统3、齿轮组4、保温被、卷被杆组5、牵引杆组6、卷被支架系统7。

（2）将保温被用卷被杆卷起放置在温室一侧，另一侧固定在牵引杆上，由在固定在另一侧支架的卷被杆缠绕牵引绳带动运动实现铺盖；反之，逆向转动实现收起。动力通过电机和齿条实现动力的供给。

（3）所述联动齿条2设置在温室的山墙两侧，齿条长度L与跨度d、拱架弧跨比α满足以下关系式：$L=d+\alpha+L_0$，，L_0为常数确保因安装误差造成的齿条短缺。

（4）所述轨道系统3设置在山墙轨道外侧，由拱架轨道31、槽轮32、滚轮33，牵引绳34组成。

（5）所述槽轮32固定在拱架轨道下方为牵引绳的运行节点，由L型支架321和深槽轮322组成。

（6）所述槽轮32优选的设置7个，均布在两侧山墙拱架的下方，作为牵引杆运动的变向节点，减小牵引绳的张力，增加保温被的拉铺流畅程度。其中两个优选的放置在近拱架底部处，便以将保温被贴紧温室棚面。

（7）所述滚轮33设置在牵引杆两端，在轨道杆上限位滚动运行。

（8）所述齿轮组4固定在卷被杆的两端，通过卷被杆在支架和齿条上转动。

（9）保温被放置在天沟上，其延温室长边一侧卷在卷被杆组5上，一侧固定在牵引杆组6上。优选的，保温被在东西向温室上从北侧向南侧铺盖，在南北向温室上从东侧向西侧铺盖。保温被放置地面支架上，对跨度和保温被的厚无特殊要求。

（10）拱架弧长S与卷被杆直径D、保温被厚度a满足以下关系式：$S=\pi[nD+n(n+1)a]$，n为保温被在卷被杆上卷起层数。

（11）所述卷被杆组有两根卷被杆分别放置在单栋温室的两侧天沟上的卷被支架系统上，一根卷被杆缠绕牵引绳带动牵引杆运动，一根卷被杆卷动保温被实现保温被的铺收。

（12）所述支架系统7由支座71和轴承72组成。支座高度稍高于保温被半径，具体高度根据保温被的半径确定。

因此该用于大棚的落地式铺卷保温被的设备具有可以减轻大棚荷载并增加大棚采光等优点，适合大规模推广。

4 备用方案

为了保证电机的动力可以顺利无误的输送到每一个传动部件，额外提供2种备选方案：

（1）将齿轮齿条系统改为链轮链条系统，试验两种系统的动力传输效果，是否可以保证每一个传动部件都可以同步稳定的转动。

（2）采用万向节和锥齿轮传输动力，克服卷放过程的起伏和错位，提高容错率，保证动力的稳定传输。

5 说明书附图（图1至图5）

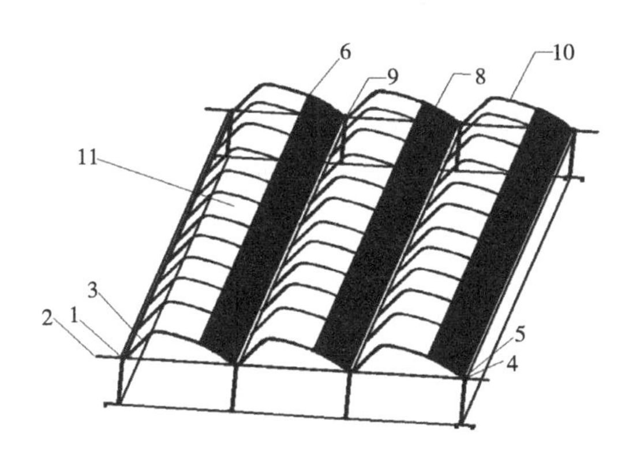

图1 连栋温室全屋面外保温立面

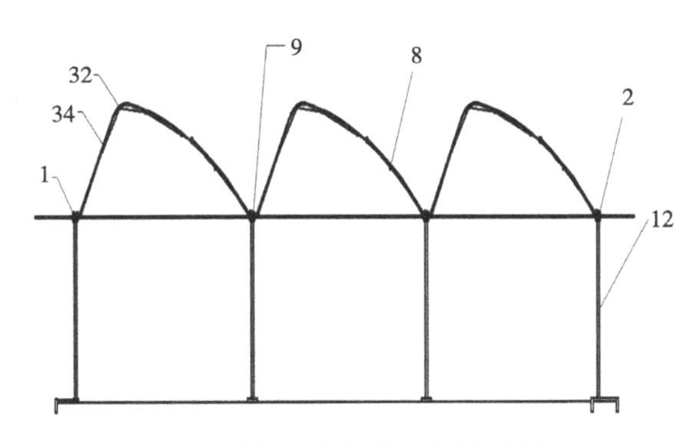

图2 连栋温室全屋面外保温侧视

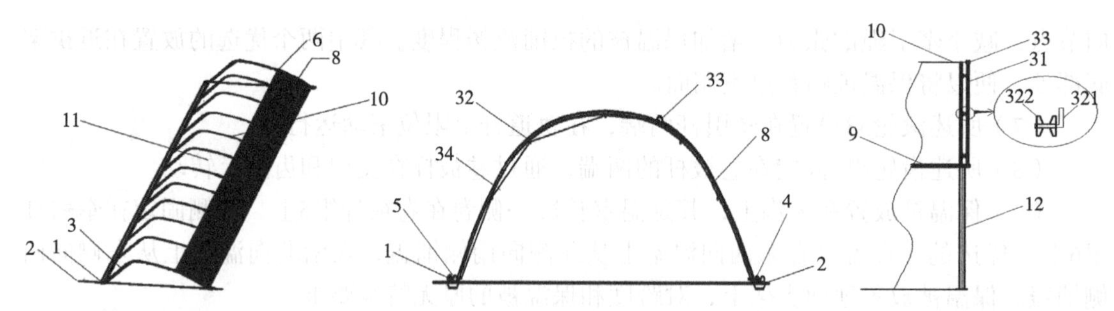

图 3　单栋温室全屋面外保温立面　　**图 4　单栋温室全屋面外保温侧视**　　**图 5　山墙端轨道架**

参考文献

陈海珍，张放军，楼杰. 2010. 新型农用保温被结构设计与保温性能研究［J］.产业用纺织品（243）：12-15.

程思霖.一种温室保温毯［P］.中国专利，201020524851.3，2011-06-22.

胡瑶玫，李娇，吴松. 2014. 几种日光温室保温被的保温性研究［J］.农机化研究（1）：199-201.

李萍萍，胡永光. 2002. 冬季塑料大棚多重覆盖及电加热增温效果研究［J］.农业工程学报，18（2）：76-79.

李胜利，霍颖君，孙治强. 2008. 不同层次简易覆盖的巨型塑料大棚温度特征研究［J］.河南农业大学学报，42（6）：621-624.

李天来，孙周平，黄文永，等. 2014. 齿轮式滑动覆盖温室传动装置［P］.中国专利，201320320284.3，4-9.

凌坚，马承伟，林聪，等. 2002. 温室缀铝膜保温幕节能性能的实验研究初报［J］.农业工程学报，18（1）：89-93.

刘晨霞，马承伟，王平智，等. 2015. 日光温室保温被传热的理论解析及验证［J］.农业工程学报，31（2）：170-177.

马承伟，黄之栋，林聪，等. 2000. 连栋温室保温节能技术探索［J］.发展中的中国工厂化农业，261-266.

乔正卫，邹志荣，杨双晓. 2008. 一种日光温室保温被的保温性研究［J］.机化研究（6）：131-133.

沈军，李亚，温祥珍，武英霞. 2007. 非对称连跨式节能温室的温度性能分析［J］.四川农业大学学报，25（3）：322-327.

宋明军，郭晓东. 2005. 双连栋节能日光温室的设计建造及其温光性能观测［J］.河南农业科学，9：79-82，84.

宋卫堂，金銮，顾帆，等. 2017. 一种拱棚保温被全棚面卷放装置及其使用方法［P］.中国专利，201611206803.8，4-26.

陶国富，崔绍荣，吴小兰. 2003. 连栋塑料温室保温幕保温性能的应用研究［J］.农机化研究，4：155-158.

佟国红，Christopher David M.，李天来，等. 2010. 日光温室保温被卷放位置对温度环境的影响［J］.农业工程学报，26（10）：253-259.

王海波，郝志强，刘凤之，等. 2016. 一种设施园艺用保温被导轨式卷放装置［P］.中国专利，201620603339.5，11-23.

王纪章，赵青松，李萍萍. 2012. 温室多层覆盖的冬季保温效果研究［J］.中国蔬菜（18）：106-110.

吴松、李娇，刘文玺，等. 2014. 轨道式日光温室卷铺机构［P］.中国专利，201420373166.3，11-5.

禹夏青，张亚红. 2017. 保温被外设PE黑膜对日光温室保温性的影响［J］.江苏农业科学，45（8）：187-191.

周升，张义，程瑞锋，等. 2016. 大跨度主动蓄能型温室温湿环境监测及节能保温性能评价［J］.农业工程学报，32（6）：218-223.

番茄茎秆不同部位声发射频谱响应特征差异性分析

张佳[1, 2]，郭文忠[1]，余礼根[1]，李灵芝[2]，秦渊渊[1, 2]，梁贝贝[1, 2]，乔晓军[1]，李海平[2]
（1. 北京农业智能装备技术研究中心，农业智能装备技术北京市重点实验室，北京　100097；
2. 山西农业大学园艺学院，太谷　030801）

摘要：本文以“佳丽14号”番茄为试材，通过番茄负水头灌溉盆栽试验，采集番茄盛果期茎秆不同部位（上部、中部、下部）的声发射信号，并对其时域波形图、波形特征参数和频谱特征参数的变化进行差异性分析。结果表明：番茄声发射信号的时域信号振幅和振幅变化持续时间随番茄茎秆自上部到下部呈先减小后增大的趋势，下部时域信号振幅最大为0.25V，持续时间最大为350μs；番茄茎秆自顶端到中部、根部，声发射信号波形特征参数呈逐渐上升的趋势，下部声发射活动最为活跃，其声发射计数、幅值、能量、上升时间、持续时间和峰值频率分别达到57.7、30.7dB、6.3、24.4μs、169.7μs、4.7kHz；番茄频谱特征参数分析表明，共振峰频率和功率谱特征参数随番茄茎秆自上部到下部呈先减小后增大的趋势，上部和下部共振峰频率相差较小，下部的主频能量、中心频率、加权功率谱频率、有限频带能量面积、功率谱面积和功率谱方差为最大值，共振峰幅值随番茄茎秆自上部到下部呈逐渐减小的趋势，上部幅值明显高于中部和下部。综合分析番茄不同部位时域波形图、声发射波形特征参数和频谱特征参数可知，番茄茎秆不同部位声发射信号差异显著，番茄茎秆下端的声发射活动最为强烈，可将番茄茎秆下部作为声发射信号采集的最佳位置。

关键词：番茄茎秆；声发射信号；频谱分析；波形图；高度

Spectrum Analysis of Acoustic Emissions for Tomato Plants at Different Height

Zhang Jia[1, 2], Guo Wenzhong[1], Yu Ligen[1], Li Lingzhi[2], Qin Yuanyuan[1, 2],
Liang Beibei[1, 2], Qiao Xiaojun[1], Li Haiping[2]
(*1. Beijing Research Center of Intelligent Equipment for Agriculture, Key Laboratory of Agricultural Intelligent Equipment Technology, Beijing* 100097; *2. College of Horticulture, Shanxi Agricultural University, Taigu* 030801)

Abstract: Took tomato variety of ‘Jiali14’ as experimental materials with the potted experiment under negative pressure irrigation. The study collected three kinds of acoustic emissions signals at the full fruit stage of tomato stem, upper part, middle part, lower part, respectively, And studied the difference analysis of time domain waveforms, waveform parameters, spectral characteristics. The results showed that the time-domain amplitude and amplitude duration time of acoustic emissions of tomato were tended to decrease firstly and then

基金项目：“十三五”国家重点研发计划（2017YFD0201503）；国家自然科学基金青年科学基金（51509005）；山西省研究生联合培养基地人才培养项目（2016JD22）

作者简介：张佳（1992—），女，山西晋城人，在读硕士，主要从事蔬菜栽培与生理研究
通信作者：余礼根（1985—），男，安徽岳西人，助理研究员，博士，主要从事动植物声信息感知技术研究
李海平（1970—），男，山西朔州人，副教授，博士，硕士生导师，主要从事蔬菜栽培与生理研究

increase with the change from upper part to lower part of the tomato stem. The lower part of time domain amplitude and amplitude duration time were 0.25 V, 350 μs; With the change of tomato stem from top to bottom, the waveform parameters of acoustic emissions signal gradually increased, and the bottom acoustic emissions signal was the most active. The acoustic emissions counts, amplitude, energy, rise time, duration and peak frequency were 57.7, 30.7 dB, 6.3, 24.4 μs, 169.7 μs and 4.7 kHz, respectively. Analysis of spectral characteristics of tomato showed that with the change of tomato stem from upper part to lower part, the formant frequency and spectral characteristics were tended to first decrease and then increase. The difference of the upper part and lower part formant frequency was small, and the dominant frequency energy, frequency, weighted power spectrum frequency, finite frequency band energy area, power spectrum area and power spectrum variance of lower part of tomato stem were the maximum values. The amplitude of formant gradually decreased with the change of tomato stem from upper part to lower part, and the amplitude of upper part was significantly higher than that of middle part and lower part. According to the analysis of time-domain waveform, acoustic emissions waveform parameters and spectral characteristics of different parts of tomato stem, the acoustic emissions signals of different parts of tomato stem are different and the lower part of acoustic emissions activity is the strongest. The lower part of tomato stem can be regarded as acoustic emissions signal the best place to collect.

Key words: Tomato stem; Acoustic emissions signal; Spectrum analysis; Waveform diagram; Height

1 引言

木质部是番茄植株水分运输的主要通道，番茄植株耗水过大，致使木质部导管的张力随之增大，导管内水柱断裂而出现空穴化现象，此时，张力突然释放产生冲击波，即植株声发射现象[1-3]。声发射技术作为一种无损检测方法，已经广泛应用于材料工程[4-5]、建筑学[6]、临床医学[7]和金属检测[8]等领域。随着声发射技术的不断完善，其逐渐应用于木材裂纹及干旱检测[9-10]。对于声发射技术在番茄上的应用，主要包括病害胁迫[11-12]和水胁迫[13-14]的声发射检测及分析。目前，有大量关于植物不同高度处叶片水分利用效率[15]、水势变化[16]、微量元素变化[17]和糖分含量[18]等方面的研究，张永娥等[15]以北京西山的侧柏林为研究对象，研究冠层不同高度处叶片的水分利用率，其随林冠高度的变化规律一致，均表现为上层>中层>下层；桑永青[16]以苹果树为研究对象，研究不同灌水处理对苹果园SPAC（土壤—植物—大气）系统水势的影响，同一棵苹果树叶片水势从植株低部到植株高部逐渐降低。目前，基于声发射信号采集的定位研究主要应用于管道泄漏[19]、泥岩断裂[20]、光纤损伤[21]及植株病害区域定位[12]等。王秀清[12]等将声发射传感器分别固定于番茄植株茎部占整体高度的1/3和2/3处，通过比较不同位置的声发射信号频次判断病害胁迫声源定位。但针对番茄不同高度处的声发射信号特征规律及信号测定位置的选取报道较少。本文通过温室盆栽试验，研究盛果期番茄声发射信号波形参数与频谱特征参数随茎秆不同高度的变化规律，并通过比较不同高度的声发射信号差异选取番茄声发射信号采集的最佳位置。

2 试验材料与方法

2.1 试验地点与材料

试验于2017年3—7月在北京市农林科学院试验温室内进行（N39° 56′，E116° 16′）。试验采用负水头灌溉盆栽试验，如图1所示。试验设置负压值为60hPa，对应的土壤体积含水量为19.7%，每个处理重复4次。番茄品种选用“佳丽14号”，于3月15日每盆定植长势一致的植株1株，花盆的规格为：上口直径34.2cm、底直径18.5cm、盆高22.3cm，每盆装供试土壤11.0kg。负水头灌溉试验装置主要由塑料盆、陶瓷盘、供水桶、控压管和导气管构成；当装置运行时，控压管及供水桶上的阀门属于关闭状态，进水管的阀门属于打开状态，供水桶内水分进入陶瓷盘，陶瓷盘空腔内气体全部排到供水桶，由于土壤基质势的作用陶瓷盘内水分缓慢进入土壤，致使陶瓷盘内压强逐渐减小，供水桶内水分再次进入陶瓷盘，如此不断循环，使供水桶内水分在负压控制下连续不断进入土壤以供土壤蒸发及番茄植株生长发育。

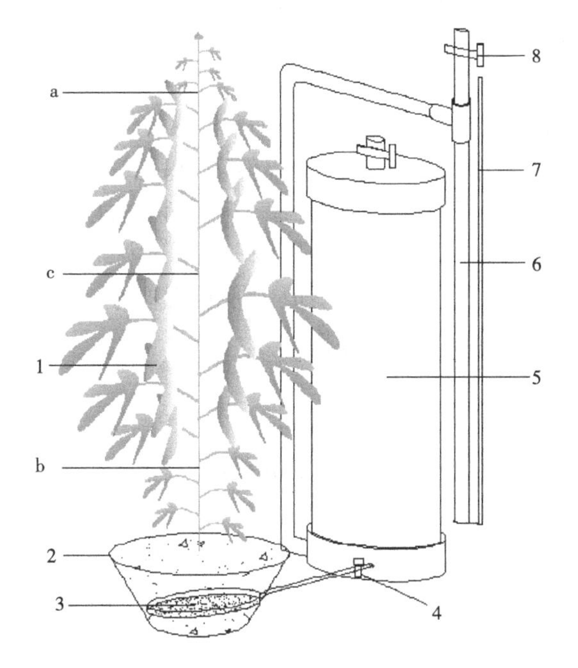

1.番茄；2.塑料盆；3.陶瓷盘；4.进水管；5.供水桶；6.控压管；7.导气管；8.阀门

1.Tomato；2.Plastic pot；3.Ceramic plate；4.Water inlet tube；
5.Water supply barrel；6.Pressure control tube；7.Airway tube 8.Valve

图 1 负水头灌溉试验装置

Figure 1 The experiment device with negative pressure irrigation

2.2 声发射信号测定

声发射信号采集选用性能稳定可长时间连续运行的MICROⅡ监测系统（Physical Acoustic Corporation，Princeton，New Jersey，USA）、配备的声发射采集卡为PCI-2（8

通道同步采集、18位A/D分辨率、40M/s采样率、1kHz-3MHz频率范围）、声发射传感器选用Nano30（响应频率为125kHz-750kHz、灵敏度为62［-72］dBref.1V/（m/s）［1V/μbar］）、放大器选用2/4/6型（20kHz-1 200kHz，20/40/60dB可选），实行24h连续采集，每隔1d存储为1个数据文件。为了研究60hPa负压值下番茄声发射信号随植株不同高度变化规律，试验选取番茄盛果期（5月16日至6月20日）进行声发射监测。

声发射信号采集时，将声发射传感器分别固定在番茄茎秆不同高度处，试验选取3个部位进行测定，如图1所示：番茄植株顶端生长点以下第五叶位置的茎秆部位（上部，a）、番茄植株根部倒五叶位置的茎秆部位（下部，b）、上部与下部之间中心点位置的茎秆部位（中部，c），为防止声发射传感器固定的茎秆部位脱水，在传感器与茎部之间涂上凡士林。

2.3 数据处理

选用MICROⅡ声发射监测系统配套的声发射信号分析软件AE win-2（Physical Acoustic Corporation，Princeton，New Jersey，USA），从有效的声发射信号中提取出典型声发射波形特征参数，包括幅值、计数、能量、上升时间、持续时间、峰值频率共6个波形特征参数及声发射信号时域波形图；在此基础上，应用LabVIEW 2014（National Instruments，Austin，TX，USA）编程计算频谱特征参数，得出第一共振峰频率、第一共振峰幅值、第二共振峰频率、第二共振峰幅值、第三共振峰频率、第三共振峰幅值、主频、主频能量、中心频率、加权功率谱频率、有限频带能量面积、功率谱面积、功率谱方差共13个频谱特征参数。统计分析番茄茎秆上部、中部、下部的声发射信号（幅值>40dB）的波形参数和频谱特征参数（以均值±标准偏差的方式表示），并对番茄茎秆不同高度声发射信号时域波形图进行比较分析。

3 结果与分析

3.1 番茄茎秆不同高度声发射信号时域波形图的差异

声发射信号时域波形图可直观反映幅值随时间的变化情况[22]，不同胁迫条件、不同采集位置与不同使用环境下往往产生不同类型的模态声波。番茄茎秆不同高度声发射信号时域波形图如图2所示。为了研究番茄茎秆不同高度声发射时域波形图变化，选取番茄盛果期温室环境温度、相对湿度、光照基本相同的3个晴天，对这3天的声发射时域波形图进行比较及分析。由图2可知，从信号幅值的角度分析，其振幅变化范围在0V～0.25V之间，随番茄茎秆由上部至下部的高度变化，时域信号最大振幅表现为：茎秆下部>茎秆上部>茎秆中部，第2、3天的茎秆下部时域信号最大振幅分别达到0.25V、0.125V。番茄茎秆不同高度的时域信号振幅变化持续时间位于250μs～600μs，随番茄茎秆由上部至下部高度变化时，时域信号振幅变化持续时间表现为：茎秆下部>茎秆上部>茎秆中部，第2天的茎秆下部时域信号振幅变化持续时间位于250μs～600μs之间，第1、2、3天的茎秆上部时域信号振幅变化持续时间均位于250μs～350μs范围内。由于随叶片叶龄的增加水分利用效率呈先上升后下降的趋势[23]，茎秆中部叶片水分利用效率最高，声发射信号时域信号振幅、振幅变化持续时间及功率谱特征参数最小，而茎秆下部叶片叶龄较大，水分利用效率较低，以致随着番茄茎秆由上部至下部的高度变化，下部的时域信号最大振幅值及振幅变化持续时间均为最大。

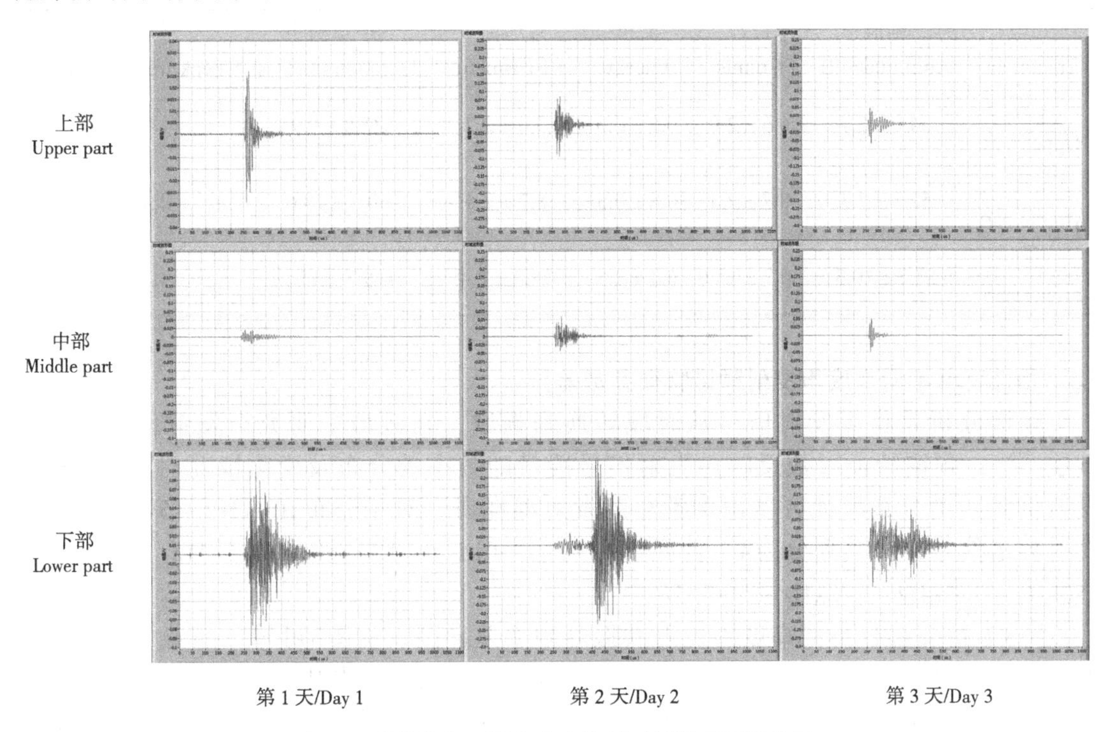

图 2　番茄茎秆不同高度声发射时域波形图的差异

Figure 2　The differences of acoustic emission time domain waveforms at different height of tomato stem

3.2　番茄茎秆不同高度声发射信号波形特征参数差异

声发射波形特征参数用以表征声发射信号特征，主要包括声发射信号幅值、计数、能量、上升时间、持续时间及峰值频率共6个特征参数。幅值反映声发射事件的能量，用于鉴别是否是声发射信号；计数用于反映声发射发生的剧烈程度；能量反映声发射信号强度；上升时间用来鉴别是否是声发射事件；持续时间反映声发射信号的频率，用于声发射源信号识别；峰值频率反映波形振幅峰值的频率[24]。由表1可知，番茄茎秆不同高度声发射信号有一定的差异，随着番茄茎秆由上部至下部的高度变化，幅值表现为茎秆下部>茎秆上部>茎秆中部，番茄茎秆下部幅值最大达到57.7dB，计数、能量、峰值频率、上升时间和持续时间均表现为茎秆下部>茎秆中部>茎秆上部，番茄茎秆下部上升时间和持续时间分别为24.4μs、169.7μs。据胡清[25]研究表明，松针水势变化趋势为：树冠下层>树冠中层>树冠上层，树冠下层松针水势最高；水势可反映植株水分运输的能力，水分运输过程中番茄植株底端水分运输阻力较小，水势偏高，植株顶端水分运输阻力增大，水势降低，致使随着番茄茎秆由上部至下部的高度变化，声发射信号计数、能量、上升时间、持续时间、峰值频率呈逐渐增大的趋势，茎秆下部的幅值、计数、能量、上升时间、持续时间和峰值频率均为最大，表明番茄茎秆下部声发射信号最为活跃。

表 1　番茄茎秆不同高度声发射信号波形特征参数差异

Table 1　The differences of acoustic emission waveform parameters at different height of tomato stem

处理 Treatment	幅值 Amplitude/dB	计数 Counts	能量 Energy	上升时间 Rise time/μs	持续时间 Duration time/μs	峰值频率 Peak frequency/kHz
上部 Upper part	54.0 ± 7.7	6.6 ± 2.0	0.5 ± 0.3	5.0 ± 2.2	35.8 ± 9.9	1.7 ± 0.8
中部 Middle part	52.5 ± 7.3	8.7 ± 2.4	0.8 ± 0.5	8.7 ± 2.4	60.3 ± 18.6	2.4 ± 1.3
下部 Lower part	57.7 ± 10.3	30.7 ± 13.0	6.3 ± 2.5	24.4 ± 2.4	169.7 ± 75.5	4.7 ± 2.2

3.3　番茄茎秆不同高度声发射频谱特征参数差异

频谱分析是将声发射信号从时域转换到频域，在频域中研究声发射信号特征的方法[26]，频谱特征参数反映声发射信号能量随频率的分布及变化情况[14]，通过声发射信号频谱分布特征可以识别番茄声发射源性质。对于分布在0kHz ~ 500kHz的声发射频谱信号，将其划分为4个区间，分别是：0kHz ~ 125kHz，125kHz ~ 250kHz，250kHz ~ 375kHz，375kHz ~ 500kHz。

图3为番茄茎秆不同高度声发射共振峰频率差异图，由图3可知，随着番茄茎秆由上部至下部的高度变化，第一、二、三共振峰频率均表现为先减小后增大，茎秆上部的共振峰频率最大；番茄茎秆上部、中部、下部共振峰频率分别位于130kHz ~ 150kHz、95kHz ~ 121kHz、114kHz ~ 140kHz范围内；茎秆上部的第一、二、三共振峰频率均出现在125kHz ~ 250kHz区间，茎秆中部的第一、二、三共振峰频率均位于0kHz ~ 125kHz区间，茎秆下部的第一、二共振峰频率在0kHz ~ 125kHz区间，第三共振峰频率在125kHz ~ 250kHz区间。由图4可知，随着番茄茎秆由上部至下部的高度变化，茎秆上部第一、二、三共振峰幅值均明显高于中部和下部，茎秆上部的第一、二、三共振峰幅值分别为2.8×10^{-6}V、3.4×10^{-6}V、4.8×10^{-6}V。由于番茄茎秆上部靠近生长点，与番茄茎秆中部和下部相比是木质化程度最低的部位，细胞组织对于水分运输较为敏感，致使茎秆上部共振峰频率和幅值为最高。

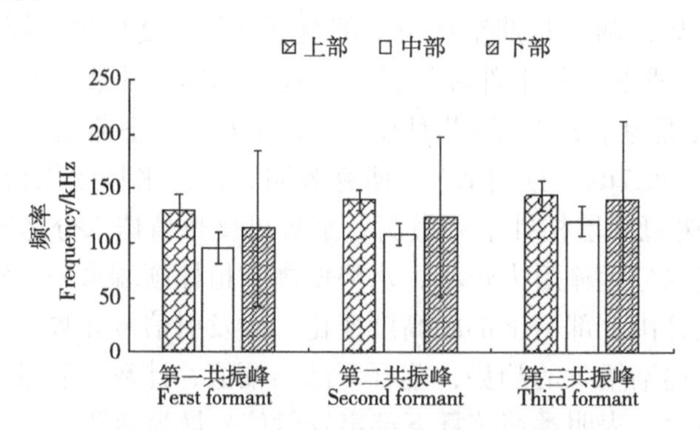

图 3　番茄茎秆不同高度声发射共振峰频率的差异

Figure 3　The differences of acoustic emission formant frequency at different height of tomato stem

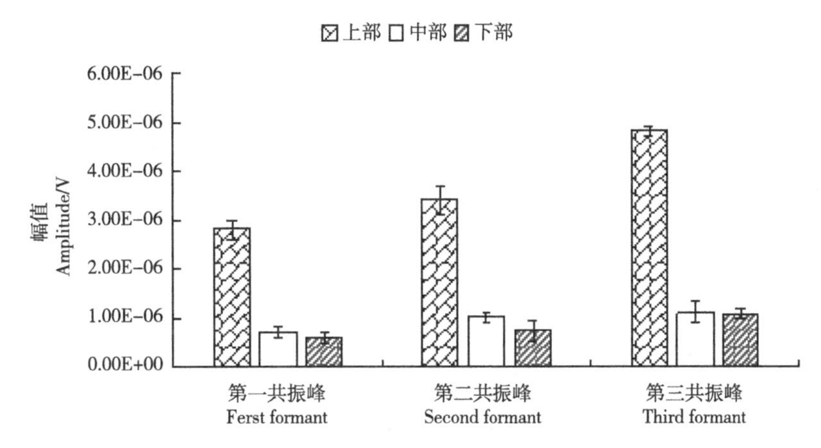

图 4 番茄茎秆不同高度声发射共振峰幅值的差异

Figure 4 The differences of acoustic emission formant amplitude at different height of tomato stem

表2是番茄茎秆不同高度声发射功率谱特征参数。由表2可知，随着番茄茎秆上部、中部、下部的高度变化，主频表现为：茎秆上部>茎秆下部>茎秆中部，番茄茎秆下部的主频为211.8kHz；从主频能量角度分析，不同高度主频能量表现为：茎秆下部>茎秆中部>茎秆上部，番茄茎秆下部的主频能量最大；中心频率与加权功率谱频率在数值方面没有差异，随着番茄茎秆高度的变化，中心频率表现为：茎秆下部>茎秆上部>茎秆中部，番茄茎秆下部的中心频率和加权功率谱频率最大达到226.6kHz；有限频带能量面积和功率谱面积在数值上相同，随着番茄茎秆高度的变化，茎秆上部至下部有限频带能量面积先减小后增大，茎秆下部的有限频带能量面积和功率谱面积最大，茎秆中部的有限频带能量面积和功率谱面积最小，茎秆下部的有限频带能量面积为78.3V^2；从功率谱方差角度分析，随着番茄茎秆由上部至下部的高度变化，功率谱方差先减小后增大，茎秆下部的功率谱方差最大，茎秆中部的功率谱方差最小，茎秆下部的功率谱方差为1.2×10^{-6}。据陈凯利[23]研究表明：水分利用效率随叶片叶龄的增加呈先上升后下降的趋势，与番茄茎秆中部叶片水分利用效率相比，番茄茎秆上部和下部叶片水分利用效率较低，因此，主频、中心频率、加权功率谱频率、有限频带能量面积、功率谱面积随番茄茎秆由上部至下部高度变化呈先减小后增大的趋势，茎秆下部的主频能量、中心频率、加权功率谱频率、有限频带能量面积、功率谱面积和功率谱方差均为最大。

表 2 番茄茎秆不同高度声发射功率谱特征参数差异

Table 2 The differences of acoustic emission spectral characteristics at different height of tomato stem

处理 Treatment	主频 Main frequency /kHz	主频能量 Main frequency energy/V^2	中心频率 Center frequency/ kHz	加权功率谱频率 Weighted power spectral frequency/ kHz	有限频带能量面积 Limited band energy area/V^2	功率谱面积 Power spectral area	功率谱方差 Power spectral variance
上部 Upper part	221.7 ± 77.3	$(7.7\pm0.03)\times10^{-4}$	213.6 ± 63.4	213.6 ± 63.4	36.3 ± 0.1	36.3 ± 0.1	$1.5\times10^{-7}\pm3.9\times10^{-13}$

（续表）

处理 Treatment	主频 Main frequency /kHz	主频能量 Main frequency energy/V^2	中心频率 Center frequency/ kHz	加权功率谱频率 Weighted power spectral frequency/ kHz	有限频带能量面积 Limited band energy area/V^2	功率谱面积 Power spectral area	功率谱方差 Power spectral variance
中部 Middle part	170.1 ± 90.4	（1.8 ± 0.02）× 10^{-3}	190.7 ± 78.6	190.7 ± 78.6	16.3 ± 0.04	16.3 ± 0.04	$1.1 \times 10^{-7} \pm 2.1 \times 10^{-13}$
下部 Lower part	211.8 ± 80.9	（3.2 ± 0.1）× 10^{-3}	226.6 ± 56.9	226.6 ± 56.9	78.3 ± 2.2	78.3 ± 2.2	$1.2 \times 10^{-6} \pm 5.3 \times 10^{-10}$

4　结论

（1）本试验通过分析番茄茎秆不同部位声发射信号表明番茄声发射时域波形振幅、振幅变化持续时间及功率谱特征参数随番茄茎秆自上部到下部呈先减小后增大的趋势；距离番茄根部越近，声发射信号的幅值、计数、能量、上升时间、持续时间、峰值频率越高；番茄茎秆上部共振峰频率和幅值为最高。

（2）综合分析番茄茎秆不同高度声发射波形特征参数、频谱特征参数及时域波形图可知，番茄茎秆下部声发射活动最为强烈，可将番茄茎秆下部作为声发射信号采集的最佳位置。

参考文献

［1］ 左力翔，李俊辉，李秧秧，等. 散孔材与环孔材树种枝干、叶水力学特性的比较研究［J］.生态学报，2012，32（16）：5 087-5 094.

［2］ 赵平，孙谷畴，倪广艳，等. 成熟马占相思水力导度对水分利用和光合响应的季节性差异［J］.应用生态学报，2013，24（1）：49-56.

［3］ Gagliano M. Green symphonies：a call for studies on acoustic communication in plants［J］. Behavioral Ecology，2013，24（4）：789-796.

［4］ Nosov V V. Acoustic-emission quality control of plastically deformed blanks［J］. Russian Journal of Nondestructive Testing，2017，53（5）：368-377.

［5］ Pochaps’kyi E P，Klym B P，Rudak M O，et al. Application of the magnetoelastic acoustic emission for the corrosion investigations of steels［J］. Materials Science，2017，52（5）：742-745.

［6］ Khandelwal M，Ranjith P G. Study of crack propagation in concrete under multiple loading rates by acoustic emission［J］. Geomechanics and Geophysics for Geo-Energy and Geo-Resources，2017，（2）：1-12.

［7］ Illanes A，Schaufler A，Maldonado I，et al. Time-varying acoustic emission characterization for guidewire coronary artery perforation identification［C］. IEEE Computers in Cardiology，2017.

［8］ Botvina L R，Soldatenkov A P，Tyutin M R，et al. On interrelation of damage accumulation in structural steels and physical parameters estimated by methods of acoustic emission and metal magnetic memory［J］. Russian Metallurgy，2017，2017（1）：10-17.

［9］ Diakhate M，Bastidas A E，Pitti R M，et al. Cluster analysis of acoustic emission activity within wood material：towards a real-time monitoring of crack tip propagation［J］. Engineering Fracture Mechanics，2017，180：254-267.

[10] Roo L D, Vergeynst L, Baerdemaeker N D, et al. Acoustic emissions to measure drought-induced cavitation in plants [J]. Applied Sciences, 2016, 6 (3): 1-10.

[11] 王秀清，游国栋，杨世凤. 基于作物病害胁迫声发射的精准施药 [J]. 农业工程学报，2011，27 (3)：205-209.

[12] 王秀清，张春霞，杨世凤. 番茄病害胁迫声发射信号采集与声源定位 [J]. 农业机械学报，2011，42 (4)：159-162.

[13] Kageyama K, Mori O. Deficit irrigation of miniature tomato based on estimation of embolism risk by measurements of acoustic emission and stress wave at stem [J]. Journal of Experimental Mechanics, 2014, 13: 85-91.

[14] 余礼根，李长缨，陈立平，等. 番茄声发射信号功率谱特征分析 [J]. 农业机械学报，2017，48 (10)：189-194.

[15] 张永娥，余新晓，陈丽华，等. 北京西山侧柏林冠层不同高度处处叶片水分利用效率 [J]. 应用生态学报，2017，28 (7)：2 143-2 148.

[16] 桑永青. 蓄水坑灌下不同灌水处理对苹果园SPAC系统水势影响研究 [D]. 太原：太原理工大学，2016.

[17] 杨延杰，谢春玲，林多，等. 弱光对番茄茎叶中铜含量的影响 [J]. 华北农学报，2009，24 (b12)：180-182.

[18] 张华文，秦岭，王海莲，等. 甜高粱茎秆糖分含量的变化分析 [J]. 华北农学报，2009，24 (s2)：69-71.

[19] 杨丽丽，谢昊飞，李帅永，等. 气体管道泄漏声发射单一非频散模态定位 [J]. 仪器仪表学报，2017，38 (4)：969-976.

[20] 杨健锋，梁卫国，陈跃都，等. 不同水损伤程度下泥岩断裂力学特性试验研究 [J]. 岩石力学与工程学报，2017，36 (10)：2 431-2 440.

[21] Sai Y, Zhao X, Hou D, et al. Acoustic emission localization based on FBG sensing network and SVR algorithm [J]. Photonic Sensors, 2017, 7 (1): 48-54.

[22] 张强，张石磊，王海舰，等. 基于声发射信号的煤岩界面识别研究 [J]. 电子测量与仪器学报，2017，31 (2)：230-237.

[23] 陈凯利. 水分对番茄不同叶龄叶片光响应特性与果实品质和产量的影响 [D]. 杨凌：西北农林科技大学，2012.

[24] 郑冰环. 基于声发射的机械断裂定位研究 [D]. 北京：中国矿业大学，2015.

[25] 胡清，吕军，李水冰，等. 旱季云南松松针水势变化规律 [J]. 云南大学学报（自然科学版），2014，36 (3)：433-438.

[26] 黄晓红，李莎莎，张艳博，等. 砂岩声发射信号的功率谱分析 [J]. 矿业研究与开发，2013，33 (2)：38-42.

负水头灌溉对番茄不同生育期生长特性及水分利用效率的影响

张佳[1]，秦渊渊[1]，郭文忠[2]，余礼根[2]，李灵芝[1]，李海平[1*]

（1. 山西农业大学园艺学院，太谷　030801；2. 北京农业智能装备技术研究中心，北京　100097）

摘要：通过温室盆栽试验，研究负水头不同供水吸力对番茄各生育期耗水量和光合特性、植株根系活力和产量的影响，为负水头在番茄生长生产中的合理应用提供理论依据。采用负水头灌溉装置，设置4个不同吸力值，分别为50hPa（T1）、60hPa（T2）、70hPa（T3）、80hPa（T4），测定不同吸力值下番茄各生育期内的耗水量和光合特性、植株根系活力及产量，进行比较分析。通过比较各处理间番茄不同生育期耗水量可知，T1、T2、T3、T4处理的番茄初果期耗水量均最高，分别为22.05kg·pot^{-1}、18.89kg·pot^{-1}、21.11kg·pot^{-1}、18.46kg·pot^{-1}；各处理的番茄叶片净光合速率和蒸腾速率随番茄植株的生长呈现先增大后减小的趋势，与番茄各生育期耗水量的变化趋势基本相同；与T1、T3、T4处理相比，T2处理的番茄植株根系活力、产量和水分利用效率为最高，分别为1.08mg·g^{-1}·h^{-1}、0.71Kg·pot^{-1}、17.8g·Kg^{-1}。负水头吸力值60hPa（T2）处理的番茄植株耗水量最小、植株根系活力及产量最高，表明吸力值60hPa的土壤含水量最有利于番茄生长，为温室番茄生长节水灌溉和负水头灌溉合理应用于番茄等温室作物优质高产提供一定的理论依据。

关键词：负水头灌溉；番茄；耗水量；光合特性；根系活力

Effect of Tomato Growth Characteristics and Water Use Efficiency at Different Growth Stages with Negative Pressure Irrigation

Zhang Jia[1]，Qin Yuanyuan[1]，Guo Wenzhong[2]，Yu Ligen[2]，Li Lingzhi[1]，Li Haiping[1*]

（1.*College of Horticulture*，*Shanxi Agricultural University*，*Taigu*　030801；2. *Beijing Agricultural Intelligent Equipment Technology Research Cente*，*Beijing*　100097）

Abstract：The aim of test was to study the effects of negative pressure value on the water consumption，photosynthesis，root activity and the the yield of tomato during different growing stages through greenhouse pot experiment to provide a theoretical basis for the rational application of negative pressure irrigation in tomato growth and production.The negative pressure irrigation device was used to set the four negative pressure values of 50 hPa（T1），60hPa（T2），70hPa（T3）and 80 hPa（T4）respectively. Measured water consumption and photosynthetic characteristics during different growth stages，and plant root activity and yield，and comparative

---

作者简介：张佳（1992–），女（汉），山西晋城人，硕士研究生，研究方向为蔬菜栽培与生理

*通讯作者：李海平，男，副教授，硕士生导师，Tel：18703545568；E-mail：lihp0205@163.com

基金项目："十三五"国家重点研发计划（2017YFD0201503）；国家自然科学基金青年科学基金（51509005）；山西省研究生联合培养基地人才培养项目（2016JD22） analysis.By comparing the water consumption of tomato during different growth stages, The water consumption at the initial fruit stage of T1, T2, T3 and T4 treatments was up to the highest 22.05 $kg \cdot pot^{-1}$, 18.89 $kg \cdot pot^{-1}$, 21.11 $kg \cdot pot^{-1}$, 18.46 $kg \cdot pot^{-1}$ respectively. The photosynthetic rate and transpiration rate of tomato plants increased firstly and then decreased with the growth of tomato plants, and the trend was same with that of tomato. Compared with T1, T3 and T4 treatments, the root activity , yield and water use efficiency of T2-treated tomato plants were the highest, 1.08 $mg \cdot g^{-1} \cdot h^{-1}$, 0.71 $kg \cdot pot^{-1}$, 17.8 $g \cdot kg^{-1}$ respectively.The negative pressure value of 60 hPa（T2）treated tomato plants had the lowest water consumption and the highest plant root activity and yield, indicating that the 60 hPa soil water content was the most conducive to tomato growth, This study will provide some theoretical basis of negative pressure irrigation Reasonable application of high quality and high yield of tomato and other greenhouse crops.

Key words: Negative pressure irrigation; Tomato; Water consumption; Photosynthetic characteristics; Root activity

番茄作为我国主要设施蔬菜之一，属喜温作物，其营养丰富，口感极佳；土壤水分条件对番茄植株生长影响较大[1]，当土壤水分亏缺时，水势降低，造成番茄植株水分供应减少，破坏植株体内的水分代谢平衡，直接影响番茄植株的生长发育[2-3]。因此，研究不同土壤含水量对番茄生长的影响有重要意义。负水头灌溉装置通过调节吸力值使土壤含水量维持在一个稳定范围内，可避免土壤的干湿交替[4]。余礼根[5]等研究不同供水模式对番茄生长、光合特性、耗水量等的影响，结果表明：与称重法灌溉相比，负水头灌溉显著促进番茄植株的生长，且负水头灌溉下番茄耗水量降低，产量显著增加；目前关于番茄不同生育期的研究多为番茄抗病性[6]、矿质元素含量[7]等。对于负水头灌溉下番茄不同生育期耗水量及光合特性的研究较少，本文以“佳丽14号”番茄为试验材料，采用负水头灌溉，设置4个不同吸力值，研究不同处理下番茄不同生育期耗水量、光合特性、及植株根系活力和产量的变化，以期为负水头灌溉在番茄生长生产中的合理应用提供理论依据。

1 试验材料与方法

1.1 试验材料

试验于2017年3—7月在北京市农林科学院试验温室内进行（39° 56′32.60″N，116° 16′53.73″E）。以“佳丽14号”番茄为试验材料，采用温室盆栽栽培，每盆装供试土壤11.0kg，供试土壤基本理化性质为：土壤容重为1.5$g \cdot m^{-3}$，最大田间持水量为25%（即体积含水率为37.5%），EC为0.5$ms \cdot cm^{-1}$。

1.2 试验设计

番茄幼苗于育苗室内培养至五叶一心，2017年3月15日定植于试验温室塑料花盆中，盆的上口直径为34.2cm，底直径18.5cm、盆高22.3cm，每盆定植长势一致的番茄幼苗1株，共定植20株，对所有供试番茄浇透水缓苗15d，至4月2日起设置不同吸力值分别为50hPa（T1）、60hPa（T2）、70hPa（T3）、80hPa（T4）进行供水控水处理，对应的土壤含水量分别为22.7%、19.7%、14.4%、11.9%。为研究不同生育期耗水量和光合特性的变化规律，将番茄生育期划分为幼苗期（3月15日至4月1日）、开花期（4月2—14日）、初果期（4月15日至5月15日）和盛果期（5月16日至6月20日）。

1.3 试验主要仪器与设备

负水头灌溉盆栽试验装置如图1所示。负水头灌溉试验装置主要由塑料盆、陶瓷盘、首部、控压管和导气管构成；当装置运行时，控压管及首部上的阀门属于关闭状态，进水管的阀门属于打开状态，首部内水分进入陶瓷盘，陶瓷盘空腔内气体全部排到首部，由于土壤基质势的作用陶瓷盘内水分缓慢进入土壤，致使陶瓷盘内压强逐渐减小，首部内水分再次进入陶瓷盘，如此不断循环，使首部内水分在负压控制下连续不断进入土壤以供土壤蒸发及番茄植株生长发育[8]。

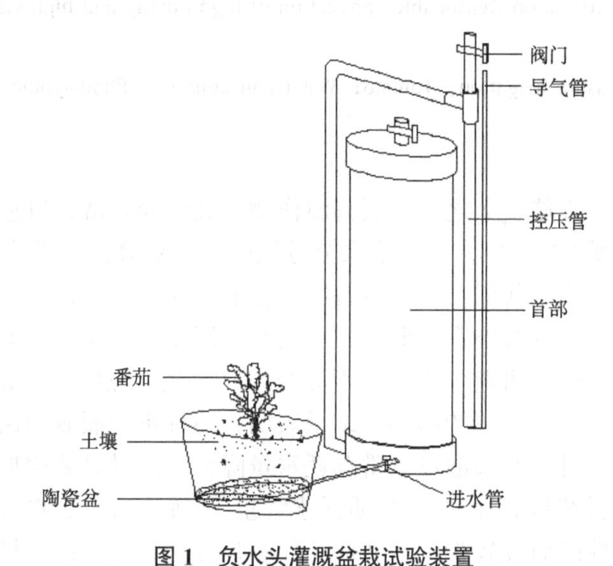

图1 负水头灌溉盆栽试验装置

Figure 1 Pot experiment device with negative pressure irrigation

2 测定指标及方法

2.1 耗水量的测定

在番茄生育期内每天9:00使用非接触式管道液位红外传感器（WS03A，CAEA electrical appliance co.，ltd，Beijing，China）读取负水头灌溉盆栽试验装置首部内的水位高度h。

日耗水量计算公式为：$V=A\times(h_2-h_1)$

式中：V为日耗水量（cm^3），A为负水头灌溉盆栽试验装置首部内径对应的横截面积，h_2（cm）为测量当天的水位高度值，h_1（cm）为测量前一天水位高度值；各生育期耗水量为各生育期内每日耗水量之和。

2.2 光合特性的测定

选取番茄各生育期中的一个晴天9:00—11:00测定，采用LI-6400XT光合作用测量系统（LI-COR Biotechnology，Nebraska，USA）测定番茄叶片净光合速率和蒸腾速率等光合参数。每个处理测定3个重复，选取生长点下第4片叶片为所测量的叶片。

2.3 根系活力和产量的测定

番茄生长后期分别选取不同处理的植株根系（根尖部分）测定根系活力，采用TTC法测定。从果实成熟开始，每2d采摘1次，每次采摘完后，分别称出不同处理当天采摘的番茄质量。产量统计用15kg电子称称量，保留两位小数。

2.4 数据处理

采用Excel2017计算番茄不同生育期耗水量、叶片净光合速率和蒸腾速率等光合特性、植株根系活力及产量并绘制图表；采用SPSS19.0进行差异显著性分析。

3 结果与分析

3.1 不同生育期番茄耗水量的变化

不同负水头吸力值下番茄各生育期耗水量变化如图2所示。由图2可知，番茄不同生育期耗水量有一定差别，随番茄生育进程的变化，T1、T2、T3、T4处理的番茄各生育期耗水量均表现为：初果期>盛果期>开花期>幼苗期，T1、T2、T3、T4处理的番茄初果期耗水量分别为22.05kg·pot^{-1}、18.89kg·pot^{-1}、21.11kg·pot^{-1}、18.46kg·pot^{-1}；与番茄初果期耗水量相比，T1、T2、T3、T4处理的番茄盛果期耗水量分别减少了21.5%、31.3%、48.4%、11.9%；与番茄初果期和盛果期耗水量相比，T1、T2、T3、T4处理的番茄幼苗期和开花期耗水量明显减少，幼苗期与开花期各处理间的耗水量相差较小。

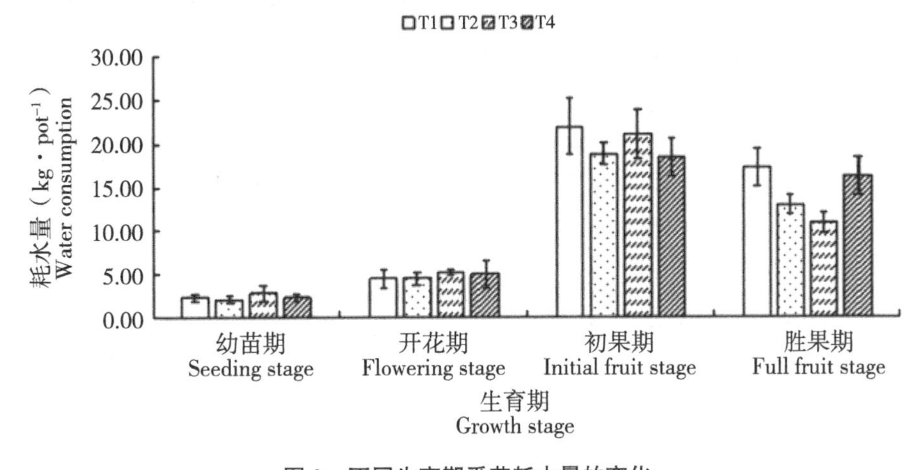

图2 不同生育期番茄耗水量的变化

Figure 2 Changes of water consumption of tomato in different growth stages

3.2 不同生育期番茄光合特性的变化

图3（a）是不同负水头吸力值下各生育期番茄叶片净光合速率的变化情况。由图3（a）可知，T1、T2、T3、T4处理的番茄叶片净光合速率随番茄植株的生长呈现先增大后减小的趋势；T2、T3、T4处理的番茄叶片净光合速率在番茄初果期达到最大，分别为14.80μmol·m^{-2}·s^{-1}、14.42μmol·m^{-2}·s^{-1}、15.53μmol·m^{-2}·s^{-1}；在番茄幼苗期和盛果期，T1、T2、T3、T4处理的番茄净光合速率相差较小。

不同负水头吸力值下各生育期番茄叶片蒸腾速率的变化情况如图3（b）所示。由

图3（b）可知，T1、T2、T3、T4处理的番茄叶片蒸腾速率随番茄植株生育期进程的变化先增大后减小，T1、T2、T3、T4处理的番茄初果期叶片蒸腾速率达到最大，分别为11.12mmol·m^{-2}·s^{-1}、11.46mmol·m^{-2}·s^{-1}、10.06mmol·m^{-2}·s^{-1}、12.20mmol·m^{-2}·s^{-1}；在番茄幼苗期，T2处理的番茄叶片蒸腾速率高于其他处理；在番茄开花期，T2、T3处理的番茄叶片蒸腾速率相差较小。

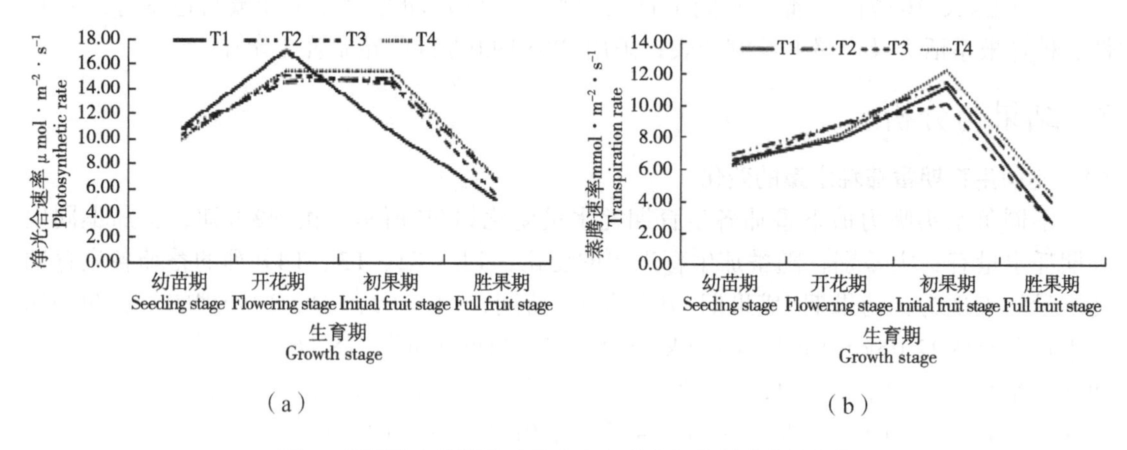

图3　不同生育期番茄叶片净光合速率和蒸腾速率的变化

Figure 3　Changes of photosynthetic rate and transpiration rate of tomato leaf in different growth stages

3.3　不同吸力值处理的番茄根系活力和产量变化

不同处理对番茄根系活力和产量的影响见表1。由下表可知，T1、T2、T3、T4处理的番茄根系活力先增大后减小，T2处理的番茄根系活力最高，达到1.08mg·g^{-1}·h^{-1}。T1、T2、T3、T4处理的番茄产量表现为：T2>T1>T3>T4，T2处理的番茄产量最高，达到0.71kg·pot^{-1}，与T1、T3、T4处理相比，增加幅度分别为4.3%、14.4%、30.1%。

表　不同处理对番茄根系活力和产量的影响

Table 1　Effects of different treatments on plant yield of tomatoes

	T_1	T_2	T_3	T_4
根系活力/mg·g^{-1}·h^{-1} Root activity	0.97 ± 0.06ab	1.08 ± 0.08b	0.99 ± 0.05b	0.94 ± 0.04c
产量/kg·pot^{-1} Yield	0.68 ± 0.07a	0.71 ± 0.05a	0.62. ± 0.08ab	0.55 ± 0.03b

注：不同小写字母表示处理间差异达0.05显著水平

Note：Different lowercase letters indicate a significant difference of 0.05 between treatments

3.4　不同吸力值处理对番茄水分利用效率的影响

图4是不同处理的番茄水分利用效率的变化情况。由图4可知，T1、T2、T3、T4处理的番茄水分利用效率呈先增大后减小的趋势，T2处理的番茄水分利用效率最高，为17.8g·kg^{-1}；与T2处理的番茄水分效率相比，T1、T3、T4处理的番茄水分利用效率分别减少17.4%、12.9%、34.8%。

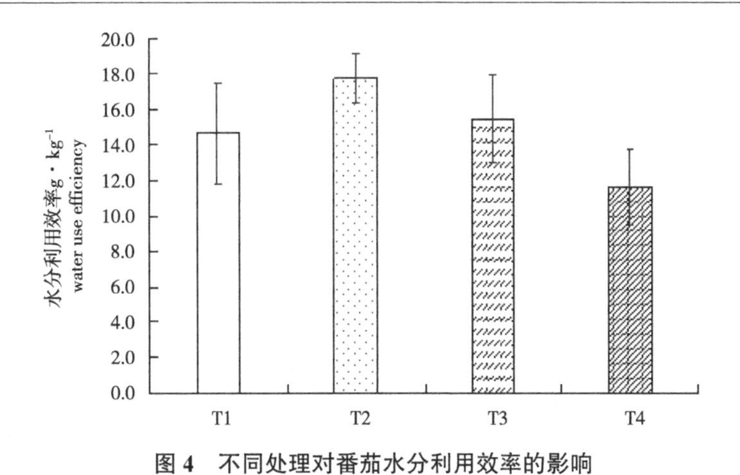

图 4　不同处理对番茄水分利用效率的影响

Figure 4　Effects of different treatments on water use efficiency of tomatoes

4　讨论与结论

负水头灌溉作为一种新型节水灌溉方式，主要是通过设置不同的吸力值来控制土壤的不同含水量，实现对土壤含水量的精准和持续控制[9]。本试验结果表明：不同负水头吸力值下番茄各生育期耗水量有一定差异，随番茄生育期进程的变化，各处理的番茄耗水量均表现为：初果期>盛果期>开花期>幼苗期，幼苗期和开花期耗水量明显低于初果期和盛果期，由于初果期是番茄植株生长的重要时期，需同时满足番茄果实生长和叶片花序发育对水分的需求[10]，因此，初果期和盛果期耗水量明显高于幼苗期和开花期。

卫如雪[11]等研究不同供水吸力对番茄花期光合特性的影响，结果表明：番茄叶片在上午10:00净光合速率和蒸腾速率随吸力值的增加先减小后增大，本试验结果表明：在番茄开花期叶片净光合速率和蒸腾速率随吸力值的增加先减小后增大，与卫如雪[11]等的研究结果基本一致。50hPa、60hPa、70hPa、80hPa处理的番茄叶片净光合速率和蒸腾速率随番茄植株的生长呈现先增大后减小的趋势，与番茄各生育期耗水量的变化趋势基本相同，说明番茄各生育期耗水量与番茄叶片净光合速率和蒸腾速率之间有一定的相关性。余礼根[10]等以番茄为试材，研究不同供水吸力对其生长、产量及水分利用效率的影响，结果表明：60hPa吸力值处理的番茄株高增长率最高为31.5%，50hPa和60hPa处理的番茄产量较高，60hPa处理的番茄水分利用效率为23.4g·kg^{-1}，本试验结果表明，50hPa、60hPa处理的番茄产量高于70hPa、80hPa处理，分别达到0.68kg·pot^{-1}、0.71kg·pot^{-1}，T2处理的番茄水分利用效率最高为17.8g·kg^{-1}，与余礼根[5]等的研究结果基本一致。本文试验结果表明：各处理间番茄根系活力有显著差异（$P<0.05$），且60hPa处理的番茄根系活力最高，因此，负水头吸力值为60hPa的土壤含水量更适合温室番茄植株根系的生长发育。

综合分析不同处理的番茄耗水量、光合特性、根系活力、产量和水分利用效率可知，负水头吸力值为60hPa的土壤含水量更适合温室番茄植株的生长发育，为负水头灌溉技术合理应用及番茄生长最佳灌溉决策提供理论依据。

参考文献

［1］ 李霞，解迎革，薛绪掌，等.不同基质含水量下盆栽番茄蒸腾量、鲜物质积累量及果实产量的差异［J］.园艺学报，2010，37（5）：805-810.

［2］ 赵平，孙谷畴，倪广艳，等.成熟马占相思水力导度对水分利用和光合响应的季节性差异［J］.应用生态学报，2013，24（1）：49-56.

［3］ 张志焕，韩敏，张逸，等.水分胁迫对不同抗旱性砧木嫁接番茄生长发育及水气交换参数的影响［J］.中国农业科学，2017，50（2）：391-398.

［4］ Cai Y，Wu P，Zhang L，et al. Effects of designed flow rate and soil texture on infiltration characteristics of porous ceramic irrigation emitters［J］.Transactions of the Chinese Society of Agricultural Engineering，2017，33（7）：100-106.

［5］ 余礼根，卫如雪，白红武，等.不同供水模式对番茄生长及声发射的影响［J］.江苏农业科学，2017，45（17）：105-108.

［6］ 孔涛，姜飞，丁永发，等.嫁接番茄不同生育期抗病性及其与根际环境的关系［J］.山东农业科学，2014（1）：94-96.

［7］ 李娟，田萍，李建设，等.微咸水灌溉方式对不同生育期设施番茄矿质元素含量的影响［J］.华北农学报，2017，32（2）：200-210.

［8］ 张芳，张建丰，薛绪掌，等.气象因子对负水头供液下番茄日耗液量的敏感性分析［J］.农业机械学报，2017，48（8）：229-238.

［9］ Li Y，Xue X，Qian Z，et al. Automatic measurement of greenhouse tomato evapotranspiration based on negative pressure irrigation system［J］Transactions of the Chinese Society of Agricultural Engineering，2017，33（10）：137-144.

［10］ 张佳，李海平，李灵芝，等.空间电场对番茄初果期生长及生理特性的影响［J］.山西农业大学学报（自然科学版），2017，37（11）：785-788.

［11］ 卫如雪，余礼根，郭文忠，等.不同供水吸力对番茄光合特征与声发射特性的影响［J］.北方园艺，2017（20）：93-99.

CO_2增施对四川弱光区设施黄瓜叶片光系统功能的影响

张泽锦[1,2]，唐丽[1,2]，李跃建[2]，刘小俊[1,2]

（1. 四川省农业科学院园艺研究所　蔬菜种质与品种创新四川省重点实验室，成都　610066；
2. 农业部西南地区园艺作物生物学与种质创制重点实验室，成都　610066）

摘要：为探究CO_2增施对弱光区设施黄瓜叶片光系统II、光系统I、跨膜质子梯度及生长的影响，以‘川翠3号’黄瓜为材料，利用吊袋式CO_2气肥进行黄瓜的CO_2增施处理，采用分光光度计、LI-6400便携式光合测定仪和Dual-PAM-100双通道荧光测定仪分别测定处理和对照黄瓜叶片的光合色素含量、叶片气体交换参数、光系统II和光系统I荧光参数及跨膜质子梯度的550～515nm差示信号等参数；此外，测定黄瓜叶片的叶面积、单果重、单株结果数、株高等生长指标。结果表明，CO_2增施使黄瓜叶片叶面积、叶绿素b、总叶绿素含量、表观量子效率（AQE）增高，而叶绿素a/b下降；PSII和PSI的最大电子传递速率和跨膜电位（$\Delta\psi$）增高导致pmf增高。此外，CO_2增施使得黄瓜单果重、单株结果数及株高均有增加。由此可见，叶片接受的光能更多的用于CO_2同化作用，同时提高了黄瓜叶片的弱光利用率和抗光氧化的能力，最终增加单果重和单株结果数。

关键词：CO_2升高；PSII；PSI；跨膜质子梯度；黄瓜；生长

Effects of CO_2 Enrichment on Functions of Photosystem in Cucumber Leaves in Poor Light Region of Sichuan Basin

Zhang Zejin[1,2], Tang Li[1,2], Li Yuejian[2], Liu Xiaojun[1,2]

（1. *Horticulture Research Institute, Sichuan Academy of Agricultural Sciences / Vegetable Germplasm Innovation and Variety Improvement Key Laboratory of Sichuan Province, Chengdu*　610066；2. *Key Laboratory of Horticultural Crop Biology and Germplasm Enhancement in Southwest, Ministry of Agriculture, Chengdu*　610066）

Abstract: In this paper, the effect of CO_2 enrichment on the photosystem II（PS II）, photosystem I（PS I）and proton motive force in cucumber leaves and its growth were investigated in poor light region by using hanging bag type CO_2 fertilizer. The chlorophyll content, gas exchange parameters, photosystem II and photosystem I fluorescence parameters of the treated and controlled cucumber leaves were measured by spectrophotometer, LI-6400 portable photosynthesis measuring instrument and Dual-PAM-100 fluorometer. In addition, the leaf area, fruit weight, fruit number per plant, plant height of cucumber leaves were measured. The results showed that elevated CO_2 concentration increased the cucumber leaf area, chlorophyll b content, total chlorophyll content and apparent quantum efficiency（AQE）, decreased the chlorophyll a/b. Elevated CO_2 concentration increased maxium electric transportation rate of PSII and PSI, membrane potential（$\Delta\psi$）and proton motive force（pmf）. Elevated CO_2 concentration increased fruit weight, fruit number per plant. The results indicated that the light

作者简介：张泽锦，男，四川成都人，博士，助理研究员，主要从事蔬菜设施栽培研究。E-mail：zhangzj127@163.com received by the leaves is more used for CO_2 assimilation, and the weak light utilization and anti-photooxidation of cucumber leaves are improved under CO_2 enrichment condition. Finally, the fruit weight and fruit number per plant is increased.

Key words: Elevated CO_2 Concentration; Photosystem II; Photosystem I; Proton motive force; Cucumber; Growth

蔬菜作物干物质产量的90%～95%则是由光合作用所制造，刘保才等（1995）用通径分析证明，影响日光温室蔬菜产量的三大环境因素中，CO_2浓度对产量形成贡献最大。位于中国的西南地区的四川盆地全年的平均太阳辐射小于1 500MJ·m^{-2}，属于中国光照资源5类地区。黄瓜是该地区设施栽培的主要品种之一，约占四川设施栽培面积25.9%。关于黄瓜的光合作用前人分别从气体交换、光合关键酶及生长发育等方面做了大量工作，但其中关于叶片光系统功能的研究主要集中在光系统Ⅱ（PSⅡ）进行叶绿素荧光的探讨，对于光系统Ⅰ（PSⅠ）和跨膜质子梯度的影响却鲜有报道，CO_2浓度升高对黄瓜叶片2个光合系统影响的机理尚未明确；此外，有关CO_2增施对黄瓜生长等方面的研究主要集中在北方密闭性较好且光照充足的日光温室中，但针对南方弱光地区生产大量采用密闭性较差的塑料大棚中黄瓜生产类似研究较少，经济适用且易于推广的CO_2施肥方式缺乏相应推广应用的理论支撑。为了进一步明确CO_2增施对弱光区黄瓜叶片光系统功能的作用机制和吊袋式CO_2增施技术在设施黄瓜促成生产中的应用价值，本试验研究CO_2增施对黄瓜叶片PSI、PSII、跨膜质子梯度以及生长的影响，为四川弱光区设施黄瓜促成高产栽培技术提供理论依据。

1 材料和方法

1.1 试验材料与试验设计

试验于2016年1月至7月在四川省农业科学院园艺研究所塑料大棚进行。以黄瓜（品种：川翠3号）为实验材料，育苗于2月1日进行，待黄瓜苗长到三叶一心时，定植于塑料大棚中，1周后进行实验处理。CO_2增施期间光照强度最高为1 212μmol·m^{-2}·s^{-1}。CO_2增施采用吊袋式CO_2气肥的增施方法，每包CO_2发生剂重110g，1个塑料大棚均匀吊挂16个CO_2施肥袋，每个施肥袋吊挂位置距地面1.5m处。另外，1个大棚无CO_2吊袋设为CK，每隔15d更换1次CO_2增施包。由于气温的逐渐升高，CO_2增施于4月30日结束。3月和4月根据天气状况10:00左右进行开启大棚天窗进行放风除湿。2个棚内黄瓜施肥按照常规处理进行。

1.2 测定项目及方法

1.2.1 大棚内CO_2浓度的测量

为了准确监测温室内CO_2浓度动态变化情况，自主集成多通路CO_2连续监测系统，可以连续长期同时且多点记录CO_2浓度变化。本系统主要由5个CO_2浓度传感器（Esense，SenseAir，瑞典）和1个记录系统（RX400A，美控，中国）构成，CO_2浓度传感器与记录系统通过数据线相连。经校正，5个CO_2浓度传感器之间无显著差异。

1.2.2 叶片叶绿素含量及叶面积测定

本试验于CO_2增施时（3月18日和4月3日）和CO_2增施结束后（5月11日和5月14日），选取植株中部的黄瓜叶片参照杨劲峰等（2002）方法进行叶面积测量；叶片叶绿素含量采用丙酮法进行测定。

1.2.3 气体交换参数的测定

本试验于4月上旬，选择晴天对黄瓜叶片光合特性进行测定。黄瓜叶片光合速率测定用便携式光合作用测量系统（LI-6400，LI-COR公司，美国），对试验区内选取的5株黄瓜，测定植株中部叶片的光响应曲线。

1.2.4 叶片荧光参数的测定

（1）光系统诱导荧光曲线的测量。黄瓜叶片暗适应2h的叶片，利用双通道荧光仪dual-PAM100（Walz，德国）中诱导曲线测量程序测定诱导荧光曲线，Actinic light设置为8档。

（2）光响应荧光曲线的测量。在使用光适应的叶片，利用双通道荧光仪Dual-PAM-100（Walz，德国）中light curve测量程序测定黄叶片PSII和PSI快速荧光响应曲线。

（3）快速叶绿素荧光诱导动力学曲线的测量。利用双通道荧光仪Dual-PAM-100（Walz，德国）自动测量程序，在叶片暗适应180min后测定黄瓜叶片快速叶绿素荧光诱导动力学曲线，JIP-test参数根据Strasser等（2000）计算得到。

（4）暗光暗诱导的550～515nm差示信号测定。根据Schreiber and Klughammer（2008）方法，利用双通道荧光仪dual-PAM-100的P515/535模块系统及其自动程序测定。

1.3 数据分析

试验数据的处理及分析利用Microsoft Excel 2010和SPSS 18.0软件完成。分别对各试验区的试验数据进行单因素方差分析。

2 结果与分析

2.1 吊袋式CO_2增施方式对塑料大棚内CO_2浓度日变化的影响

晴天CO_2增施棚内CO_2浓度最高约为550mmol·mol^{-1}，最低浓度约为320 mmol·mol^{-1}，对照棚内最高CO_2浓度为480mmol·mol^{-1}，最低CO_2浓度为310mmol·mol^{-1}；阴天CO_2增施棚内CO_2浓度最高约为510mmol·mol^{-1}，最低浓度约为400mmol·mol^{-1}，对照棚内最高CO_2浓度为470mmol·mol^{-1}，最低CO_2浓度为380mmol·mol^{-1}（图1）。由此可见，设施内CO_2浓度变化呈现“U”型，午后随着植物光合作用的加快，设施内CO_2浓度降低，但夜晚，植物主要是呼吸作用，所以设施内CO_2浓度增高。吊袋式CO_2增施方法在晴天主要在光照强度和温度上升后，CO_2增施加快；阴天则是全天缓慢释放CO_2，但是释放量低于晴天。

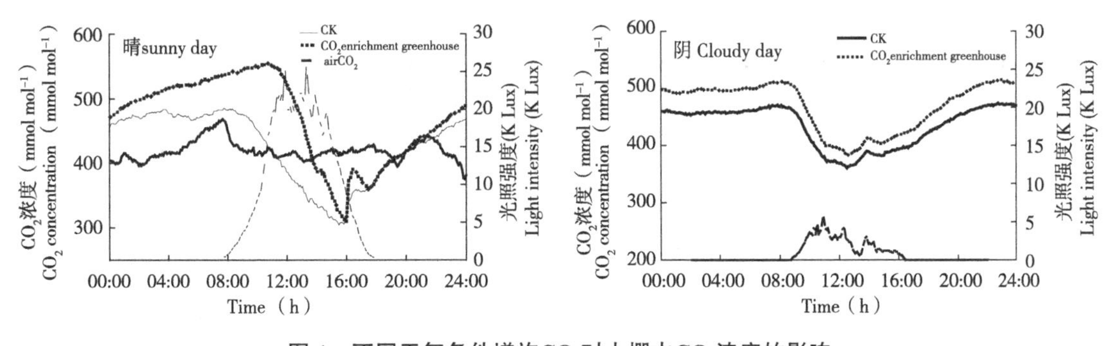

图1 不同天气条件增施CO_2对大棚内CO_2浓度的影响

Figure 1 Effect of CO_2 enrichment on CO_2 concentration in greenhouse under different weather conditions

2.2 CO_2增施对叶片光合系统I和光合系统II的影响

CO_2增施处理中PSI的量子产量［Y（I）］、相对电子传递速率［ETR（I）］以及由于受体侧限制引起的PSI处非光化学能量耗散的量子产量［Y（NA）］与对照黄瓜叶片无显著差异，但CO_2增施处理供体侧引起的由于供体侧限制引起的PSI处非光化学能量耗散的量子产量［Y（ND）］显著低于对照，为对照处理的97.2%。PSII的量子产量［Y（II）］、相对电子传递速率［ETR（II）］以及光保护和光损伤的两个参数PSII处非调节性能量耗散的量子产量［Y（NO）］和PSII处调节性能量耗散的量子产量［Y（NPQ）］在CO_2增施处理和对照间均无显著差异（表1）。

表1　CO_2增施对黄瓜叶片PSI和PSII量子产量及电子传递速率的影响（n=12）

Table 1　Effect of CO_2 enrichment on PSI and PSII quantum yield and relative electron transport rate of cucumber leaves（n=12）

	Y（Ⅰ）	ETR（Ⅰ）	Y（ND）	Y（NA）
CO_2增施 CO_2 enrichment	0.196 ± 0.002	76.100 ± 0.954	0.729 ± 0.005	0.076 ± 0.016
CK	0.193 ± 0.003	76.000 ± 2.722	0.750 ± 0.004	0.087 ± 0.028
ANOVA	NS	NS	*	NS
	Y（Ⅱ）	ETR（Ⅱ）	Y（NO）	Y（NPQ）
CO_2增施 CO_2 enrichment	0.118 ± 0.005	45.800 ± 1.997	0.341 ± 0.014	0.541 ± 0.010
CK	0.127 ± 0.016	45.800 ± 2.252	0.321 ± 0.007	0.551 ± 0.013
ANOVA	NS	NS	NS	NS

通过荧光响应曲线的测量，采用Eilers and Peeters（1998）的公式进行光响应曲线的拟合得到快速光曲线的初始斜率（α）、最大电子传递速率（ETRmax）及最小饱和光强（I_K）。从表2可以看出，CO_2增施提高了PSII和PSI的ETRmax以及PSI的I_K，分别比对照提高了40.2%，28.4%及20.4%。CO_2增施未提高PSII和PSI的快速荧光初始斜率和PSII的最小饱和光强（I_K）。

表2　CO_2增施对黄瓜叶片光响应荧光参数的影响（n=12）

Table 2　Effect of CO_2 enrichment on light response curve fluorescence parameters in cucumber leaves（n=12）

	PSⅡ		
	α（electrons/photons）	ETRmax（$\mu mol \cdot m^{-2} \cdot s^{-1}$）	I_K（$\mu mol \cdot m^{-2} \cdot s^{-1}$）
CO_2增施 CO_2 enrichment	0.46 ± 0.05	56.95 ± 1.91	113.90 ± 7.48
CK	0.41 ± 0.02	40.63 ± 6.65	100.93 ± 2.40
ANOVA	NS	*	NS

（续表）

	PSI		
	α（electrons/photons）	ETRmax（$\mu mol \cdot m^{-2} \cdot s^{-1}$）	I_K（$\mu mol \cdot m^{-2} \cdot s^{-1}$）
CO_2增施 CO_2 enrichment	0.50 ± 0.03	72.27 ± 11.25	143.67 ± 14.40
CK	0.47 ± 0.02	56.27 ± 8.96	119.37 ± 13.80
ANOVA	NS	*	*

2.3　CO_2增施对叶片光合系统Ⅱ能量分配的影响

初始荧光（F_o）是PSII作用中心完全开放即所有的电子受体QA、QB和PQ等最大程度氧化态时的荧光产量，此时的荧光强弱与天线色素含量的多少及作用中心的活性状态有关。天然色素降解和非光化学能量耗散易造成F_o的降低，而光合机构的可逆失活又使其升高。黄瓜CO_2增施处理的初始荧光F_o与未增施的对照F_o分别为0.20和0.19，无显著差异，最大光化学效率Fv/Fm反应了潜在的量子效率，CO_2增施和对照分别为0.80和0.79，无显著差异。

从PSII中能量分配来看，CO_2增施处理黄瓜叶片反应中心捕获的用于电子传递的能量（ET_o/RC）为1.70，比对照显著增加了7.6%；而处理与对照的黄瓜叶片单位反应中心吸收的光能（ABS/RC），单位反应中心捕获的用于还原QA的能量（TR_o/RC）以及单位反应中心热耗散的能量（DI_o/RC）无显著差异。以上结果表明，CO_2增施黄瓜叶片PSII反应中心捕获的用于电子传递的能量显著增加（ET_o/RC）可能是导致光反应最大电子传递速率的提高的原因之一。

PI_{ABS}指以吸收光能为基础的性能指数，由捕获的激子将电子传递到电子传递链中超过QA的其他电子受体概率（ψ_o）、光合活性反应中心的数目（RC/ABS）和最大光化学效率（ϕP_o）三个参数组成。CO_2增施处理的RC/ABS比对照处理增加25%；ψo和ϕPo两个参数处理与对照之间无显著差异。CO_2增施处理的黄瓜叶片PI_{ABS}比CK显著提高了11.0%。因此，CO_2增施使黄瓜叶片单位面积上光合活性反应中心的数目显著增加可能是提高其叶片光合的性能指数增高的主要原因（表3）。

表 3　CO_2增施对黄瓜叶片叶绿素荧光响应参数的影响（n=12）

Table 3　Effect of CO_2 enrichment on chlorophyll fluorescence parameters in cucumber leaves（n=12）

		CO_2增施 CO_2 enrichment	CK	ANOVA
PSⅡ能量分配 PSⅡ energy distribution	F_o	0.20 ± 0.01	0.19 ± 0.01	NS
	Fv/Fm	0.80 ± 0.01	0.79 ± 0.01	NS
	ABS/RC	3.73 ± 0.08	3.55 ± 0.21	NS
	TR_o/RC	2.93 ± 0.06	2.78 ± 0.15	NS
PSⅡ能量分配 PSⅡ energy distribution	DI_o/RC	0.80 ± 0.02	0.77 ± 0.06	NS
	ET_o/RC	1.70 ± 0.03	1.58 ± 0.02	*

（续表）

		CO_2增施 CO_2 enrichment	CK	ANOVA
光合性能指数 Photosynthetic performance index	RC/ABS	0.35 ± 0.01	0.28 ± 0.02	*
	ψo	0.58 ± 0.01	0.57 ± 0.02	NS
	ϕ_{Po}	0.79 ± 0.01	0.78 ± 0.01	NS
	PI_{ABS}	1.51 ± 0.05	1.36 ± 0.21	*

2.4　CO_2增施对叶片跨膜质子动力势的影响

在光化光关闭后，暗—光—暗诱导的550 ~ 515nm差示信号先下降后上升，从550 ~ 515nm差示信号曲线变化可得出跨膜质子梯度（ΔpH）、跨膜电位（Δψ）、跨膜质子动力势（pmf）及可逆玉米黄素（Zeaxanthin，Z）相对含量的变化。

黄瓜叶片暗适应时与经过照光关灯后测量结束时550 ~ 515nm差示信号的差值代表叶片可逆玉米黄素相对含量。如图2和表4所示，CO_2增施处理后“暗基线”的增加显著高于对照，玉米黄素相对含量为16.63，比对照增加34.7%，CO_2增施处理促进了可逆玉米黄素的形成。以上结果说明CO_2处理的黄瓜叶片比对照具有更强减轻黄瓜叶片光氧化胁迫的能力。

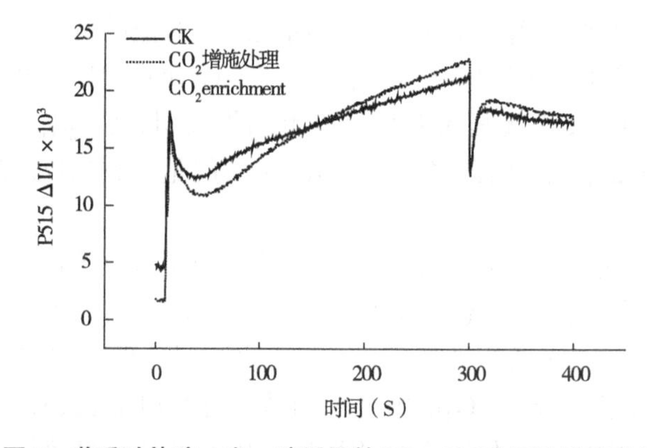

图2　黄瓜叶片暗—光—暗诱导的550 ~ 515差示信号的变化

Figure 2　Effect of CO_2 enrichment on slow dark-light-dark induction transients of the 550 ~ 515 nm difference signal of cucumber leaves

表4　CO_2增施对黄瓜叶片跨膜质子参数的影响（n=12）

Table 3　Effect of CO_2 enrichment on membrane potential（Δψ），proton gradient（ΔpH），proton motive force（pmf）and relative content of xeaxanthin（n=12）

	Δψ	ΔpH	pmf	玉米黄素相对含量 relative content of xeaxanthin
CO_2增施 CO_2 enrichment	4.97 ± 0.38	5.13 ± 0.76	10.10 ± 0.46	16.63 ± 1.97
CK	3.83 ± 0.25	4.63 ± 0.80	8.47 ± 0.95	12.35 ± 1.65
ANOVA	*	NS	*	*

光化光关闭后的550～515nm差示信号曲线变化，可反映叶片Δψ及ΔpH大小。光化光照射达到稳定时550～515nm差示信号与关闭光化光并达到稳态后550～515nm差示信号的差值，反映叶片类囊体膜Δψ相对大小；关闭光化光达到稳态后P515信号与关闭光化光后信号下降的最低点差值反映类囊体膜ΔpH相对大小[1, 2]。CO_2增施显著提高了Δψ和pmf，分别比对照处理提高29.8%和19.2%；两处理的ΔpH无显著差异。pmf增高的主要原因是由于Δψ的增高，因为较高的Δψ提高了类囊体膜上的代谢活性及蛋白运转，同时提高了膜的渗透性。

2.5　CO_2增施对叶片光响应气体交换参数的影响

由表5可以看出，CO_2处理显著提高了黄瓜叶片的表观量子效率（AQE），比对照提高了16.7%；暗呼吸速率（Rd）同时也显著提高，是对照处理的1.24倍。光饱和点，光补偿点及最大光合速率则处理与对照无显著差异。

表 5　CO_2增施对黄瓜叶片光响应气体交换参数的影响（n=6）

Table 5　Effect of CO_2 enrichment on light response curve parameters in cucumber leaves（n=6）

	Ls（μmol·m^{-2}·s^{-1}）	Lc（μmol·m^{-2}·s^{-1}）	AQE（μmol·$μmol^{-1}$）	Rd（μmol·m^{-2}·s^{-1}）	Pmax（μmol·m^{-2}·s^{-1}）
CO_2增施 CO_2 enrichment	954 ± 203	27 ± 2	0.084 ± 0.001	2.24 ± 0.13	19.6 ± 1.4
CK	1 119 ± 118	27 ± 3	0.072 ± 0.003	1.00 ± 0.14	20.9 ± 0.8
ANOVA	NS	NS	*	*	NS

2.6　CO_2增施对黄瓜叶面积和叶绿素含量的影响

由表6可以看出，4月3日测定CO_2增施处理黄瓜叶片的叶绿素b和总叶绿素含量分别为1.37mg·g^{-1}和5.23mg·g^{-1}，显著高于对照处理，分别是对照处理的1.3和1.1倍。但CO_2增施处理黄瓜叶片的叶绿素a/b为2.00，显著低于对照处理，是对照处理的74.3%。5月14日测定黄瓜叶片叶绿素含量各指标CO_2增施处理与对照无显著差异。

表 6　CO_2增施对黄瓜叶片叶绿素含量的影响（n=15）

Table 6　Effect of CO_2 enrichment on chlorophyll content parameters in cucumber leaves（n=15）

时间 Time	处理 Treatment	叶绿素a（mg·g^{-1}）Chlorophyll a content	叶绿素b（mg·g^{-1}）Chlorophyll b content	总叶绿素（mg·g^{-1}）Total chlorophyll content	叶绿素a/b Chlorophyll a/b
4月3日 3-April	CO_2增施 CO_2 enrichment	2.75 ± 0.09	1.37 ± 0.04	5.23 ± 0.15	2.00 ± 0.04
	CK	2.78 ± 0.08	1.03 ± 0.02	4.64 ± 0.05	2.69 ± 0.12
	ANOVA	NS	*	*	*
5月14日 14-May	CO_2增施 CO_2 enrichment	2.72 ± 0.07	1.34 ± 0.13	5.13 ± 0.26	2.03 ± 0.20
	CK	2.61 ± 0.19	1.37 ± 0.08	5.07 ± 0.27	1.91 ± 0.14
	ANOVA	NS	NS	NS	NS

由图3可以看出，3月15日CO_2增施和对照处理的黄瓜叶面积分别为45.0m^{2}和24.0m^2，CO_2增施处理的黄瓜叶面积比对照增加了87.5%。5月11日测量CO_2增施处理和对照处理的黄瓜叶面积分别为221.7m^{2}和200.5m^2，无显著差异。4月底以后由于气温的上升，塑料大棚均未闭棚，CO_2增施棚的CO_2浓度与对照棚内CO_2浓度均为大气浓度，因此5月CO_2增施和对照处理之间黄瓜中部叶片的叶绿素含量与叶面积无显著差异。

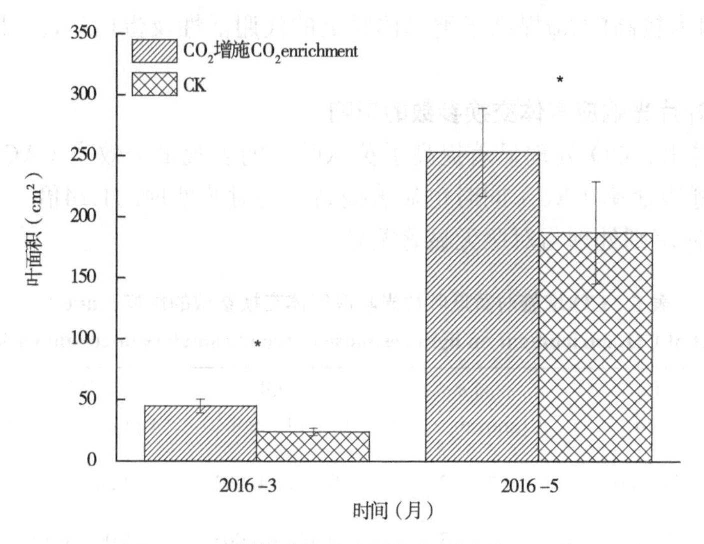

图3　CO_2增施对黄瓜叶面积的影响（n=15）

Figure 3　Effect of CO_2 enrichment on leave area of cucumber（n=15）

2.7　CO_2增施对黄瓜生长的影响

由表7可以看出，CO_2增施提高了黄瓜单果重，收获期平均单果重为0.32kg，显著高于对照处理的0.24kg。收获期结束后，测量黄瓜植株长。CO_2增施处理黄瓜植株长和单株果数分别为257.5cm和7.2，显著高于对照处理的185.8cm和5.4，是对照处理的1.4倍和1.3倍。

表7　CO_2增施对黄瓜生长的影响（单果重、单株果数及植株长n=12）

Table 6　Effect of CO_2 enrichment on growth of cucumber（Fruit weight，Fruit number per plant and Vine length n=12）

	单果重（kg）Fruit weight	单株果数 Fruit number per plant	植株长（cm）Vine length
CO_2增施 CO_2 enrichment	0.32 ± 0.06	7.2 ± 0.2	257.5 ± 43.4
CK	0.24 ± 0.05	5.4 ± 0.3	185.8 ± 15.9
ANOVA	*	*	*

3　讨论

光合作用中较强的CO_2传导能力，较高的光反应和暗反应活性是叶片提高光合速率的生理基础[3]。CO_2不仅是植物光合作用的主要原料，还是作物产量的形成的重要来源。短期提高CO_2浓度不但增加了植物光合作用的碳源，还提高了C_3植物的光合速率[4]。温室内CO_2浓度由200μmol·mol^{-1}提高到400μmol mol^{-1}时，设施内黄瓜、番茄和辣椒的净光合速率提高55%～130%[5]。本研究发现，CO_2增施时期使得黄瓜叶片叶面积和叶绿素含量增高，而叶绿素a/b下降，更好地利用散射光，在弱光条件下提高叶片净光合速率[6]。CO_2增高使得黄瓜叶片表观量子效率（AQE）增高。表观量子效率（AQE）值为弱光阶段的光响应曲线的斜率，其值大小与色素蛋白复合体数目相关。CO_2升高后的黄瓜叶片总叶绿素含量和叶绿素b含量的增高则是导致其利用弱光的能力增强的主要原因。叶片光利用率的提高，有助于减轻西南地区早春的弱光环境对黄瓜生长产生的胁迫。

与植物叶片“表观性”的气体交换参数净光合速率相比，光合荧光参数具有反应光合作用“内在性”特点。本研究发现，CO_2升高提高了黄瓜叶片内的两个光系统的最大电子传递速率，降低了由于供体侧限制引起的PSI处非光化学能量耗散的量子产量（Y（ND）），PSI具有接受更多的电子在供体侧用于光合作用电子传递的能力，从而引起叶片的PSI的最小饱和光强（I_K）增大。此外，PSⅡ和PSI的最大电子传递速率的提升，更多的光能用于电子链中的电子传递，为叶片暗反应提供更多的能量。

跨膜电势（Δψ）和跨膜质子势（ΔpH）不仅可以驱动类囊体膜上ATP酶合成ATP[7]，跨膜质子势（ΔpH）还能通过qE机制耗散过多光能，下调光合效率，从而起到调节光合能力的作用[8]。本试验中，CO_2升高导致黄瓜叶片跨膜电位（Δψ）增高，这说明促进了玉米黄质的形成，通过类囊体ATP酶从腔内流向基质的H^+的量增多。光合作用中另外提供能量的则是内囊体薄膜上的跨膜质子梯度所构成，CO_2增施显著的提高了跨膜质子动力势（pmf），而跨膜电位（Δψ）则是pmf提高的根本原因。更多的氢离子通过跨膜质子动力势的增多而进入内囊体膜，从而通过ATPase流出，产生更多的ATP用于暗反应的能量供应。

与前人报道相比单相比株结果数增加，但单果重并无显著差异是苗期CO_2增施黄瓜增产的主要原因有所不同[9]，本实验黄瓜除了单株果数增加，单果重也有较大幅度增加。由于本实验在黄瓜定植一周后增施CO_2，导致前期叶片弱光捕获能力增加，同时叶片PSII和PSI活性增强，将捕捉到的光能高效的转化为化学能，产生更多的ATP用于暗反应的能量供给，促进了黄瓜叶片CO_2同化能力的增加，为前期黄瓜生长奠定了良好的物质基础；此外，CO_2增施增加了黄瓜花芽的分化，增加了开花数，从而导致单株坐果数的增加。这与番茄叶片在CO_2增加的环境中具有较高的CO_2同化和运转效率，同化产物运转至果实促进了坐果和果实发育类似[10，11]。由此可见，在春季黄瓜塑料大棚促成栽培过程中从定植后开始增施CO_2能显著的促进其生长。

4 结论

CO_2增施增加了黄瓜的叶面积、光合色素量、2个光系统的电子传递效率及类囊体的跨膜质子梯度，叶片接受的光能更多的用于CO_2同化作用，同时提高了黄瓜叶片的弱光利用率和抗光氧化的能力，最终增加单果重和单株结果数。此外，本实验结果可以看出，操作简单且成本较低的吊袋式施肥方式，虽然增加的CO_2浓度不多，但其促进生长效果明显，在四川弱光地区春季黄瓜促成栽培中具有较好的应用前景。

参考文献

[1] Kramer D.A and C.A Sackstedera. Diffused-optics flash kinetic spectrophotometer（DOFS）for measurements of absorbance changes in intact plants in the steady-state［J］. Photosynthesis Research，1998，56（1）：103-112

[2] Cruz J.A.，T.J. Avenson，A. Kanazawa，et al., Plasticity in light reactions of photosynthesis for energy production and photoprotection［J］. Journal of Experimental Botany，2004，56（411）：395-406

[3] 乜兰春，陈贵林. 西瓜嫁接苗生长动态及生理特性研究［J］. 西北农业学报，2000，9（1）：100–103

[4] Drake B.，M. Gonzàlez-Meler，S. Long. More efficient plants：A consequence of rising atmospheric CO_2［J］. Annual review of Plant Physiology and Plant Molecular Biology，1997，48（1）：609-639

[5] Nederhoff E. M.，Vegter J. G. Photosynthesis of stands of tomato，cucumber and sweet pepper measured in greenhouse under various CO_2 concentrations［J］. Annals and Botany，1994，73（4）：353-361

[6] 许大全. 光合作用学［M］. 北京：科学出版社，2013

[7] Dzbek J. Korzeniewski B. Control over the contribution of the mitochondrial membrane potential（Δψ）and proton gradient（ΔpH）to the protonmotive force（Δp）. In silico studies［J］. Journal of Biological Chemistry，2008，283（48）：33 232-33 239

[8] Ioannidis N E，J.A. Cruz，K. Kotzabasis，et al.，Evidence that putrescine modulates the higher plant photosynthetic proton circuit［J］. 2012，Plos one，7（1）：e29864

[9] 董乔，宋阳，孙潜，等. 不同光强和CO_2浓度对温室嫁接黄瓜光合作用及叶绿素荧光参数的影响［J］. 北方园艺，2015，39（22）：1-6

[10] Tripp K. E.，M. M Peet，D. M. Pharr. CO_2- enriched yield and foliar deformation among tomato genotypes in elevated CO_2 environment［J］. 1991. Plant Physiol.，96（3）：713-719

[11] Fyrdrych J. Factors affecting photosynthetic productivity of sweet pepper and tomato in CO_2 enriched atmosphere［J］. 1984. Act. Hort.，162：255-264

贮藏条件对蔬菜种子质量和贮藏寿命的影响

朱怡航，潘学勤，赵颖雷，黄丹枫*
（上海交通大学农业与生物学院，上海　200240）

摘要： 蔬菜种子通常对贮藏温度和湿度非常敏感。商品蔬菜种子一般以编织袋包装，贮藏于温度和相对湿度（RH）分别在8～15℃和30%～55%之间的仓库中。丸粒化种子具有比正常种子更好的出苗特性，但是对贮藏条件更加敏感，表现比正常种子差。本研究以大葱、生菜、结球甘蓝种子为材料，探讨了温度和RH对种子商品质量和种子寿命的影响。通过饱和盐浓度加速老化（SSAA）和模拟贮藏试验研究了贮藏温度和RH对种子发芽率和种子含水量的影响。另外，通过试验数据模拟Ellis种子寿命预测公式，计算了不同蔬菜种子的生活力常数和不同商品仓库条件下的种子寿命。结果表明，贮藏温度的升高、RH的降低，种子含水量随之降低，并在一般种子仓库的温湿度范围内呈较好的线性关系。相同贮藏条件下，丸粒化的甘蓝种子具有高于正常种子的含水量，且贮藏寿命较正常种子短。此外，根据不同蔬菜种子的生活力常数和贮藏寿命，就一般蔬菜种子仓库的温湿度设置而言，生菜种子应尽可能贮藏在8℃冷库中，其他3种种子可根据需求贮藏在8℃仓库或15℃仓库中，且丸粒化甘蓝种子应贮藏在湿度控制较好的仓库中。

关键词： 蔬菜种子；温度；湿度；丸粒化；种子寿命

The Influence of Storage Conditions on Vegetable Seeds Qualities and Seed Longevity

Yihang Zhu，Xueqin Pan，Yinglei Zhao，Danfeng Huang*
（*School of Agriculture & Biology*，*Shanghai Jiao Tong University*，*Shanghai*　200240）

Abstract: Vegetable seeds are usually sensitive to storage temperature and humidity. Commodity vegetable seeds are usually stored extensively in woven bags in warehouses with steady temperature and relative humidity（RH）around 8～15℃ and 30%～55% respectively. Pelleted seeds perform better in seedling yet worse in storage than normal seeds. This study investigate the effect of temperature and RH on seed commodity qualities and seed longevity of vegetable seeds. Germination and moisture content were investigated via the saturated salt accelerated aging（SSAA）tests. In addition，the viability constants of different vegetable seeds and the seed longevity under different commodity warehouse conditions were estimated by simulating the experimental data based on Ellis' seed longevity predict equation. The results show that with storage temperature rising and RH falling，seed moisture content decreases and appears linearity in the temperature and RH range of general warehouses. The pelleted cabbage seeds have higher moisture content and shorter seed longevity than normal seeds. Moreover，in terms of viability constants of different vegetable seeds and the temperature and RH range of general warehouses，lettuce seeds should be stored in 8℃ warehouses prior to other 3 and the RH should be paid more attention to where pelleted cabbage seeds are stored.

通讯作者：黄丹枫（1956—），女，博士，教授，研究方向：设施蔬菜与蔬菜生理生态，E-mail：hdf@sjtu.edu.cn

Key words：Vegetable seed；Storage condition；Temperature；Relative humidity；Seed longevity

1 引言

常见的商品蔬菜种子在种子分类上均属于中命或短命种子，其室温条件的贮藏年限一般不超过3年，而实际贮藏时间还受到贮藏环境温湿度和种子本身的贮藏特性影响[1]。评价蔬菜种子商品质量最主要的指标是种子的发芽率，它能影响蔬菜种子发芽的数量与整齐度。蔬菜种子的贮藏寿命是指一定贮藏条件下（温度、湿度），蔬菜种子的发芽率随时间不断下降，最终达到最低发芽率标准值所经过的贮藏时间[2]。影响种子发芽率的外界环境因素主要是温度和湿度，这两个因素之间还存在耦合关系。一般来说，低温（0℃以上）、低湿的条件可延缓蔬菜种子发芽率的下降。根据Eills提出的种子寿命预测公式，可以利用公式中的生活力常数衡量并预测蔬菜种子的贮藏寿命，该公式认为降低贮藏温度和种子含水量可以延长蔬菜种子的贮藏寿命[3]。

种子本身的特性也会影响到蔬菜种子质量，最主要的因素有种子含水量和种子丸粒化，后者的材料一般包括农药、化肥、微量元素等活性成分，多为亲水物质，在长期贮藏的过程中会增加种子吸湿能力，影响贮藏寿命[4]。相比包衣技术，丸粒化种子还具有整齐的规格，但其种子间更易成团，使局部吸湿甚至霉变，影响种子寿命。许多研究表明丸粒化种子的发芽率、成苗率要显著高于一般种子，但在长期贮藏中，即便经过干燥的种子仍表现出含水量过高、发芽率下降快等问题[5]。

评价蔬菜种子商品质量的另一个指标是种子含水量。种子含水量是一个较为复杂的影响因素，降低种子含水量有利于种子贮藏，但贮藏环境的温度、湿度均会改变种子含水量，一般认为环境温度升高、湿度降低会降低种子含水量[6-8]。有许多研究把一定温度、湿度下的种子含水量成为平衡含水量，并绘成曲线[9-11]，如图1所示。

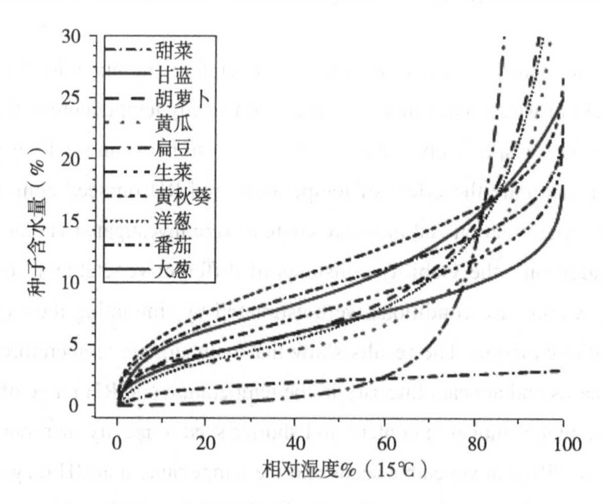

图1　不同蔬菜种子的平衡含水量曲线

综上，本文选取4种蔬菜种子（大葱、生菜、结球甘蓝、丸粒化结球甘蓝）作为材料。利用恒温箱控制温度，选取较高的温度（30、45℃）和湿度（60%、75%、80%）组合，进行一定温湿度下的模拟贮藏试验，并定期测量试验种子的含水量、发芽率。通过试验评价贮藏温度、湿度对蔬菜种子商品质量的影响。

2 材料与方法

2.1 种子材料

试验选取如下4种蔬菜种子作为试验材料：长悦大葱、玛丽娜生菜、欧克结球甘蓝（非丸粒化）、欧克甘蓝（丸粒化）。种子材料在试验前存放于15℃，40%RH的种子仓库，其初始发芽率分别为98%、99%、99%和93%。

2.2 各指标的测定

种子发芽率种子发芽采用纸上发芽，温度25℃，测试100粒种子一重复，重复4次；记试验开始后第7天的发芽种子数，并计算发芽率。

种子含水量每次种子发芽试验同时去相同处理条件的种子样品，重复3次，称重，放置于130℃烘箱2h，取出后再称重，以烘前重量为基数计算种子含水量。

2.3 模拟贮藏试验方法

分别将4种蔬菜种子放置于不同的温度、湿度组合处理条件下模拟贮藏，利用恒温箱控制温度，采用饱和盐溶液法控制湿度，选取温度（30、45℃）和湿度（60%、75%和80%）交叉组合。于试验开始后每7d测定种子发芽率和种子含水量。

利用种子寿命预测公式[12]，可以更好的研究环境条件对种子贮藏寿命的影响。公式如下：

$$\log\frac{P}{K_i-V}=K_E-C_w\log m-C_H t-C_Q t^2$$

式中：P表示贮藏的天数（period），K_i表示初始生活力（芽率概率单位probit germ.），V表示最终生活力（同上），m表示种子含水量（moisture/%），t表示贮藏温度（temp./℃），K_E、C_W、C_H、C_Q分别为不同类别种子的参数。

根据试验数据，参照公式做多元回归分析，得到每种种子的4个参数。将参数代回公式，以初始发芽率的概率值K_i，标准发芽率的最终生活力V，计算不同温度、湿度、种子含水量下蔬菜种子的贮藏寿命P。

3 结果与讨论

3.1 温湿度对蔬菜种子含水量的影响

贮藏环境的温湿度会影响蔬菜种子的含水量，试验开始后每7d测量1次试验种子的含水量，发现在第7天以后，4种种子的含水量均到达最高值且保持稳定。比较4种种子不同处理和试验初始的含水量，结果如图2所示。可以看到，试验种子的含水量在不同的温湿度处理下有显著的差异，贮藏环境湿度越低，种子含水量越低；除生菜种子在相对湿度75%以上处理外，温度越高，种子含水量越低。高温高湿条件下的生菜种子可能由于种子外表皮的结构导致其含水量随温湿度的变化趋势不同。

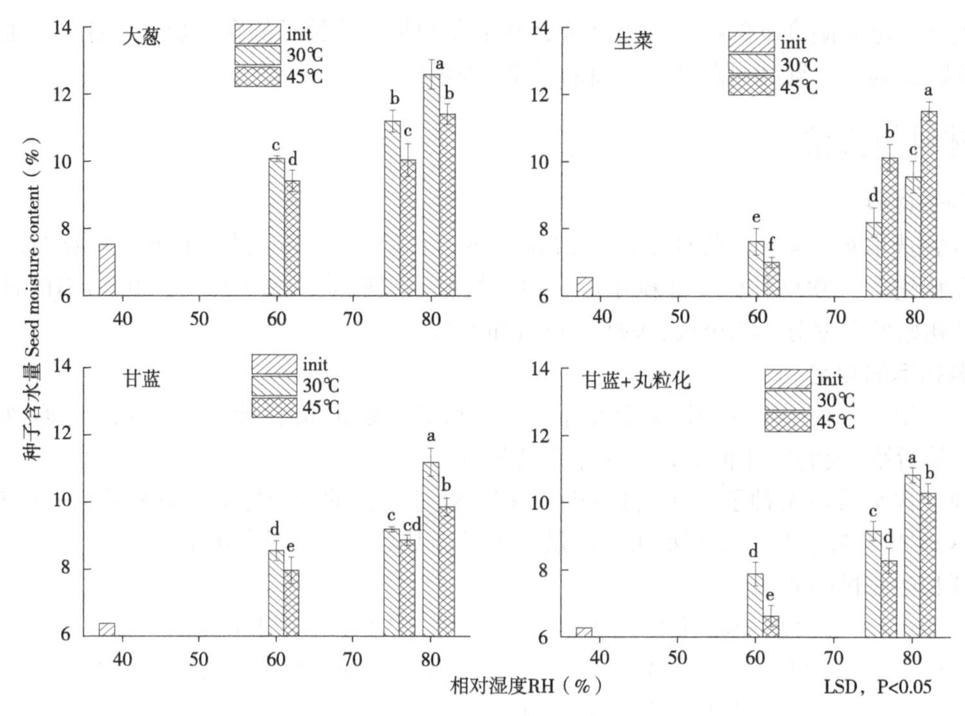

图 2　不同温湿度处理对4种蔬菜种子含水量的影响

进一步方差分析发现，温度和湿度的交互作用对4种种子含水量的影响不显著。根据图1中蔬菜种子吸湿平衡曲线，一定温度下，种子含水量对环境相对湿度（RH）呈反S型曲线，但当RH在30%～60%，温度在5～45℃的范围内，种子含水量与其的关系近似为线性关系。由于一般种子仓库的温湿度均处在以上范围内，因此可以采用多元线性回归拟合不同贮藏温度、湿度对种子含水量的关系公式，结果如表1所示。可以看出在此范围内，贮藏温度、湿度对种子含水量有较好的线性关系，并可用于仓库蔬菜种子水分的快速估计。

表 1　种蔬菜种子含水量与贮藏环境温湿度关系近似公式

种子种类	公式*	R-Square
大葱	mc=0.997−0.047 161 0・t+0.176・RH	0.990 02
生菜	mc=3.574−0.044 347 8・t+0.089・RH	0.982 36
甘蓝	mc=0.611−0.041 278 3・t+0.152・RH	0.976 95
丸粒化甘蓝	mc=2.410−0.043 986 8・t+0.111・RH	0.972 58

*式中mc表示种子含水量（%），t表示温度（5～45℃），RH表示相对湿度（30%～60%）；代入贮藏环境的温度和湿度，即可得出该条件下平衡状态的种子含水量。

3.2　丸粒化对结球甘蓝种子含水量及发芽率的影响

对两种甘蓝种子在不同处理下的含水量、发芽率进行检测。不同处理的种子含水量结果如图3所示，可以看到，在30℃条件下，一般种子与丸粒化种子的含水量无显著差异；在45℃条件下，较高的湿度则会显著提高丸粒化种子含水量。通过进一步方差分析发现，环境湿度与种子丸粒化之间存在交互作用，丸粒化甘蓝种子的含水量相较非丸粒化种子对环境湿度更为敏感。在2个温度条件下，湿度较高处理的丸粒化种子含水量均显著高于湿度较低处理的种子含水量，而非丸粒化种子在60%和75%湿度下的含水量差异不显著。这说明在相同贮藏条件下，丸粒化的甘蓝种子具有高于正常甘蓝种子的含水量，从而影响到其种子贮藏期间的表现。

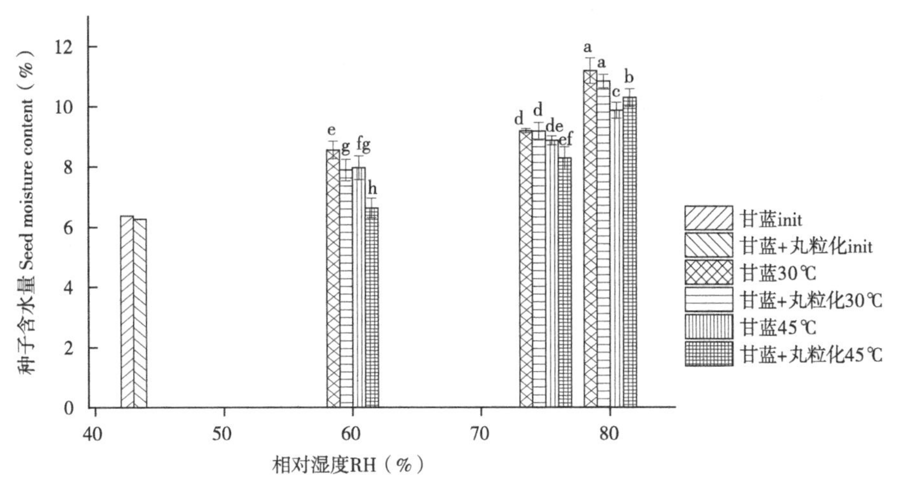

图 3　丸粒化对甘蓝种子含水量的影响

模拟贮藏试验开始后每7d进行1次发芽试验，测定2种甘蓝种子的发芽率，并绘成曲线，如图4所示。结果显示，高温和高湿均会显著加速种子发芽率的下降，而丸粒化并没有直接影响。通过对试验结束时芽率的方差分析发现，除温度和湿度对种子发芽率的影响存在交互作用外，环境湿度和丸粒化对种子发芽率的影响也存在交互作用，具体表现为：正常甘蓝种子在30℃，75%和80%条件下的芽率有显著差异，而丸粒化甘蓝种子在这些条件下无显著差异。因此，在研究贮藏温度、湿度对种子贮藏寿命的影响时，应分别计算两种甘蓝种子的生活力常数及贮藏寿命。

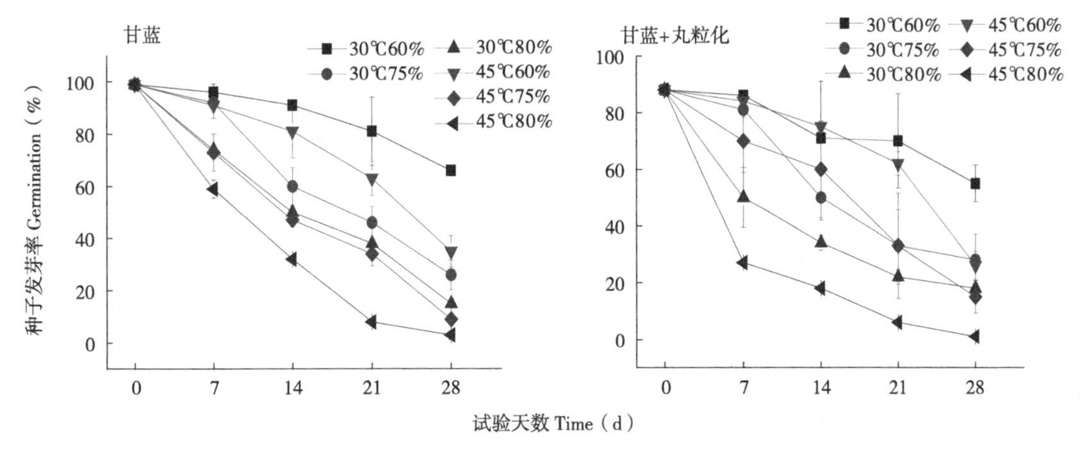

图 4　丸粒化对甘蓝种子发芽率的影响

3.3 不同温度、湿度对4种蔬菜种子贮藏寿命的影响

种子贮藏寿命即种子从进入贮藏阶段一直到其发芽率降至标准规定值的时间长度。为研究温度和湿度对种子贮藏寿命的影响，模拟贮藏试验开始后每7d进行一次发芽试验，测量4种蔬菜种子的发芽率，并绘成曲线，如图5所示。结果显示，高温和高湿均会显著加速种子发芽率的下降。通过进一步的方差分析发现，温度和湿度对种子贮藏寿命有显著影响，并存在交互作用。例如，30℃下，75%和80%处理大葱种子试验结束时芽率有显著差异，而45℃处理下的芽率无显著差异；45℃下，60%和75%处理生菜种子试验结束时芽率有显著差异，而30℃处理下的芽率无显著差异。

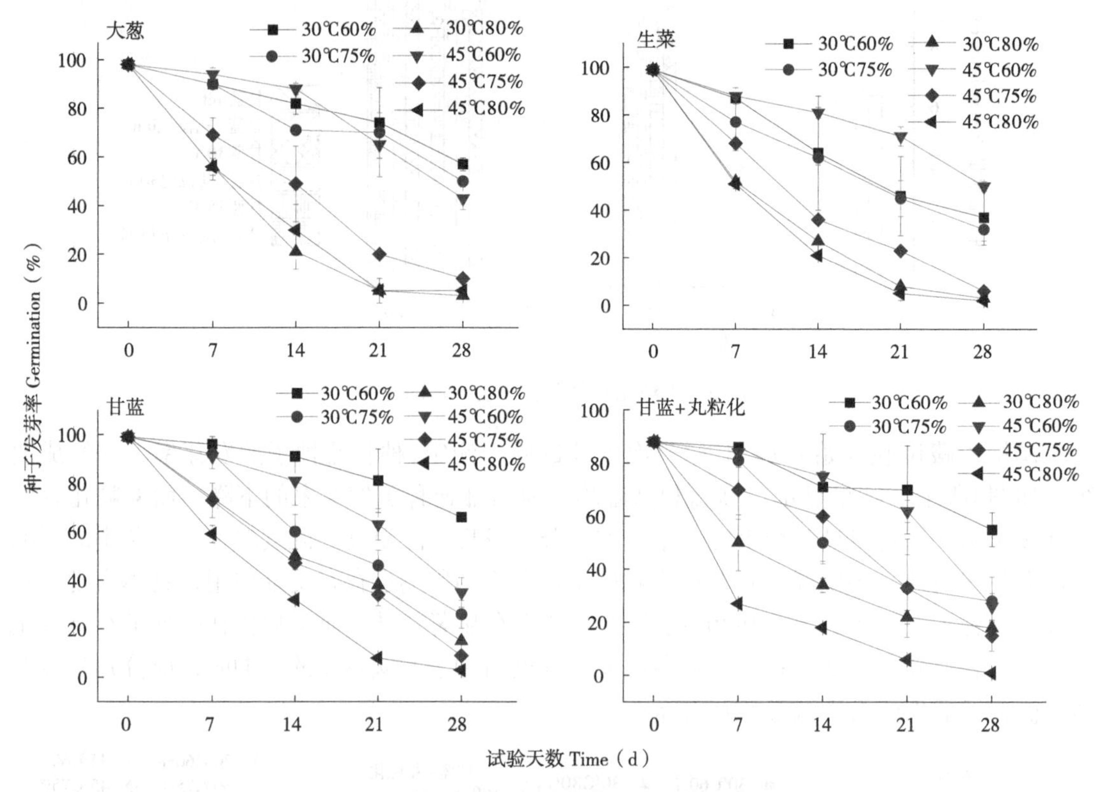

图5　不同环境温湿度对4种蔬菜种子贮藏寿命的影响

根据种子寿命预测公式（Ellis　et al.，1988）计算得到4种蔬菜种子的生活力常数，如表2，其中RH表示相对湿度（%），其他字母与公式中含义相同。

表2中最后一列（P）即为相应条件下的种子贮藏寿命，可以看到在15℃仓库中，大葱和甘蓝的贮藏寿命都有2年，而生菜和丸粒化甘蓝种子的贮藏寿命仅有1.5年；在8℃冷库中，生菜种子的贮藏寿命大约有1.5年，其他3种蔬菜种子的贮藏寿命都超过4年。就一般蔬菜种子仓库的温湿度设置而言，生菜种子应尽可能贮藏在8℃冷库中，其他3种种子可根据需求存放在8℃仓库或15℃仓库中。另外，由于丸粒化甘蓝种子的C_W参数高于普通甘蓝种子，随着温度的降低，其贮藏寿命增幅较普通甘蓝种子小，因此需要贮藏于湿度控制较好的仓库中。

表 2　4种蔬菜种子的生活力常数及不同贮藏条件下的贮藏寿命的预测

种子种类	K_E	C_W	C_H	C_Q	K_i	V	m	t	RH	P
大葱	7.199 84	3.592 1	0.079 73	3.33E-04	0.98	0.8	0.073 296	15	40	804
生菜	4.458	1.955 26	0.015 88	2.22E-04	0.99	0.8	0.064 83	15	40	568
甘蓝	5.316 03	1.470 7	0.070 91	1.33E-03	0.99	0.85	0.060 718	15	40	816
丸粒化甘蓝	5.632 03	2.069 81	0.049 92	6.31E-04	0.93	0.85	0.061 902	15	40	557
大葱	7.199 84	3.592 1	0.079 73	3.33E-04	0.98	0.8	0.076 597	8	40	2 806
生菜	4.458	1.955 26	0.015 88	2.22E-04	0.99	0.8	0.067 934	8	40	727
甘蓝	5.316 03	1.470 7	0.070 91	1.33E-03	0.99	0.85	0.063 608	8	40	3 913
丸粒化甘蓝	5.632 03	2.069 81	0.049 92	6.31E-04	0.93	0.85	0.064 981	8	40	1 422

4　结论

蔬菜种子在不同的贮藏温度、湿度条件下，其种子含水量受到显著的影响，并随贮藏温度的升高、湿度的降低而降低；丸粒化与非丸粒化甘蓝种子相比，在相同贮藏条件下，其种子含水量显著升高，贮藏过程中发芽率下降更快。同时，影响蔬菜种子贮藏寿命的因素有贮藏温度和种子含水量，而种子含水量又受贮藏温度、湿度影响。因此，本文通过较高的温湿度试验设计，探究蔬菜种子贮藏温度、湿度对蔬菜种子贮藏寿命的影响，并根据种子寿命预测公式计算相应蔬菜种子的生活力常数，以预测蔬菜种子在不同贮藏温度、湿度条件下的贮藏寿命。以上结果不仅可以更好地指导蔬菜种子的贮藏，还有助于根据不同的蔬菜种子进行贮藏条件的分类，减少蔬菜种子仓库发生种子劣变的风险。

参考文献

[1] George R A T. Vegetable seed production [M]. 第3版. CABI，2009.

[2] Copeland L O，Mcdonald M B. Principles of Seed Science and Technology [M]. Boston，MA：Springer Science & Business Media，2012.

[3] Ellis R H，Hong T D. Quantitative response of the longevity of seed of twelve crops to temperature and moisture in hermetic storage [J]. Seed Science and Technology，2007，35（2）：432–444.

[4] Schwember A R，Bradford K J. Oxygen interacts with priming，moisture content and temperature to affect the longevity of lettuce and onion seeds [J]. Seed Science Research，Cambridge，UK：Cambridge University Press，2011，21（3）：175–185.

[5] Mandal A B，Mondal R，Dutta P M S. Seed Enhancement Through Priming，Coating and Pelleting for Uniform Crop Stand and Increased Productivity [J]. Journal of the Andaman Science Association Vol，2015，20（1）：26–33.

[6] Reddy B S，Chakraverty A. Equilibrium Moisture Characteristics of Raw and Parboiled Paddy，Brown Rice，and Bran [J]. Drying Technology，Taylor & Francis Group，2004，22（4）：837–851.

[7] CHEN C. Moisture sorption isotherms of pea seeds [J]. Journal of Food Engineering，2003，58（1）：45–51.

［8］ Correa P C，Reis M F T，Oliveira G H H de，et al. Moisture desorption isotherms of cucumber seeds：modeling and thermodynamic properties［J］. Journal of Seed Science，ABRATES - Associação Brasileira de Tecnologia de Sementes，2015，37（3）：218–225.

［9］ Alhamdan A M，Alsadon A A. Moisture sorption isotherms of four vegetable seeds as influenced by storage conditions［C］// XXVI International Horticultural Congress：Issues and Advances in Transplant Production and Stand Establishment Research 631. 2002：63–70.

［10］ Gazor H R. Moisture isotherms and heat of desorption of canola［J］. Agricultural Engineering International：CIGR Journal，2010，12（2）：79–84.

［11］ Alhamdan A M，Alsadon A A. Moisture sorption isotherms of four vegetable seeds as influenced by storage conditions［J］. Acta Horticulturae，2004，631：63–70.

［12］ Ellis R H，Roberts E H. Improved Equations for the Prediction of Seed Longevity［J］. Ann. Bot.，1980，45（1）：13–30.

不同围护结构日光温室环境性能比较

李纯青[1]，王传清[1]，魏珉[1，2]，王秀峰[1，2]，隋申利[3]，赵利华[3]，李艳玮[3]
（1. 山东农业大学园艺科学与工程学院，泰安　271018；2. 农业部黄淮海设施农业工程科学观测实验站，泰安　271018；3. 山东寿光欧亚特菜有限公司，寿光　262700）

摘要：以3种不同围护结构的日光温室为观测对象，对2—3月的室内光温环境进行了比较研究，主要结果如下：保温性能以下挖式厚土墙日光温室最优，大跨度拱棚型日光温室次之，聚苯板异质复合墙体日光温室最差。下挖式厚土墙日光温室气温空间分布均匀性优于聚苯板异质复合墙体和大跨度拱棚型日光温室。日光温室采光性能以大跨度拱棚型日光温室最优，作物冠层高度处的光照强度由南向北逐渐减弱，聚苯板异质复合墙体日光温室内光照分布更均匀，大跨度拱棚型日光温室北侧的光照条件明显变差。土地利用率以大跨度拱棚型日光温室最大，聚苯板异质复合墙体次之，下挖式厚土墙日光温室最低。
关键词：日光温室；墙体结构；气温；光照强度；土地利用率

Comparison of Environmental Factors in Solar Greenhouses with Different Envelope Retaining enclosure structure

Li Chunqing[1]，Wang Chuanqing[1]，Wei Min[1.2]，
Wang Xiufeng[1.2]，Sui Shenli[3]，Zhao Lihua[3]，Li Yanwei[3]
（1. *College of Horticultural Science and Engineering*，*Shandong Agricultural University*，*Tai' an*　271018；
2. *Scientific Observation and Experimental Station of Environment Controlled Agricultural Engineering in Huang-Huai-Hai Region*，*Ministry of Agriculture*，*Tai' an*　271018，3. *Shouguang Eurasian Vegetables Co. Ltd.*，*Shouguang*　262700）

Abstract: Light and temperature conditions in 3three kinds of wall structures greenhouses with different envelope structures were observed selected as research object .At the same time，a comparative study of the indoor light and temperature condition in the months of from February- to March in Shouguang of Shandong province was conducted. The main results are as follows: The sinkingunk solar greenhouse with thick earth wall has best heat preservation，followed by onglarge span tunnel-type shedsolar greenhouse，and solar greenhouse with polystyrene heterogeneityheterogeneous composite polystyrene board wall solar greenhouse is the worst. The spatial distribution uniformity of air temperature in sunk solar greenhouse with thick earth wall sinking earth wall solar greenhouse under the air temperature spatial distribution uniformity is better than the heterogeneous composite polystyrene board wall solar greenhouse and the long-span shed solar greenhouseothers. Large span tunnel-type solar greenhouse has the best The lighting performance，however light intensity in north side decrease obviously，when thelong-span shed greenhouse lighting performance is the best.T light distribution in solar greenhouse with polystyrene heterogeneity wall light is distributedis more evenly in the heterogeneous composite. Land utilization rate of large span tunnel-type solar greenhouse is the highest，while sunk solar greenhouse with thick earth wall is the lowest.

polystyrene board wall solar greenhouse.T the long-span shed solar greenhousesignificantlyThe long-span shed solar greenhouse land utilization rate is the largest，the heterogeneous composite polystyrene board wall solar greenhouse is the second，and the sinking earth wall solar greenhouse is the lowest.

Key words: Solar greenhouse；Wall structure；Air temperature；Light intensity；Land utilization rate

日光温室是中国北方地区冬春季节的主要栽培设施类型，墙体是日光温室的重要组成部分。土质墙体具有良好的蓄热保温性能，但建造费工，占地面积大[1-3]；砖墙占地面积虽小，但成本上升，且有时保温性能较差[4-6]。随着科技和产业的发展，人们越来越关注便于标准化建造、机械化作业和智能化管理的新型墙体结构日光温室，如用聚苯板、草砖、棉被等材料建造的组装式温室[7-10]。为了探明不同围护结构日光温室的环境性能，本试验选取生产中3种日光温室进行了跟踪观测，以期为结构和材料优化创新提供依据。

国家“十二五”科技支撑计划课题（2014BAD05B03）；山东省农业重大应用技术创新课题［鲁财农指（2015）16号］；国家现代农业产业技术体系建设专项（CARS-23）

1　材料与方法

1.1　试验温室

供试日光温室位于山东省寿光市蔬菜高科技示范园内，下挖式厚土墙日光温室、聚苯板异质复合墙体日光温室和大跨度拱棚型日光温室进行比较研究。不同围护结构日光温室结构参数如表1所示。

表 1　不同围护结构日光温室结构参数

Table 1　The parameters of solar greenhouse structure with different retainingenclosure structures

日光温室	下挖深度	长度	跨度	脊高	采光屋面角	后墙高度	后墙结构
Solar greenhouse	Sunken depth（m）	Length（m）	Span（m）	Ridge Height（m）	Front roof angle（°）	Back wall Height（m）	Wall structure
1号温室	0.80	90	12	6	29（流线型）	5	土墙底厚6.8m、顶厚1.5m
2号温室	—	90	12	6	29（抛物线型）	5	5cm聚苯板+20cmGFB保温板+5cm聚苯板
3号温室	—	90	12+12	7.5	32（抛物线型）	—	塑料薄膜+3cm保温被

注：1号温室为下挖式厚土墙日光温室；2号温室为聚苯板异质复合墙体日光温室；3号温室为大跨度拱棚型日光温室。Note: No. 1 ismeans tsunk solar greenhouse with thick earth wall he sinking earth wall solar greenhouse；No. 2 imeans solar greenhouse with polystyrene heterogeneity wall the heterogeneous composite polystyrene board wall solar greenhouse；No. 3 imeans large span tunnel-type solar greenhouse.the long-span shed solar greenhouse

1.2　测定方法

2016年2—3月采用Hobo温湿光照记录仪连续测定，间隔15 min自动采集数据1次。下挖式厚土墙日光温室和聚苯板异质复合墙体日光温室的温度测点在温室长度方向的中部，南、中、北3个测点分别在温室跨度由南向北的1/6、3/6、5/6处，离地面高度1.6 m；气温空间分布测点在上述3个位置的垂直方向上离地面高度分别为0.10 m、1.5 m、3.0 m和4.5 m。光照测点位于上述3个位置的植株冠层上方，离地面高度1.8 m。大跨度拱棚型日光温室温度测点在温室中部，由南向北1/12、3/12、5/12、7/12、9/12、11/12处，离地面高度1.6 m；气温空间分布测点在上述六个位置的垂直方向上离地面高度分别为0.10 m、1.5 m、3.0 m和4.5 m；光照测点位于上述六个位置的植株冠层上方，离地面高度1.8 m；大跨度拱棚型日光温室南部温度或光照测点选取为1/12、3/12处测点处温度或光照的平均值，大跨度拱棚型日光温室温室中部测点选取为5/12、7/12测点处温度或光照的平均值，大跨度拱棚型日光温室温室北部测点选取为9/12、11/12测点处温度或光照的平均值。

夜间平均气温为每天夜间18:00 ~ 8:00的气温平均值，白天平均气温为每天9:00 ~ 17:00的气温平均值。透光率为室内光照强度/室外光照强度 × 100%，光照强度取10:00 ~ 16:00平均值。

1.3 数据处理与分析

试验数据采用Microsoft Excel 2003整理，绘图采用CAD 2010和Excel 2003软件。

2 结果与分析

2.1 不同围护结构日光温室保温性能

从表2可以看出，2号温室的平均夜间气温、平均昼间温度、平均最高气温、平均最低气温分别比1号温室低4.2℃、2.0℃、1.7℃、3.6℃；3号温室分别比1号温室低2.8℃、1.4℃、1.5℃、2.4℃。

表 2 不同围护结构日光温室温度比较（℃）

Table 2 Comparison of air temperature in solar greenhouses with different retainingenvelope structures（℃）

项目 Item	夜平均气温 Night mean temperature	昼平均气温 Day mean temperature	平均最高气温 Average highest temperature	平均最低气温 Average lowest temperature
室外	1.3	6.8	11.4	−2.4
1号温室	18.4	22.5	29.1	16.6
2号温室	14.2	20.5	27.4	13.0
3号温室	15.6	21.1	27.6	14.2

2.2 不同围护结构日光温室气温日变化

2.2.1 晴天日变化

图1为2月26日、27日、29日晴天条件下温室内的气温日变化，3个日光温室的变化规律基本一致，以1号温室温度最高，3号温室次之，2号温室最低。日光温室内最高气温出现在12:30左右，1号、2号和3号温室分别达26.8℃、25.5℃和25.8℃；最低温度出现在早晨揭开保温被之前，1号、2号和3号温室分别达15.1℃、10.9℃和12.8℃。

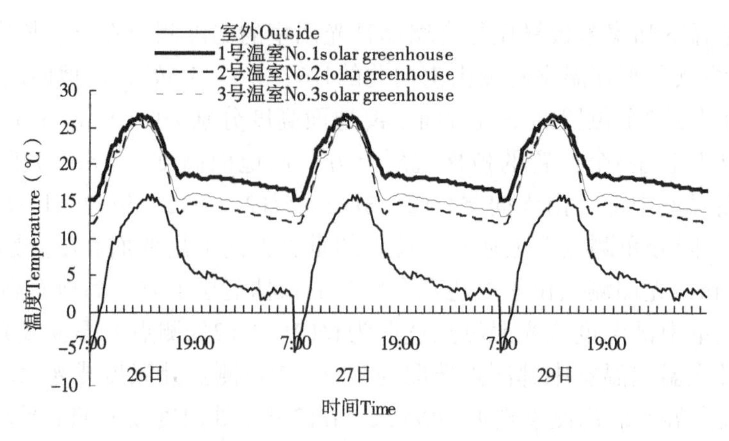

图1　晴天不同围护结构日光温室温度日变化

Figure 1　Temperature variation in solar greenhouses with different retainingenvelope structures in sunny days

2.2.2　阴天日变化

图2为2月24日、25日和28日阴天温室内气温的日变化，3个温室变化趋势相同，最高气温出现在13:00左右，分别为14℃、11.9℃和12.2℃，最低气温出现时间在早晨揭开保温被之前，分别为10.1℃、8.3℃和8.1℃。

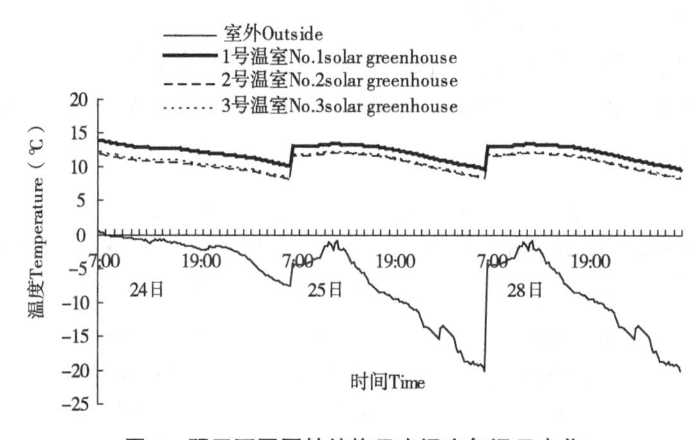

图2　阴天不同围护结构日光温室气温日变化

Figure 2　Temperature variation in solar greenhouses with different retainingenvelope structures in sunnycloudy days

2.3　不同围护结构日光温室气温空间分布

2.3.1　晴天

由表3可知，水平方向上，昼间最高温度以中部最高，南部次之，北部最低，相差0.4～1.6℃；夜间最低温度以北部最高，中部次之，南部最低，相差0.1～1.0℃。不同围护结构日光温室昼间最高温度均随着高度升高而上升，相差1～2℃；夜间最低温度变化趋势相反，1号温室近地面层高于作物冠层0.3℃，2号温室高于作物冠层0.5～0.6℃，3号温室高于作物冠层0.8～1.0℃。

表 3 晴天不同围护结构日光温室气温空间分布（℃）

Table 3 Spatial distribution of air temperature in solar greenhouses with different envelope structures with different retaining structures inon sunny days（℃）

离地面高度 Height from the ground	1号温室 No.1 solar greenhouse			2号温室 No.2 solar greenhouse			3号温室 No.3 solar greenhouse		
夜间平均最小值 Nighttime average lowest value	南部	中部	北部	南部	中部	北部	南部	中部	北部
10cm	17.8	18.0	18.2	16.2	16.5	16.6	16.4	17.0	17.2
1.5m	17.3	18.0	18.0	15.6	15.8	16.0	14.8	15.8	16.0
3.0m	—	17.8	18.2	—	15.9	15.9	—	15.9	15.9
4.5m	—	—	17.8	—	—	15.9	—	—	16.1
昼间平均最大值 Daytime average highest value									
10cm	30.4	31.5	30.2	25.3	26.0	25.0	25.3	25.3	23.9
1.5m	30.4	31.6	29.4	28.4	28.6	28.6	27.2	27.6	27.0
3.0m	—	31.6	29.0	—	28.7	28.7	—	28.3	28.9
4.5m	—	—	32.0	—	—	28.8	—	—	28.6

注：观测日期2016年3月19日

Note: Observation date was March 19，2016

2.3.2 阴天

由表4可知，阴天在水平方向上昼间最高温度同样以中部最高，南部次之，北部最低，相差0.2～1.2℃；夜间最低温度以北部最高，中部次之，南部最低，相差0.1～0.7℃。垂直方向上，1号温室不同高度间温度差异较小，仅0.1～0.3℃，而2号和3号温室存在明显的气温空间分布不均匀现象。昼间最高气温随着高度升高而增大，夜间最低气温则是近地面层气温高于作物冠层，相差0.5～0.6℃。

表 4 阴天不同围护结构日光温室气温空间分布（℃）

Table 4 Spatial distribution of air temperature in solar greenhouses with different envelope structure with different retaining structures inon cloudy days（℃）

离地面高度 Height from the ground	1号温室 No.1 solar greenhouse			2号温室 No.2 solar greenhouse			3号温室 No.3 solar greenhouse		
夜间最低温度	南部	中部	北部	南部	中部	北部	南部	中部	北部
10cm	18.5	18.6	18.6	14.0	14.3	14.4	17.7	17.4	17.6
1.5m	18.1	18.5	18.6	12.6	12.8	13.3	16.3	16.3	16.3

（续表）

离地面高度 Height from the ground	1号温室 No.1 solar greenhouse			2号温室 No.2 solar greenhouse			3号温室 No.3 solar greenhouse		
	南部	中部	北部	南部	中部	北部	南部	中部	北部
夜间最低温度									
3.0m	—	18.5	19.0	—	13.0	13.2	—	16.5	16.6
4.5m	—	—	18.7	—	—	13.1	—	—	16.6
昼间最高温度									
10cm	27.7	28.5	27.6	28.9	28.9	28.0	26.6	27.4	26.2
1.5m	27.9	28.2	27.7	30.1	31.3	30.1	30.5	31.1	30.0
3.0m	—	28.9	28.6	—	31.2	31.0	—	31.4	30.6
4.5m	—	—	29.8	—	—	31.1	—	—	31.4

注：观测日期2016年3月18日

Note: Observation date was March 18，2016

2.4 不同围护结构日光温室光照特点

2.4.1 透光率

1号和2号温室采光屋面角相同，但采光屋面形状不同，分别为流线型和抛物线型，2号温室比1号温室透光率增加4.7%，说明抛物线形更好； 2号和3号温室的采光屋面形状均为抛物线型，但后者屋面角增加，透光率提高10.1%。

2.4.2 时空变化

图3为晴天条件下（2月29日）不同围护结构日光温室内的光照强度，以12:00左右室内光照强度最高，之后逐渐下降。2号温室光照强度的空间分布较1号温室和3号温室均匀，3号温室侧光照条件明显变差，与南侧光照强度相差638μmol·m^{-2}·s^{-1}。

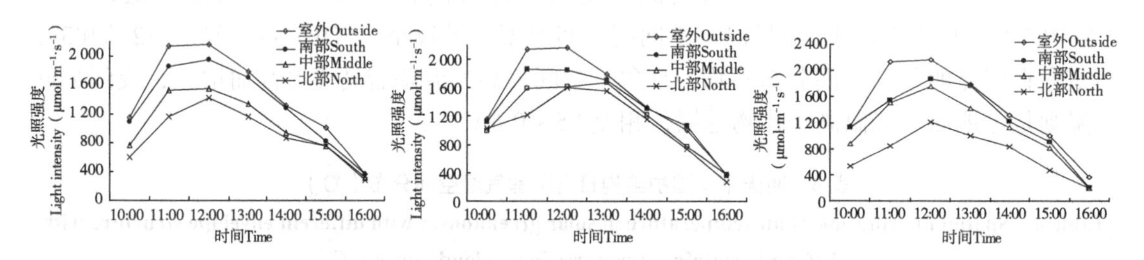

图 3　不同围护结构温室日光温室光照强度日变化

Figure 3　DiurnavVariation of light intensity ofin solar greenhouses with different envelope structure with different retaining structures

2.5 不同围护结构日光温室土地利用率

不同围护结构日光温室土地利用率以大跨度拱棚型日光温室最高，聚苯板异质复合墙体日光温室次之，下挖式后土墙日光温室最低，结果如表5所示。

表 5　不同围护结构日光温室土地利用率

Table 4　Land utilization rate of solar greenhouses with different envelope structures

日光温室 Solar greenhouse	种植面积 Cultivation area /m^2	墙体占地面积 Wall covering area/m^2	温室间面积 Area between greenhouses/m^2	占地面积 Total area/m^2	土地利用率 Land utilization rate/%
1号温室	1080	636	540	2256	48
2号温室	1080	39	540	1659	65
3号温室	2160	48	540	2748	78

注：土地利用率=种植面积/占地面积，占地面积为种植面积、墙体占地面积、温室间距面积的总和

Note: Land Utilization rate= cultivation area / totalr area，total area include cultivation area，wall covering are and area between greenhouses

3　讨论与结论

日光温室墙体结构影响保温性能。陈瑞生、周长吉[115]提出日光温室较理想的墙体内侧应由蓄热能力较强的材料组成蓄热层，外层由导热、放热能力较差的材料组成保温层，中间为隔热层。本试验中不同围护结构日光温室相比，聚苯板异质复合墙体日光温室、大跨度拱棚型日光温室的平均夜温分别比下挖式厚土墙日光温室低4.2℃和2.8℃，说明了室内温度状况与墙体的蓄热保温能力密切相关。李丽平[12]研究表明，温室热容量大，有利于保温。大跨度拱棚型日光温室室内空间大，缓冲性能强，可能是夜温高于聚苯板异质复合墙体日光温室的原因。

采光屋面形状和角度大小是影响日光温室采光的两个关键因素。下挖式厚土墙日光温室和聚苯板异质复合墙体日光温室采光屋面角均为29°，仅形状不同，前者为流线型，平均透光率达79.8%，而后者为抛物线型，平均透光率达84.5%，两者相差4.7%，说明抛物线型更优。聚苯板异质复合墙体日光温室和大跨度拱棚型日光温室的采光屋面形状均为抛物线型，但屋面角度分别为29°和32°，但导致后者平均透光率比前者增加10.1%，说明屋面角度大小比屋面形状对温室采光的影响更大。

总之，不同围护结构日光温室的采光保温性能不同，保温性能以下挖式厚土墙日光温室最优，大跨度拱棚型日光温室次之，聚苯板异质复合墙体日光温室最差。下挖式厚土墙日光温室内部气温空间分布均匀性优于聚苯板异质复合墙体和大跨度拱棚型日光温室。日光温室采光性能以大跨度拱棚型日光温室最优，聚苯板异质复合墙体日光温室次之，下挖式厚土墙日光温室最差，但是聚苯板异质复合墙体日光温室较下挖式厚土墙日光温室和大跨度拱棚型日光温室光照强度的空间分布更均匀，大跨度拱棚型日光温室北侧光照强度明显降低。土地利用率以大跨度拱棚型日光温室最大，聚苯板异质复合墙体次之，下挖式厚土墙日光温室最低。

参考文献

[1]　杨建军，邹志荣，张智等.西北地区日光温室土墙厚度及其保温性的优化[J].农业工程学报，2009，25（8）:180-185

[2] 张纪涛，林琭，闫万丽等.山西省日光温室结构问题的调查研究[J].中国蔬菜，2013(4):90-94.
[3] 黎贞发，于红.持续低温及低温连阴天气下几种典型日光温室保温性能评价[J].中国农学通报，2013，29(23):123-128.
[4] 李明，魏晓明，周长吉等.发泡水泥对日光温室黏土砖墙保温蓄热性能的改善效果[J].农业工程学报，2014，30(24):187-192.
[5] 佟国红，David M. Christopher.墙体材料对日光温室温度环境影响的CFD模拟[J].农业工程学报，2009，25(3):153-157.
[6] 张泽平，李珠，董彦莉. 建筑保温节能墙体的发展现状与展望[J].工程力学，2007，24(增刊II):121-128.
[7] 李小芳，陈青云.墙体材料及其组合对日光温室墙体保温性能的影响[J].中国生态农业学报，2006(4):185-189.
[8] 周长吉.一种以涤棉轻质保温材料为墙体和后屋面的组装式日光温室[J].农业工程技术，2017，37(13):51-54
[9] 赵丽玲，樊东隆，杨爱华，赵贵宾.冬季辽阳型与白银型日光温室的温、湿度特性比较[J].北方园艺，2014(15):40-43.
[10] 蒋锦标，姜兴胜，乔军，等. 对我国蔬菜温室的评价及新型日光温室的研发[J].中国蔬菜，2011(11):8-10.
[11] 陈端生，郑海山，刘步洲.日光温室气象环境综合研究Ⅰ.墙体、覆盖物热效应研究初报[J].农业工程学报，1990(2):77-81.
[12] 李丽平，张亚红.阴阳棚共用墙体不同季节的吸放热状况分析.北方园艺，2010(15):80-84.

高效能日光温室被动式建筑设计方法探讨

陈超，杨枫光，李印，韩枫涛，于楠，李亚茹，姜理星
（北京工业大学绿色建筑环境与节能技术北京市重点实验室，北京　100124）

摘要：利用被动式建筑设计方法，提高日光温室冬季“反季节”蔬菜作物生产太阳能利用率，实现日光温室不加温或少加温生产，具有非常重要意义。本研究结合现有日光温室建筑构造特点以及冬季“反季节”蔬菜作物生产期间温室建筑热负荷动态变化特性，运用建筑热工、建筑气候、建筑构造与太阳能利用等学科的原理和方法，采用被动式建筑设计方法，重点研究朝向、建筑空间形态对提高日光温室太阳能被动利用率的影响规律，提出了不同地理纬度地区日光温室最佳建筑朝向以及建筑空间形态特征参数设计计算模型。分析计算结果表明，朝向以及建筑空间形态特征参数优化后的日光温室建筑，冬季需热量可减少15.7%。

关键词：高效能日光温室；被动式建筑设计方法；朝向；建筑空间形态特征参数；设计计算简化模型

Discussion on the Design Method of Passive Building for High Efficiency Solar Greenhouse

Chen Chao，Yang Fengguang，Li Yin，Han Fengtao，Yu nan，Li Yaru，Jiang Lixing
（*Green building environment and energy saving technology Beijing Key Laboratory*，*Beijing University of Technology*，*Beijing*　100124）

Abstract：Using passive building design method to improve solar energy utilization rate of in winter，and realize solar greenhouse production without heating or less heating is of great significance. This paper based on the existing building structure characteristics of solar greenhouse and the dynamic heat load characteristics of solar greenhouse in winter the key period of “anti-season” vegetable crop production. Using the principles and methods of building thermal engineering，building climate，building structure and solar energy utilization，the passive building design method is also adopted. Focus on the influence of orientation and the key parameters of solar greenhouse morphology on increasing the passive utilization of solar energy. The optimum design model of the orientation and the key parameters of solar greenhouse morphology in different geographical latitude areas is put forward. The results of analysis show that the heat demand in winter can be reduced by 15.7% in the solar greenhouse building following the optimization of the orientation and the parameters of the building space.

Key words：High efficiency solar greenhouse；The design method of passive building ；Orientation；Key

---

基金项目：国家自然科学基金资助项目（No. 51578012）；“十三五”国家重点研发计划（No. 2016YFC0700206）

通信作者：陈超（1958—），女，湖南人，教授，博士生导师，主要从事相变蓄热技术与可再生能源技术研究。北京工业大学建筑工程学院，100124，E-mail: chenchao@bjut.edu.cn parameters of solar greenhouse morphology; Simplified calculation model of design

1 引言

日光温室属于典型的直接受益式太阳房建筑。然而，太阳辐射与气象要素的双重周期性热作用，直接影响日光温室的光照与光热特性。太阳能随时间、随季节的冬天变化特性，直接影响不同朝向日光温室前屋面获得的太阳光照量以及光照时长；而日光温室建筑空间形态特征参数（跨度、脊高、北墙高度、后屋面水平投影长度）的不同组合，决定了白天通过日光温室前屋面进入温室太阳能量与通过温室围护结构向外散失热量（特别是夜间）的平衡程度。如何根据不同地理纬度地区太阳辐射与气象参数的变化规律，合理的确定日光温室建筑最佳朝向、建筑空间形态特征参数，对提高日光温室冬季“反季节”蔬菜作物生产太阳能被动利用率、降低日光温室供热需求，具有不可忽视的重要作用，而这正是被动式建筑设计理念的体现。

自20世纪80年代日光温室技术出现后，国内一些学者就开始了关于日光温室建筑朝向与空间形态的研究。李军等[1]根据西北型节能日光温室采光设计理论中的温室建筑朝向和前屋面仰角的设计原理，给出了西北地区温室建筑最佳朝向为正南或者南偏西5°～8°的研究结果。白义奎等[2]以沈阳地区为例，研究了日光温室建筑朝向对进光量的影响，研究结果表明沈阳地区温室建筑朝向南偏西5°～6°时进光量最大。关于日光温室建筑空间形态的研究，李天来[3]、白义奎等[4]以温室越冬作物健康生长对温光的最低需求、以及冬至日日光温室前屋面截获的太阳能等于春分日地平面截获之太阳能为控制条件，给出了节能日光温室建筑空间形态特征参数设计值确定方法。周长吉等[5]基于所提出的确保蔬菜作物获得充足光照、保证冬至日正午前后4 h内透过温室前屋面的太阳辐照度衰减率不超过2%、且栽培区最后一排作物的冠层全天可接受到太阳照射为控制条件，并利用三角函数关系，给出了日光温室建筑空间形态特征参数的取值方法。

本研究将基于被动式建筑设计理念，运用建筑热工、建筑气候、建筑构造与太阳能利用等学科的原理和方法，重点研究朝向、建筑空间形态对提高日光温室太阳能被动利用率的影响规律，提出不同地理纬度地区日光温室最佳建筑朝向以及建筑空间形态特征参数设计计算模型，为高效能日光温室建筑的优化设计，提供方法参考。

2 材料与方法

2.1 日光温室建筑动态热负荷计算分析

本节采用EnergyPlus能耗模拟软件，以北京地区一长度为80m、跨度为8m的典型日光温室为计算对象，对其冬季热负荷动态变化特性进行了计算分析。该计算温室墙体均为240 mm砌块砖墙、墙体外侧采用100mm聚苯板保温材料；温室前屋面采用0.12 mm的EVA薄膜、夜间加盖40 mm保温覆盖物；后屋面采用内夹100mm聚苯板保温材料彩钢板。温室朝向为南偏西6°；主要建筑空间特征参数为：脊高3.5m、北墙高度2.8m、后屋面水平投影长度为0.9m。

由图1计算结果表明，在冬季“反季节”蔬菜作物生产关键期（12月1日至1月31日）

确保温室内温度不低于8℃的计算条件下，冬季需要向日光温室补充供热量（图1a））， 而其中需要在夜间补充的热量占冬季总供热负荷的78.1%（图1b））。

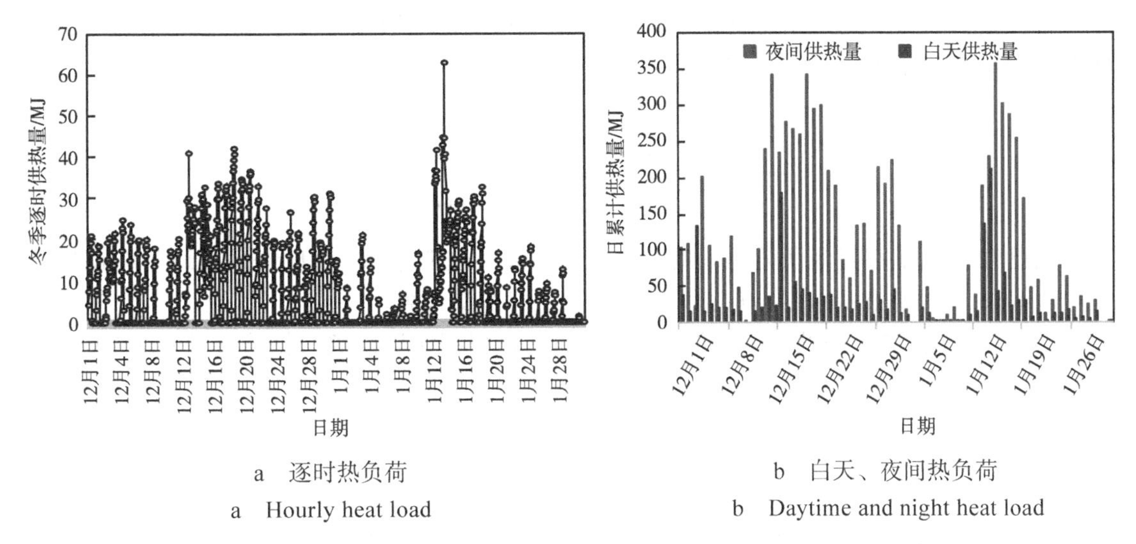

a　逐时热负荷

a　Hourly heat load

b　白天、夜间热负荷

b　Daytime and night heat load

图1　日光温室冬季“反季节”蔬菜作物生产关键期建筑热负荷随时间变化

Figure 1　Heat load of building in solar greenhouse varies with time in winter the key period of “anti season” vegetable crop production

2.2　日光温室建筑朝向优化设计

日光温室建筑朝向不但直接影响日光温室1d可接受的日照时间，同时还影响有效日照时间内日光温室内墙体和地面上被阳光照射的面积（也称光斑面积）。由于太阳辐射和室外气象参数的周期性动态变化特性，科学确定不同地理纬度日光温室建造最佳朝向，对确保日光温室内蔬菜作物各生长期获得足够的光热能，具有非常重要的意义。受全年太阳运动轨迹动态变化的影响，太阳高度角h和方位角α的动态变化特性直接影响日光温室前屋面接受太阳辐射照度的大小。在北半球随着纬度的增加，日出时间逐渐延后，日落时间逐渐提前，一天的日照时长逐渐减小。因此，不同地理纬度地区的日光温室，由于其可接收太阳辐射的有效日照时长不同，其前屋面截获太阳辐射的能力也不同。

对于日光温室建筑朝向设计，保证其前屋面可最大化接收太阳能十分重要。在太阳辐射有效照射时间段内，日光温室前屋面累计截获的太阳总辐射量最大时的温室朝向即为最佳朝向。因此，所谓日光温室最佳朝向的问题，实际上是计算期间内温室前屋面截获太阳直射辐照量强度累计量最大值的问题。

蔬菜作物关键生长期间（*n*天）透过日光温室前屋面累计太阳辐照量*S*可表示为式（1）。

$$S=\sum_{i=1}^{n}\int_{t_1}^{t_2}\tau_{tn}\times[I_0P^{\operatorname{csch}_t}(\cos\theta\times\sinh_t+\sin\theta\times\cosh_t\times\cos(\alpha_t-\gamma))+\frac{1}{2}\times I_0\sinh_t\times\frac{1-P^{\operatorname{csch}_t}}{1-1.4\ln P}\cos^2\frac{\theta}{2}]\times Adt \qquad (1)$$

式中，n为蔬菜作物关键生长期天数，取11月1日至次年2月28日；τ_{tn}为正南向日光温室*t*时刻前屋面塑料薄膜透过率（为简化计算，本研究日光温室前屋面塑料薄膜透过率仅按正南向简化处理），%。

求解日光温室建筑最佳朝向的问题，实际上是求解日光温室前屋面累计截获太阳总辐照量的最大值问题。因此，可以式（1）为目标函数，以日照时间和日照质量为约束条件，对日光温室前屋面累计太阳辐照量S求最值。根据极值原理，可令$dS/d\gamma=0$，即可得到日光温室建筑最佳朝向γ_{max}计算式（2）。

$$\gamma_{max}=\arctan\frac{\sum_{i=1}^{n}\int_{t_1}^{t_2}\tau_{tn}P^{\operatorname{csch}_t}\cosh_t\sin\alpha_t dt}{\sum_{i=1}^{n}\int_{t_1}^{t_2}\tau_{tn}P^{\operatorname{csch}_t}\cosh_t\cos\alpha_t dt} \tag{2}$$

当已知日光温室建筑所处地理纬度ϕ，当地大气透明系数P及日光温室蔬菜作物越冬生产期间n，即可根据式（2）确定所建日光温室建筑最佳朝向γ_{max}。

实际上，式（2）中的大气透明系数P值通常是根据实测气象数据统计及式（3）计算得到的[6]。

$$P=\sqrt[\frac{1}{\sinh}]{\frac{I_{DN}}{I_0}} \tag{3}$$

式中，I_0为太阳辐射常数，W/m^2；h为太阳高度角，°；I_{DN}为实测气象数据中到达地面上的法向太阳辐射强度，W/m^2。

设定冬季“反季节”蔬菜作物生产关键期为11月1日至次年2月28日，根据式（2）可以计算得到不同地理纬度地区日光温室建筑最佳朝向，如图2所示。

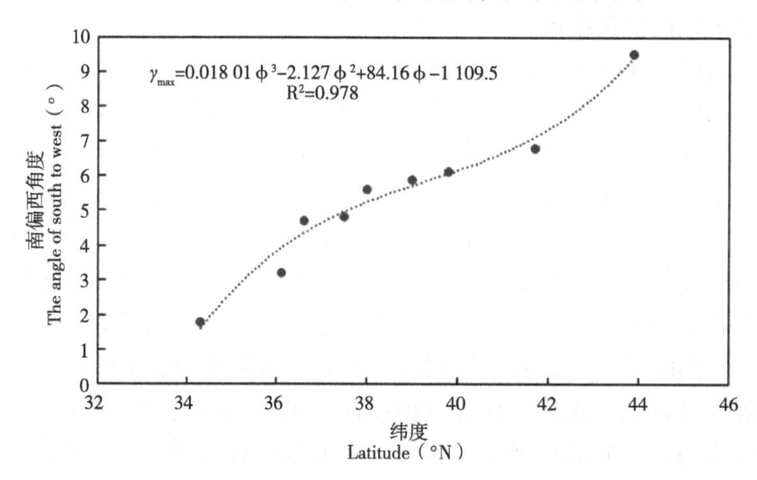

图 2　不同纬度地区日光温室建筑最佳朝向

Figure 2　Optimum orientation values of solar greenhouse in different latitudes

根据图2计算结果，可得到日光温室建筑最佳朝向与地理纬度的拟合关系式，也即，可建立关于不同地理纬度地区日光温室建筑最佳朝向简化模型式［式（4）］。

$$\gamma_{max}=0.01801\phi^3-2.127\phi^2+84.16\phi-1109.5 \quad (R^2=0.978) \tag{4}$$

2.3　日光温室建筑空间形态特征参数优化设计

反映日光温室建筑空间形态特征的参数主要有跨度、脊高、后屋面水平投影长度、北墙高度等。这些参数的合理匹配与否，一方面关乎白天通过日光温室透明前屋面进入温室内的太阳光热量是否足够大；二方面决定了集集热、吸热和蓄热于一体的温室北墙体面积大小；三方面也决定了日光温室非透明墙体与透明屋面的面积比，而该屋面与墙体面积的比则决定了白天进入日光温室太阳能量与通过围护结构向外散失热量的博弈程度。显然，前屋面越大，白天可进入的太阳光热能亦越大；但由此向外流失的热量特别是在夜间也随之增大。大量实测与分析结果表明，夜间前屋面既使加盖保温覆盖物，但由于其热阻明显小于墙体的，导致约有60%～70%的热损失是从前屋面流失的；另外，脊高过高，也将导致温室建筑空间过大，维持温室必要热环境需要补充的热量特别是夜间的也越大。

对于高效能日光温室，冬季“反季节”蔬菜作物生产期，全部利用太阳能或外加少量能源为日光温室营造其必要的热环境是最高目标。根据建筑热工设计理论，可以将向温室提供的累积补充供热量最小作为日光温室建筑空间形态特征参数优化设计的控制目标，即有式（5）成立。

$$Q_{min}=min\sum_{i=1}^{n}\left(\sum_{t_1}^{t_2}\tau_{ti}I_{ti}\sqrt{(L-C)^2+(\lambda L)^2}\times l-\sum_{t_1}^{t_2}K_f\times(t_{wti}-t_n)\times\sqrt{(L-C)^2+(L)^2}\times l-\sum_{t_2}^{t_1}K_f'\times(t_{wti}-t_n)\sqrt{(L-C)^2+(L)^2}\times l-\sum_{t_1}^{t_2}K_b(t_{wti}-t_n)\sqrt{(L-H_w)^2+C^2}\times l-\sum_{0:00}^{24:00}u_{1bti}\right) \quad (5)$$

式中，τ_{ti}为前屋面薄膜透过率，%；t_1为保温覆盖物卷起时间，t_2为保温被覆盖时间。I_{ti}为投射在前屋面的太阳总辐射强度（可查阅当地气象数据），W/m^2；F_f为日光温室前屋面面积，m^2。K_f为前屋面白天（保温覆盖物卷起）的传热系数，W/（m^2·℃），K_f'前屋面夜间（保温被覆盖）的传热系数，W/（m^2·℃）；t_{wti}为室外空气设计计算温度，℃；t_n为维持温室热环境必要的温室空气设计计算温度，℃；式中，K_b为后屋面传热系数，W/（m^2·℃）；F_b为温室后屋面面积，m^2。u_{1bti}温室后屋面向外界流失热量，W。

作为目标函数式（5）的约束条件：①确保种植区最后一排作物的冠层大暑日～翌年小满日都能接受到太阳照射，即有式（6）成立；②冬至日至大寒日期间，日光温室北墙可以接收到太阳照射，即有式（7）。

$$\sin\theta_1=\frac{\lambda L-L_P}{\sqrt{(\lambda L-L_P)^2+(C-P)^2}}\geq\sin h \quad (6)$$

$$\tan\theta_2=\frac{\lambda L-H_w}{C}\geqslant\tan h \quad (7)$$

式中，L_p为植株高度，一般取2m；P为温室走道宽度，一般取0.8m；h为太阳高度角；式中，L为日光温室的跨度，m；λ为日光温室高跨比，λ=H/L；C为日光温室的后屋面水平投影长度，m；H_w为日光温室北墙高度，m；l为日光温室的长度，m。

考虑到北半球大暑日至翌年小满日期间，大暑日正午的太阳高度角最高，式（8）中太阳高度角h可取当地大暑日正午时的；同理，冬至日至大寒日期间，大寒日正午的太阳高度角最高，式（9）中太阳高度角h取当地大寒日正午时的。

根据式（7）至式（9），影响温室最小累积补充供热量Q_{min}的主要参数有太阳辐射强度I、室外空气温度t_{out}、温室高跨比λ、温室跨度L、后屋面水平投影长度C和北墙高度H_w等参数（如式（8））。

$$Q_{min}=f(\mathrm{t_{out}},I,\lambda,L,\mathrm{C},H_w) \quad (8)$$

2.3.1　高跨比简化计算模型

结合EnergyPlus能耗模拟软件、并应用单一变量控制方法，可发现温室高跨比与当地太阳辐射量强度以及室外空气温度具有强相关性[7]。基于该分析结果，并利用Matlab曲线拟合工具箱cftool，可建立关于不同纬度地区冬季反季节蔬菜作物关键生产期，温室高跨比λ与当地室外空气平均温度和日平均太阳辐射量的关联拟合式（9）[7]。

$$\lambda = 0.4301 + 0.0008249 \times \bar{t}_s + 0.007702 \times \bar{I}_S \quad (9)$$

式中，$\bar{t}_s$为冬季蔬菜作物生产关键时期的当地室外平均温度，℃；$\bar{I}_s$为冬季蔬菜作物生产关键时期的当地日平均太阳辐射量，MJ/（m^2•day）。

根据式（9），对于越冬生产的日光温室，室外空气平均温度和日平均太阳辐射量越低的地区，相应的日光温室高跨比λ也应降低。

2.3.2　后屋面水平投影长度简化计算模型

式（10）为式（6）的变形。日光温室后屋面的水平投影长度越长，相应地越冬生产期间需要向温室提供的累计补充供热量也将随之减小。为了简化计算，可考虑取其临界值作为温室建筑后屋面水平投影长度的计算值，即有式（11）成立。

$$C \le \sqrt{\left(\frac{\lambda L - L_P}{\sin h_s}\right)^2 - (\lambda L - L_P)^2} + P \quad (10)$$

$$C = \lambda\sqrt{\left(\frac{\lambda L - L_P}{\sin h_s}\right)^2 - (L - L_P)^2} + P \quad (11)$$

式中，h_s为当地大暑日正午太阳高度角，°。

2.3.3　北墙高度简化计算模型

式（12）为式（7）的变形。随着北墙高度的不断增高，越冬生产期间需要向温室提供的热量也随之减小。为了简化计算，可考虑取其临界值作为温室建筑北墙体高度的计算值，即有式（13）成立。

$$H_w \le \lambda L - C \times \tan\ h_c \quad (12)$$

$$H_w = \lambda L - \left\{\sqrt{\left(\frac{\lambda L - L_P}{\sin h_s}\right)^2 - (\lambda L - L_P)^2 + P}\right\} \times \tan h_c \quad (13)$$

式中，h_c为当地大寒日正午太阳高度角，°。

综上分析，当日光温室的建设场地确定，查阅当地冬季蔬菜作物生产关键期的室外空气平均温度和日平均太阳辐射量，即可根据式（9）、式（11）和式（13）给出所在地日光温室建筑空间形态特征参数的设计推荐值。

3　计算案例节能效果比较分析

为了评价第2.2节和第2.3节给出的不同地理纬度地区日光温室建筑最佳朝向、空间形态特征参数简化设计计算方法的有效性，仍然2.1节的北京地区日光温室计算对象为比较对象，并假定优化温室的跨度、长度，以及墙体热工性能同2.1节，根据第2节优化方法的计算结果为：温室朝向为南偏西6°；主要建筑空间特征参数为：脊高4.2m、北墙高度

3.1m、后屋面水平投影长度1.8m。

同样在日光温室越冬生产关键期（12月1日至1月31日）确保温室内温度不低于8℃的计算条件下，根据EnergyPlus能耗模拟软件可计算得到优化前后两个温室冬季需要补充的供热量（图3）。由图3可见，优化后日光温室冬季累积补充供热量为8150MJ，较2.1节现行温室的（9670MJ）减少了15.7%的供热量需求。而这种能耗的减少，仅仅是通过日光温室建筑的科学设计得以实现的。由此也说明了，被动式建筑设计方法在建筑节能中的重要性。

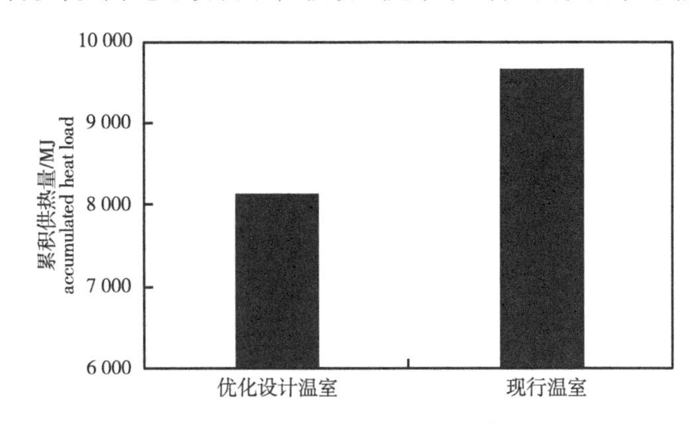

图 3　计算案例节能效果比较

Figure 3　Comparison of energy saving results in calculating cases

4　结论

本研究采用被动式建筑设计方法，运用建筑热工、建筑气候、建筑构造与太阳能利用等学科的原理和方法，研究并提出了不同地理纬度地区日光温室最佳建筑朝向以及建筑空间形态特征参数设计计算模型。

EnergyPlus能耗模拟软件比较计算结果表明，在确保日光温室内温度不低于8℃的计算条件下，在日光温室越冬生产关键期（12月1日至1月31日），优化后日光温室冬季累积补充供热量为8150MJ，较现行温室的（9670MJ）减少了15.7%的供热量需求，节能效果明显。

参考文献

［1］ 李军，邹志荣，杨旭，等.西北型节能日光温室采光设计中方位角和前屋面角的分析、探讨与应用［J］. 西北农业学报，2003，12（2）：105-108.

［2］ 白义奎，刘文合，王铁良，等.日光温室朝向对进光量的影响分析［J］.农业机械学报，2005，36（2）：73-75，84.

［3］ 李天来. 日光温室蔬菜栽培理论与实践［M］.北京：中国农业出版社，2013.

［4］ 白义奎，周东升，曹刚，等. 北方寒区节能日光温室建筑设计理论与方法研究［J］.新疆农业科学，2014，51（6）：990-998.

［5］ 魏晓明，周长吉，曹楠，等. 基于光照的日光温室总体尺寸确定方法研究［J］. 北方园艺，2010（15）：1-5.

［6］ 彦启森. 建筑热过程［M］.北京：中国建筑工业出版社，1986.

［7］ 陈超. 现代日光温室建筑热工设计理论与方法［M］. 北京：科学出版社，2017.

设施栽培理论和技术

PROTECTED CULTIVATION THEORY
AND TECHNOLOGY

Evaluation of The Aeroponics, Hydroponics and Vermiculite Systems for Greenhouse Tomato Production for Antarctic and Long-Duration Spaceflight

Dong Chen[1, 2], Wang Minjuan[1, *], Fu Yuming[3], Gao Wanlin[1, *]

(1. *College of Information and Electrical Engineering*, *China Agricultural University*, 100083, *Beijing*; 2. *Higher Education Evaluation Center of the Ministry of Education*, *Beijing*; 3. *School of Biological Science and Medical Engineering*, *Beihang University*, *Beijing*, 100191)

Abstract: Nutrient delivery system is one of the most important key technologies for tomato (*Lycopersicon esculentum* Mill.) growth and development in both space such as Bio-regenerative Life Support Systems (BLSS) and unban agriculture fields. To realize the high space mission fidelity, the Antarctic station is usually selected for the location of the mobile test facility, because of its logistical, infrastructural and environmental properties. The objective of this study is to investigate the influences of different nutrient delivery systems (aeroponics, hydroponics and vermiculite) on the tomato growth, photosynthetic characteristics, antioxidant capacity, biomass yield and quality during their life cycle. The results showed that the chlorophyll a content of tomato plants grown in aeroponics system had the top advantages from initial stage 1.47 mg/g to late stage 1.21 mg/g. Both tomato photosynthesis and stomatal conductance maximized at the development stage and then decreased later in senescent leaves. At the initial stage and the development stage, POD activities in the aeroponics treatment were higher than other two treatments, reached 3.6 U/mg prot and 4.6 U/mg prot, respectively. The fresh yield 431.3 g/plant of hydroponics treatment group was significantly lower than others. At the same time, there were no significant differences among nutrient delivery systems in the per fruit fresh mass, which was 14.2~17.5 g/fruit.

Key words: Bioregenerative life support systems; Aeroponics; Hydroponics; Vermiculite

1 Introduction

Sustained human presence in space requires the development of bio-regenerative life support systems (BLSS) (Dong et al., 2017; Sirko et al., 1994; Tikhomirov et al., 2011; Wheeler et al., 1993). When fully integrated into crewed habitats these systems decrease resupply requirements by (re-) generating human resources through biological processes (Dempster et al., 2009). Within a BLSS, higher plants take a central role as they can provide food production, CO_2 reduction, O_2 production, water recycling and waste management (Fu et al., 2016). It has also been suggested through both anecdotal and scientific means that plants can also positively influence crew psychological well-being (Wu and Wang, 2015). These aspects play a key role for future space crews, who may spend a significant part of their life away from Earth.

To realize the high space mission fidelity, the Antarctic station is usually selected for the location of the mobile test facility, because of its logistical, infrastructural and environmental properties. The crew size and composition during winter is also similar to a crewed spacecraft. The winter period, with duration of 9 to 10 months, is foreseen for the validation of the systems. The long-duration of the test campaign allows several cultivation cycles of plants and therefore several possibilities for testing various system settings and operational modes. The research initiative focuses on BLSS, especially greenhouse modules, and how these technologies can be integrated in future human-made space habitats.

Nutrient delivery is the provision of water and nutrients in the amount necessary for optimal plant growth over all plant development stages. The method of delivery of the nutrient/water mix (hydroponic solution) can be conducted in various ways. These include soil-based, nutrient media based hydroponics systems as well as systems requiring no substrate, such as aeroponics. Each configuration can provide certain advantages, but for space-based BLSS, aeroponics can provide the benefit that no soil or substrate is required (minimizing waste) while potentially producing higher plant yields.

The basic principle of aeroponic systems is to grow plants suspended in a closed or semi-closed environment by spraying the plant's dangling roots with an atomized nutrient-rich water solution. Aeroponic equipment involves the use of sprayers, misters, foggers, or other devices to create a fine mist of solution to deliver nutrients to plant roots. No soil or grow media is needed for the whole life cycle. Furthermore, the plant's nutrient uptake can be improved by the exact control of plant root environment. Through this innovative irrigation principle, a general reduction in nutrient solution throughput, decrease of water loss, higher plant density (than traditional grow procedures), limitation of disease transmission, and potentially higher plants yields can be achieved.

However, utilization capacity of the soilless system in BLSS has not yet expanded at all on the Antarctic stations and long-duration spaceflight scale due to higher capital investment. Any system failure in BLSS could result in a catastrophic event culminating with mission abort or, even worse, loss of crew. It is necessary to investigate the different soilless system availability effects on the space agricultural production and evaluate different consequences caused by the soilless system disturbance in the artificial condition. Furthermore, as one of the most popular vegetable plants, tomato has been investigated for the long-term space missions during these years (Lu et al., 2005; Nechitailo et al., 2004). Therefore, we cultivated tomato (*Lycopersicon esculentum* Mill.) plants in aeroponics, hydroponics and vermiculite systems and investigated the influences of different systems on the tomato growth, photosynthetic characteristics, antioxidant capacity, biomass yield and quality during their life cycle.

2 Materials and Methods

2.1 Plant material and cultivation conditions

Tomato (*Lycopersicon esculentum* Mill.) plants were grown in a controlled environment cabinet under a light intensity of 250 $\mu mol \cdot m^{-2} \cdot s^{-1}$ at the top of the plants containing the fruit. The tomato planting density was 5 plants per m^2. The temperature ranged from 25±3℃ during the light period (14h) to 18±3℃ in the dark (10h) , and the relative humidity was 70±10%. The temperature was controlled using cool air following into cabinet from central cooler. The CO_2 level was the same as that of atmosphere outside. The modified Hoagland nutrient solution was the basic culture medium (Table 1) .

Table 1 Nutrient compositions

Compositions	Concentration (mg/L)
$Ca(NO_3)_2 \cdot 4H_2O$	945
KNO_3	607
$NH_4H_2PO_4$	115
$MgSO_4 \cdot 7H_2O$	493
FeEDTA	28
H_3BO_3	2.86
$MnSO_4 \cdot 4H_2O$	2.13
$ZnSO_4 \cdot 7H_2O$	0.22
$CuSO_4 \cdot 5H_2O$	0.08
$(NH_4)_6Mo_7O_{24} \cdot 4H_2O$	0.02
pH	5.8 ~ 6.3

2.2 Culture systems

To investigate the influence of different culture methods on the tomato plants during ontogenesis, experiments were divided into 3 groups according to different systems: aeroponics, hydroponics and vermiculite systems (Figure 1) . The three systems constructed primarily of relatively inert materials such as glass, Teflon and stainless steel were used for growth of tomato plants. Each plant growing area was 1 m^2.

An aeroponics system was designed by China Agricultural University. The aeroponic tubing and misters are designed so that they can be easily installed within the top of the growth trays. The architecture includes pairs of misters spaced at 20 cm intervals. The misters are angled downwards to ensure good coverage of the plants roots over the entire growth cycle of the respective crops. The tubing and misters are easily installed within the top of the plant channels during the preparation of each growing cycle. Excess aeroponic water is collected in the bottom of the slightly angled channels and flows out drainage tubes back to the respective petal reservoir. The solution is analyzed (e.g. ion-selective, EC and pH sensors) and composition adjustments are then made accordingly.

A hydroponics system was used with the modified Hoagland nutrient solution as listed in Table 1. The nutrient solution was poured into the cultivating plate that was covered by cystosepiments where there were some holes for plant cultivation.

An vermiculite system was designed and specific details on chamber design and control have been described previously (Dong et al., 2016; Dong et al., 2015b) . Plants were grown in vermiculite, continuously maintained under optimal irrigation, and supplied with 1X strength Hoagland solution every 3 days.

The tomato seeds were germinated in the equipment at 24℃ air temperature for 10 days and the tomato seedlings were transplanted to different systems. According to the growth characteristics of tomato plants, the tomato life cycle was divided into four stages: Initial stage (0~30 days) , Mid-stage (30~60d) , Development stage (60~90d) and Late stage (90~120d) . Fruit were harvested every week after 100 days after transplanting. Early yield, fruit number and single fruit mass were calculated from fruits in the first three harvests.

2.3 Photosynthetic characteristics analyses

2.3.1 Chlorophyll contents

At the initial stage, mid-stage, development stage and late stage, the content of chlorophyll a and chlorophyll b in the tomato leave was measured with an ultraviolet spectrophotometer(SP-75, Shanghai spectrum instruments co., LTD, China) (Macknney, 1941) . Leaf samples were frozen in liquid nitrogen and stored at −80℃ until measured.

2.3.2 Photosynthetic efficiency

Portable photosynthesis instrument (Li-6400XT, Li-Cor, USA) was used for the determination of photosynthetic characteristics. Leaf gas-exchange parameters included photosynthetic rate (A) , stomatal conductance (gs) and intercellular CO_2 concentration (C_i) using the second leaf at the tomato terminal bud. Water use efficiencies (A/gs) were calculated by dividing A by gs and the instantaneous carboxylation efficiencies(A/C_i)were also calculated(Dong et al., 2015a) .

2.4 Antioxidant capacity analyses

2.4.1 Peroxidase (POD) activity

POD activity was analyzed spectrophotometrically at 470 nm using guaiacol as a phenolic substrate with hydrogen peroxide (Díaz et al., 2001) . The reaction mixture contained 0.15 ml of 4% (v/v) guaiacol, 0.15 mol/L of 1% (v/v) H_2O_2, 2.66 mL of 0.1 mol/L phosphate buffer (pH= 7.0) and 40 μmol/L of enzyme extract. Blank sample contained the same mixture without enzyme extract.

2.4.2 Catalase (CAT) activity

CAT activity was determined according to the method described by Kumar and Knowles (Kumar and Knowles, 1993) . CAT reaction solution consisted of 100 mmol/L Na_2HPO_4-

NaH_2PO_4 buffer solution（pH=7.0）and 0.1mol/L H_2O_2. The optical density was determined every 1 min at λ=240 nm.

2.4.3 Malonaldehyde（MDA）content

Determination of MDA depended on the method of Stewart and Bewley（Stewart and Bewley，1980）. Briefly，10mol/L 0.1% trichloroacetic acid（TCA）pestle homogenate was used to centrifuge wheat leaves（0.5 g）at 4 000 rpm for 10 min. 2mol/L supernatant was added to 4mol/L 5% thiobarbituric acid（TBA）which was made up by 20% TCA. The mixture was heated at 95℃ for 30 min and then cooled in ice-bath rapidly. The supernatant was obtained by centrifuging at 3 000 rpm for 10 min. The absorbency of the supernatant was recorded at 532 nm. The value for non-specific absorption at 600 nm was subtracted. The MDA content was calculated using its extinction coefficient of 155 $mM^{-1}\cdot cm^{-1}$.

2.5 Biomass yield analyses

2.5.1 Lycopene & β-carotene

The lycopene and β-carotene（mg/100 mol/L）were evaluated in hydroponically grown tomatoes（Nagata and Yamashita，1992）. One gram of tomato sample was taken in a test tube; poured acetone: hexane（4:6）in the test tube and then the mixture was homogenized. The optical density of the homogenized mixture was measured at 663，645，505 and 453 nm. The values of lycopene and β-carotene were calculated by following formula:

Lycopene（mg/100 ml）$=-0.045\,8A_{663}+0.204A_{645}+0.372A_{505}-0.080\,6A_{453}$

β-carotene（mg/100 ml）$=0.216A_{663}-1.22A_{645}-0.304A_{505}+0.452A_{453}$

where: A is the absorbance at 663，645，505 and 453 nm.

2.5.2 Edible and inedible biomass

Fruit weight(g), total yield(g/plant) and different part weight were recorded according to(Dong et al., 2014; Rosen, 1999).

2.6 Data statistics

The experiment was setup in a completely randomized design. All experiments were performed in triplicate. The average value of total 6 measurements ± standard deviation was regarded as the final result. All statistical analyses were performed using SPSS 18.0. P values less than .05 were considered statistically significant.

3 Results and discussion

Different nutrient delivery systems had significant effects on the growth and development of the tomato plants. From Figure 2a，chlorophyll a concentrations fluctuated during the whole life cycle. The aeroponics system had the top advantages from initial stage 1.47 mg/g to late stage 1.21 mg/g and the hydroponics system group had a lower level at both mid-stage 1.56 mg/g and development stage 1.62 mg/g. About the chlorophyll b，only the development stage was the sensitive stage for different nutrient delivery systems，the value was 0.79 mg/g，0.59 mg/

g and 0.78 mg/g, respectively (Figure 2b) . Aeroponics system and vermiculite system had a significant higher value than that for the hydroponics system. The value added of chlorophyll b was greater than that of chlorophyll a for both of the aeroponics and vermiculite systems, especially at the development stage. And thus the chlorophyll a/b decreased (Figure 2c) . Throughout the senescence period, particularly at the late stage, there was no significant difference either chlorophyll a or b contents from the tomato leaves. Senescence typically involves cessation of photosynthesis and degeneration of cellular structures, with strong losses of chlorophyll and the (Ougham et al., 2008) .

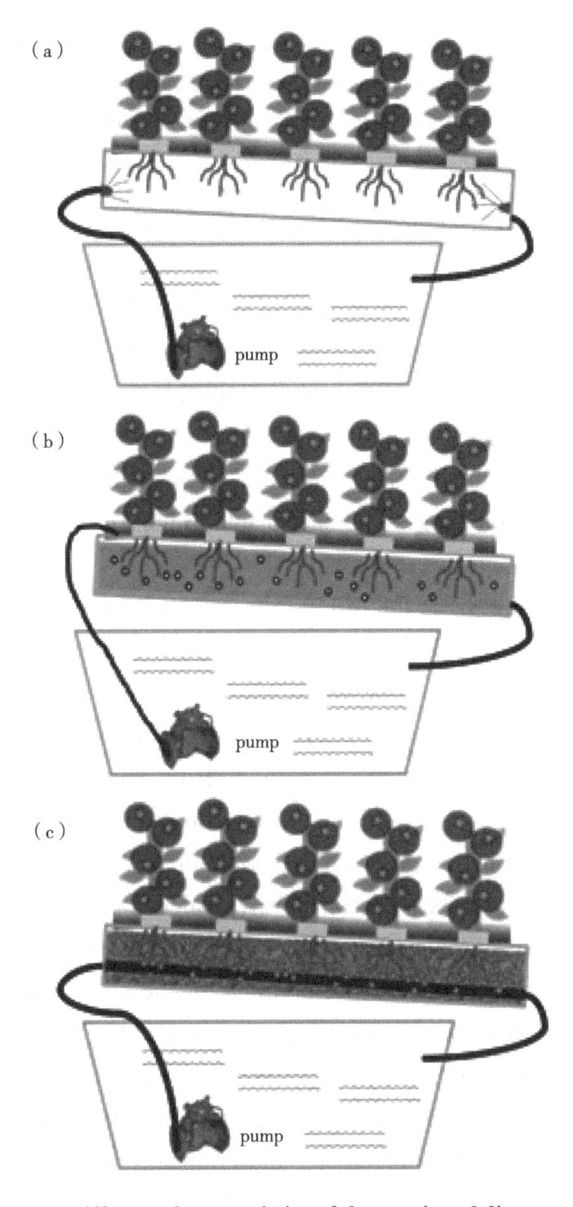

Figure 1　Different characteristics of the nutrient delivery systems

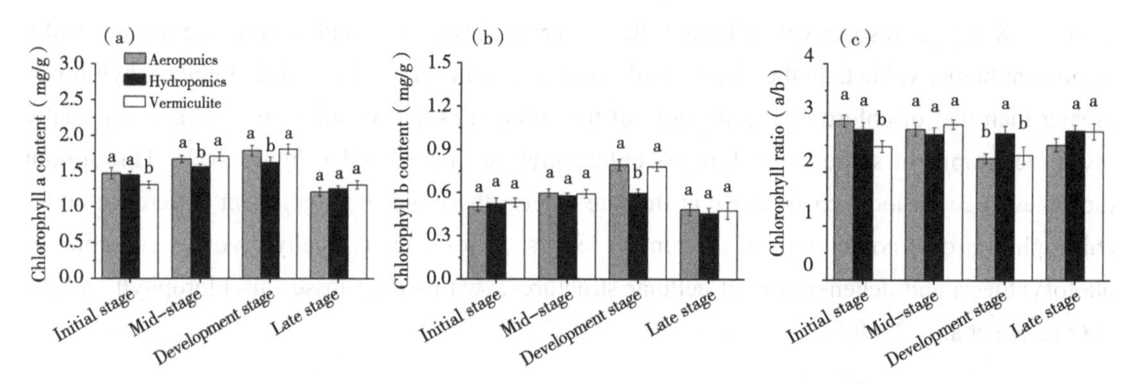

Figure 2 Response of chlorophyll a (a) and chlorophyll b (b) and chlorophyll ratio (c) of tomato leaves at different stages of ontogenesis to different treatments. Vertical bars are means ± SD. Within each graph, bars labeled with lowercase letters are significantly different at $p \leq 0.05$.

Tomato plants sense and respond to different nutrient delivery systems through modified photosynthesis (A) and stomatal conductance (gs) (Figure 3a-b). Both A and gs maximized at the development stage and then decreased later in senescent leaves. At the development stage, the tomato plants under vermiculite treatment reached 16.4 μmol CO_2 $m^{-2}\cdot s^{-1}$. Despite closing the stomata, the transpiration rate of tomato leaves was increased at the mid-stage using hydroponics treatment (Figure 3c), reflecting an imbalance between water uptake and transpiration rate. In particular, the increase in A at the development stage, was smaller than the increase in gs, so A/gs decreased (Figure 4d), allowing the tomato plants to more efficiently use water for the high yield in finally.

Stimulation of photosynthesis is the driving force for increased growth and yield of tomato in aeroponics and vermiculite groups. Over prolonged periods of exposure to different nutrient delivery systems (days to weeks), the stimulation of photosynthesis gradually diminished. However, at the mid-stage and development stage, the effect of nutrient delivery systems became more significant, which may be because the stages from vegetative growth to reproductive growth are more sensitive. The water and the ion concentration played more important roles in the plant growth, which led to the photosynthesis better or not. Hydroponics, managed as a closed system, enables nutrient to be well controlled, which is necessary in order to avoid hyper-accumulation and the consequent toxic effects. It can also be used for improving the quality and the shelf-life of fruit with adequate nutrient concentrations for the human diet. Nutrient sensing is an intrinsic property of guard cells, which are thought to respond to the intercellular carbon dioxide concentration (C_i) rather than CO_2 concentration at the leaf surface. Ion and organic solute concentrations mediate the turgor pressure in the guard cells that determines stomatal aperture. Stomata closure implies lower CO_2 availability, and ultimately lower CO_2 fixation by Rubisco, which may finally result in lower biomass production (Leakey et al., 2009).

We also studied the activity of the antioxidant enzymes POD and CAT and the production of MDA in tomato leaves during ontogenesis. POD and CAT activities increased during early leaf development, reaching their maximal levels at the development stage and decreasing later in senescent leaves. Also, these antioxidant enzymatic activities were at low level in the plants grown under vermiculite condition throughout leaf development (Figure 4a-b) . At the initial stage and the development stage, POD activities in the aeroponics treatment were higher than other two treatments, reached 3.6 U/mg prot and 4.6 U/mg prot, respectively. As shown in Figure 4c, MDA production increased with leaf growth and development, especially at the mid-stage and the development stage, suggesting that these two stages may be more important role in regulating leaf senescence in tomato plants by increasing reactive oxygen species (ROS) production. Alternatively, lower ROS production may decrease the incidence of oxidative stress, translating as a reduced stimulus for antioxidant production. Such reduced oxidative pressure would also correspond well with a smaller POD activity. Greater C_i may translate into greater CO_2/O_2 ratio in the chloroplast, causing lower oxygenation rates by Rubisco and therefore lower rates of photorespiration and associated ROS production in tomato cultivar. However, the mechanism needs to be further investigated. Moreover, responses of plant organs to environmental factors should be different among different growth stages. Plants have adaptability to environmental changes.

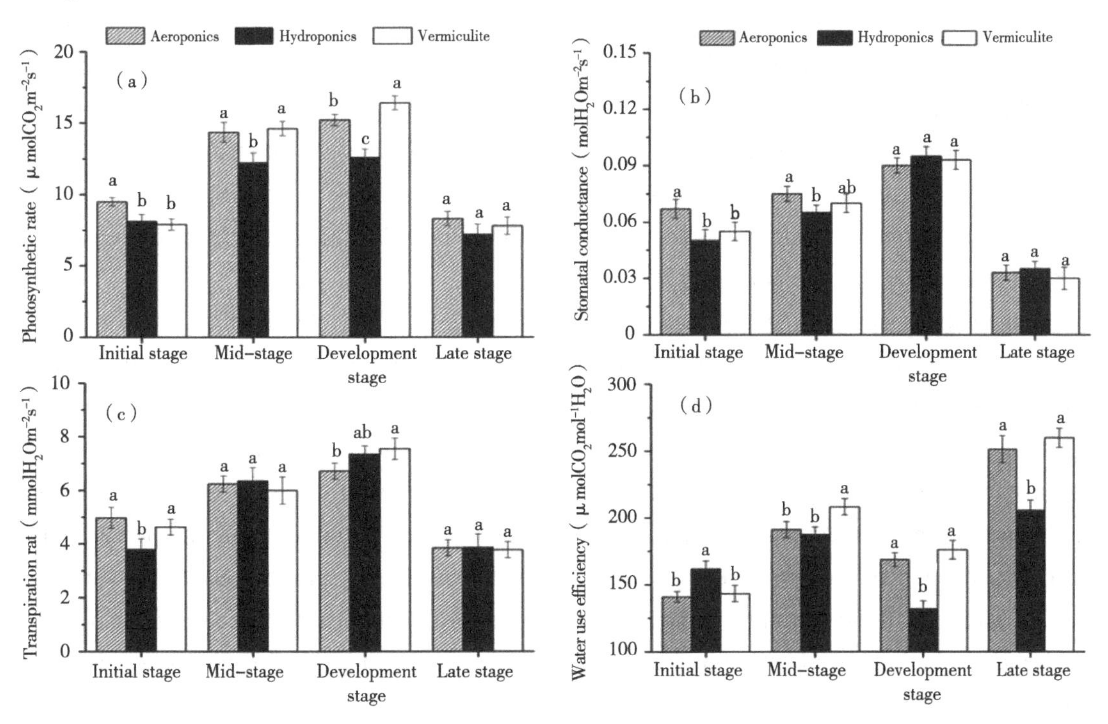

Figure 3 Response of photosynthetic rate (a) , stomatal conductance (b) , transpiration rate (c) and water use efficiency (d) of tomato leaves at different stages of ontogenesis to different treatments. Vertical bars are means ± SD. Within each graph, bars labeled with lowercase letters are significantly different at $p \leq 0.05$.

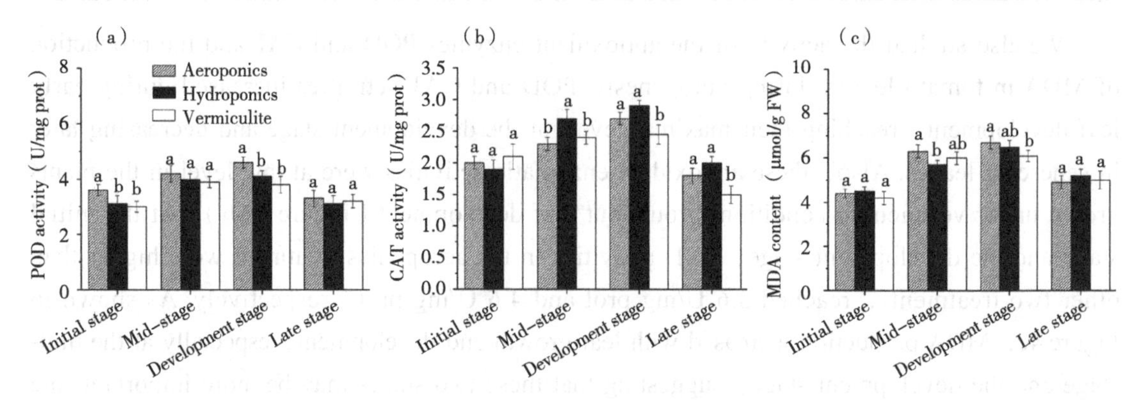

Figure 4 Response of POD activity (a), CAT activity (b) and MDA content (c) of tomato leaves at different stages of ontogenesis to different treatments. Vertical bars are means ± SD. Within each graph, bars labeled with lowercase letters are significantly different at $p \leq 0.05$.

The lycopene (mg/100 g) values were recorded from different tomato plants under aeroponics, hydroponics and vermiculite systems those advocated already measured values range from 4.57 to 5.01 mg/100 g. The lycopene content showed red colored tomato (Baranska et al., 2006), and the degree of redness is directly proportional to the concentration of lycopene, while orange or yellow color shows less concentration of lycopene. The lycopene content of tomato plants grown in the aeroponics and vermiculite systems was higher than the other group. But the tomato β-carotene content was no significant difference among different treatments. The results are in the range of the findings of recorded 0.23 ~ 4.00 mg/100 g of β-carotene in tomato fruits (Baranska et al., 2006; Hyman et al., 2004).

In tomatoes, the content of β-carotene has been shown to increase from the green to the fully ripe stage (Fraser et al., 1994). Since β-carotene accumulation is a ripening-related event in tomato and its formation from lycopene appears to be principally under transcription regulation (Fraser et al., 2007; Lee et al., 2012)

In the experiment treatments, there was no difference in fruit yield per plant and fresh mass per fruit between aeroponics and vermiculite grown plants (Figure 6a). However, the yield of hydroponics treatment group was significantly lower than others. The fresh weight reached only 431.3 g/plant. At the same time, there were no significant differences among nutrient delivery systems in the per fruit fresh mass (Figure 6b), which was 14.2~17.5 g/fruit. The plant yield depends on the weight and number of fruit. The number of tomato fruit was lower in the hydroponics treatment group. Therefore, the fruit weight was directly proportional to the yield of a plant. In addition, the dry weight of the tomato plants in different parts was also the same result, which means the tomato plants in hydroponics treatment may not so thick and strong compared with aeroponics and vermiculite treatments (Figure 6c).

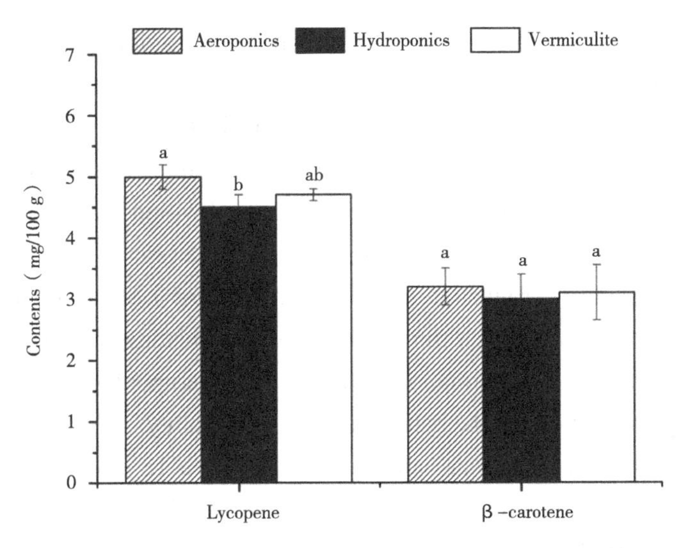

Figure 5 Response of Lycopene and β-carotene content of tomato plants to different treatments. Vertical bars are means ± SD. Within each graph, bars labeled with lowercase letters are significantly different at p≤0.05.

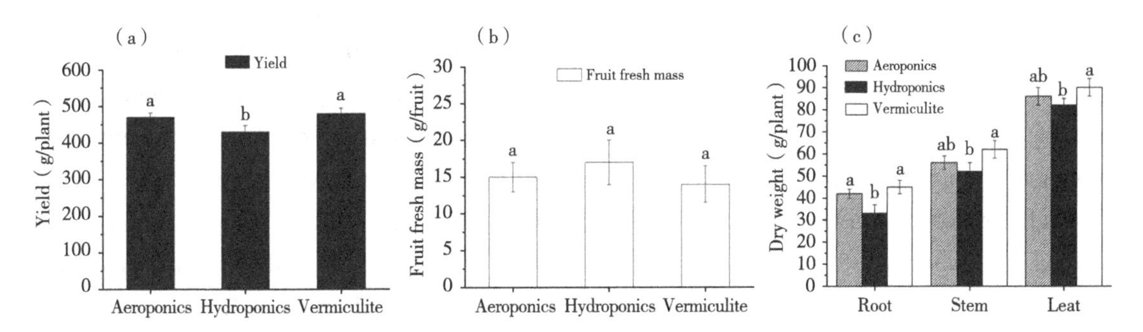

Figure 6. Response of yield (a) , fruit fresh mass (b) and different part dry weight (c) of tomato plants to different treatments. Vertical bars are means ± SD. Within each graph, bars labeled with lowercase letters are significantly different at p≤0.05.

Commercial scale production of tomato and other solanaceous vegetables is significantly hindered by attacks of soil-borne diseases and sudden temperature fluctuation under the open field conditions. To cope with these challenges, the soil-less technique is considered a promising tool for commercial vegetable productions (Mavromatis et al., 2003) . Hydroponics is the most intensive method for crop production in the agricultural industry (Lópezmillán et al., 2009) . This enables the plants to achieve higher growth of the shoot system with more vegetation, larger fruits, flowers and other edible parts. Plants in hydroponics grow up to two times faster with higher yields than with conventional soil farming methods due to higher oxygen levels around the root system, optimum pH along with increased nutrient and water uptake (Ghazvini et al., 2007) . The hydroponic culture of tomato and other susceptible vegetable crops can facilitate their successful and profitable production. Therefore, a precise detection and management of biotic and abiotic stresses should be taking into consideration for immense production.

The incorporation of aeroponics, as described, presents several advantages over other hydroponic nutrient delivery system options, in particular for missions to the lunar surface where transporting growth substrate can itself present challenges from a mass and waste perspective. That said, aeroponics is a technology that although discussed within the hydroponics literature for some time remains one that requires further investigation (Clawson et al., 2000; Weathers and Zobel, 1992). In particular, production improvements (if any) of aeroponics need to be more systematically assessed as well as such things as appropriate misting frequency and clogging and cleaning requirements for a true confirmation of the utility of aeroponics (in particular in a hybrid aeroponics-other implementation) to be better known with certainty.

It is well established that there are anywhere from 15 to 18 essential plant nutrients that govern plant growth, development and reproduction (of which oxygen, carbon and hydrogen can be considered non-fertilizer nutrients) (Jones and Jones, 1997). If any given nutrient is deficient, plant growth can be negatively influenced. Similar effects can arise when given nutrients are present in excessive concentrations, either by direct toxic effects or more likely, by limiting the uptake of other nutrient ions. In addition to sensor technology, further work is required in the development of the control system that could subsequently control nutrient salt additions to result in the optimum solution of interest. This is inherently challenging because the operator cannot simply add one nutrient ion of interest (e.g. K^+) but instead they must also add the other components contained within the employed salt (e.g. NO_3^- from KNO_3). Thus when attempting to adjust the concentration of a given ion, other ions may go out of balance without appropriate control or without the forethought to carry additional nutrient salts options that may otherwise not have been carried. Further developments surrounding the logic behind such hydroponic solution management choices must also be advanced.

Although literature exists suggesting the benefits of aeroponics nutrient delivery systems, little experimental evidence exists that demonstrates these benefits in a quantitative manner. Controlled experiments will be conducted to assess these possible yield benefits. Experimental utilization and documentation of the benefit and operational utilization of ion-selective sensors and various UV disinfection systems will also be conducted. Consideration of the use of aeroponics, ion-selective sensors and disinfection systems in the Antarctic as well as space-based systems will also be performed.

Keeping in view the importance of tomato and need of Chinese Antarctic Great Wall Station, Antarctic Zhongshan Station and Kunlun Station, it was imperative to carry out an experiment on tomatoes under greenhouse conditions by using different nutrient delivery systems in China. The research evaluation was based on tomato morphological, qualitative and analytical parameters, which are imperative for the development of rapid screening techniques and proper selection method of different nutrient delivery systems. In addition the project will further enhance the knowledge about crew time assessments. The quantification of realistic crew time requirements over the long-term operation of this greenhouse within this mission relevant environment will have considerable benefit over laboratory extrapolations. In addition to providing a plant production facility that will benefit the station's long duration crews, the project strives to advance the technology readiness of a number of technologies that could be applied to on orbit and planetary surface plant production systems.

4 Conclusions

Our results clearly demonstrate that, nutrient delivery is the provision of water and nutrients in the amount necessary for optimal tomato growth over all plant development stages. Both tomato photosynthesis and stomatal conductance maximized at the development stage and then decreased later in senescent leaves. Stimulation of photosynthesis is the driving force for increased growth and yield of tomato in aeroponics and vermiculite groups. There was no difference in fruit yield per plant between aeroponics and vermiculite grown plants, but the hydroponics treatment group was lower. Fully understanding how the biochemical mechanisms are directly or indirectly affected by water and nutrients in fruit tissues at ripening should help considerably in optimizing the treatment procedures and protocols. These findings are useful for designing and operating growth technologies for enhanced tomato production.

Research in this area should continue as in addition to enhancing our understanding of fundamental plant biology it will continue to enhance tomato yields and overall system reliability. This improved knowledge of plant nutrition will further the ability of the lunar base operators to also garner scientific data related to the influence of the reduced gravity environment of the lunar surface on crop growth.

5 Acknowledgements

The authors would like to thank their colleagues for their support of this work. The detailed comments from the anonymous reviewers were gratefully acknowledged. This work was supported by Project of Scientific Operating Expenses from Ministry of Education of China (Grant #: 2017PT19).

References

Baranska, M., Schütze, W., Schulz, H., 2006. Determination of lycopene and beta-carotene content in tomato fruits and related products: Comparison of FT-Raman, ATR-IR, and NIR spectroscopy. Analytical Chemistry, 78: 8456.

Clawson, J.M., Hoehn, A., Stodieck, et al., 2000. Re-examining Aeroponics for Spaceflight Plant Growth. Environment.

Dempster, W.F., Nelson, M., Silverstone, S., et al., 2009. Carbon dioxide dynamics of combined crops of wheat, cowpea, pinto beans in the Laboratory Biosphere closed ecological system. Advances in Space Research, 43: 1 229-1 235.

Díaz, J., Bernal, A., Pomar, F., et al., 2001. Induction of shikimate dehydrogenase and peroxidase in pepper (Capsicum annuum L.) seedlings in response to copper stress and its relation to lignification. Plant Science, 161: 179-188.

Dong, C., Fu, Y., Liu, G., et al., 2014. Growth, Photosynthetic Characteristics, Antioxidant Capacity and Biomass Yield and Quality of Wheat (*Triticum aestivum* L.) Exposed to LED Light Sources with Different Spectra Combinations. Journal of Agronomy & Crop Science, 200: 219–230.

Dong, C., Fu, Y., Xie, B., et al., 2017. Element Cycling and Energy Flux Responses in Ecosystem Simulations Conducted at the Chinese Lunar Palace-1. Astrobiology, 17.

Dong, C., Liu, G., Fu, Y., et al., 2016. Twin studies in Chinese closed controlled ecosystem with humans: The effect of elevated CO 2 disturbance on gas exchange characteristics. Ecological Engineering, 91: 126-130.

Dong, C., Shao, L., Liu, G., et al., 2015a. Photosynthetic characteristics, antioxidant capacity and biomass yield of wheat exposed to intermittent light irradiation with millisecond-scale periods. Journal of plant physiology, 184: 28-36.

Dong, C., Shao, L., Wang, M., et al., 2015b. Wheat Carbon Dioxide Responses in Space Simulations Conducted at the Chinese Lunar Palace-1. Agronomy Journal, 108.

Fraser, P.D., Enfissi, E.M.A., Halket, J.M., et al., 2007. Manipulation of Phytoene Levels in Tomato Fruit: Effects on Isoprenoids, Plastids, and Intermediary Metabolism. Plant Cell, 19: 4 131-4 132.

Fraser, P.D., Truesdale, M.R., Bird, C.R., et al., 1994. Carotenoid Biosynthesis during Tomato Fruit Development (Evidence for Tissue-Specific Gene Expression). Plant Physiology, 105: 405-413.

Fu, Y., Li, L., Xie, B., et al., 2016. How to Establish a Bioregenerative Life Support System for Long-Term Crewed Missions to the Moon or Mars. Astrobiology, 16: 925.

Ghazvini, R.F., Peyvast, G., Azarian, H., 2007. Effect of Clinoptilolitic Zeolite and Perlite on the Yield and Quality of Strawberry in Soilless Culture. International Journal of Agriculture & Biology, 885-888.

Hyman, J.R., Gaus, J., Foolad, M.R., 2004. rapid and accurate method for estimating tomato lycopene content by measuring chromaticity values of fruit puree. Ophthalmology, 106: 1 156–1 165.

Jones, J.B., Jr, Jones, J.B., Jr, 1997. Hydroponics: a practical guide for the soilless grower. Horttechnology, 15.

Kumar, G., Knowles, N.R., 1993. Changes in Lipid Peroxidation and Lipolytic and Free-Radical Scavenging Enzyme Activities during Aging and Sprouting of Potato (Solanum tuberosum) Seed-Tubers. Plant Physiology, 102: 115-124.

Lópezmillán, A.F., Sagardoy, R., Solanas, M., et al., 2009. Cadmium toxicity in tomato (Lycopersicon esculentum) plants grown in hydroponics. Environmental & Experimental Botany, 65: 376-385.

Leakey, A.D.B., Ainsworth, E.A., Bernacchi, C.J., et al., 2009. Elevated CO_2 effects on plant carbon, nitrogen, and water relations: six important lessons from FACE. Journal of Experimental Botany, 60: 2 859-2 876.

Lee, J.M., Joung, J.G., Mcquinn, R., et al., 2012. Combined transcriptome, genetic diversity and metabolite profiling in tomato fruit reveals that the ethylene response factor SlERF6 plays an important role in ripening and carotenoid accumulation. Plant Journal for Cell & Molecular Biology, 70: 191.

Lu, J.Y., Liu, M., Xue, H., et al., 2005. [Random amplified polymorphic DNA analysis of tomato from seeds carried in Russian Mir space station]. Space Medicine & Medical Engineering, 18: 72-74.

Macknney, G., 1941. Absorption of light by chlorophyll solutions. J. Biol. Chem, 140: 315-322.

Mavromatis, A.G., Athanasouli, V., Vellios, E., et al., 2003. Characterization of Tomato Landraces Grown under Organic Conditions Based on Molecular Marker Analysis and Determination of Fruit Quality Parameters. Journal of Agricultural Science, 5.

Nagata, M., Yamashita, I., 1992. Simple Method for Simultaneous Determination of Chlorophyll and Carotenoids in Tomato Fruit. Nippon Shokuhin Kagaku Kogaku Kaishi, 39: 925-928.

Nechitailo, G.S., Jinying, L., Huai, X., et al., 2004. Influence of long term exposure of space flight on Tomato seeds: effects on the first and second generation of plants, 35th COSPAR Scientific Assembly.

Ougham, H., Hörtensteiner, S., Armstead, I., Donnison, et al., 2008. The control of chlorophyll catabolism and the status of yellowing as a biomarker of leaf senescence. Plant Biology, 10: 4-14.

Rosen，W.G.，1999. Introduction to Plant Physiology. Aibs Bulletin，10：47.

Sirko，R.J.，Smith，G.C.，Hamlin，L.A.，et al.，1994. Lunar base CELSS design and analysis. Advances in Space Research，14：105-112.

Stewart，R.R.C.，Bewley，J.D.，1980. Lipid Peroxidation Associated with Accelerated Aging of Soybean Axes. Plant Physiology，65：245-248.

Tikhomirov，A.，Ushakova，S.，Velichko，et al.，2011. Assessment of the possibility of establishing material cycling in an experimental model of the bio-technical life support system with plant and human wastes included in mass exchange. Acta Astronautica，68：1 548-1 554.

Weathers，P.J.，Zobel，R.W.，1992. Aeroponics for the culture of organisms，tissues and cells. Biotechnology Advances，10：93.

Wheeler，R.M.，Corey，K.A.，Sager，et al.，1993. Gas exchange characteristics of wheat stands grown in a closed，controlled environment. Crop science，33：161-168.

Wu，R.，Wang，Y.，2015. Psychosocial interaction during a 105-day isolated mission in Lunar Palace 1. Acta Astronautica，113：1-7.

不同坐果方式对温室番茄糖含量的影响

张培钰，温祥珍，李亚灵

（山西农业大学园艺学院，太谷　030801）

摘要：针对消费者反映温室番茄激素坐果风味差的问题，试验设计了三种坐果方式方式，即震动坐果，自然坐果和激素喷花坐果，测定了影响品质要素的果实中可溶性糖、淀粉、果糖、蔗糖和维生素C含量。结果表明：试验中番茄果径的生长速率，激素喷花的要高。糖含量中，激素喷花较震动和自然坐果方式果实的可溶性糖含量下降一半左右；第二穗和第三穗果实中震动坐果、自然坐果均比激素坐果的淀粉、果糖、蔗糖和维生素C含量高。相对而言震动比自然糖含量略高，可认为与授粉效果较好有关，风味物质的增加是品质较好的内在要素，因此，震动坐果提高果实风味的效果较好。

关键词：番茄；坐果方式；糖含量

The Effect of Different Fruits on the Content of Greenhouse Tomato Sugar

Zhang Peiyu，Wen Xiangxhen，Li Yalin

（*Shanxi Agricultural University*，*School of horticulture*，*Taigu*　030801）

Abstract：In view of the problem of poor taste of tomato fruit in greenhouse. Three fruit-setting methods were designed，That is shaking fruit，natural fruit setting and hormone spray fruit setting. The contents of soluble sugar，starch，fructose，sucrose and VC in the fruit which affected the quality factors were determined. The results showed that the growth rate of tomato fruit diameter was higher than that of hormone spray. In the sugar content，the soluble sugar content of the fruit by hormone spraying was about half lower than that of the shaker and the natural fruit-setting mode. The content of starch，fructose，sucrose and VC in the fruit of the second ear and the third ear were higher than that of the steroid fruit. The content of vibration is slightly higher than that of natural sugar，which can be considered to be related to better pollination effect. The increase of flavor is the internal factor of better quality. Therefore，the effect of shaking fruit setting on improving fruit flavor is better.

Key words：Tomato；Fruit setting；Sugar content

基金项目：国家自然科学基金重点项目“温室环境作物生长模型与环境优化调控”（编号：61233006），山西省煤基重点科技攻关项目“设施蔬菜高效固碳技术研究与示范”（项目编号：FT201402-05）

已申请国家发明专利“叶菜类蔬菜智能控制生产系统”（专利申请号：201710908963.5）

作者简介：张培钰（1993—），女（汉），山西临汾人，硕士，研究方向：栽培生理

通讯作者：李亚灵（1962—），女（汉），山西灵石人，教授，博士生导师，研究方向：栽培生理，E-mail：yalingli1988@163.com

1 引言

番茄在日常消费中占据很大比例，是我国栽培主要蔬菜。露地栽培条件下，通常通过昆虫和风的摇动授粉坐果，在设施中，因通风较差，昆虫少，不利于传粉，造成番茄授粉不良，不利坐果从而减少产量[1]（郜玉江，2016），这些成为技术难题。随着激素保果技术的应用，设施番茄面积迅速扩展，成为主栽蔬菜之一。但是消费者反映设施番茄果实果肉硬化、炒食不烂，风味变差，生食无味[2]（霍建勇等，2005）。除品种因素外，生产者反映同一品种自然坐果的番茄风味会好很多。为证实这一现象，笔者进行了如下试验。

2 材料和方法

本研究所采用番茄（Lycopersicon esculentum）品种为“阿粉162”。试验时间为2017年8月8日至11月12日。地点在山西农业大学设施农业工程中心的日光温室内。试验时建造了3个自然光照人工气候室，其结构坐北朝南，长3m、宽2.4m、前屋面高1.5m，后屋面高2.3m，番茄8月10日定植到自然光照人工气候室。

试验设计三种处理：A：震动授粉，用“小蜜蜂授粉器”，产于山东嘉禾科技有限公司。震动时将袋去掉，震动后立即套袋；B：自然授粉，即对照，不做任何处理；C：激素保果，采用“番茄果霉康”300倍液喷花穗，喷时将袋去掉，对其柱头喷一次，喷完立即套袋。在试验处理之前，三个室的植株都在花蕾期套袋，防止其他因素的干扰，待开花时对其进行处理，番茄坐果后去袋。

试验测定项目及方法

试验测定了第二穗果和第三穗果中可溶性总糖、淀粉果糖、蔗糖维生素C含量。可溶性总糖和淀粉用蒽酮—硫酸法[3]（Hans-Peter Kläring et al 2015）；果糖和蔗糖用间苯二酚显色法[4]（张志良等，2003）；维生素C含量用2，6-二靛酚法测定[5]（王元秀等，2002）。在花后10天取第一次样，之后每隔10天1次，3次重复。

果径变化量，用游标卡尺定株测量。每处理选5株生长一致的植株，7天测一次果径。果实生长速率（V）计算：

$$V=(l2-l1)/d。$$

其中，l2——本次测定果径值，单位cm；l1—上次测定果径值，单位cm；d—测定间隔期，单位：天。

文中的数据处理采用Microsoft Excel 2003软件处理。

3 结果与分析

3.1 不同处理对番茄果实生长速率的影响

从图1和表1中可看出，果实生长均呈“S”型曲线，处理前2周生长量均小，在2～6周期间处理快速生长期，随后生长趋缓。但不同处理间的差异有所不同，震动授粉处理前期生长较快，5周后进入缓慢生长过程；自然授粉处理前期生长稍慢，6周后进入缓慢生长过程；激素处理的果实生长接近自然授粉处理。认为这与授粉效果、种子发育有关。即A处理授粉效率，初期激素产生多，生长较快高，随种子发育，消耗较多养分，生长转缓早。B处理授粉正常，果实发育正常，C处理前期达到正常果实激素水平，保证了坐果及相应的长势，但果实种种子少，维持了较快的生长速度。总体看激素比其他两种授粉方式果实生长率快。果实成熟后时，C处理比A、B处理快了15.88%和3.14%。

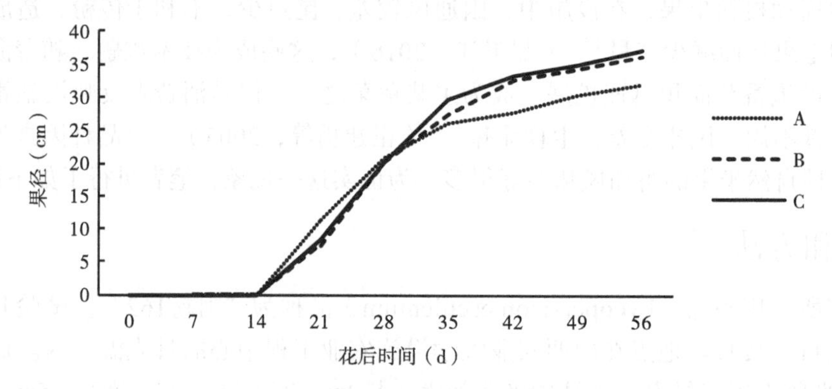

图1　不同处理对果径变化的影响

Figure 1　influence of different treatments on the change of fruit diameter

表1　果实生长率

Table 1　growth rate of fruit

时间（d）	不同处理方式		
	A	B	C
14～21	1.6	1.07	1.2
21～28	1.38	1.82	1.71
28～35	0.76	1.01	1.35
35～42	0.23	0.76	0.49
42～49	0.34	0.23	0.24
49～56	0.28	0.26	0.34

3.2　不同处理对果实中可溶性糖的影响

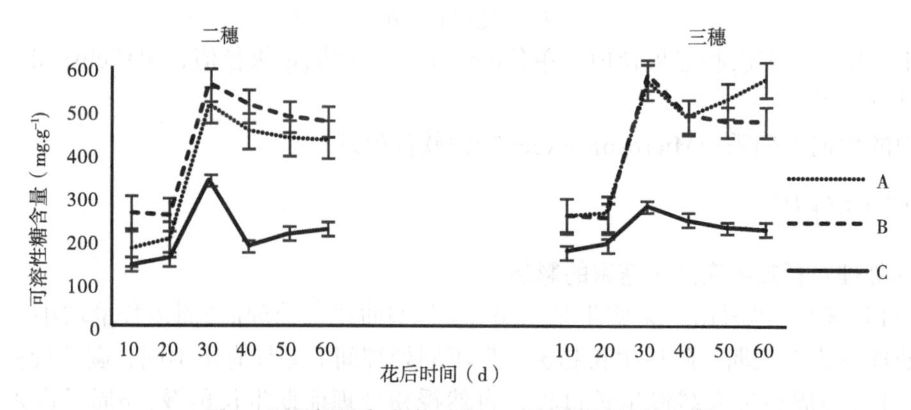

图2　不同授粉方式中二穗、三穗果可溶性糖含量

Figure 2　the soluble sugar content of two or three ears in different pollination method

番茄果实中可溶性糖含量是决定番茄风味的主要因素之一。测定第二、三穗果果实发育过程中可溶性糖含量如图2所示。在果实膨大期直至成熟过程，3种处理可溶性糖含量的变化趋势基本一致，都是先上升后下降最终趋于稳定。但处理间有显著差异，初期差异相对较小，在花后30d直至果实成熟，A、B处理显著高出C处理2倍以上，二穗果A处理高于B处理，三穗果前期无明显差异，后期A处理有升高趋势。

3.3 不同处理对果实中对淀粉含量的影响

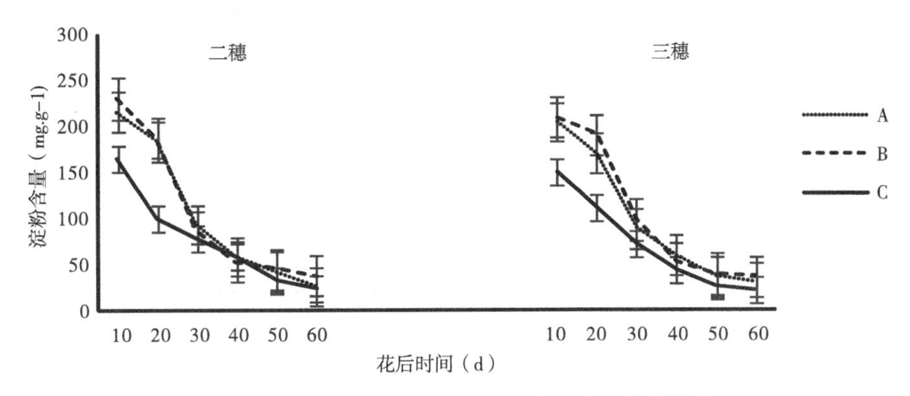

图 3　不同授粉方式中二穗、三穗果淀粉含量变化

Figure 3　changes in starch content of two and three panicles in different pollination methods

测定表明：番茄果实中淀粉的含量开花后10d含量最多，随生长到成熟快速减低，如图3所示。也就是番茄坐果后作为贮藏物质的淀粉积累最多，后逐渐分解转化为其他果实内含物。不同处理间变化趋势一致，差异相对较小，成熟期淀粉含量更加接近。在花后10天到30d中C处理比A、B处理淀粉的含量明显较低，第二穗果，果实成熟后，A、B处理分别比C处理增加了14.13%和60.9%。第三穗果增加的量分别为38.55%和69.32%。

3.4 不同处理对果实中对果糖含量的影响

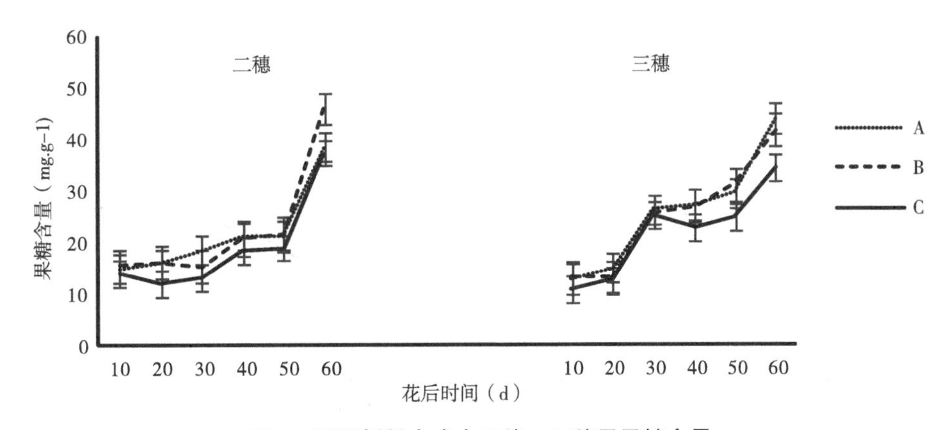

图 4　不同授粉方式中二穗、三穗果果糖含量

Figure 4　the content of two or three ear fruit fructose in different pollination methods

果糖是甜度较高的一种糖，也是决定果实风味的主要糖类之一。测定结果表明随果实成熟度提升果糖含量均呈上升趋势，尤其在后期（图4）。各处理间的差异相对较小，但C处理的果糖含量还是要比其他两种处理要低。第二穗果中，A、B处理比C处理果糖含量分别增加了2.94%和22.9%；第三穗果，花后30d前，三种处理差异较小，30d后C处理的果糖含量要比其他2处理显著低，成熟时分别增加了27.28%和20.87%。

3.5 不同授粉方式对蔗糖含量的影响

蔗糖在果实中是长距离运输的一种物质，是果实中糖积累的主要形式，是果实品质形成的因子。表5中2和3分别代表二穗果和三穗果。第二、三穗果实中蔗糖变化趋势基本是一致的，都是先增后降，在果实膨大期（花后30d）3种处理均达到最大值，之后下降（表2）。第二穗果实成熟后，A、B处理比C处理的蔗糖含量分别增加了39.67%和17.46%；第三穗果A、B处理比C处理分别增加了34.51%和12.78%。根据计算表明A处理蔗糖含量显然高于其他两种处理，C处理果实中蔗糖含量偏低。

表 2　不同处理番茄中蔗糖含量

Table 2　Different processing of sucrose in tomatoes

花后时间（d）	不同处理方式中蔗糖含量 单位（$mg·g^{-1}$）					
	A2	B2	C2	A3	B3	C3
10	41.06	46.69	39.00	78.98	70.90	54.63
20	61.02	57.45	45.59	71.67	57.38	49.64
30	104.64	88.98	60.94	237.06	149.98	133.72
40	74.44	70.56	62.79	62.90	81.73	75.09
50	56.55	67.79	48.21	71.06	63.36	52.93
60	44.96	37.80	32.19	64.93	54.44	48.27

3.6 不同授粉方式对维生素C含量的影响

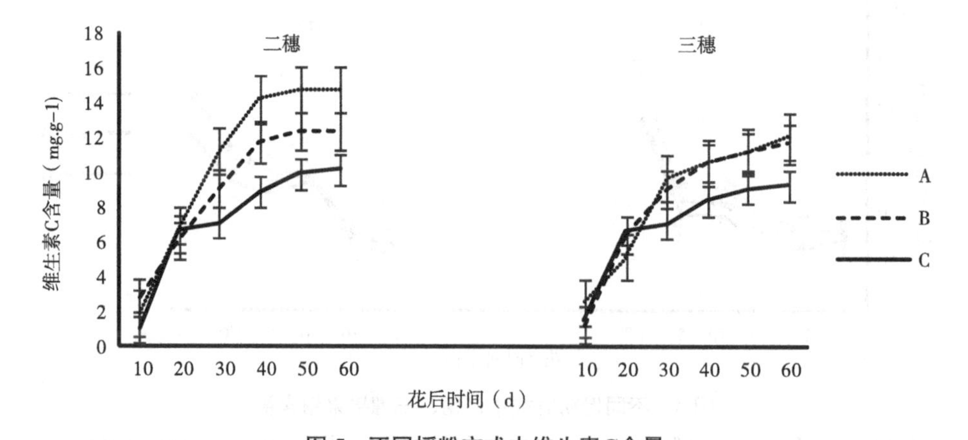

图 5　不同授粉方式中维生素C含量

Figure 5　Vc content in different pollination methods

番茄果实中维生素C含量也是判断果实风味的主要因素之一。图5是第二、三穗果实中维生素C含量变化趋势曲线，可明显看出A、B处理比C处理的要高，坐果20d三种处理差异不大增速均快，之后A、B处理保持了快速增长的势头，C处理则趋缓，差距加大。而三种处理的变化趋势都基本一致。在第二穗果中，A、B处理分别增长44.7%和21.32%，且A处理与C处理之间存在显著性差异；第三穗果中A、B处理比C处理分别增加30.29%和25.67%，但无显著性差异。A处理比B处理果实维生素C含量略高。

同时，从图5可看出，维生素C含量的变化曲线图类似于抛物线。经计算知，在第二穗果中，A、C处理均在花后53天时存在最大值分别为15.07mg·g^{-1}和10.05mg·g^{-1}，B处理在花后56天时有极大值为12.55mg·g^{-1}；第三穗果的A处理在花后57d时存在极大值为12.06mg·g^{-1}，B处理在花后52d有极大值为11.85mg·g^{-1}，C处理在花后50d有极大值9.35mg·g^{-1}。即它们都在花后50d到60天这个时间段内存在最大值，而这个时间段正是果实成熟阶段，所以在果实成熟阶段维生素C含量最高。

4 讨论

本试验所测的糖类（可溶性糖、果糖、蔗糖和淀粉）从番茄小果到果实成熟，变化规律与刘以前（2006）测定的趋势是一致的。不同坐果方式对维生素C含量也有影响，震动坐果和自然坐果中维生素C含量相差不大，但用激素喷花的番茄维生素C含量要比其他两种偏低，这个结论与前人所研究的结论一致[6]（邢艳红等，2005）。

（1）从番茄品质上来说，积累不同的糖对果实品质及风味也不同。可溶性糖是测品质的一个重要指标，积累可溶性糖含量高则品质好[7]（刘以前，2006）。本试验3种坐果方式中激素坐果积累的可溶性糖含量明显比前两种坐果方式减少一半左右，可能是由于激素坐果导致果实营养物质无法及时填充。在二穗的激素坐果和三穗震动坐果的40d后可溶性糖含量都在增加，初步认为在40d前果实内部的营养物质前期由于受其他因素的抑制，影响物质代谢，而在40d后抑制因素消失，物质代谢恢复，此时可溶性糖含量也在增加。淀粉是作为一种转化物质暂时储存，后逐步分解为果糖和葡萄糖[8]（Arthur A et al 1999）。在番茄果实风味中，番茄粉质化也可能与淀粉含量有关。激素坐果的淀粉含量比其他两种坐果方式要低，所以激素坐果的番茄吃起来没有粉质化的口感。就番茄的甜度来说，由于果糖的甜度是葡萄糖的两倍，因此改变果实中果糖与葡萄糖的比率就可以改变果实的风味[9]（周长久等，1995）。葡萄糖和果糖二者所占的比例相似，总和可以占到可溶性固形物的75%[10]（陈俊伟等，2000）。本文中的果糖含量，二穗果在花后50d前，三种坐果处理含量较明显，三穗果是在花后30d以后三种处理才有明显区别，猜想可能由于二三穗果位置的差异使得营养物质先到达二穗果，在到达三穗果，使果实内部营养物质的转化存在时间上的差异。同时二三穗激素坐果的果糖含量要比震动和自然坐果的果糖含量低，这大大降低了番茄果实风味，使番茄吃起来索然无味。蔗糖的积累对番茄的作用就更重要了，蔗糖降解后为果糖和葡萄糖且可在植物体内快速改变其含量[11]（刘凤军等，2014）。本试验中蔗糖含量存在一个有趣的现象，即在三穗果中震动的蔗糖含量在20d后先急速上升在第30d最高之后急速下降，比同一时期的二穗果含量的2倍还多，这可能是由于番茄内转化蔗糖的酶此时含量少使蔗糖含量无法及时降解或降解量少，导致蔗糖的积累量过多，而之后酶含量急剧增加同时蔗糖含量急剧下降。这些都需要进一步验证。

（2）目前番茄的坐果方式多样化，可以采用授粉器进行震动授粉，也可以使用激素进行人工喷花、点花、蘸花，生物坐果主要包括蜜蜂坐果、熊蜂坐果等昆虫坐果，利用风授粉也是坐果方式中的一种[12]（Muhammad Amjad Bashir et al. 2017）。目前普通温室中，人们会选择震动坐果和激素坐果来提高坐果率。激素坐果，是通过花期喷施外源激素刺激果实膨大，但是使用激素授粉易造成果实品质变差，畸形果率高。从整体来看，激素坐果中各种糖含量都要比震动和自然坐果的含量低，且激素坐果可使番茄内部的糖类及其他一些营养物质无法及时填充，更重要的是还会造成激素残留，影响消费者的身体健康。这一方法在一些农业发达国家已被禁止，我国出口番茄也禁止使用植物激素类物质[13]（胡帆，2012）。所以随着人们对食品安全性的日益重视，绿色环保坐果技术越来越受到人们重视[14]（Farra J et al. 2000）。

5 结论

番茄风味主要有内含物决定，试验中不同坐果方式内含物变化趋势基本一致，但数量及比例存在显著差别，是引起口感不同主要原因。通过对温室番茄坐果方式对品质指标的影响，可以看出激素确实影响了果实的生长率、大小、果实内部的糖含量及维生素C含量，使用激素坐果的果实生长率确实比震动授粉的和自然授粉的要快，但内部的营养物质无法及时提供造成了果实风味的下降。激素喷花较震动和授粉方式果实的可溶性糖含量下降一半左右；第二穗果实中震动授粉、自然授粉比激素授粉的淀粉含量增加了14.13%和60.9%，果糖、蔗糖分别增加了2.94%、22.9%和39.67%、17.46%，维生素C含量增加了44.7%和21.32%；第三穗果淀粉含量增加了38.55%、69.32%、果糖、蔗糖的增加量分别为27.28%、20.87%和34.51%、12.78%，维生素C含量增加了30.29%和25.67%。总的来说激素除了可以刺激番茄坐果提高生长率外，无太多的好处。震动坐果和自然坐果在对于糖类含量无明显差别，但经前人的研究震动坐果的坐果率要比自然坐果的要高。所以对于普通温室和农民来说震动坐果是不错的选择。

参考文献

[1] 郜玉江. 不同授粉方式对大棚番茄的影响效果初探［J］. 石河子科技，2016，12，（6）：1-2.

[2] 霍建勇，刘静，冯辉，等. 番茄果实风味品质研究进展［J］. 中国蔬菜，2005（2）：34-36.

[3] Hans-Peter Kläring，Yvonne Klopotek，Angelika Krumbein，et al., The effect of reducing the heating set point on the photosynthesis，growth，yield and fruit quality in greenhouse tomato production［J］. Agricultural and Forest Meteorology . 2015，178-188.

[4] 张志良，瞿伟菁. 植物生理学实验指导［M］. 北京：高等教育出版社，2003.

[5] 王元秀，庄海燕. 微量滴定法测定猕猴桃中维生素C的含量［J］. 广西农业科学，2002（4）：188-189.

[6] 邢艳红，彭文君，安建东. 不同蜂授粉对设施番茄产量和品质的影响［J］. 试验研究报告，2005，25（7）：8-10.

[7] 刘以前，沈火林，石正强. 番茄果实生长发育过程中糖的代谢［J］. 华北农学报，2006，21（3）：51-56.

[8] Arthur A，Schaffer，Marina petreikov，et al. Modification of Carbohydrate Content in Developing Tomato Fruit［J］. Hort

Science，1999，34（6）：1 024-1 027.

［9］ 周长久，王鸣. 现代蔬菜育种学［M］. 北京：科学技术文献出版社，1995.

［10］ 陈俊伟，张良诚，张上隆. 果实中的糖分积累机理. 植物生理学通讯，2002，36（6）：497-503.

［11］ 刘凤军，李军，徐君，等. 外源物质叶面喷施对番茄果实品质的影响. 中国农学通报［J］.2014，30（31）：164-168.

［12］ Muhammad Amjad Bashir a，Abid Mahmood Alvia，Khalid Ali Khan b，Muhammad Ishaq Asif Rehmani c. Mohammad Javed Ansari b，Role of pollination in yield and physicochemical properties of tomatoes（*Lycopersicon esculentum*）［J］.Saudi Journal of Biological Sciences. 2017.

［13］ 胡帆. 授粉方法对番茄坐果率和产量的影响［J］. 长江大学学报，2012，9（7）：3.

［14］ Farra J，Pollock C，Gallagher J. Sucrose and the integration of metabolism in vascular plants［J］. Plant Sci，2000，154：1-11.

光质对西兰花芽苗菜营养品质的影响

龚春燕，苏娜娜，陈沁，张晓燕，崔瑾*
（南京农业大学生命科学学院，南京　210095）

摘要：研究光质对西兰花芽苗菜生长和营养品质的影响。在白光（W）、蓝光（B）、红光（R）和紫外光B（UV-B）处理下培养西兰花芽苗菜。与对照白光相比，蓝光和UV-B处理后子叶中可溶性蛋白和花青苷以及下胚轴中维生素C和可溶性蛋白的含量均显著地提高。UV-B处理的西兰花芽苗菜生长受到抑制，但其全株鲜重、可食鲜重和可食率与白光相比并无明显差异；下胚轴以及子叶中游离氨基酸含量均显著提高；花青苷、类黄酮以及总酚的含量在下胚轴中也提高得较为显著；子叶中可溶性糖的含量比白光有较显著性的提高。UV-B处理还显著增加了西兰花芽苗菜子叶中萝卜硫苷（Glucoraphanin，GRA）以及下胚轴中总硫苷的含量。UV-B是提高西兰花芽苗菜营养品质的最佳光质。
关键词：光质；西兰花芽苗菜；营养品质

Effects of Light Quality on Nutritional Quality in Broccoli Sprouts

Gong Chunyan，Su Nana，Chen Qin，Zhang Xiaoyan，Cui Jin*
（*College of Life Sciences*，*Nanjing Agricultural University*，*Nanjing*　210095）

Abstract：Objective The effects of light quality on the growth and nutritional quality of broccoli sprouts were studied. Methods Broccoli sprouts were cultured under white light（W），blue light（B），red light（R）and ultraviolet light B（UV-B）. Results The contents of soluble protein and anthocyanin in cotyledons，vitamin C and soluble protein in hypocotyls were significantly increased by compared with that control white light. The growth of broccoli sprouts treated with UV-B was inhibited，but the fresh weight of whole plant，edible fresh weight and edible rate did not show significant difference compared with white light；the content of free amino acids were increased significantly in hypocotyls and cotyledons. The content of total phenols，flavonoids and anthocyanins in hypocotyls were also increased significantly under white light. The content of soluble sugar in cotyledons treated with monochromatic light was significantly higher than that under white，among which the content of soluble sugar in cotyledons was the highest after UV-B treatment. UV-B treatment also significantly increased the content of glucoraphanin（GRA）in the cotyledons and the total glucosinolates in hypocotyls. Conclusion UV-B is the best quality to improve the quality of broccoli sprouts.
Key words：Light quality；Broccoli sprouts；Nutritional quality

研究表明，经常食用西兰花（*Brassica oleracea* L.var.*italic* Planch）可以降低癌症和

作者简介：龚春燕（1991—），女，硕士研究生，E-mail:2015116010@njau.edu.cn
通讯作者：崔瑾，E-mail:cuijin@njau.edu.cn

心血管疾病的风险。这些有益效果归因于一类重要的含硫次级代谢产物，称为芥子油苷。与成熟植物相比，由于其幼嫩的生理状态，芸薹属植物芽苗菜含有更高含量的芥子油苷（20倍以上）。芥子油苷在人体内的分解产物是异硫氰酸酯，对各种癌症具有防御性功能。异硫氰酸酯由葡萄糖醛酸通过黑芥子酶的水解作用形成，其对许多类型癌细胞的活性较强，是已知最有希望的抗癌剂[1]。

近年来，人们希望通过外源添加如植物激素，硒，$CaCl_2$等的处理来增加植物组成中的芥子油苷，维生素C，总酚，类黄酮，花青苷等的含量以此提高芽苗菜的营养品质。维生素C在预防动脉粥样硬化，神经和心血管疾病治疗方面也具有重要的医学价值。研究发现酚类化合物具有抗氧化，抗肿瘤，抗病毒和抗菌等生物活性[2]。类黄酮是植物的重要次生代谢产物，具有清除自由基，延缓衰老和调节免疫力作用。花青苷是高等植物中主要的水溶性色素，在许多植物生理过程中起关键作用[3]。西兰花中检测到的主要的芥子油苷是萝卜硫苷（GRA），3-吲哚基甲基硫苷（GBS）和1-甲氧基-3-吲哚甲基硫苷（NGBS），其中GRA含量最高。此外，萝卜硫苷（GRA）是最重要的硫苷，其占西兰花中总硫苷含量的50%以上。

光作为一种节能，环保，易于操作的调控能源，广泛应用于植物设施栽培领域。目前关于西兰花光调控的研究仅限于不同强度的UV-B辐射对西兰花芽苗菜疾病和害虫的抗性研究以及光质对羽衣甘蓝芽苗菜健康品质的研究。关于单色光对西兰花芽苗菜子叶和下胚轴营养品质的研究尚未见报道。

本文研究了在西兰花芽苗菜生长中，不同单色光质对其活性物质积累和抗氧化活性的影响。探讨光调控对西兰花芽苗菜营养品质的影响，为光调控应用于芽苗菜工厂化生产提供理论依据。

1 材料与方法

1.1 培养条件

将经过次氯酸钠消毒后洗至中性的西兰花种子30℃蒸馏水中浸泡4h后均匀铺撒在纱布上，黑暗25℃催芽24h。将发芽的种子播种并置于黑暗25℃培养36h，再转移至不同光照培养箱中，箱内相对湿度为（75±5）%，温度为（25±2）℃，16h照光，8h黑暗处理5d后收苗。

1.2 光质处理

将西兰花芽苗菜放在在冷光源培养箱，箱内顶置LED光源，可发出白光（W）、蓝光（B）及红光（R）；另外使用紫外窄谱灯管作UV-B光源，以白光培养（W）作为对照（图1）。表1为光谱能量分布主要技术参数，调节光量为30μmol/（$m^{-2}·s^{-1}$）左右。

表1 光谱能量分布主要技术参数

Table 1 Major technique parameters of light spectral energy distribution

光质	光谱能量分布	峰值波长	波长半宽	光强
Light quality	Light spectral energy distribution	Λp /nm	Δλ /nm	Light intensity
W	100%白光100%White	380～750	5	30±3
B	100%蓝光100%Blue	460	5	30±3

（续表）

光质	光谱能量分布	峰值波长	波长半宽	光强
R	100%红光100%Red	637	5	30 ± 3
UV-B	100%紫外100%UV-B	280	—	5

注：W（白光）、B（蓝光）、R（红光）光强单位为μmol/（$m^{-2}\cdot s^{-1}$），UV-B光强单位为W/m^{-2}

1.3 测定指标

1.3.1 不同光质处理下西兰花芽苗菜生长、可溶性糖、可溶性蛋白以及游离氨基酸的测定

生长指标：测定西兰花芽苗菜全株鲜重，可食鲜重，下胚轴长。鲜样置于105℃杀青15min，85℃烘干至恒重后测定全株干重及其可食干重。

可溶性糖含量的测定：参照王学奎[4]的方法称取0.5g鲜样放入玻璃试管中，加入10ml去离子水，沸水浴30min（重复提取2次），合并提取液并过滤到25ml的容量瓶中，定容至刻度。取1ml提取液加入到20ml试管中，加入10ml浓硫酸溶液和1ml 2%蒽酮—乙酸乙酯溶液，沸水浴1min，室温下测定630nm波长下的吸光值。

可溶性蛋白含量测定：采用王学奎[4]的方法称取1g鲜样，加2ml蒸馏水研磨成匀浆，转移到离心管中，再用6ml蒸馏水多次洗涤，在室温下放置1h后8 000g离心15min，上清液转入10ml容量瓶，定容至刻度。提取液加入蛋白试剂后静置3min，在595nm波长下测定吸光值。

游离氨基酸含量的测定：提取方法同可溶性糖。取0.05ml上清液，加0.5ml茚三酮酒精溶液，沸水浴后加5ml 95%乙醇，于570nm波长下测定吸光值。

1.3.2 不同光质处理下西兰花芽苗菜总酚、类黄酮、花青苷与维生素C的测定

总酚含量测定：用50%甲醇提取芽苗菜的总酚含量。室温下12 000g离心10min。使用福林—酚试剂测定酚类化合物，并在765nm读取其吸光值。样品中总酚含量用没食子酸当量（GAE）/100gFW表示。

类黄酮含量测定：用0.25ml的80%甲醇提取芽苗菜后加入1.25ml蒸馏水稀释后向混合物中加入75ml 5%$NaNO_2$溶液，7min后，加入150ml 10% $AlCl_3\cdot 6H_2O$溶液，反应混合物加入0.5ml NaOH（1mol/L）后在510nm处测定混合物的吸光值。

花青苷含量测定：根据Su[3]的方法将0.5g样品于室温黑暗中用10ml含1%HCl的甲醇溶液萃取24h，10 000g离心15min。子叶花青苷需要在测量前使用氯仿萃取。读取A530和A657的吸光值。差值每增加0.01定义为一个单位Unit。

维生素C含量测定：取0.5g鲜样，用5ml 2%草酸研磨提取后用HPLC测定。色谱柱：SB-C18色谱柱；流速：0.8ml/min；检测波长：254nm；柱温：30℃；进样量：20μl；流动相：0.1%草酸：甲醇=95%：5%。

1.3.3 不同光质处理下西兰花芽苗菜硫苷含量的测定

硫苷含量测定：取1g鲜样，加入6ml 75%甲醇煮沸25min灭酶，研磨匀浆后，再用2ml 75%甲醇萃取残余物，80℃恒温水浴中浸提25min，12 000g离心15min，收集上清液。将2ml上清液加到1ml DEAE Sephadex A-25柱（乙酸活化）中，加入500μl硫酸酯酶，30℃下反应16h。用2ml去离子水洗脱后过0.22μm滤膜。通过LC-MS系统（G2-XS QTof，

Waters）分析样品。将2μl溶液注入UPLC柱，流速为0.4ml/min。使用Masslynx4.1进行数据采集和处理。色谱柱：石英毛细柱（30mm × 0.25mm，0.25μm）；柱温32℃；进样量0.5μl；检测波长为226nm，流速为1.5ml/min，内标为烯丙基硫苷。

1.4 数据处理

数据处理采用SPSS 19.0软件，Origin 8.0软件作图，每个样品做3次重复分析。

2 结果与分析

2.1 光质对西兰花芽苗菜生长的影响

如图1所示（bar=1cm），与白光相比，红光和蓝光处理促进下胚轴的伸长；尤其是红光下伸长的较为显著。UV-B处理的西兰花芽苗菜生长受到抑制，但表2结果表明，其全株鲜重、可食鲜重和可食率与白光相比并无明显差异。

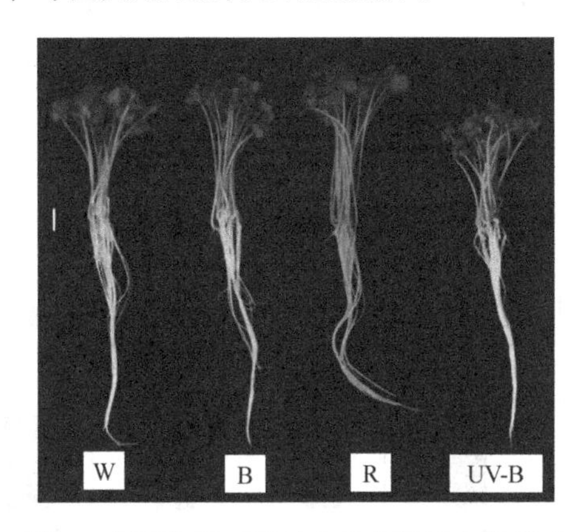

图 1 光质处理对西兰花芽苗菜形态的影响

Figure 1 Effects of light quality treatment on morphology on broccoli sprouts

如表2所示，与对照白光相比，红光下西兰花芽苗菜下胚轴显著伸长，而UV-B对下胚轴伸长明显抑制，其次是白光和蓝光；在红光培养下西兰花芽苗菜全株鲜重最高，与其他光质相比，其可食鲜重与对照相比增加约19%；蓝光和UV-B处理后全株鲜重以及可食鲜重与白光相比并没有提高，可食干重和可食率均没有显著变化；各光质处理下的全株干重也没有显著性差异。

表 2 光质对西兰花芽苗菜生长的影响

Table 2 Effects of light quality on the growth of broccoli sprouts

光质 Light quality	下胚轴长（cm）Hypocotyl length（cm）	全株鲜重（g）Total fresh mass（g）	全株干重（g）Total dry mass（g）	可食鲜重（g）Edible fresh mass（g）	可食干重（g）Edible dry mass（g）	可食率（%）Edible rate（%）	含水率（%）Water content（%）
W（Con）	3.23 ± 1.42^{b}	0.072 ± 0.002^{ab}	$0.004\,6 \pm 0.000\,2^{a}$	$0.047 \pm 0.001V4^{bc}$	0.004 ± 0.002^{a}	65.25 ± 2.94^{a}	93.59 ± 0.22^{a}
B	3.43 ± 0.41^{b}	0.072 ± 0.007^{ab}	$0.004\,8 \pm 0.000\,7^{a}$	$0.048 \pm 0.004\,6^{b}$	0.004 ± 0.001^{a}	66.47 ± 0.51^{a}	93.44 ± 0.27^{a}

（续表）

光质 Light quality	下胚轴长（cm） Hypocotyl length（cm）	全株鲜重（g） Total fresh mass（g）	全株干重（g） Total dry mass（g）	可食鲜重（g） Edible fresh mass（g）	可食干重（g） Edible dry mass（g）	可食率（%） Edible rate（%）	含水率（%） Water content（%）
R	4.58 ± 0.27[a]	0.078 ± 0.003[a]	0.004 9 ± 0.000 5[a]	0.056 ± 0.002 2[a]	0.0044 ± 0.001[a]	71.74 ± 3.69[a]	93.77 ± 0.40[a]
UV-B	2.19 ± 0.33[c]	0.065 ± 0.008[b]	0.005 0 ± 0.000 4[a]	0.042 ± 0.002[c]	0.004 ± 0.002[a]	65.13 ± 3.99[a]	92.26 ± 1.03[b]

注：表中数据均以鲜样质量计。小写字母不同表示差异显著（$P<0.05$）

2.2 光质对西兰花芽苗菜可溶性糖、可溶性蛋白和游离氨基酸含量的影响

如图2A所示，单色光处理后子叶中可溶性糖的含量均比白光显著性提高，其中UV-B处理后的最高，提高了69.49%。图2B结果表明，与对照相比，蓝光和UV-B下可溶性蛋白的含量在子叶中分别显著提高了18.26%和8.7%；不同光照条件下，蓝光和UV-B处理的芽苗菜下胚轴中可溶性蛋白含量高于红光和白光的培养，其中UV-B培养下的西兰花芽苗菜与白光和红光相比具有显著性提高。如图2C所示，UV-B照射后的游离氨基酸含量在子叶和下胚轴中与白光相比分别提高67.95%和12.51%。

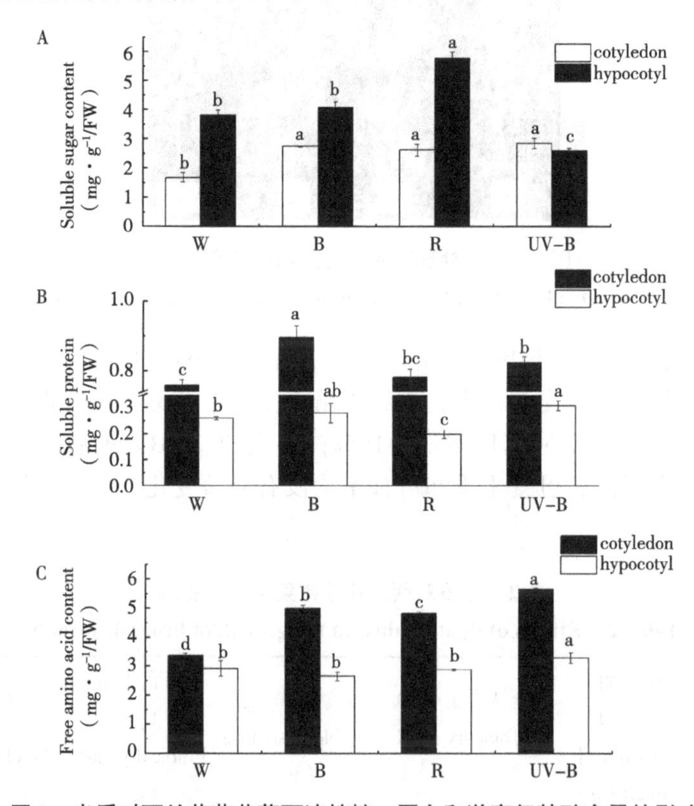

图 2　光质对西兰花芽苗菜可溶性糖、蛋白和游离氨基酸含量的影响

Figure 2　Effects of light quality on soluble sugar，soluble protein and free amino content in broccoli sprouts

2.3 光质对西兰花芽苗菜总酚、类黄酮、花青苷和维生素C含量的影响

经光质处理后，西兰花芽苗菜下胚轴中总酚含量在UV-B处理下显著性提高，比白光处理平均增加了26.04%（图3A）。

与其他光质相比，UV-B显著增加了西兰花芽苗菜下胚轴类黄酮含量。经蓝光、红光照射后的下胚轴中类黄酮与对照白光相比均没有提高，其中蓝光和白光处理下类黄酮含量相差甚微，没有显著性差异（图3B）。

本文也研究了不同光质处理下西兰花芽苗菜中花青苷含量。与对照相比，子叶中花青苷含量经蓝光和UV-B处理后分别显著地上升65.62%和55.35%，下胚轴中分别提高7.58%和74.75%（图3C）。

与白光相比，西兰花芽苗菜下胚轴经蓝光处理后，维生素C含量显著增加；而在子叶中，不同光照条件下，红光处理的子叶中维生素C含量高于其他单色光（图3D）。

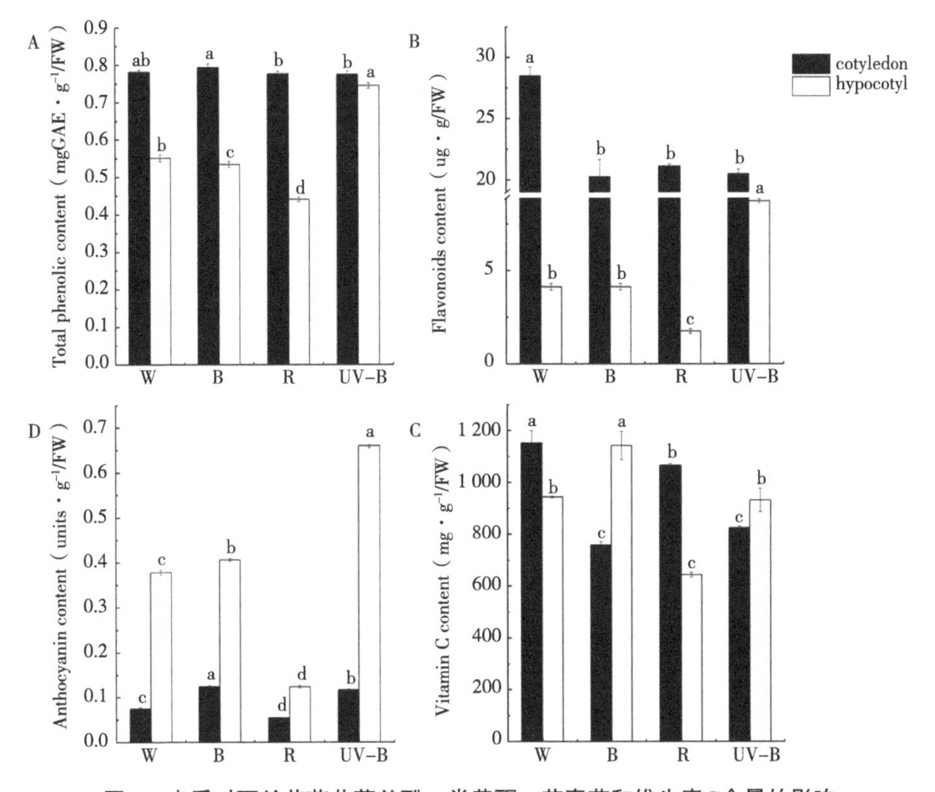

图3 光质对西兰花芽苗菜总酚、类黄酮、花青苷和维生素C含量的影响

Figure 3 Effects of light quality on total phenolic，flavonoid，anthocyanin and vitamin C contents in broccoli sprouts

2.4 光质对西兰花芽苗菜硫苷含量的影响

在西兰花芽苗菜的子叶和下胚轴中定量测定总芥子油苷含量以及单个芥子油苷含量（表3和表4）。本研究中，在西兰花芽苗菜中鉴定出了5种脂肪族芥子油苷和4种吲哚族芥子油苷。如表3所示，与白光处理相比，子叶经蓝光，红光和UV-B照射后均显著地提高了

GRA的含量。此外，相比于对照，GBS和4HGBS含量在蓝光培养后分别提高了约56%和35%。

表 3 不同光质处理下西兰花芽苗菜子叶中硫苷含量的变化

Table 3 Effect of different lights on glucosinolate composition and content（μmol/g FW）in cotyledons of broccoli sprouts

		W	B	R	UV-B
脂肪族	GRA	3.41 ± 0.20^{b}	4.43 ± 0.22^{a}	4.61 ± 0.45^{a}	4.03 ± 0.08^{ab}
	SIN	0.15 ± 0.01^{a}	0.17 ± 0.03^{a}	0.14 ± 0.01^{a}	0.13 ± 0.04^{a}
	GAL	0.48 ± 0.02^{b}	0.55 ± 0.06^{ab}	0.63 ± 0.07^{a}	0.42 ± 0.03^{b}
	GER	1.33 ± 0.12^{a}	0.68 ± 0.04^{c}	1.12 ± 0.05^{b}	0.80 ± 0.04^{c}
	GNA	7.38 ± 0.48^{a}	6.33 ± 0.64^{ab}	5.28 ± 0.37^{b}	8.51 ± 0.28^{ab}
吲哚族	GBS	0.075 ± 0.01^{b}	0.17 ± 0.01^{a}	0.08 ± 0.01^{b}	0.05 ± 0.01^{c}
	NGBS	0.11 ± 0.01^{b}	0.12 ± 0.01^{ab}	0.14 ± 0.01^{a}	0.10 ± 0.01^{b}
	4HGBS	0.29 ± 0.01^{b}	0.45 ± 0.03^{a}	0.35 ± 0.06^{b}	0.18 ± 0.01^{c}
	4MGBS	0.11 ± 0.01^{a}	0.07 ± 0.01^{bc}	0.09 ± 0.01^{ab}	0.05 ± 0.01^{c}
总硫苷		13.33 ± 0.86^{a}	12.86 ± 0.56^{a}	12.89 ± 1.07^{a}	11.76 ± 0.27^{a}

注：GRA：4-甲基亚磺酰基丁基硫苷（萝卜硫苷）；SIN：2-丙烯基硫苷；GAL：5-甲基亚磺酰基戊基硫苷；GER：4-甲硫基丁基硫苷；GNA：3-丁烯基硫苷；GBS：3-吲哚基甲基硫苷；NGBS：1-甲氧基吲哚基-3-甲基硫苷；4HGB：4-羟基吲哚基-3-甲基硫苷；4MGB：4-甲氧基吲哚基-3-甲基硫苷；下同。不同小写字母表示在$P<0.05$下的差异显著

表 4 不同光质处理下西兰花芽苗菜下胚轴中硫苷含量的变化

Table 4 Effect of different lights on glucosinolate composition and content（μmol/g FW）in hypocotyls of broccoli sprouts

		W	B	R	UV-B
脂肪族	GRA	5.42 ± 0.33^{a}	5.06 ± 0.84^{a}	5.00 ± 0.52^{a}	4.03 ± 0.24^{a}
	SIN	0.31 ± 0.01^{a}	0.29 ± 0.01^{a}	0.3 ± 0.03^{a}	0.29 ± 0.01^{a}
	GAL	0.28 ± 0.04^{ab}	0.34 ± 0.01^{a}	0.18 ± 0.01^{b}	0.37 ± 0.06^{a}
	GER	0.27 ± 0.01^{c}	0.11 ± 0.01^{d}	0.47 ± 0.04^{b}	0.86 ± 0.03^{a}
	GNA	1.33 ± 0.18^{c}	2.16 ± 0.14^{b}	2.20 ± 0.04^{b}	3.47 ± 0.01^{a}
吲哚族	GBS	0.25 ± 0.15^{b}	0.14 ± 0.06^{b}	0.67 ± 0.03^{a}	0.08 ± 0.01^{b}
	NGBS	1.23 ± 0.02^{a}	0.78 ± 0.01^{b}	0.70 ± 0.01^{c}	0.71 ± 0.01^{c}
	4HGBS	0.32 ± 0.08^{b}	0.71 ± 0.01^{a}	0.40 ± 0.04^{b}	0.18 ± 0.02^{c}
	4MGBS	0.71 ± 0.37^{a}	0.37 ± 0.04^{c}	0.62 ± 0.02^{b}	0.08 ± 0.01^{d}
总硫苷		9.74 ± 0.31^{bc}	8.57 ± 0.08^{c}	10.94 ± 0.67^{b}	13.42 ± 1.39^{a}

表4反映了西兰花芽苗菜下胚轴中硫苷含量的变化。蓝光和UV-B促进了GAL含量的积累，分别提高21.43%和32.14%。红光显著促进了GBS的提高，其含量约为对照白光的2.7倍。GER和GNA在UV-B处理下分别提高了约3.1和2.6倍。蓝光处理的下胚轴中4HGBS含

量是白光处理下的2.21倍。脂肪族芥子油苷的含量在红光处理下没有显著增加，但都高于白光下的处理。UV-B处理后芽苗菜下胚轴中总芥子油苷含量在整个处理中最高，与对照白光相比提高了约37.78%。

3 讨论

本研究发现，与对照相比，在子叶和下胚轴中，UV-B均能显著提高西兰花芽苗菜可溶性蛋白的含量，这与王虹等[6]的研究结果基本一致。这可能是由于UV-B促进了西兰花芽苗菜的生理代谢过程，提高了与代谢相关的酶活性。本研究发现，与白光对照处理相比，UV-B处理也显著增加了西兰花芽苗菜子叶中游离氨基酸的含量。

与其他光质相比，UV-B处理后的西兰花芽苗菜下胚轴中的酚类物质显著性增加。这种差异可能与UV-B促进参与酚类化合物合成酶的产生有关[7]。西兰花芽苗菜下胚轴中类黄酮含量在UV-B处理后显著高于白光，其次是蓝光>白光>红光。西兰花芽苗菜的子叶和下胚轴的生理功能不同，两者的代谢途径不同，所以西兰花芽苗菜子叶和下胚轴中的黄酮含量存在差异。本研究发现，UV-B辐射都能有效地诱导西兰花芽苗菜子叶和下胚轴中的花青苷积累，具有在实际生产中加以应用以提高西兰花芽苗菜营养品质的可能性。据报道，在番茄中，环境胁迫可以诱导与维生素C合成相关基因的表达，增加维生素C含量。在本研究中，蓝光照射后下胚轴中维生素C含量与对照白色相比提高的最为显著，这与徐茂军等[8]等研究结果相似。

芥子油苷是主要在十字花科作物中发现的一类含氮和硫的次级代谢产物，因其在抗癌活性及植物防御微生物病原体和食草昆虫方面具有重要作用，其分解产物已经成为被广泛研究的对象，其中萝卜硫苷的分解产物萝卜硫素是目前发现的抗癌活性最强的成分。我们的结果表明，西兰花芽苗中含有脂肪族芥子油苷的比例较高，这与前人研究结果类似[9]。植物将芥子油苷和黑芥子酶（myrosinase，MYR）储存在不同位置，一般情况下两者不能互相接触，但当植物受到胁迫造成组织损伤时，MYR与芥子油苷接触后被水解生成异硫氰酸盐、腈类、硫氰酸盐等物质。与白光处理相比，子叶经蓝光，红光和UV-B照射后均显著地提高了GRA的含量；在西兰花芽苗下胚轴中，蓝光和UV-B促进了GAL含量的积累；GER和GNA在UV-B处理下极显著提高，这可能与这3种光质提高了芽苗菜中黑芥子酶的活性相关。此外，相比于对照，子叶中GBS和4HGBS含量在蓝光培养后分别升高，红光处理后中显著促进下胚轴GBS的提高；蓝光处理的下胚轴中4HGBS含量是白光处理下的两倍多，这可能与植物内源茉莉酸诱导吲哚族芥子油苷的提高有关。UV-B处理后芽苗菜下胚轴中总芥子油苷含量在整个处理中最高，与对照白光相比提高的极为显著，这可能与不同的防御机制有关。此外，在UV-B下生长的芽苗菜下胚轴中总芥子油苷含量在所有处理中最高。硫苷侧链的结构很大程度上直接决定了硫苷的种类及水解产物的活性，其侧链的二次修饰是通过羟基化、甲基化、氧化和去饱和等作用，对侧链进行不同的修饰，从而导致硫苷种类的多样性。单一硫苷及总硫苷含量子叶不同于下胚轴，可能原因是在不同的植物中硫苷种类差异较大，因此造成器官的差异性。

4 结论

本研究结果表明：紫外光B、蓝光和红光照射的西兰花芽苗菜具有更高的营养品质。蓝光和UV-B主要提高了芽苗菜中游离氨基酸、可溶性蛋白和花青苷的含量；红光主要提高了可溶性糖的含量；UV-B还提高了总酚的含量。蓝光，红光和UV-B主要提高了子叶中GRA的含量；蓝光和UV-B促进下胚轴中GAL含量的积累；UV-B处理还提高了下胚轴GER和GNA的含量以及下胚轴中总芥子油苷含量。综合考虑，紫外光B（UV-B）是西兰花芽苗菜实际生产中较理想的光照条件。

参考文献

［1］ Zhang C，Su Z Y，Khor T O，et al. Sulforaphane enhances Nrf2 expression in prostate cancer TRAMP C1 Cells through Epigenetic Regulation［J］. Biochemical Pharmacology，2013，85（9）：1398. PAJAK P，SOCHA R，GALKOWSKA D，et al. Phenolic profile and antioxidant activity in selected seeds and sprouts［J］. Food Chemistry，2014，143（1）：300..

［2］ Su N，Lu Y，Wu Q，et al. UV-B-induced anthocyanin accumulation in hypocotyls of radish sprouts continues in the dark after irradiation.［J］. Journal of the Science of Food & Agriculture，2016，96（3）：886.

［3］ 王学奎. 植物生理生化实验原理和技术［M］. 北京：高等教育出版社，2006.

［4］ 刘金，魏景立，刘美艳，等. 早熟苹果花青苷积累与其相关酶活性及乙烯生成之间的关系［J］. 园艺学报，2012，39（7）：1235-1242.

［5］ 王虹，姜玉萍，师恺，等.光质对黄瓜叶片衰老与抗氧化酶系统的影响［J］. 中国农业科学，2010，43（3）：529-534

［6］ NA Y L，LEE M J，KIM Y K，et al. Effect of light emitting diode radiation on antioxidant activity of barley leaf［J］. Journal of the Korean Society for Applied Biological Chemistry，2010，53（6）：685-690.

［7］ 徐茂军，朱睦元，顾青. 光诱导对发芽大豆中半乳糖酸内酯脱氢酶活性和维生素C合成的影响［J］. 营养学报，2002，24（2）：212-215.

［8］ Tian L，Li X，Yang R，et al. NaCl treatment improves reactive oxygen metabolism and antioxidant capacity in broccoli sprouts［J］. Horticulture Environment & Biotechnology，2016，57（6）：640-648.

S-腺苷甲硫氨酸对黄瓜断根扦插苗生长及生理代谢的影响

刘鑫，李晓彤，荆鑫，王硕硕，巩彪，魏珉，史庆华*

（山东农业大学园艺科学与工程学院，作物生物学国家重点实验室，农业部黄淮海设施农业工程科学观测实验站，泰安 271018）

摘要：以黄瓜品种“津研四号”为试验材料，研究了S-腺苷甲硫氨酸（SAM）对断根后黄瓜苗期不定根发生、植株生长和生理代谢的影响。结果表明，断根黄瓜幼苗基部用50μmol·L^{-1}的SAM溶液浸泡5min，能显著促进不定根的生长发育、叶绿素的积累、光合作用的增强以及提高植株对N、P、K的吸收能力，促进了黄瓜幼苗的生长。进一步分析表明，SAM处理上调了黄瓜下胚轴内多胺和生长素相关基因的表达水平及物质积累，可能是SAM促进黄瓜不定根发生和幼苗生长发育的机理。

关键词：S-腺苷甲硫氨酸；黄瓜；扦插；不定根；生理代谢

Effect of S-adenosylmethionine on Growth and Physiological Metabolism of Cucumber Cutting Seedlings

Liu Xin，Li Xiaotong，Jing Xin，Wang Shuoshuo，Gong Biao，Wei Min，Shi Qinghua*

（*College of Horticulture Science and Engineering*，*Shandong Agricultural University*；*State Key Laboratory of Crop Biology*；*Huang-Huai-Hai Region Scientific Observation and Experimental Station of Environment-Controlled Agricultural Engineering*，*Ministry of Agriculture*，*Tai'an*，*Shandong* 271018）

Abstract：In the present experiment，‘Jinyan NO.4’ cucumber was used as experimental material to study the effect of application exogenous S-adenosylmethionine（SAM）on adventitious root generation，seedlings growth and physiological metabolism. The results showed that the adventitious root growth，chlorophyll accumulation，photosynthesis were significantly improved in cucumber cutting seedlings treated with 50 μmol/L SAM solution for 5 min，and N，P，K contents and cucumber seedlings growth were promoted. Further investigations indicated that SAM treatment up-regulated the expression polyamines and auxin-related genes as well as polyamine and auxin accumulation in cucumber hypocotyls，which was the possible mechanism of SAM promoting adventitious root generation and plant growth of cucumber cutting seedlings.

Key words：S-adenosylmethionine；cucumber；Cutting；Adventitions root；Physiological metabolism

研究表明，黄瓜幼苗断根扦插后也具有较强的生长势（李应超，2011）。因此，研究

基金项目：“十三五”国家重点研发项目（2017YFD0201600）；山东省现代农业产业技术体系项目（SDAIT-05-10）；山东省农业科学院农业科技创新工程项目（CXGC2016B06）

作者简介：刘鑫（1991—），男，山东泰安人，硕士，主要从事植物栽培与生态研究

* 通信作者 Author for correspondence（E-mail：qhshi@sdau.edu.cn）

促进黄瓜扦插苗生根及生长的方法具有重要的意义。在促进扦插生根方面，人们普遍使用吲哚乙酸、萘乙酸等激素类物质，但近年来植物生长调节剂残留问题存在较大的争议。因此，研究开发安全有效的新型调节物质，对农业的可持续生产非常重要。

S-腺苷甲硫氨酸（S-adenosy1-L-methionine，SAM）最早是由Cantoni在1952年发现（Cantoni，1952），对治疗肝病、抑郁症、胃癌等疾病上有显著效果（程兮等，2015）。研究表明，SAM在调控植物对缺铁、干旱和盐碱的适应性以及在抵抗病原菌中发挥重要作用，但在调控植物根系发育方面未见报道。以“津研四号”黄瓜为试验材料，研究了SAM对黄瓜断根扦插苗根系发育、植株生长和生理代谢的影响，研究结果一方面可为黄瓜育苗新途径的研发提供技术支撑，具有一定的应用价值，另一方面对丰富SAM的生理功能具有一定的理论意义。

1　材料与方法

试验于2017年3月至2017年9月在山东农业大学作物生物学国家重点实验室植物培养室内进行，供试材料为天津市蔬菜研究所提供的黄瓜品种‘津研四号’。将催好芽的种子，播于盛有育苗基质（草炭：蛭石：珍珠岩=3:1:1）的50空穴盘中。植物培养室环境设定：白天12h，夜晚12h，温度为白天28℃，夜晚18℃，白天光强为600μmol·m^{-2}·s^{-1}，空气湿度为60%。

待黄瓜子叶展平后，选取长势一致的黄瓜幼苗从下胚轴基部切除主根，并进行如下处理：①CK：自根苗；②T1：将切除主根的黄瓜苗基部浸于去离子水中5min后插于新的盛有相同育苗基质的穴盘中，扦插深度为1.5cm；③T2：将切除主根的黄瓜苗基部浸于浓度为50μmol/L的SAM溶液中5min，之后扦插方法同T1（各处理幼苗总株数为50株）。将处理后的黄瓜幼苗放于原培养室内培养，培养期间只浇清水，不再添加肥料。

2　结果与分析

2.1　SAM对黄瓜幼苗植株生长的影响

由表1可得，与CK相比，T1与T2的株高和地上部鲜重都有不同程度的增加，且T2增长幅度较大，株高和地上部鲜重分别比CK增加28.22%和44.51%（P<0.05）。T1和T2的茎粗以及T1的地上部干重和全株干重与CK无显著差异，T2地上部干重相比CK增加了13.16%，全株干重增加了11.24%（P<0.05）。

表1　SAM对黄瓜幼苗植株生长的影响

Table 1　Effect of SAM on plant growth vigor of cucumber

处理 treatment	株高 Plant height（cm）	茎粗Stem diameter（cm）	地上部鲜重FW of shoot(g)	地上部干重DW of shoot(g)	全株干重DW of shoot and root(g)
CK	20.52 ± 2.95b	4.83 ± 0.27a	9.48 ± 1.02c	0.76 ± 0.09b	0.89 ± 0.08b
T1	22.14 ± 1.62b	4.53 ± 0.38a	12.17 ± 0.16b	0.75 ± 0.04b	0.86 ± 0.05b
T2	26.31 ± 2.37a	4.99 ± 0.49a	13.7 ± 0.26a	0.87 ± 0.04a	0.99 ± 0.06a

注：不同小写字母表示不同处理间差异显著（P＜0.05）（下同）

2.2 SAM对黄瓜幼苗植株叶片色素和净光合速率的影响

由表2可以看出，与CK相比，T1，T2的光合色素、净光合速率均显著提高，其中，T1的叶绿素a、叶绿素b、类胡萝卜素和净光合速率分别比对照提高了47.33%、46.30%、36.00%和16.08%（P<0.05）；T2的这4项指标分别比CK提高了51.91%、48.15%、36.00%和38.92%（P<0.05）。

表 2　SAM对黄瓜幼苗植株叶片色素和净光合速率的影响

Table 2　Effect of SAM on chlorophyll contents and photosynthesis of cucumber

处理 treatment	叶绿素a Chlorophyll a（$mg \cdot g^{-1}$ FW）	叶绿素b Chlorophyll b（$mg \cdot g^{-1}$ FW）	类胡萝卜素 Carotenoids（$mg \cdot g^{-1}$ FW）	净光合速率Pn（$\mu mol \cdot m^{-2} \cdot s^{-1}$）
CK	1.31 ± 0.09b	0.54 ± 0.04b	0.25 ± 0.02b	12.00 ± 0.30c
T1	1.93 ± 0.01a	0.79 ± 0.02a	0.34 ± 0.03a	13.93 ± 0.32b
T2	1.99 ± 0.05a	0.80 ± 0.03a	0.34 ± 0.01a	16.67 ± 0.21a

2.3 SAM对黄瓜幼苗植株根系的影响

由表3得出，黄瓜幼苗断根扦插后根系得到快速生长，尤其是在SAM处理下根系生长速度和根系活力增幅更为显著。在断根扦插第15d时，T1和T2单株总根长分别为66.64cm和108.52cm，而30d时分别为670.58cm和845.99cm。单株根系表面积也呈现类似的变化。T1和T2根系活力在断根扦插15d和30d后均显著高于CK，在15d和30d分别比CK增加8.26%，31.15%和9.35%，14.11%。30d相比15d根系活力各处理整体均呈现下降趋势，这可能是因为穴盘空间限制了根系生长。在断根扦插后30d，T1根系鲜重与CK无显著差异，T2的根系鲜重比对照增加了12.12%；而根系干重在各处理与CK之间无显著差异。

表 3　SAM对黄瓜幼苗植株根系的影响

Table 3　Effect of SAM on root of cucumber

处理 treat-ment	总根长cm Total root length	根系表面积cm^2 Total root area	总根长cm Total root length	根系表面积cm^2 Total root area	根系活力 root activity（μg·g-1FW·h）		根鲜重 FW of root（g）	根干重 DW of root（g）
	15d		30d		15d	30d	30d	30d
CK	169.65 ± 9.18a	48.55 ± 5.81a	789.28 ± 37.56a	86.48 ± 2.95a	65.34 ± 1.17c	59.90 ± 1.76b	1.32 ± 0.10b	0.13 ± 0.09a
T1	66.64 ± 3.65c	11.06 ± 1.50c	670.58 ± 40.66b	71.41 ± 1.53b	70.74 ± 1.40b	65.50 ± 1.91ab	1.31 ± 0.03b	0.11 ± 0.04a
T2	108.52 ± 7.82b	20.72 ± 0.65b	845.99 ± 21.27a	85.96 ± 3.42a	85.69 ± 3.96	68.35 ± 4.73a	1.48 ± 0.10a	0.12 ± 0.04a

2.4 SAM对黄瓜幼苗下胚轴多胺含量及相关基因表达量的影响

*ADC*精氨酸脱羧酶（*ADC*）基因、硫代腺苷甲硫氨酸脱羧酶（*SAMDC*）基因和亚

精胺合酶（*SPDS*）是编码多胺合成关键酶的基因。由图1和图2可知，黄瓜幼苗断根后，*ADC*基因（图1A）的表达在24h内不同程度下调，腐胺含量（图2A）也相应减少，SAM处理使其在12h时下降幅度进一步增大，而腐胺减少量也增大，这表明SAM降低了黄瓜幼苗下胚轴内腐胺含量的积累。而*SAMDC*（图1B）和*SPDS*（图1C）基因的表达受SAM处理显著诱导，使亚精胺（图2B）和精胺含量（图2C）在处理12h时显著增加。

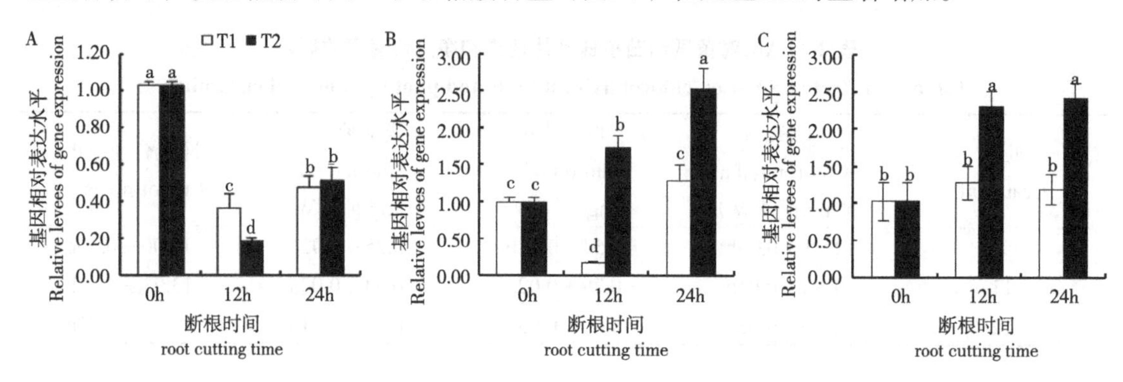

图 1　外源SAM对黄瓜幼苗下胚轴ADC（A）、SAMDC（B）和SPDS（C）表达的影响

注：T1：去离子水处理，T2：SAM处理（下同）。

Figure 1　Effect of exogenous SAM on the expression of *ADC*（A）、*SAMDC*（B）and *SPDS*（C）in the hypocotyl of Cucumber

Note：T1：Deionized water treatment，T2：SAM treatment（the same below）.

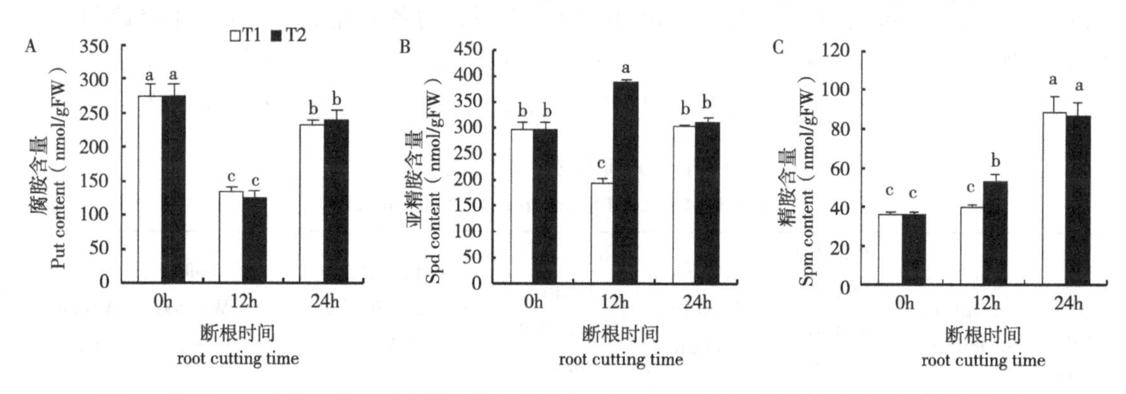

图 2　外源SAM对黄瓜幼苗下胚轴游离态腐胺（A），亚精胺（B）和精胺（C）含量的影响

Figure 2　Effect of exogenous SAM on free Put（A），Spd（B）and Spn（C）content in cucumber Hypocotyl

2.5　SAM对黄瓜幼苗下胚轴生长素含量及相关基因表达量的影响

CsARL1（*Adventitious rootless 1*）为不定根特异表达基因，荧光定量PCR分析表明，SAM处理显著诱导了*CsARL1*基因的表达（图3A），尤其是断根处理后12h，SAM处理中*CsARL1*表达量为不处理的4.82倍。进一步测定黄瓜下胚轴基部生长素含量（图3B）的变化，在处理12h时SAM的添加使黄瓜下胚轴生长素积累量增加31.01%。

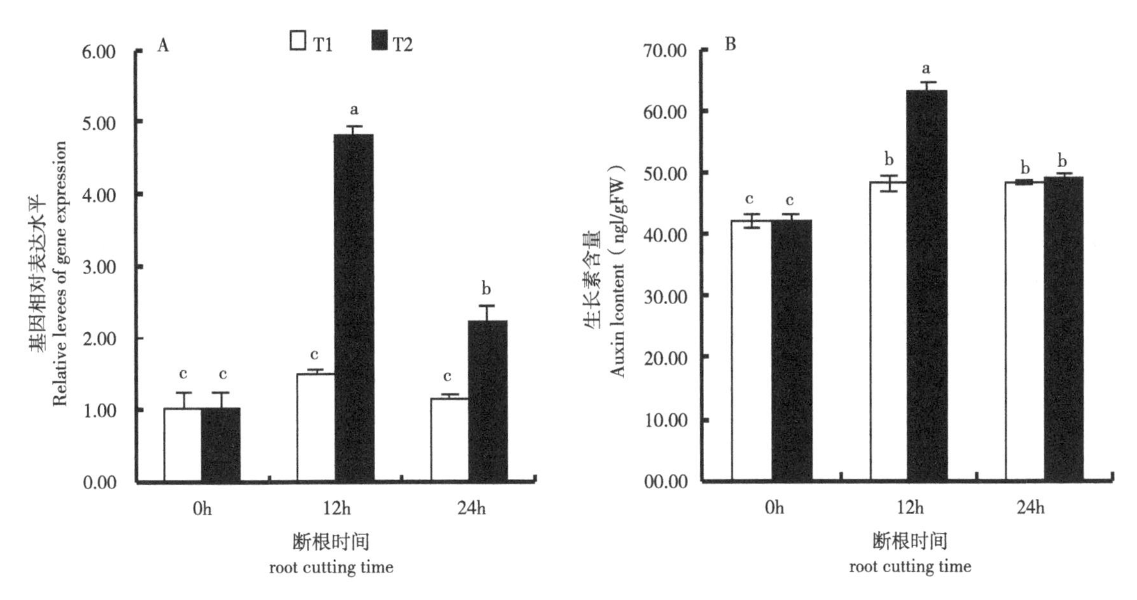

图3　外源SAM对黄瓜幼苗下胚轴基部不定根发生过程基因表达水平（A）及生长素含量（B）的影响
Figure 3　Effect of exogenous SAM on the gene expression levels（A）and Auxin（B）during adventitious root formation on the base of cucumber Hypocotyl

3　讨论与结论

由结果分析可以看出，SAM可有效促进黄瓜断根后不定根的生成。相比清水处理，SAM溶液处理下*ADC*基因表达下降，使黄瓜下胚轴内腐胺含量减少，同时SAM的添加上调了*SAMDC*和*SPDS*基因的表达。*SAMDC*基因和*SPDS*基因的上调，加速了SAM脱羧形成脱羧硫代腺苷甲硫氨酸（dcSAM），并与黄瓜下胚轴中氨丙基的反应，促进了在亚精胺合酶和精胺合酶的作用下将腐胺转化为亚精胺和精胺的过程。徐璐等（2014）研究表明多胺可以促进生长素的积累，而生长素是调控不定根发生最重要的植物激素，它可诱导中柱鞘上的某些细胞分化及分裂，并逐渐形成根原基，最终形成侧根，之后不再需要生长素的作用。本研究结果与其一致，在不定根发生过程中，黄瓜下胚轴生长素含量呈先升高后降低的趋势。在12h和24h时，SAM的添加使得黄瓜下胚轴生长素含量均比清水处理高，尤其是12h增加幅度最大，调控生长素含量的*CsARL1*基因表达量可被SAM诱导，可能是其增加生长素的重要原因。

根系活力反映植物根系对水分和养分的吸收能力。SAM促进了根系的生长及活力的提升，从而促进了黄瓜幼苗的叶绿素含量，叶绿素含量在一定范围内的增加可以有效增强光合作用（顾骏飞等，2016）。SAM的添加显著增强了黄瓜幼苗的净光合速率，这与其提高了叶绿素含量有密切关系。光合作用的增强促进了光合产物向根部的供应（罗凡等，2014），提高了根系活力和生长速率，发达的根系又反过来促进植物对养分的吸收，从而促进了黄瓜幼苗的生长。

References

[1] Birnbaum K D. How many ways are there to make a root. Current Opinion in Plant Biology [J]. 2016, 34: 61-67.

[2] Cantoni G.L. The nature of the active methyl donor formed enzymatically from l-methionine and adenosine triphosphate . Journal of the American Chemical Society [J] .1952, 74. 2 942-2 943.

[3] 程兮，杨卫平，马丁，等. S腺苷甲硫氨酸对肝细胞癌手术病人的疗效 [J]. 外科理论与实践，2015，20（5）：403-407.

[4] 顾骏飞，周振翔，李志康，等. 水稻低叶绿素含量突变对光合作用及产量的影响 [J] .作物学报，2016，42（4）：551-560.

[5] 李应超. 一种培育黄瓜实生苗强大根系的栽培方法：中国. 102138405A [P]. 2011，8-3.

[6] 罗凡，张厅，龚雪蛟，等. 不同施肥方式对茶树新梢氮磷钾含量及光合生理的影响 [J]. 应用生态学报，2014，25（12）：3 499-3 506.

[7] 徐璐，邢树堂，孙宪芝，等. 多胺对菊花激素含量与花芽分化的影响.植物生理学报 [J]，2014，50（8）：1 195-1 202.

不同补光灯对设施草莓光合生长及产量品质的影响

钱舒婷，李建明

（西北农林科技大学　园艺学院，农业部西北设施园艺工程重点实验室，杨凌　712100）

摘要：通过研究不同补光灯对草莓植株光合生长及产量品质的影响，以期为草莓设施栽培补光提供参考。以‘红颜草莓’为供试材料，试验设置了5个补光处理，分别为T1：激光生长灯；T2：红蓝比为4.9:1的LED灯；T3：荧光灯；T4：红蓝比为3:1的LED灯；T5：高压钠灯；以不补光为对照。测定不同补光灯处理下草莓植株的形态指标、光合色素、光合特性以及果实的产量和品质。红蓝比为4.9的LED灯和红蓝比为3的LED灯更有利于提高草莓植株的光合速率，促进植株的营养生长和草莓的提前成熟，对于增产的效果更明显；果实可溶性固形物和可溶性糖含量则以高压钠灯处理下最大。LED灯有利于促进草莓植株的光合生长，提高果实的产量和品质，高压钠灯虽然有利于提高果实的品质，但因为产量不高且耗电量大等弊端，其经济效益远不及LED，因此，综合考虑LED灯更适合作为冬春季节寡日照地区大棚草莓的补光光源。

关键词：补光；草莓；光合特性；产量；品质

Effects of Different Supplemental Lighting on Photosynthesis, Growth, Yield and Quality of Strawberry in Greenhouse

Qian Shuting, Li Jianming

(*The Agriculture Ministry Key Laboratory of Protected Horticultural Engineering in Northwest* , *College of Horticulture* , *Northwest A&F University* , *Yangling* , *Shaanxi*　712100)

Abstract: The study investigated the effects of different supplemental lighting on photosynthesis, growth, yield and quality of strawberry, in order to provide reference of Supplemental Lighting on strawberry. Taking Benihoppe strawberry variery as experimental material, different supplemental lighting were tested to examine the effects on morphology, chlorophyll content, Photosynthesis, Yield and Quality of Strawberry. Four treatments were set: T1, Plant Laser Light; T2, LED (R/B=4.9) ; T3, Fluorescent Lamp; T4, LED (R/B=3) ; T5, High Pressure Sodium Lamp; CK, without lamp. LED plant light source with the ratio of red/ blue photons was 4.9:1 and 3 : 1 are better on improving photosynthetic rate, promoting vegetative growth, ripening strawberries ahead of time and increase production; The soluble solid content and soluble sugar content under High Pressure Sodium Lamp treatment was the highest. LED was beneficial to the improvement of photosynthetic rate, the increase of fruit production and some qualities of strawberry. High Pressure Sodium Lamp was beneficial to the improvement of qualities of strawberry, but it consumed much electricity and the yield was not high. Therefore, using LED as the

作者简介：钱舒婷（1993—），女，安徽六安人，硕士，主要从事设施环境工程研究。E-mail：986421256@qq.com light source for the greenhouse strawberry is the best choice，beacause of the higher economic benefits .

Key words：Supplemental lighting；Strawberry；Photosynthetic characteristics；Yield；Quality

草莓由于含有丰富的营养物质和较高的药用价值，所以深受人们的喜爱。近年来，随着草莓经济效益的增加，其栽培面积也越来越大，目前已经成为设施栽培中的主要栽培作物之一。

光是设施栽培过程中的主要限制因素，因此研究光照对草莓生长的影响具有重要意义。目前，对草莓的研究主要集中在光强、光质和光周期等方面，光照强度影响植株的生长发育，同时对植物体内营养物质的含量也有显著影响[1-2]；光质对植物的影响要更加复杂，它不仅可以调控植株的光合作用[3]，还作为触发信号影响植株的生长[4]；光周期则影响植物的花性分化和成花诱导[5-9]，以及植物的营养生长。

关于不同补光灯对植物生长影响的研究也有少量的报道，研究发现，高压钠灯补光更适用于植株的开花、结果阶段[10-12]。荧光灯由于其光谱成分主要是红橙光和蓝紫光，所以它的光能利用率也较高。LED灯则是最近几年发展起来的新型光源，与荧光灯相比，LED光源对于提高叶用莴苣植株光合能力更有优势[13]。

总体来看，关于不同补光灯补光效果的综合对比研究还较少，因此本试验以草莓作为供试材料，探究荧光灯、高压钠灯、LED灯以及市面上新出现的激光生长灯等对植物光合生长和果实产量品质的影响，综合分析各补光灯的补光效果，以期为冬春季节寡日照地区大棚草莓的补光光源选择提供参考。

1　材料与方法

1.1　试验材料

选用“红颜”草莓作为供试材料，于2016年12月至2017年2月在陕西杨凌设施草莓产业园中的一座大跨度钢架棚内对草莓进行补光处理，供试材料设200株为一个小区，试验随机选取8株，重复3次。

1.2　试验设计

供试大棚的总面积约1 600m^2，在棚内靠北部弱光区共设6个小区，各小区面积50m^2，分别安装以下五种类型的补光灯：T1激光生长灯、T2 LED生长灯（R/B=4.9:1）、T3荧光灯、T4 LED生长灯（R/B=3:1）和T5高压钠灯，以同区域不补光植株为对照（CK）。各补光灯的属性如下表1，除激光生长灯的补光时间为16h外，其他各处理均补光4h。

表 1　不同处理光源的属性和补光时间

Table 1　property and the opening time of the supplemental lighting

处理 Treatment	光源 Light souce	紫外线强度 （uw/cm^2） Uitraviolet intensity	冠层光照强度（umol/（m^2·s） Light intensity of strawberry canopy	红蓝比 Red and blue ratio	补光时间 （h） Opening time
CK	不补光 Without lamp				

（续表）

处理 Treatment	光源 Light souce	紫外线强度 （uw/cm^2） Uitraviolet intensity	冠层光照强度（umol/（m^2·s） Light intensity of strawberry canopy	红蓝比 Red and blue ratio	补光时间 （h） Opening time
T1	激光生长灯 Plant Laser Light	0.4	1	5:1	16
T2	LED生长灯 LED light	2.1	15	4.9:1	4
T3	荧光灯 Fluorescent Lamp	0.7	8	1.93:1	4
T4	LED生长灯 LED light	4.7	22	3:1	4
T5	高压钠灯 High Pressure Sodium Lamp	40.4	215	8.5:1	4

1.3 测定项目指标及方法

1.3.1 物理性状的测定

选取5个处理下8个生长状态大致相同的新叶挂牌标记，分别在补光40d、46d、52d、58d、64d后测定草莓的叶片长、叶片宽和节间长，每个处理重复测定3次；补光15d后，从标记过的8棵植株中选择8个生长状态大致相同的幼果进行标记，记录果实从幼果到果实发白再到成熟所需要的时间。补光结束后，从各处理中随机采摘10颗成熟果实，并用天秤称出每颗果实的重量求出平均单果重，统计每平方米的产量和最大单果重。用直尺测量株高，并统计各处理下的平均叶片数。

1.3.2 光合指标的测定

采用Li-6800便携式光合仪测定叶片的光合参数，于2017年1月3日9:00—11:00分别对各试验区内选取的8株草莓生长点下的第3片完全展开功能叶片进行第一次测定，以后每隔10d测1次光合数据，一共测了6次。叶绿素和类胡萝卜素含量的测定参考赵世杰等[14]的方法，每处理重复3次。

1.3.3 品质指标的测定

果实可溶性固形物含量的测定用手持式固形物含量测试仪测出；果实可溶性糖含量的测定用蒽酮比色法测出[15]，果实总酸含量用NaOH滴定法测出[15]，果实可溶性蛋白质含量用考马斯亮蓝法测出[16]，并记录数据。试验重复 3 次，取平均值。

1.4 数据统计与分析

采用SPSS 20.0和Excel 2003软件对数据进行统计分析，采用Duncan新复极差法进行显著性检验（α=0.05），利用Excel 2003软件作图.

2 结果与分析

2.1 不同补光灯对草莓叶片光合色素含量的影响

由图1中可以看出，除了T1（激光生长灯）处理下的叶片光合色素含量与对照差异不显著外，其他各处理的光合色素含量均比CK高，其中T3（荧光灯）处理下的叶片光合色素的含量最大，荧光灯对于促进草莓叶片光合色素合成的效果最显著。

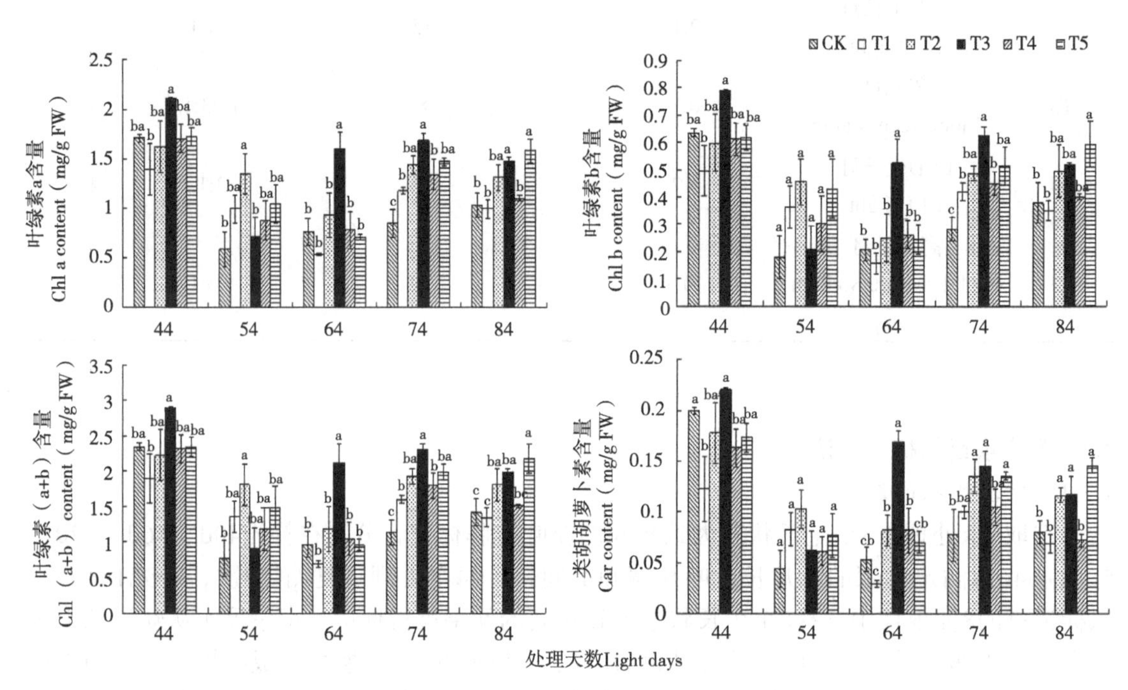

图柱上标不同小写字母表示差异显著（P<0.05）。下同Different letters indicate significant difference at 0.05 level.The same below.

CK：不补光；T1：激光生长灯；T2：红蓝比为4.9的LED灯；T3：荧光灯；T4：红蓝比为3的LED灯；T5：高压钠灯

CK：Without light；T1：Plant Laser Light；T2：LED（R/B=4.9）；T3：Fluorescent Lamp；T4：LED（R/B=3）；T5：High Pressure Sodium Lamp.

图1 不同补光灯对草莓叶片光合色素含量的影响

Figure 1 Effects of different Supplemental Lighting on pigment content of strawberry leaves

2.2 不同补光灯对草莓叶片光合特性的影响

由图2中可以看出，不同光源处理下草莓叶片净光合速率大小的顺序依次为T4（红蓝比为3:1的LED灯）>T5（高压钠灯）>T2（红蓝比为4.9:1的LED灯）>T1（激光生长灯）>T3（荧光灯）>CK，各处理间的蒸腾速率、气孔导度和胞间CO_2相差不大。可见，T4处理对于提高草莓叶片净光合速率的效果最显著。

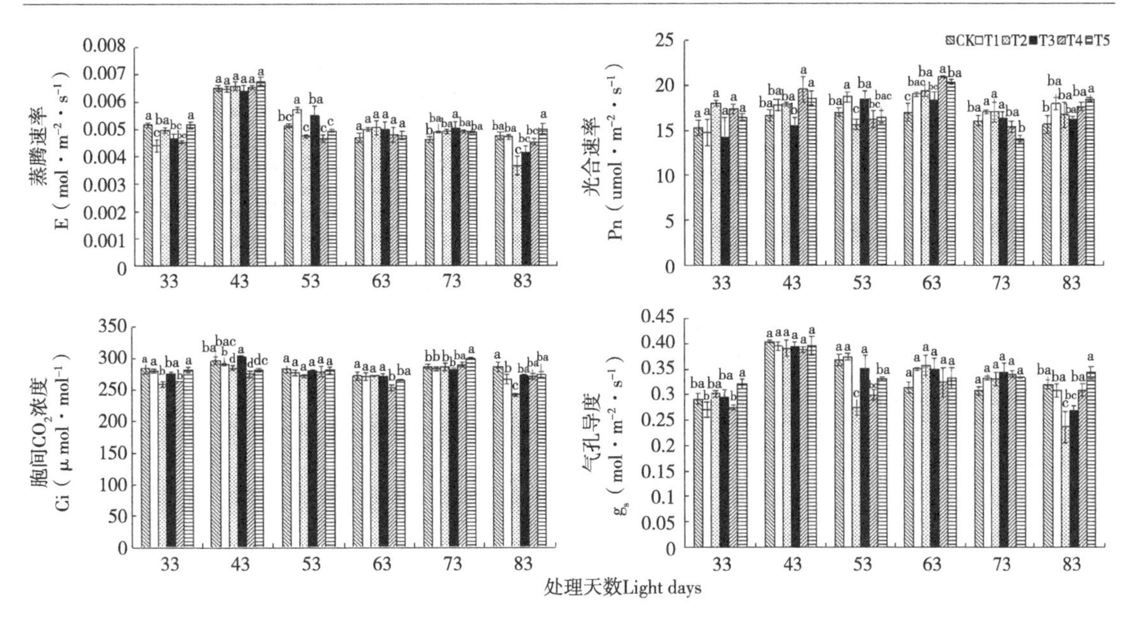

图 2　不同补光灯对草莓叶片光合特性的影响

Figure 2　Effects of different Supplemental Lighting on photosynthetic characteristic of strawberry leaves

2.3　不同补光灯对草莓株高和叶片生长的影响

由表2可以看出，T2（红蓝比为4.9:1的LED灯）、T3（荧光灯）和T4（红蓝比为3:1的LED灯）处理下，草莓的株高、叶片数和叶片大小均高于T1（激光生长灯）、T5（高压钠灯）处理及CK。说明在草莓的开花结果阶段，T2、T3、T4处理更有利于促进大棚草莓的光合作用以及植物营养物质的积累。

表 2　不同补光灯对草莓叶纵径、叶横径、节间长、叶片数和株高的影响

Table 2　Effects of different Supplemental Lighting on Leaf longitudinal diameter，Leaf transverse diameter，internode length，leaf number and plant height

	处理	处理后天数Days after treatment（d）				
		27	33	39	45	51
叶纵径/cm Leaf longitudinal diameter	CK	5.78a	5.92a	5.92a	6.06a	6.08a
	T1	5.78a	5.88a	5.88a	5.98a	6.05a
	T2	5.53a	5.73a	5.83a	5.9a	5.98a
	T3	5.77a	5.9a	5.93a	6.13a	6.23a
	T4	5.8a	5.93a	5.95a	6.07a	6.13a
	T5	5.64a	5.76a	5.8a	5.88a	5.98a
叶横径/cm Leaf transverse diameter	CK	4.78a	4.86a	4.94a	5.02a	5.1a
	T1	4.88a	4.98a	5.05a	5.12a	5.18a
	T2	4.83a	5a	5.03a	5.13a	5.25a
	T3	4.83a	5.07a	5.17a	5.37a	5.43a
	T4	4.87a	4.97a	5a	5.08a	5.13a
	T5	4.66a	4.88a	4.9a	5a	5.08a

（续表）

	处理	处理后天数Days after treatment（d）				
		27	33	39	45	51
节间长/cm Internode length	CK	10.08a	10.34a	10.38a	10.54a	10.62a
	T1	10.33a	10.48a	10.53a	10.63a	10.75a
	T2	11a	11.15a	11.18a	11.35a	11.48a
	T3	11.3a	11.5a	11.63a	11.9a	12a
	T4	10.67a	10.88a	10.95a	11.13a	11.2a
	T5	10.16a	10.38a	10.46a	10.64a	10.72a
叶片数 Leaf number（个/m^2）	CK			681		
	T1			591		
	T2			663		
	T3			729		
	T4			643		
	T5			675		
株高 Plant height（cm）	CK			14		
	T1			15		
	T2			20		
	T3			25		
	T4			19		
	T5			14		

2.4 不同补光灯对草莓果实生长的影响

由表3可以看出，与T1（激光生长灯）、T3（荧光灯）处理以及CK相比，T2（红蓝比为4.9:1的LED灯）、T4（红蓝比为3:1的LED灯）和T5（高压钠灯）处理下草莓果实从幼果到长大成熟所需要的时间更短，更有利于促进果实的提前成熟。

表 3　不同补光灯处理对草莓果实生长的影响

Table 3　Effects of different Supplemental Lighting on the growth of strawberry 月/日

处理 Treatment	幼果期 young fruit period	浆果变色期 Berry Color changing period	成熟期 Mature period
CK	12/16	1/29	2/5
T1	12/16	1/25	1/31
T2	12/16	1/28	2/4
T3	12/16	1/25	2/1
T4	12/16	1/24	1/31
T5	12/16	1/26	2/2

2.5 不同补光灯对草莓果实产量的影响

由表4可以看出，T2（红蓝比为4.9:1的LED灯）和T4（红蓝比为3:1的LED灯）处理下，草莓的产量要明显高于其他处理，且T2（红蓝比为4.9:1的LED灯）处理下草莓的单果重、结果数和产量均为最高，T5（高压钠灯）处理下的产量最低。可见，T2（红蓝比为4.9:1的LED灯）处理可以显著提高果实产量。

表 4 不同补光灯对草莓果实产量的影响

Table 4 Effects of Different Supplemental Lighting on the production of strawberry

处理 Treatment	单果重 Fruit weight（g）	结果数 Number of fruit（个/m²）	产量 Yield（g/m²）
CK	11.95b	305	3 646
T1	13.10ab	310	4 060
T2	16.22a	381	6 179
T3	12.34b	337	4 160
T4	16.06a	356	5 717
T5	11.26b	329	3 704

同列不同小写字母表示差异显著（P<0.05）Different small letters in the same column meant significant difference at 0.05 level

2.6 不同补光灯对草莓果实品质的影响

由表5可以看出，T5（高压钠灯）处理下的果实可溶性固形物和可溶性糖含量均为最高，T3（荧光灯）处理下的果实蛋白质含量最高，各处理间可滴定酸含量相差不大。由此可见，高压钠灯较其他处理更有利于提高果实的品质。

表 5 不同补光灯对草莓果实品质的影响

Table 5 Effects of Different Supplemental Lighting on the fruit quality of strawberry

处理 Treatment	可溶性固形物（%） Soluble solids content	可溶性糖（μg/g FW） Soluble sugar	可滴定酸（%） Titratable acid	可溶性蛋白（mg/g FW） Protein
CK	9.73cd	2.72ab	1.64a	1.40bc
T1	10.67ab	2.28bc	1.71a	0.60d
T2	10.23abc	2.45abc	1.52a	0.79cd
T3	10.17bc	2.17c	1.44a	2.81a
T4	9.4d	2.37abc	1.59a	1.57b
T5	10.73a	2.78a	1.54a	0.93cd

同列不同小写字母表示差异显著（P<0.05）Different small letters in the same column meant significant difference at 0.05 level

3 讨论

3.1 不同补光灯对草莓光合色素含量的影响

对于大多数作物而言，蓝光会降低叶片的叶绿素含量，而红光则有利于提高叶片的叶绿素含量[17-18]。有研究表明，蓝光荧光灯下的宝粉番茄幼苗叶片的叶绿素含量最低，红光下最高[19]；近年来，刘庆等[20]研究也发现，草莓叶片光合色素的含量在红/蓝/黄（7/2/1）处理下最高，蓝光下最低。本研究结果显示，荧光灯处理下的光合色素含量最高，这可能是因为荧光灯更有利于草莓叶片光合色素的合成。

3.2　不同补光灯对草莓光合特性的影响

在植物可以利用的光源中，红、橙光是被叶绿素吸收最多的，蓝、紫光也可以被叶绿素和胡萝卜素等强烈吸收[21]。刘庆等[20]研究发现，红光处理下的草莓叶片光合速率最大，其次是红蓝黄与红蓝混合光处理，蓝光处理下的叶片光合速率最小。本研究结果显示，LED灯处理下的叶片光合速率较其他处理要高，这可能是因为LED灯更有利于提高叶片的光合能力，李雯琳等[13]研究发现，LED灯对于提高叶用莴苣叶片光合能力的效果比荧光灯更显著，LED灯有利于提高叶片的*PSII*活性和*QA*的还原速率。荧光灯处理下的草莓叶片光合速率最低，但该处理下的叶片光合色素含量最大，这可能是因为荧光灯抑制了叶片光合产物的输出，增加了叶片的淀粉积累，而淀粉的积累不利于植物光合作用的进行。

3.3　不同补光灯对草莓株高和叶片生长的影响

不同光源对于草莓植株生长的影响显著，近年来谢景等[22]研究发现，LED光源对于提高叶用莴苣叶片光合能力的效果比荧光灯更显著，因为LED灯的光能利用率高，植物在LED灯下的生长也更为旺盛。本研究结果也显示，T3（荧光灯）处理下的叶片光合速率最低，但该处理下的草莓植株生长最旺盛，叶绿素含量最高，这可能是因为荧光灯处理不利于草莓植株同化产物的输出，植物合成的光合产物主要用于营养生长。

3.4　不同补光灯对草莓生育期的影响

浦正明等[23]研究发现，光照虽然对草莓第一花序的形成和花芽分化没有直接的影响，但是增加光照强度、延长光照时间却能明显促进草莓第二花序的形成及花芽分化，虽然目前关于草莓所需的最佳光照时间和光照强度还不清楚，但草莓在4～12h的光周期范围内的花芽分化更明显。各处理中，除激光生长灯的补光时间为16h，其余处理下的草莓光照时间均在4～12h的光周期范围内，激光生长灯的光强又因为最弱，所以该灯下草莓成熟所需要的时间最长。

3.5　不同补光灯对草莓产量的影响

对于草莓而言，影响产量的主要因素是花芽分化的时间早晚、数量多少以及质量的好坏。余红等[24]研究发现，光照强度虽然能影响草莓的花芽分化，但它的作用效果比光质和光周期要小得多。本研究中，除激光生长灯外的各处理间补光时间均一致，所以影响草莓产量的主要因素是光质量，各处理中，T2（红蓝比为4.9:1的LED灯）和T4（红蓝比为3:1的LED灯）处理下的草莓产量最高，这可能是因为LED灯下的草莓光合速率较高，LED灯又因为光能利用率高的优势，所以较其他灯而言更有利于提高草莓的产量。

3.6　不同补光灯对草莓果实品质的影响

光强和光质对于调控果实的品质都有一定的影响，光质可以调控果实的糖、酸代谢，王英利等[25]研究发现，红光可以提高番茄的糖酸含量，红光与低剂量的*UV-B*复合可以提高果实番茄红素以及糖、酸的含量；蓝光可以提高番茄果实可溶性蛋白以及维生素C的含量[26]。光照强度对于提高果实的品质也有一定影响，钟霈霖等[27]研究发现，光照强度对草莓的维生素C、果胶和总糖的含量都有显著的影响，对草莓单果重、硬度、总酸含量以及可溶性固形物含量等没有显著影响。本研究中，高压钠灯处理下的果实可溶性固形物含量最高，可能是因为高压钠灯的红光比例最高，红光有利于植物碳水化合物的合成，从而提高果实可溶性固形物含量[20]。各处理间果实可溶性糖含量虽有差异，但都相差不大，这可能是因为光强对红颜草莓可溶性糖含量的影响不大。本研究中荧光灯处理下的可溶性糖含量最低的原因，可能是因为荧光灯的红光比例最低；而该处理下蛋白质含量高的原因，可能是因为荧光灯的蓝光比例较高，蓝光促进了草莓果实蛋白质的积累[20]。

4 结论

不同补光灯对草莓植株的生长发育、光合特性、果实产量以及品质等都有显著的影响。本试验中，红蓝比为4.9的LED灯和红蓝比为3的LED灯更有利于提高草莓植株的光合速率，促进植株的营养生长以及果实的提前成熟，对增产的效果显著；荧光灯处理下叶片的叶绿素含量和果实的蛋白质含量最高；而高压钠灯处理下的果实可溶性固形物和可溶性糖含量则为最高。综合结果分析，红蓝比为4.9的LED灯对草莓植株的生长发育、光合特性、果实产量以及促进草莓提前成熟的效果都要比其他处理好，高压钠灯虽然有利于提高果实的品质，但是由于高压钠灯的耗电量大且对于提高果实产量的效果不显著等原因，使得高压钠灯的经济效益远不及LED灯，所以将LED灯作为冬春季节寡日照地区大棚草莓的补光光源最好，将红蓝比为4.9的LED灯作为红颜草莓的补光光源最好。

参考文献

[1] Vergeer L H T，Aarts T L，Degroot J D.The wasting disease and the effect of abiotic factors（light-intensity，temperature，salinity）and infection with labyrinthula-zosterae on the phenolics content of zostera-marina shoots［J］.Aquatic Botany，1995，52：35-44.

[2] Keller M，Hrazdina G. Interaction of nitrogen availability during bloom and light intensity during veraison. II. Effects on anthocyanin and phenolic development during grape ripening［J］. American Society for Enology and Viticulture，1998，49：341-349.

[3] 郑洁，胡美君，郭延平.光质对植物光合作用的调控及其机理［J］.应用生态学报，2008，19（7）：1 619-1 624.

[4] 胡阳，江莎，李洁，等.光强和光质对植物生长发育的影响［J］.内蒙古农业大学学报，2009，30（4）：296-303.

[5] 李程，裴忠孝，甘林叶，等.光周期对春石斛开花及多胺含量的影响［J］.植物生理学报，2014，50（8）：1 167-1 170.

[6] 杨娜，郭维明，陈发棣，等.光周期对秋菊品种“神马”花芽分化和开花的影响.园艺学报，2007，34（4）：965-972.

[7] 田素波，郭春晓，郑成淑.光周期诱导植物成花的分子调控机制［J］.园艺学报，2010，37（2）：325-330.

[8] Hayama R，Coupland G. The molecular basis of diversity in the photoperiodic flowering responses of arabidopsis and rice［J］. Plant Physiol，2004，（135）：677-684.

[9] Yamasaki S，Fujii N，Takahashi H. Photoperiodic regulation of CS-ACS2，CS-ACS4 andCS-ERS gene expression contributes to the femaleness of cucumber flowers through diurnal ethylene production under short-day conditions［J］. Plant Cell Environ，2003，（26）：537-546.

[10] 杜宁，韩哲.高压钠灯用于园艺光照明的改进设想［J］.牡丹江教育学院学报，2008（1）：143-144.

[11] Runkle E S，Padhye S R，Wook Oh，et al.Replacing incan-descent lamps with compact fluorescent lamps may delay flow-ering.

Scientia Horticulturae，2012，143（16）：56-61.

［12］Blom T J，Zheng Y B. The response of plant growth and leaf gas exchange to the speed of lamp movement in a greenhouse. Scientia Horticulturae，2009，119（2）：188 -192.

［13］李雯琳，郁继华，张国斌，等. LED光源不同光质对叶用莴苣幼苗叶片气体参数和叶绿素荧光参数的影响［J］.甘肃农业大学学报，2010，45（1）：47-51，115.

［14］赵世杰，史国安，董新纯.植物生理学实验指导［M］.北京：中国农业科学技术出版社，2002.

［15］高俊凤.植物生理学实验指导［M］.北京：高等教育出版社，2005.

［16］Li H-S，Sun Q，Zhao S-J，et al. The Experiment Principle and Technique on Plant Physiology and Biochemistry. Beijing：Higher Education Press，2000（in Chinese）

［17］童哲，赵玉锦，王台，等.植物的光受体和光控发育研究［J］.植物学报，2000，42（2）：111-115.

［18］许莉，刘世琦，齐连东，等.不同光质对叶用莴苣光合作用及叶绿素荧光的影响［J］.中国农学通报，2007，23（1）：96-100.

［19］蒲高斌，刘世琦，刘磊，等.不同光质对番茄幼苗生长和生理特性的影响［J］.园艺学报，2005，32（3）：420-425.

［20］刘 庆，连海峰，刘世琦，等.不同光质LED光源对草莓光合特性、产量及品质的影响［J］.应用生态学报，2015，26（06）：1 743-1 750.

［21］杨世杰. 植物生物学［M］.北京：科学出版社，2000.

［22］谢 景，刘厚诚，宋世威，等.光源及光质调控在温室蔬菜生产中的应用研究进展［J］.中国蔬菜，2012（2）：1-7.

［23］浦正明，王华君，谢伟峰，等.光照对草莓花芽分化影响的初探［J］.上海农业科技，2004（4）：73.

［24］余 红，马华升，方献平，等.草莓花芽分化机理及调控技术研究进展［J］. 江西农业学报，2011，23（1）：58-61+67.

［25］王英利，王勋陵，岳明.UV-B 及红光对大棚番茄品质的影响［J］.西北植物学报，2000，20（4）：590-595.

［26］陈 强，刘世琦，张自坤，等.不同LED光源对番茄果实转色期品质的影响［J］.农业工程学报，2009，25（5）：156-161.

［27］钟霈霖，杨仕品，乔 荣，等.光照强度对草莓主要品质的影响［J］. 西南农业学报，2011，24（3）：1 219-1 221.

不同红蓝配比LED光源对生菜资源利用效率的影响

王君[1, 2]，仝宇欣[1, 2]，杨其长[1, 2]，魏灵玲[1, 3*]

（1. 中国农业科学院农业环境与可持续发展研究所，北京　100081；2. 农业部设施农业节能与废弃物处理重点实验室，北京　100081；3. 农业部休闲农业重点实验室，北京　100081）

摘要： 通过调查红蓝LED光源下生菜的资源利用效率，以期确定人工光植物工厂栽培环境下优化的红蓝光配比（R/B）参数。本试验红蓝LED光源的光强为200μmol·m^{-2}·s^{-1}，光质处理为R、R/B=12、R/B=8、R/B=4、R/B=1和B，并以荧光灯（FL）作为对照，生菜定植30天后收获。结果表明，①电能利用效率（EUE）和光能利用效率（LUE）随着R/B增加而升高，在R/B=12处理下达到最大，EUE和LUE分别为0.49%和1.78%，是FL处理的5.54倍和1.76倍，但与R处理间无显著性差异；②水利用效率（WUE）也随着R/B增加而升高，在R处理下达到最大，为3.32μmol CO_2·mmol·H_2O^{-1}，是FL处理的1.33倍；③随着R/B增加单叶片的光合速率显著下降，引起EUE和LUE升高的主要原因为随着R/B增加而显著增大的叶面积指数。尽管R处理下具有较高的EUE、LUE和WUE，但收获时其形态表现出节间和叶柄较长的问题，因此，推荐红蓝配比为12作为光强为200μmol·m^{-2}·s^{-1}条件下生菜生长较优的红蓝光光环境参数。

关键词： 人工光植物工厂；红光；蓝光；光能利用效率；电能利用效率；水利用效率

Effects on Resource Use Efficiency for Lettuce Exposed to Different Ratios of Red Light to Blue Light

Wang Jun[1, 2]，Tong Yuxin[1, 2]，Yang Qichang[1, 2]，Wei Lingling[1, 3*]

（1. *Institute of Environment and Sustainable in Agriculture*，*Chinese Academy of Agricultural Science*，*Beijing* 100081；2. *Key Laboratory. for Energy Saving and Waste Disposal of Protected Agriculture*，*Ministry of Agriculture*，*Beijing* 100081；3. *Key Laboratory. for Leisure Agriculture*，*Ministry of Agriculture*，*Beijing* 100081）

Abstract： The objective of this study was to determine the optimal ratio of red light and blue light（R/B）in the plant factory with artificial lighting by investigating the resource use efficiency for lettuce. In this experiment，lettuce plants（*Lactuca sativa* L.）were exposed to 200 μmol·m^{-2}·s^{-1} irradiance for a 16 h·d^{-1} photoperiod under the following seven treatments：monochromatic red light（R），monochromatic blue light（B）and the mixture of R and B with different R/B ratios of 12，8，4，and 1，fluorescent lamps（FL）. Lettuce plant under FL treatment were set as the control. The results showed that ① electric-energy use efficiency（EUE）and light use efficiency（LUE）both increased with the increased R/B ratio. EUE and LUE both reached the highest values of

收稿日期：2017-12-9　修订日期：2018—

项目资助：科技部科技伙伴计划（KY201702008），用于设施农业生产的LED关键技术研发与应用示范（2017YFB0403900）

作者简介：王君（1989—），女，主要从事设施农业环境工程方面的研究。中国农业科学院农业环境与可持续发展研究所，100081。Email：wangjun112209@163.com

*通讯作者：魏灵玲（1973—），女，博士，研究员，博士生导师，主要从事设施园艺环境工程方面的研究。中国农业科学院农业环境与可持续发展研究所，100081。Email：weilingling@agri-garden.com

0.49% and 1.78% under R/B=12 treatment，respectively，which were 5.54 times and 1.76 times in comparison with EUE and LUE under FL treatment，respectively. However，no significant differences were found between R/B=1 and R treatments. ② Water use efficiency（WUE）also increased with increasing R/B ratio with the maximum value of 3.32 μmol CO_2·mmol H_2O^{-1} under R treatment，which was 1.33 times in comparison with WUE under FL treatment. ③ Although photosynthetic rate of single leaf decreased with increasing R/B ratio，the increased EUE and LUE were mainly caused by the increased leaf area index with the increasing R/B ratio. Although lettuce plants under R treatment had the highest EUE，LUE and WUE，long internode and petiole were also found under R treatment before harvest. Based on the above results，it is concluded that R/B=12 with the light intensity of 200 μmol·m^{-2}·s^{-1} was recommended as the optimal R/B ratio for lettuce growth.

Key words：Plant factory with artificial light；Red light；Blue light；Light use efficiency；Rlectric-energy use efficiency；Water use efficiency

1 引言

近年来，随着人们日益增长的对安全食物和周年稳定供应的需求，人工光植物工厂作为重要的实现途径已经获得都市居民的广泛认可。并且，随着科研工作者对其应用技术的不断探索，人工光植物工厂获得了蓬勃发展。在人工光植物工厂领域，日本、荷兰和中国发展迅猛[1, 2]。据报道，在日本，商业化生产的人工光型植物工厂已经有近250处[2]，而我国人工光植物工厂也在向商业化模式转变。

但是，人工光植物工厂在推广应用过程中仍存在初期投入成本和运行成本高等问题。随着技术进步，行业标准地不断制定完善，初期投入作为必要投入可以大幅降低，运行成本也可以通过环境调控手段实现缩减。已经有大量的研究致力于通过环境调控来减少人工光植物工厂的运行成本，其中主要包括：①对植物工厂内部冷凝水进行收集，循环再利用，提高水的利用效率[3]；②将人工光植物工厂的照明期设置到外界空气温度较低的夜间，利用风机引进外界低温空气与空调协同降温的方法，以低功率的风机减少高功率空调的运行时间，减少了降温耗电量[4]；在此基础上，以零浓度差法供给CO_2，使植物工厂内外CO_2浓度保持一致，减少了CO_2的施放量[5]；③通过提高CO_2浓度，优化LED光强参数提高作物产量和光能利用效率[6-8]。目前，节能LED作为波宽窄、光强可调的优质光源，不断取代电光转化率低的荧光灯作为人工光植物工厂的植物生长光源。因此，探索最佳的LED组合光源来提高作物的资源利用效率，从而降低人工光植物工厂的运行成本，是其在人工光植物工厂中应用亟须解决的问题。

本文从叶绿素吸收最多的红蓝光入手，通过优化红蓝光配比来提高生菜的资源利用效率，为减少人工光植物工厂的运行成本提供光参数指导。并从光合速率（P_n）、叶片数和叶面积指数（LAI）等指标入手分析不同红蓝光配比对生菜光能利用效率（LUE）、电能利用效率（EUE）和水利用效率（WUE）的影响。

2 材料与方法

2.1 试验设计

以奶油生菜（*Lactuca sativa* L.）作为栽培对象，蛭石:草炭=3:1（V/V）的混合物作为育苗基质，育苗光源采用荧光灯，光强为150μmol·m^{-2}·s^{-1}，当第二片叶完全展开后，选择长势一致的幼苗进行随意分配，定植到7套完全相同的水培系统中。试验处理中的光源采用红（R）、蓝光（B）波峰分别为657nm和450nm的LED光源，以荧光灯（FL）作为对照，光质处理分别表述为R、R/B=12、R/B=8、R/B=4、R/B=1、B和FL（R/B/G=1.2:1:1.3），光谱详见图1所示。试验期间，光强和光周期分别为200μmol·m^{-2}·s^{-1}和16h·d^{-1}；明期和暗期空气温度分别为24℃和20℃，湿度为60%，CO_2浓度为400μmol·mol^{-1}；采用日本山崎营养液配方，pH值≈5.8；EC≈1.5mS.cm^{-1}。生菜定植30天后收获，该试验重复进行两次。

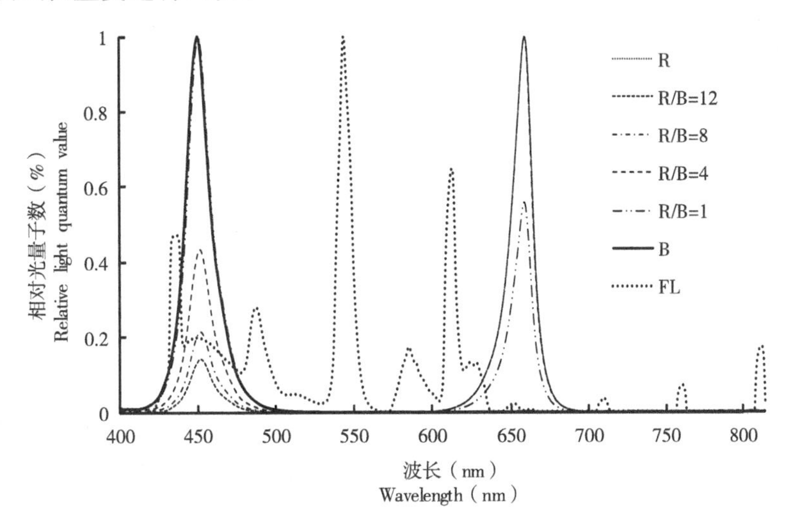

图1 不同光源的光谱分布

Figure 1 The light spectrum of different light sources

2.2 试验设备和测量方法

红蓝LED光源（东莞生物光环境科技有限公司，中国）的光强和光质均可以通过调节直流稳压电源的电流实现。采用36W荧光灯（TL-D 36W，Philips）作为对照。光强和光谱分别是通过Li-1500（Li-COR，USA）和光谱仪USB 200 spectrometer（Ocean Optics，Dunedin，FL，USA）进行测定的。光源耗电量是采用北电仪表（PowerBay-SSM，精度：0.1W，中国）进行测量的。

样品干重是将生菜鲜样在85℃下烘干至恒重后用电子天平（Sartorius BL610，精度：0.01g，德国）进行称量获得的。

叶面积指数（LAI）为叶面积总和与栽培面积的比值。叶面积测定是取每个处理长势一致的4株植株作为样品，将叶形描绘在质地均匀、形状规则的纸张上。首先对整张纸进行重量测量，然后对描绘区域进行裁剪称重，根据整张纸的既定面积计算叶片面积。

光合速率（P_n）是对样品从上往下完全展开的第二片叶进行测定。采用便携式光合仪（Li-6400，Li-COR，USA），光源选用10%蓝光和90%红光，光强为200μmol·m^{-2}·s^{-1}的红蓝LED。叶室温度为24℃，CO_2浓度为400μmol·mol^{-1}；饱和蒸汽压差VPD为1.1kPa，当P_n达到稳态时记录数据。每个处理选择一个样品进行测定，测量顺序为R、R/B=12、

R/B=8、R/B=4、R/B=1、B和FL，然后再重复此测量过程3次。

2.3 计算方法

电能利用效率（EUE）是地上累积干重的化学能与消耗电能的比值[3]。

$$EUE=\frac{\overline{DW}\cdot D\cdot S\cdot f}{W_e} \quad (1)$$

其中，$\overline{DW}$是样品干重的平均值，g；D是栽培密度，本试验中为30株·m^{-2}；S是栽培面积，本试验中为0.3m^2；f是干重与化学能之间的转换系数，取$2\times10^4 J\cdot g^{-1}$；$W_e$是光源消耗的总电能，J。

光能利用效率（LUE）是地上累积干重的化学能与植物群体冠层处接受总光能的比值[3]。

$$LUE=\frac{\overline{DW}\cdot D\cdot f}{PAR} \quad (2)$$

其中，PAR是植物群体冠层处接受光能的累积量，J·m^{-2}。

瞬时水利用效率（WUE）参考施建忠和王天铎（1996）的计算方法[9]，单位为μmol CO_2·mmol·H_2O^{-1}。

$$WUE=\frac{P_n}{T_r} \quad (3)$$

2.4 数据统计与分析

采用Excel进行数据整理，利用SAS 9.1（SAS Institute Inc. 9.1，Cary，NC，USA）软件对数据进行显著性分析，采用邓肯氏多重比较，显著水平P≤0.05。

3 结果

3.1 EUE和LUE

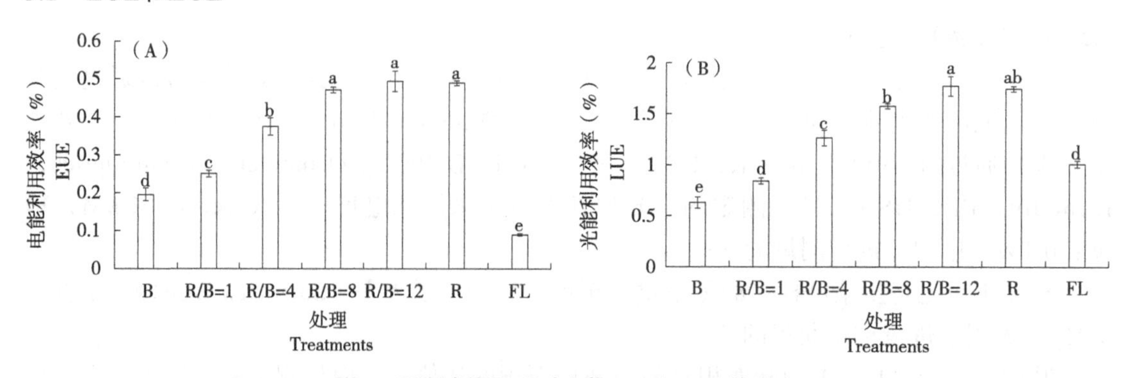

图2 不同光源处理对生菜EUE和LUE的影响

Figure 2 Effect of different lighting sources on EUE and LUE

EUE和LUE是衡量植物对人工光源所消耗电能和冠层处接受光能利用程度的重要指标[8]。从图2（A）可以看出，当R/B≤8时，EUE随着R/B增加而升高；当R/B≥8时，EUE随着R/B增加各处理间无显著性差异，R/B=12处理下EUE最大，为0.49%；R/B=12和R处理下EUE分别是B处理下EUE的2.52和2.5倍。而FL处理下生菜的EUE最小，显著低于红蓝单色光或混合光处理，B处理下生菜EUE是FL处理下EUE的2.19倍。从图2（B）可以看出，当R/B≤12时，LUE随着R/B增加而升高，R/B=12处理下LUE最大，为1.78%；但R/B=12和R处理下生菜的LUE无显著性差异，分别是B处理下LUE的2.81倍和2.76倍。当R/B≥4时，LUE随着R/B增加而显著高于FL处理，FL处理下生菜的LUE虽与R/B=1处理无显著性差异，却是B处理下生菜LUE的1.37倍。

3.2 干重

从图3可以看出，生菜地上干重随R/B增加的变化趋势与LUE一致。根部干重随R/B增加也呈现增加的趋势，R处理下根部干重达到最大。FL处理下根部干重除显著低于R处理外，显著大于R/B≤4的红蓝光处理。

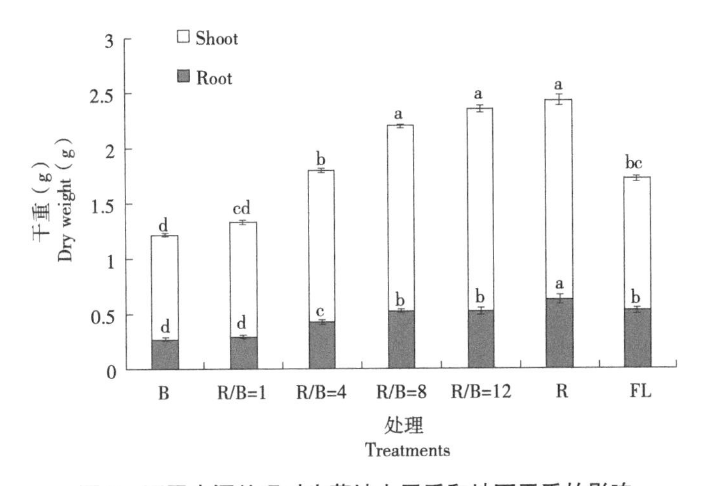

图 3　不同光源处理对生菜地上干重和地下干重的影响

Figure 3　Effects of different lighting sources on shoot dry weight and root dry weight for lettuce

3.3 LAI

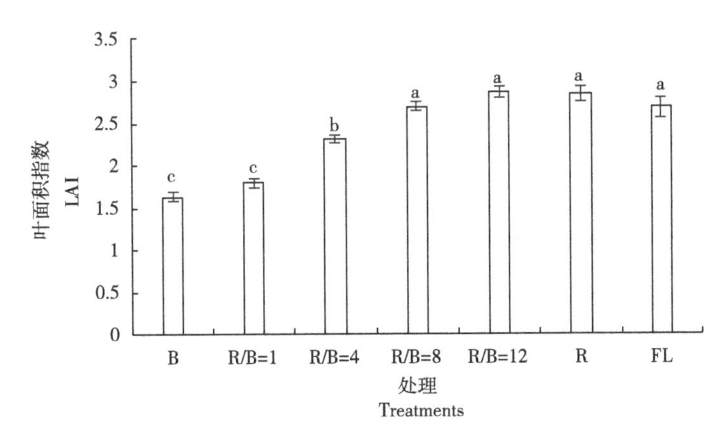

图 4　不同光源处理对生菜LAI的影响

Figure 4　Effects of different lighting sources on LAI for lettuce

从图4可以看出，当R/B≤8时，生菜LAI随着R/B增加而增加，与LUE随R/B增加的变化趋势一致；当R/B≥8时，随着R/B增加各处理间LAI无显著性差异，R/B=12和R处理下LAI分别是B处理的1.75和1.74倍。但FL处理下LAI与R/B≥8的处理间无显著性差异，且显著大于R/B≤4的红蓝光处理，是B处理下LAI的1.64倍。

3.4 P_n和叶片数

从表1可以看出，当R/B≥1时，生菜Pn随着R/B增加呈减小的趋势，在R处理下达到最小。R/B=1处理与B处理下P_n没有显著性差异。FL处理下Pn除显著低于R/B=1处理，显著高于R处理外，与其他处理间没有显著性差异。对于不同光源对生菜叶片数的影响，R/B=8处理下生菜叶片数最大，与R/B=12处理间无显著性差异。FL、B和R/B=1处理间叶片数无显著性差异，且叶片数最少。

3.5 LUE，EUE和LAI之间的关系

从图5可以看出，LUE、EUE与LAI均呈正相关关系，且具有较高的拟合系数，但该拟合曲线不包含FL处理下数据（即圈内数据）。LUE随LAI增大而增加的速度要远高于EUE。

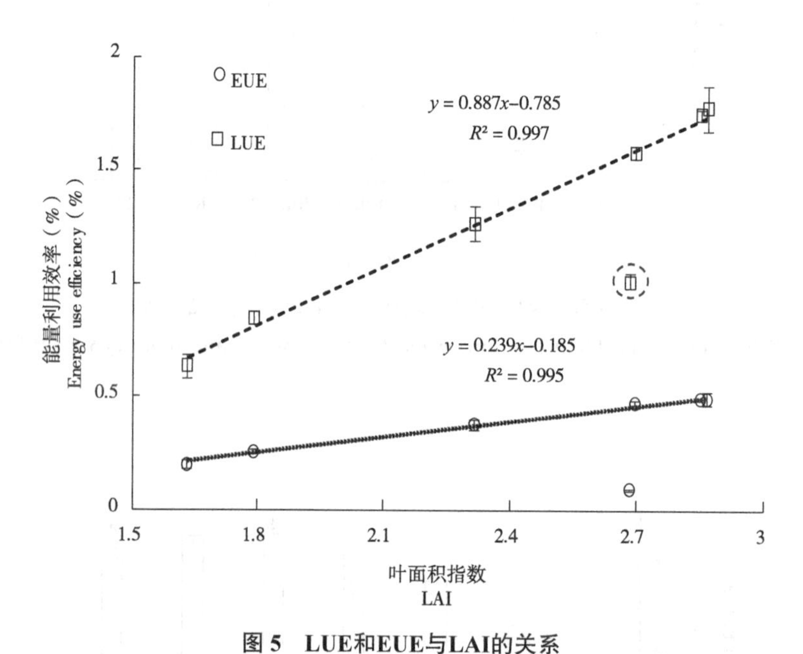

图5 LUE和EUE与LAI的关系

Figure 5 Relationship between LUE and EUE and LAI

3.6 WUE

WUE是反映植物碳同化和水分耗散关系的重要指标。WUE随着R/B增加而增大（图6），在R处理下达到最大，为3.32μmol CO_2·mmol·H_2O^{-1}，与P_n随着R/B增大的变化趋势正好相反，B和R/B=1处理下WUE无显著性差异，分别比R处理下WUE低64.2%和87.9%。而FL处理下WUE为2.49 μmol CO_2·mmol H_2O^{-1}，仅显著性低于R处理。

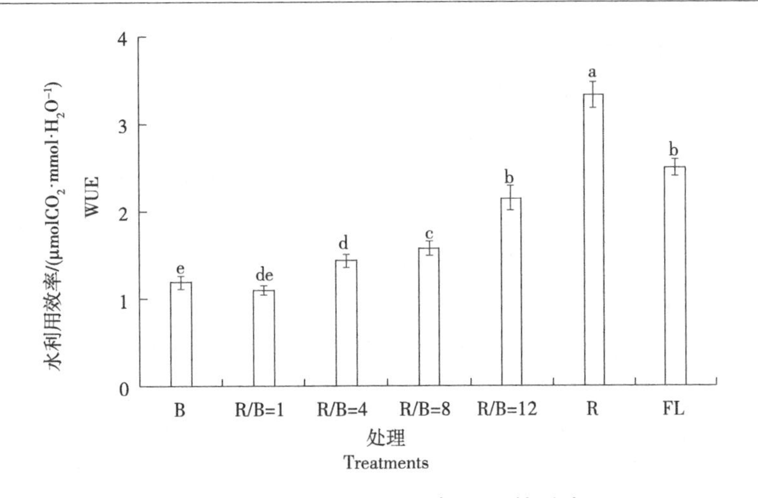

图 6　不同光源处理对生菜WUE的影响

Figure 6　Effect of different lighting sources on WUE

4　分析与讨论

4.1　不同红蓝光配比处理和FL对生菜LUE的影响（表1）

表 1　不同光源处理对生菜P_n和叶片数的影响

Table 1　Effects of different lighting sources on P_n and leaf number for lettuce

不同光源Different lighting sources	B	R/B=1	R/B=4	R/B=8	R/B=12	R	FL
P_n（$\mu mol \cdot m^{-2} \cdot s^{-1}$）	7.89ab	8.54a	7.68b	7.21bc	6.45c	4.15d	7.2bc
叶片数Leaf number	20c	22c	25b	30a	28ab	26b	22c

LUE与冠层截获的光合有效辐射和积累的碳水化合物直接相关。即使在环境可控的人工光栽培环境，大气CO_2浓度水平下，实际LUE也不高于10%[10]。本试验中获得的LUE最大值为1.78%，是光强为200$\mu mol \cdot m^{-2} \cdot s^{-1}$，R/B＝12的处理，低于王君（2016）在研究不同光强对生菜LUE影响时获得的以下结果，在R/B＝1，光强分别为200、300和400$\mu mol \cdot m^{-2} \cdot s^{-1}$处理中，光强200$\mu mol \cdot m^{-2} \cdot s^{-1}$下LUE最大，为4.17%，造成两结果矛盾的原因为后者的计算方法是将从定植至收获过程中由于栽培密度调整所收获的所有生菜的干重进行统计获得的，而前者的栽培密度始终保持不变，且显著低于后者。Yokoi et al.（2003）[6]通过将CO_2浓度从大气浓度提高到1 040 × 10^{-6}，LUE高达7.5%。因此，通过环境调控的手段可以不断提高LUE接近理论最大值。

在相同的光合有效辐射强度下，对于红蓝混合光处理，当R/B≤12时，LUE随R/B变化的趋势一致（图1B），该试验结果表明，红蓝混合光下减少蓝光的比重可以显著增加地上干重的累积（图2），引起该结果的主要原因为，减少红蓝混合光下蓝光的比例有效地增加了叶片数（表1）和LAI（图4），从而增加植物冠层截获的光合有效辐射量[11]，与Ohashi-kaneko et al.（2007）[12]发现的红光处理下生菜地上干重、叶面积、叶片数和茎长要显著高于R/B=1和B处理的研究结果一致。但同时也发现，表征叶片合成碳水化合物能力的P_n与地上干重随R/B增加变化的趋势相反，造成该结果原因可能为，①降低红蓝混合光下蓝光的比重虽然减小了单个叶片的P_n，但单叶水平上测得的光合速率与作物整株水平上的长期光同化效率可能并无必然联系[11, 13]，在今后需对不同R/B处理对植株整体同化速率的影响进行深入研究；②LAI和叶片数随着R/B增加而增加的结果抵消了蓝光比重减小对P_n降低的影响。

本文试验结果与闻婧等[14]（2011）之前研究发现的光强150μmol·m^{-2}·s^{-1}，R/B=8条件下最适宜生菜生长发育的研究结果看似矛盾，其实不然，后者使用的红蓝光源中蓝光强度为16.7μmol·m^{-2}·s^{-1}，本试验中获得最大干重的R/B＝12处理下蓝光强度为15.4μmol·m^{-2}·s^{-1}，可以分析得出，红蓝LED光源下生菜生长所需的蓝光可能并非是以红蓝光的比例决定的，而是以一定的蓝光强度作为限制条件的。Hoenecke et al.[15]（1992）也发现每天维持12h·d^{-1}的15μmol·m^{-2}·s^{-1}和30μmol·m^{-2}·s^{-1}的蓝光强度对于生菜生长是可接收的蓝光强度。

而对于FL处理，其光谱成分为R/B/G=1.2:1:1.3，与R/B=1处理的R/B值比较接近，两处理间生菜LUE、干重和叶片数均无显著性差异，但由于FL光源中包含大量的绿光成分，造成P_n减小了15.7%，LAI却提高了49.9%。由于绿光可以促进叶片伸长[13]，FL处理下生菜表现出更为松散的生长结构，截获的光合有效辐射量也随之增加，从而引起地上干重累积量比R/B=1处理提高了15.1%。并且，FL处理下生菜干重和LUE显著低于R/B≥4的各处理，大于R/B≤1处理，该试验结果表明，在光强为200μmol·m^{-2}·s^{-1}，红、蓝、绿光结合的白光下，尽管被叶片吸收的绿光驱动光合作用比红光更为有效[16]，增加红光成分比增加绿光成分显著降低了单叶片的P_n，但增加红光比重对于提高生菜整体的干重和LUE更为有效。与红蓝LED光源相比，FL处理下生菜的LUE尽管高于R/B≤1处理，由于FL光源的电光转换效率显著低于LED灯[3]，使其EUE最低（图5）。因此，在选择白光作为植物生长光源时建议优先选择白色LED。

R处理下生菜表现出较低的P_n，较高的叶片数和LAI，B处理下生菜则表现出相反的生长特征，而R/B=12的混合光处理下生菜表现出最高的干重和LUE，纯红光或蓝光不能满足生菜的生长需求，红蓝混合光才是生菜生长的较佳光源。此试验结果与前人在菠菜、萝卜[17]和黄瓜苗[18]等植物上的发现一致。与R/B=12处理相比，R处理下单个叶片的P_n显著偏低，但两处理间地上干重和LUE却无显著差异，其可能的原因为纯红光处理增加了节间和叶柄的长度[19]，从而造成R处理下地上干重的增加。

4.2　不同红蓝光和FL对生菜WUE的影响

本试验中WUE随着R/B增加而显著增加（图6），造成该结果的原因为气孔是CO_2和水分进出的通道，而叶片气孔导度随着R/B增加而显著减小[18, 20]，引起P_n显著下降（表1）。

而FL处理下生菜叶片的P_n除显著低于R/B=1的处理外，与其他处理之间无显著性差异或显著偏高，但其WUE显著高于除R处理外的其他处理，可能是由于绿光可以逆转蓝光对气孔打开的效果，导致气孔导度下降[19]，引起P_n升高的同时蒸腾速率有所下降，从而FL处理下WUE较高。

5 结论

本文通过研究不同红蓝光配比的LED光源和荧光灯下生菜对资源利用效率的影响，结果发现：

（1）相同光强下，红蓝LED混合光照射的生菜可以获得比荧光灯照射更高的干重累积量、叶面积指数和叶片数，从而获得更高的光能和电能利用效率。

（2）在光强为200$\mu mol \cdot m^{-2} \cdot s^{-1}$，红蓝光配比为12的LED光源下，生菜可以获得较优的光能利用效率和电能利用效率；并且，红蓝混合光下蓝光比重以不超过11%为宜。

（3）当红蓝光配比大于1时，在红蓝光基础上增加绿光可以提高生菜产量、光能利用效率和水利用效率。

从不同红蓝光配比光源下生菜对资源利用效率的研究结果中分析得出，推荐红蓝配比为12作为光强为200$\mu mol \cdot m^{-2} \cdot s^{-1}$条件下生菜生长较优的红蓝光光环境参数。

参考文献

［1］ 刘文科，杨其长. 设施农业照明新光源——发光二极管（LED）［J］. 科技导报（北京），2014，32（6）：12-12.

［2］ 贺冬仙. 植物工厂的概念与国内外发展现状［J］. 农业工程技术（温室园艺），2016，36（10）：13-15.

［3］ Kozai，T. Improving utilization efficiencies of electricity，light energy，water and CO_2 of a plant factory with artificial light. Technology Advances in Protected Horticulture. 2011，2-8.

［4］ 王君，杨其长，魏灵玲，等. 人工光植物工厂风机和空调协同降温节能效果［J］. 农业工程学报，2013，29（3）：177-183.

［5］ 辛敏. 引进室外冷源的植物工厂零浓度差 CO_2 施肥系统［D］. 北京：中国农业科学院，2015.

［6］ Yokoi，S.，Kozai，T.，Ohyama，K.，et al. Effects of leaf area index of tomato seedling populations on energy utilization efficiencies in a closed transplant production system ［J］. Journal of Society of High Technology in Agriculture，2003，15（4）：231-238（in Japanese）.

［7］ 沈韫赜，郭双生，艾为党，等. 红蓝 LED 光照强度对密闭生态系统中生菜生长状况及光合速率的影响［J］. 载人航天，2014，20（3）：273-278.

［8］ 王君，杨其长，仝宇欣. 红蓝光下光强对生菜电能、光能利用效率及品质的影响［J］. 中国农业大学学报，2016，21（8）：59-66.

［9］ 施建忠，王天铎. 小麦冠层不同层次叶片水分利用效率的研究——光合速率与蒸腾速率之比（P/T）的模拟［M］. 农田生态系统研究. 北京：气象出版社，1996.

［10］ Beadle，C. L. and Long，S. P. Photosynthesis is it limiting to biomass production? ［J］. Biomass，1985，8：119-168.

［11］ Kim，H. H.，Goins，G. D.，Wheeler，R. M.，et al. Green-light supplementation for enhanced lettuce growth under red-and blue-light-emitting diodes［J］. HortScience，2004，39（7）：1 617-1 622.

［12］ Ohashi-Kaneko，K.，Takase，M.，Kon，N.，et al. Effect of light quality on growth and vegetable quality in leaf lettuce，spinach and komatsuna ［J］. Environ. Control Biol.，2007，45：189-198.

［13］ Johkan，M.，Shoji，K.，Goto，F.，et al. Effect of green light wavelength and intensity on photomorphogenesis and photosynthesis in Lactuca sativa［J］. Environmental and Experimental Botany，2012，75：128-133.

［14］ 闻婧，杨其长，魏灵玲，等. 不同红蓝 LED 对生菜形态与生理品质的影响［J］. 设施园艺创新与进展——2011 第二届中

国· 寿光国际设施园艺高层学术论坛论文集，2011，232-239.

[15] Hoenecke，M. E.，Bula，R. J.，Tibbitts T W. Importance of 'Blue' Photon Levels for Lettuce Seedlings Grown under Red-light-emitting Diodes [J]. HortScience，1992，27（5）：427-430.

[16] Terashima，I.，Fujita，T.，Inoue，T.，Chow，W.S.，Oguchi，R.，2009. Green light drives leaf photosynthesis more efficiently than red light in strong white light：revisiting the enigmatic question of why leaves are green. Plant Cell Physiol. 50，684–697.

[17] Yorio，N.C.，Goins，G.D.，Kagie，H.R.，2001. Improving spinach，radish，and lettuce growth under red light-emitting diodes（LEDs）with blue light supplementation. Hort Sci 36，380–383.

[18] Hogewoning，S.W.，Trouwborst，G.，Maljaars，H.，et al.，Blue light dose-responses of leaf photosynthesis，morphology，and chemical composition of *Cucumis sativus* grown under different combinations of red and blue light. J. Exp. Bot. 2010，61，3 107–3 117.

[19] Kim，S. J.，Hahn，E. J.，Heo，J. W.，et al. Effects of LEDs on net photosynthetic rate，growth and leaf stomata of chrysanthemum plantlets in vitro [J]. Scientia Horticulturae，2004，101（1）：143-151.

[20] 王君. 红蓝光下不同光强和光质配比对生菜光合能力影响机理 [D]. 北京：中国农业科学院，2016.

[21] 王君，仝宇欣，杨其长，等. 不同红蓝配比LED光源对生菜资源利用效率的影响 [J]. 农业工程学报，2018.

潮汐灌溉液位深度对多层栽培生菜生长和品质的影响

曹晨星[1]，武占会[2, 3]，刘明池[2, 3]，季延海[2, 3]，梁浩[2, 3]，臧秋兰[2, 3]，王丽萍[1*]

（1. 河北工程大学园林与生态工程学院，邯郸　056038；2. 北京市农林科学院蔬菜研究中心，北京　100097；3. 农业部华北都市农业重点实验室，北京　100097）

摘要：采用多层潮汐灌溉设备，研究了不同潮汐灌溉液位深度对多层岩棉基质栽培生菜生长、产量和品质的影响。结果表明：灌溉液位深度为2cm时，生菜的叶片数最多，地上部干质量和单株产量显著高于其他处理，维生素C含量、可溶性蛋白含量和叶绿素含量最高；灌溉液位深度为3cm时，生菜的株高显著高于其他处理；灌溉液位深度为4cm时，生菜的硝态氮含量最低。综合来看，在灌溉间隔时间1h，灌溉维持时间3min的试验条件下，灌溉液位深度为2cm时生菜综合品质最优，产量最高。

关键词：潮汐灌溉；灌溉液位深度；生菜；多层栽培

Effects of Ebb and Flow Irrigation Depth on the Growth and Quality of Lettuce Multilayer Cultivation

Cao Chenxing[1], Wu Zhanhui[2, 3], Liu Mingchi[2, 3], Ji Yanhai[2, 3], Liang hao[2, 3], Zang Qiulan[2, 3], Wang Liping[1*]

（1. *College of Landscape and Ecological engineering*, *Hebei University of Enginneerring*, *Handan*　056001; 2. *National Engineering Research Center for Vegetables*, *Beijing Academy of Agriculture and Forestry Sciences*, *Beijing*　100097; 3. *Key Laboratory of North China Urban Agriculture*, *Ministry of Agriculture*, *Beijing* 100097）

Abstract: This experiment adopts multilayer ebb and flow irrigation equipment, studied the effects of different ebb and flow irrigation depth in rock wool on the growth, yield and quality of lettuce cultivation.The results showed that the irrigation depth of 2 cm, the leaves, dry weight and yield of lettuce were significantly higher than other treatment, the content of vitamin C, soluble protein and chlorophyll were highest; irrigation depth of 3 cm, the plant depth were significantly higher than other treatment; irrigation depth of 4 cm, nitrate nitrogen content was the lowest.In conclusion, irrigation depth is 2 cm, and under the test condition was the maintaining fluid of 3 minutes and the interval time was 1 hour, which the comprehensive quality and yeild of lettuce was the best.

Key words: Ebb and Flow irrigation; Irrigation depth; Lettuce; Multilayer cultivation

---

曹晨星，女，硕士研究生，专业方向：设施园艺与无土栽培，E-mail：495122216@qq.com

*通讯作者（Corresponding author）：王丽萍，女，教授，硕士导师，专业方向：设施园艺与无土栽培，E-mail：wlp29@163.com

基金项目：国家科技支撑计划课题（2014BAD05B05），国家大宗蔬菜产业技术体系项目（CARS-25-G-01），北京市科技计划项目（D171100002017002）

潮汐灌溉是一种从栽培基质底部给水，利用基质毛细管作用供给植物生长所需水分和养分的灌溉技术，目前已成为欧美发达国家花卉生产和蔬菜育苗的主要灌溉方式。我国潮汐灌溉起步较晚，现阶段在花卉栽培和蔬菜育苗等栽培方面研究较多（王正等，2015；高艳明等，2016）。与传统灌溉方式相比，潮汐灌溉可以实现营养液的循环利用，水份利用率高（张黎和王勇，2011）；可以降低栽培设施内环境湿度，减轻病虫害的发生（刘宏久等，2015）；易于实现水肥一体化精准控制技术，降低人工成本，有利于实现自动化管理和工厂化生产，应用前景广阔。目前，在叶菜类蔬菜生产上主要以人工浇灌、滴灌、喷灌等灌溉形式为主，潮汐灌溉应用较少。

阳台农业是农业现代化发展过程中的一种新形态，它已不再单纯的局限于阳台，屋顶、露台、办公室等闲置空间也属于阳台农业的范畴（郭迪等，2013）。阳台农业一般采用无土栽培，打破了传统土壤耕作的局限性，可以美化绿化阳台环境，为家庭提供新鲜洁净蔬菜，深受消费者欢迎（陈瑞仙等，2015）。岩棉是一种理想的阳台栽培基质，岩棉是由天然的矿物高温融化压缩而成，其无毒无菌、物理特性稳定（周长吉，2007），孔隙率高，吸水保水性良好，且质量较轻（张学军等，2012）。在荷兰园艺生产中70%以上的无土栽培基质采用岩棉（丁小涛等，2015）。因此，本试验以生菜为试验材料，以岩棉为栽培基质，研究不同灌溉液位深度对阳台多层栽培生菜生长、产量和品质的影响，以期为生菜潮汐灌溉模式的建立及生菜潮汐式多层栽培在阳台农业上的应用奠定基础。

1 材料与方法

1.1 试验材料

供试生菜品种为罗生3号（北京京研益农科技发展中心）。栽培基质采用岩棉（荷兰grodan公司），长×宽×高=10cm×10cm×6.5cm，定植孔深3.5cm。

试验设备为3层式潮汐灌溉架，长×宽×高=185cm×35cm×180cm，下层距地面57.5cm，放置营养液罐，层间距为36.5cm，栽培槽长×宽×高=160cm×30cm×8.5cm，占地面积约0.65m^2。

1.2 试验方法

试验于2016年10—11月在北京农林科学院蔬菜研究中心阳台温室内进行。采用潮汐灌溉系统供给营养液，营养液配方选用园试配方1/2剂量，微量元素采用通用配方。

设3个灌溉液位深度（营养液浸没岩棉基质块的高度）：2cm、3cm、4cm，栽培架分上层、中层、下层，共9个处理（表1），每处理定植10株，3次重复。

表1　试验设计及编号

编号	灌溉液位深度（cm）	设备层次
A1	2	上层
A2	2	中层
A3	2	下层
B1	3	上层

（续表）

编号	灌溉液位深度（cm）	设备层次
B2	3	中层
B3	3	下层
C1	4	上层
C2	4	中层
C3	4	下层

10月2日浸种催芽后播种；10月29日幼苗4叶1心时选取长势一致的植株定植于岩棉块内，置于栽培槽内。营养液从7：00～19：00循环供给，灌溉间隔时间为1h，灌溉维持时间为3min。不同处理每次供液量分别为280、520、760ml·株$^{-1}$。生菜于11月28日采收。

1.3 项目测定

于生长期内使用LI-250A光照计（美国LI-COR公司）测定不同层间光合有效辐射强度，分别在生长前期、中期和后期选取晴天和阴天各1d进行测定，试验数据取3次测定平均值。

每处理选择长势相似的10株植株测定生长指标，株高用直尺测量，叶片数按完全展开叶片数量计算，单株产量用电子天平测定，地上部干质量采用烘干法烘干至恒重后用电子天平测定，叶绿素含量采用乙醇浸提法测定。每处理再随机选取3株，取地上部分整株匀浆后进行品质测定，3次重复，维生素C含量采用2，6-二氯靛酚滴定法测定，可溶性蛋白含量采用考马斯亮蓝法测定，硝态氮含量采用水杨酸法测定［王学奎等，2006（这几种都是采用的这本书上的方法）］。

1.4 数据处理

采用Microsoft Excel 2007软件进行数据整理和作图，运用SPSS 22.0软件进行方差分析，显著性由Duncan’s新复极差法检验。

2 结果与分析

2.1 试验期间光合有效辐射强度

由图1和图2可见，不同层间光合有效辐射强度有明显差异。晴天时，各层间随着时间推移光合有效辐射强度逐渐增强，上层在12:00时光合有效辐射强度最高，之后开始逐渐降低；中层和下层在10:00时光合有效辐射强度最高，之后由于上层遮光光合有效辐射逐渐下降，至12:00后又开始上升，14:00时光合有效辐射强度达到第二次峰值后再次下降，在10:00时上层分别为中层和下层光合有效辐射强度的0.63倍和1.54倍，午时12:00上层分别为中层和下层光合有效辐射强度的3.19倍和4.67倍。阴天时，随着时间的推移，光合有效辐射强度差异逐渐增强，在12:00时上层、中层和下层光合有效辐射强度均达到最高，此时上层分别为中层和下层光合有效辐射强度的1.30倍和2.97倍。

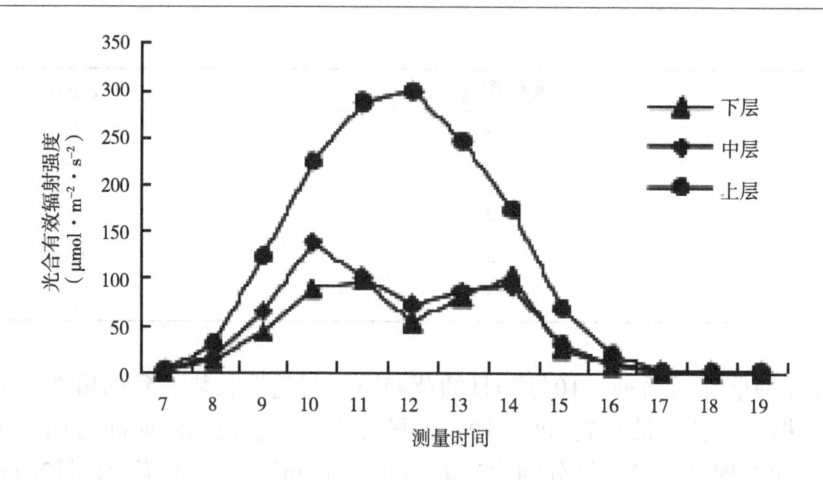

图 1　晴天光合有效辐射强度变化情况

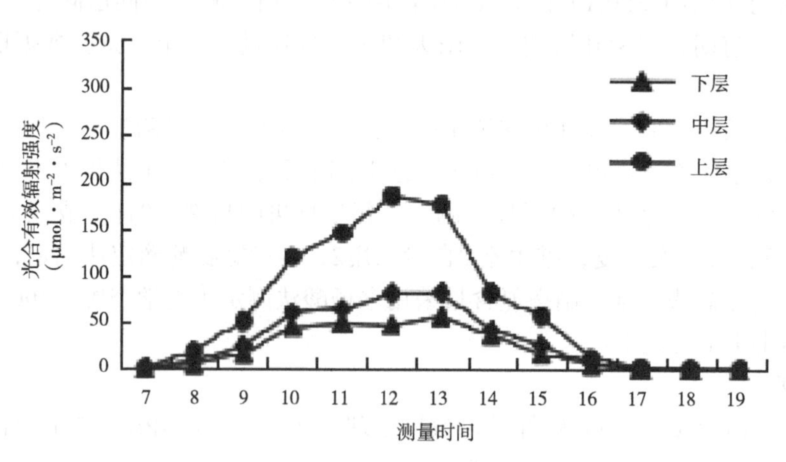

图 2　阴天光合有效辐射强度变化情况

2.2　潮汐灌溉液位深度对生菜的影响

2.2.1　潮汐灌溉液位深度对生菜生长的影响

从表2可以看出，不同潮汐灌溉液位深度对生菜株高有显著影响，B处理比A和C处理分别显著高出5.85%和6.74%；A处理的叶片数、地上部干质量显著高于B和C处理，A处理单株产量显著高于C处理，但与B处理间差异不显著；折合单个设备产量后各处理间差异显著，A处理比B和C处理分别显著增产6.13%和13.82%。

表 2　不同潮汐灌溉液位深度对生菜生长指标的影响

处理	株高/cm	叶片数	地上部干质量（g）	单株产量（g）	单个设备产量（kg）
A	17.96b	10.39a	1.08a	28.79a	1.73a
B	19.01a	10.11b	0.99b	27.23ab	1.63b
C	17.81b	10.00b	1.01b	25.36b	1.52c

注：A=（A1+A2+A3）/3，B=（B1+B2+B3）/3，C=（C1+C2+C3）/3

2.2.2　潮汐灌溉液位深度对生菜品质的影响

不同潮汐灌溉液位深度对生菜品质产生显著影响（表3）。A处理维生素C含量和可溶性蛋白含量最高，分别为0.41mg·kg^{-1}和5.43mg·g^{-1}，维生素C含量比B和C处理分别显著提高5.13%和7.89%；C处理硝态氮含量最低，为1 107.54mg·kg^{-1}，与A处理之间无显著差异。综合3项指标来看，A处理生菜品质最佳。

表3　不同潮汐灌溉液位深度对生菜品质指标的影响

处理	维生素C（mg·kg^{-1}）	可溶性蛋白（mg·g^{-1}）	硝态氮（μg·kg^{-1}）
A	0.41a	5.43a	1 183.96b
B	0.39b	5.34a	1 309.26a
C	0.38b	4.98b	1 107.54b

2.2.3　潮汐灌溉液位深度对生菜色素含量的影响

由表4可知，A处理生菜的叶绿素a、叶绿素b和胡萝卜素含量均为最高，其中叶绿素a、叶绿素b含量比B处理显著提高1.69%和3.38%；各处理间类胡箩卜素含量差异不显著。

表4　不同潮汐灌溉液位深度对生菜色素含量的影响

处理	叶绿素a（mg·g^{-1}）	叶绿素b（mg·g^{-1}）	类胡箩卜素（mg·g^{-1}）
A	5.43a	5.81a	1.51a
B	5.34b	5.62b	1.43a
C	5.42a	5.76a	1.50a

2.3　潮汐灌溉液位深度对多层栽培生菜的影响

2.3.1　潮汐灌溉液位深度对多层栽培生菜生长的影响

由表5可知，不同潮汐灌溉液位深度对多层栽培生菜株高没有显著影响，但对叶片数、地上部干质量和单株产量有显著影响。各处理不同层间单株产量差异显著，A1＞C1＞B1＞C2＞A2＞B2＞A3＞C3＞B3，表现为上层＞中层＞下层；A1处理分别比A2和A3处理显著增产69.38%和130.29%，并较最小值B3处理显著增产186.63%。地上部干质量各处理不同层间差异显著，A1处理分别比A2和A3处理显著增加82.75%和103.85%，并较最小值C3处理显著增加127.14%。

表5　不同潮汐灌溉液位深度对多层栽培生菜生长指标的影响

处理	株高（cm）	叶片数	单株产量（g）	地上部干质量（g）
A1	18.12a	11.92a	42.65a	1.59a
A2	18.03a	9.58cde	25.18e	0.87c
A3	17.73a	9.67cde	18.52g	0.78d
B1	19.50a	10.67bc	37.08c	1.53a
B2	18.93a	10.25bcd	21.08f	0.81d
B3	18.60a	9.08e	14.88i	0.58e

（续表）

处理	株高（cm）	叶片数	单株产量（g）	地上部干质量（g）
C1	17.85a	11.25ab	38.19b	1.44b
C2	17.68a	9.42de	26.66d	0.82cd
C3	17.90a	9.67cde	17.84h	0.70e

2.3.2 潮汐灌溉液位深度对多层栽培生菜品质的影响

不同灌溉液位深度和不同层栽培对生菜维生素C含量、可溶性蛋白含量和硝态氮含量具有显著影响（表6）。其中，维生素C含量和可溶性蛋白含量分别表现为A1>C1>B1>A2>B2>C2>A3>B3>C3和A1>B1>A2>B2>C1>C2>B3>C3>A3，呈现上层>下层>中层的趋势。硝态氮含量C1<C2<A1<C3<A2<A3<B1<B2<B3，总体以灌溉液位深度4cm<2cm<3cm。其中，A1比B1和C1的维生素C含量分别高出10.57%和1.95%，可溶性蛋白含量分别高出7.41%和19.8%；A1比A2和A3的维生素C含量分别高出34.79%和62.42%，可溶性蛋白含量分别高出14.73%和34.85%，品质综合表现为同层间差异小于不同层间。

表 6　不同汐灌溉液位深度对多层栽培生菜品质指标的影响

处理	维生素C（mg·kg^{-1}）	可溶性蛋白（mg·g^{-1}）	硝态氮（μg·kg^{-1}）
A1	0.523a	6.23a	1 100.00d
A2	0.388d	5.43c	1 188.54c
A3	0.322g	4.62i	1 263.33bc
B1	0.473c	5.80b	1 295.97b
B2	0.376e	5.35d	1 297.92b
B3	0.316h	4.86g	1 333.89a
C1	0.513b	5.20e	1 063.19d
C2	0.356f	4.92f	1 083.33d
C3	0.285i	4.81h	1 176.11c

2.3.3 潮汐灌溉液位深度对多层栽培生菜色素含量的影响

从表7可以看出，不同潮汐灌溉液位深度对多层栽培生菜色素含量有显著影响，叶绿素a、叶绿素b含量表现为A1>C1>C3>B1>A3>C2>B3>A2>B2，类胡萝卜素含量表现为A1>C1>A3>B1>C3>B3>C2>A2>B2，总体表现为上层>下层>中层；其中A1处理的叶绿素a、叶绿素b和类胡萝卜素含量均为最高，比B2处理分别显著提高33.38%、34.87%和19.97%。

表 7　不同潮汐灌溉液位深度对多层栽培叶生菜色素含量的影响

处理	叶绿素a（mg·g^{-1}）	叶绿素b（mg·g^{-1}）	类胡萝卜素（mg·g^{-1}）
A1	6.143 1a	6.635 5a	1.621 4a
A2	4.831 4h	5.173 7h	1.404 6h

（续表）

处理	叶绿素a（mg·g^{-1}）	叶绿素b（mg·g^{-1}）	类胡萝卜素（mg·g^{-1}）
A3	5.315 1e	5.622 1e	1.514 4c
B1	5.426 6d	5.692 2d	1.510 2d
B2	4.605 7i	4.920 0i	1.351 5i
B3	4.994 9g	5.262 0g	1.431 9f
C1	5.683 9b	6.034 5b	1.597 3b
C2	5.069 9f	5.384 6f	1.412 0g
C3	5.498 4c	5.868 3c	1.499 3e

2.3.4 潮汐灌溉液位深度对多层栽培生菜相关性状的方差分析

由表8可知，在不同的栽培层和灌溉液位深度两个因素对生菜生长和品质的影响效果不同。不同的灌溉液位深度极显著的影响了生菜的株高和硝态氮含量，而叶片数、地上部干质量、单株产量、维生素C含量、可溶性蛋白含量和叶绿素含量受不同栽培层的影响较大，均达到极显著水平。说明不同栽培层间比不同灌溉液位深度对生菜生长和品质的形成影响不同，且不同栽培层影响更大。

表 8　潮汐灌溉液位深度对多层栽培生菜相关性状的方差分析结果

变异来源	F值							
	株高	叶片数	地上部干质量	单株产量	维生素C含量	可溶性蛋白含量	硝态氮含量	叶绿素含量
栽培层	2.06	9.79*	398.86*	153.22*	68.66*	7.64*	7.35	13.07*
灌溉液位深度	19.84*	0.47	4.79	6.26	1.30	1.80	27.02*	4.10

注：*为测验达到5%显著

3　结论与讨论

灌溉方式对作物根系密度和产量有较很大影响（刘志刚等，2014），Wilson等（2003）研究得出潮汐灌溉条件下水肥利用率提高、生物量积累较多。高艳明等（2016）研究表明，在灌溉频率和灌溉时间相同的情况下，潮汐灌溉液位深度会直接影响基质含水率。Dalvi等（1999）、杨素苗等（2010）研究表明，适合的基质含水率有利于作物生长，可以显著促进作物根系生长、干物质积累和作物品质。当基质含水率较高时，基质会长期处于水分饱和状态，基质通气孔隙数量减少，通气不良，对根系造成低氧胁迫（裴云等，2015）。本试验中随着灌溉液位深度的增加，基质内水分和养分逐渐增多，通气孔隙减少。灌溉液位深度过高时，基质含水量接近饱和，使生菜根系长期浸泡在营养液内，造成的低氧胁迫抑制了生菜根系的生长，降低养分和水分的吸收量，导致产量和品质下降。所以本试验条件下灌溉液位深度2cm时有利于生菜生长和品质形成，综合品质最优、产量最高。

本试验结果表明：灌溉液位深度对多层栽培生菜的生长、产量及品质有显著影响，在灌溉间隔时间1h，灌溉维持时间3min的试验条件下，灌溉液位深度为2cm时生菜生长、产量和品质最优，是较适宜生菜生长的潮汐灌溉液位深度。

参考文献

陈瑞仙，翟云霞，解莉莉，等. 2015.阳台农业的推广应用［J］.农业工程，（5）：81-82.

丁小涛，周强，何立中，等. 2015.岩棉栽培青菜潮汐式灌溉技术［J］.中国瓜菜，28（5）：54-55.

郭迪，王晨静，陆国权.2013.我国阳台农业概况及发展前景［J］.浙江农业科学，（3）：239-241.

刘宏久，高艳明，沈富，等. 2015.番茄穴盘育苗潮汐灌溉技术研究［J］.安徽农业大学学报，42（4）：549-554.

刘志刚，王纪章，徐云峰，等. 2014.基质配方和灌溉方式对生菜根系和产量的影响［J］.农业机械学报，45（2）：156-159.

裴芸，张保才，别之龙.2015.不同土壤水分含量对生菜生长和光合特性的影响［J］.西南农业学报，28（3）：1 042-1 046.

王学奎、章文华、郝再彬、等. 2006.植物生理生化实验原理和技术（第2版）［M］.北京：高等教育出版社.

王正，武占会，刘明池，等. 2015.潮汐灌溉营养液供应时间对番茄穴盘苗质量的影响［J］.中国蔬菜，（11）：46-51.

杨素苗，李保国，齐国辉，等. 2010.根系分区交替灌溉对苹果根系活力、树干液流和果实的影响［J］.农业工程学报，26（8）：73-79.

张黎，王勇.2011.盆栽八仙花潮汐灌溉栽培试验初探［J］.北方园艺，（20）：77-79.

张学军，段静，张月红，等. 2012.温室床式潮汐灌技术及其设计方法的研究［J］.节水灌溉，（2）：33-37.

周长吉.2007.温室灌溉原理与应用［M］.北京：中国农业出版社.

Dalvi V B，Tiwarib K N，Pawade M N.1999.Response surface analysis of tomato production under microirrigation［J］.Agricultural Water Management，41：11-19.

Wilson S B，Stoffella P J，Greatz D A.2003.Compost amended media and irrigation system influence containerized perennial salvia［J］. Journal of the American Society for Horticultural Science，128（2）：260-268.

分区滴灌施氮对日光温室黄瓜产量及水氮利用效率的影响

张文东[1]，李时雨[1]，艾希珍[1, 2, 4]，魏珉[1, 2, 3, 4]，刘彬彬[2]，李清明[1, 2, 3, 41*]

（1. 山东农业大学园艺科学与工程学院，2. 作物生物学国家重点实验室，3. 农业部黄淮海设施农业工程科学观测实验站，4. 山东果蔬优质高效生产协同创新中心，泰安　271018）

摘要：为探明分区滴灌施氮对日光温室黄瓜产量及水氮利用效率的影响，本试验以‘津园578’黄瓜（*Cucumis sativus* L.）为试材，采用裂区试验设计，主区因素为滴灌方式，设分区滴灌（Partitive-area drip irrigation，P）和常规滴灌（Conventional drip irrigation，C）2种滴灌方式；副区因素为施氮量，设0（不施氮）、1（320kg N/hm^2）和2（640kg N/hm^2）3个水平，共P0、P1、P2、C0、C1、C2 6个处理，每个处理重复3次，完全随机区组排列。结果表明：①P2处理叶面积最大，较C2的增大35.5%，低氮处理下，变化趋势相反，P1较C1减小18.4%；②P1的产量较C1提高12.9%，P2的产量较C2提高11.1%；P1的水分利用效率较C1提高16.6%，P2的水分利用效率较C2提高14.7%；P1的氮素利用效率较C1提高45.4%，P2的氮素利用效率较C2提高215.7%（2.157倍）；P1与P2间产量和水分利用效率差异不显著，但P1较P2的氮素利用效率提高44.1%。综上所述，分区滴灌减施氮处理能在显著提高黄瓜产量的同时，水氮利用效率显著提升，可作为日光温室黄瓜“节水减肥”栽培的水氮管理模式。

关键词：黄瓜；分区滴灌施氮；产量；水氮利用效率

Effects of Partitive-area Drip Irrigation and Nitrogen Fertilization on Yield and Water-nitrogen Use Efficiency of Cucumber in Solar Greenhouse

Zhang Wendong[1]，Li Shiyu[1]，Ai Xizhen[1, 2, 4]，Wei Min[1, 2, 3, 4]，Liu Binbin[2*]，Li Qingming[1, 2, 3, 4*]

（1. *College of Horticulture Science and Engineering*，*Shandong Agricultural University*，*Taian*　271018；2. *State Key Laboratory of Crop Biology*，*Taian*　271018；3. *Scientific Observing and Experimental Station of Environment Controlled Agricultural Engineering in Huang-Huai-Hai Region*，*Ministry of Agriculture*，*Taian*　271018；4. *Collaborative Innovation Center of Fruit & Vegetable Quality and Efficient Production in Shandong*，*Taian*　271018）

Abstract：In order to investigate the effects of partitive-area drip irrigation and nitrogen fertilization on yield and water-nitrogen use efficiency of cucumber（Jinyuan 578）in solar greenhous， the experiment utilized the split-plot design with drip irrigation as main plots and nitrogen as subplots. The main plots was drip irrigation（Partitive-area drip irrigation，P，and Conventional drip irrigation，C），the subplots was the amount of nitrogen（0 level: 0 kg N/hm^2，1 level: 320 kg N/hm^2 and 2 level: 640 kg N/hm^2）. So there were six treatments in total，designated P0，P1，P2，C0，C1，and C2，respectively. The results showed that:①The leaf area of P2 was the largest，which

项目基金：国家自然科学基金（31471918）、山东省农业重大应用技术创新项目（鲁财农指［2016］36号）、山东省重点研发计划（2017CXGC0201）。

通讯作者：E-mail：gslqm@sdau.edu.cn

was 35.5% higher than that of C2, under low nitrogen treatments, the trend was opposite, P1 reduced 18.4%. compared with C1; ②The yield of P1 increased by 12.9% compared with C1, and the yield of P2 increased by 11.1% compared with C2; the WUE of P1 was increased by 16.6% compared with C1, the WUE of P2 increased by 14.7% compared with C2; the ANUE of P1 increased by 45.4% compared with C1, the ANUE of P2 increased by 215.7% compared with C2; there was no significant difference in yield and WUE between P1 and P2, while the ANUE of P1 increased by 44.1% compared with P2. In conclusion, P1 can significantly increase the yield and improve the use efficiency of water and nitrogen, so P1 can be used as an effective water-nitrogen management mode in solar greenhouse to achieve the goal of "water-saving and reducing fertilizer" cultivation.

Key word: Cucumber; Partitive-area drip irrigation and nitrogen fertilization; Yield; Water-nitrogen use efficiency

1 前言

2015年，农业部制定出台了《农业部关于打好农业面源污染防治攻坚战的实施意见》，明确要求到2020年，实现“一控两减三基本”的目标，明确指出要提高水分利用效率和实现化肥使用量的零增长。近年来我国设施蔬菜产业为追求经济效益，盲目的增加灌溉量和化肥用量，带来了蔬菜品质下降、土壤次生盐渍化、能源浪费等一系列问题[1]。

分区滴灌作为一种高效节水滴灌方式，源于控制性交替灌溉[2]，这种非充分灌溉方式能给予作物根系适当干旱刺激，一是调节植株气孔部分开闭，降低蒸腾耗水，提高水分利用效率[3-5]，减少冗余生长[6]；二是促进植株根系的生长，增大根际范围，提高植株对水肥的利用效率[7, 8]，已在苹果[9]、葡萄[10, 11]、番茄[12]、胡萝卜[13]、棉花[14]等多种植物上进行了试验与探索，但黄瓜（*Cucumis sativus* L.）果实含水量高，栽培中对水分依赖性强，上述节水理论是否可行还未见报道。氮肥作为植物必需养分，是蛋白质、叶绿素、酶等物质的组成部分，是生物量和产量形成的关键因素，前人以尿素作为氮源在水氮耦合方面做了大量研究[15-17]，但这与日光温室黄瓜栽培中大量使用硝态氮化肥的情况不符，以硝态氮作为氮源的水氮耦合研究也少见报道，因此，本试验以硝酸钙为氮源，研究分区滴灌施氮对日光温室黄瓜产量及水氮利用效率的影响，以期为日光温室黄瓜“节水减肥”和提质增效栽培提供技术参数和理论依据。

2 材料与方法

2.1 试验材料与设计

试验于山东农业大学科技创新园3号日光温室进行，试验材料为“津园578”（天津科润黄瓜研究所），2017年2月14日浸种催芽，3月9日定植，定植前1个月每小区施入0.5m^3腐熟鸡粪作为基肥，7月27日拉秧。试验田基础地力为：碱解氮170.6mg/kg，速效磷65.8mg/kg，速效钾606.4mg/kg，有机质38.1g/kg。采用裂区试验设计，主区因素为滴灌方式，设分区滴灌（Partitive-area drip irrigation，P）和常规滴灌（Conventional drip irrigation，C）两种滴灌方式；副区因素为施氮量，设0（不施氮）、1（320kg N/hm^2）和2（640kg N/hm^2）3个水平，共P0、P1、P2、C0、C1、C2 6个处理，每个处理重复3次，完全随机区组排列。

试验小区南北长7m，东西宽1.2m，其中栽培畦宽0.7m，过道宽0.5m，黄瓜采取双行三角形交错栽培模式，小行距30cm，大行距120cm，株距30cm，为防止水肥侧渗，每小区间预埋塑料薄膜，深度50cm。采取膜下滴灌方式，分区滴灌处理在栽培畦内布置三条滴灌带，中间滴灌带滴头间距为15cm，布置在栽培畦中间位置，两侧滴灌带滴头间距为30cm，布置在中间滴灌带左右两侧，与中间滴灌带间距为20cm；常规滴灌处理在栽培畦内布置两条滴灌带，滴灌带滴头间距为30cm，两条滴灌带到栽培畦两侧距离均为15cm；所有滴灌带滴头流量均为2.7L/h，最大工作压力均为0.25Mpa。分区滴灌灌溉时交替使用两侧两条滴灌带或中间一条滴灌带，常规滴灌灌溉时同时使用两根滴灌带。滴灌带滴头采取三角形交错方式布置，为保证分区滴灌的三根滴灌带均为三角形交错方式布置，中间滴灌带的滴头间距为15cm，使得在灌水量相同的情况下三条滴灌带滴头间的湿润锋重叠最小，最大程度的减少水分纵向运移过快造成的渗漏。浇定植水后，待黄瓜长到三叶一心开始水分处理。具体小区布置情况如图1所示。

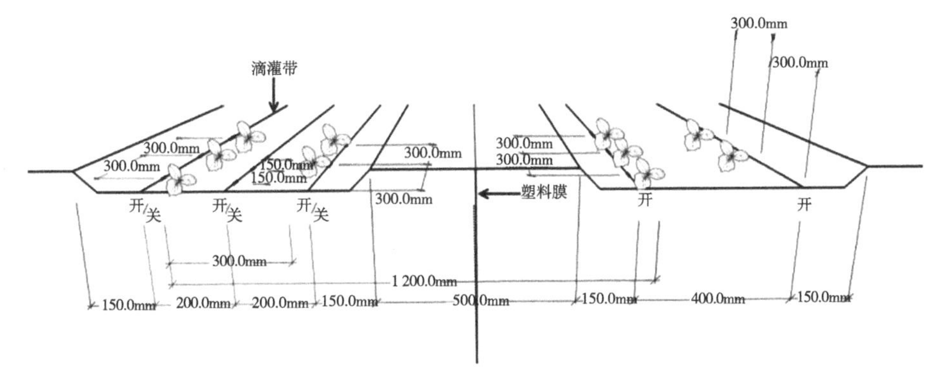

图1　滴灌方式示意

灌水量依公式$V=r\times p\times S\times h\times \theta_f\times (q_1-q_2)/\eta$计算，式中r为土壤容重，为1.15g/cm^3；p为土壤湿润比，取100%；S为滴灌润湿面积，取1.6m^2；h为灌水计划湿润层，取0.4m；θ_f为田间持水量，以土壤质量含水量表示，为40%；q_1、q_2分别为土壤水分上限（为田间持水量的90%）、土壤水分下限（为田间持水量的65%）；η为水分利用系数，滴灌取0.95[18]。灌水量由精度0.1L机械水表控制。

到黄瓜结瓜初期，使用自制肥料注入系统将氮肥分3次，每次间隔5d，从滴灌系统施入相应试验小区。

施氮量依公式m=计算，式中计划产量吸肥量为2.6kg N/1 000kg黄瓜，计划产量取10 000kg；Ns为土壤碱解氮测试值；r为养分校正系数，取0.6；氮肥种类为硝酸钙，肥料有效养分含量取15%，肥料利用率1处理取50%，2处理取25%[19]。

2.2　测定项目与方法

测定黄瓜生长指标。每处理选取10株，株高用皮尺测定，为黄瓜茎基部到生长点的绝对高度；茎粗用数显千分尺测定，选取黄瓜植株第4至第5节节间中点处为测量点；单株叶面积为第1叶起至初生展平叶叶面积之和，单叶叶面积为叶长的平方。

每小区选取10株黄瓜统计单株产量，并折算亩产量。水分利用效率（WUE）采用产量（kg）与灌溉量（m^3）比值表示[20]。氮肥农学利用效率（ANUE）采用单位施氮量所增加产量表示[21]，ANUE=（施氮处理产量-对照区产量）/施氮量。

采用HOBO公司Pro2型数采器连续测定日光温室内的温度和空气相对湿度，以日平均温度和日平均空气相对湿度作图，结果如图2所示。

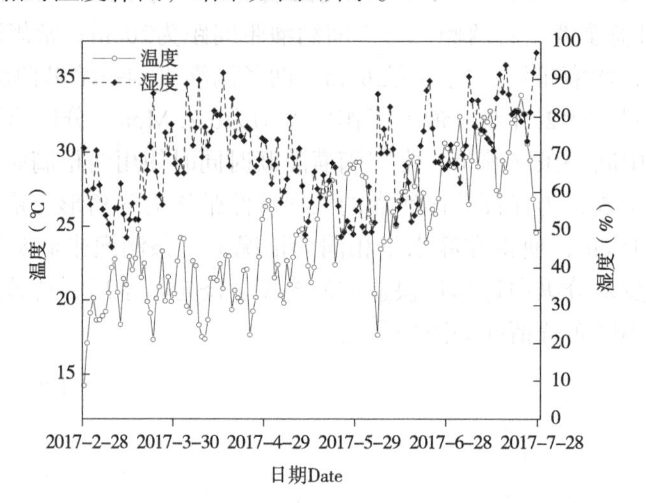

图2　日光温室内环境条件

2.3　数据处理方法

采用Excel 2010整理数据，DPS对数据进行显著性分析（显著性差异$P<0.05$），OriginPro 2017作图。

3　实验结果

3.1　分区滴灌施氮对日光温室黄瓜形态指标的影响

由表1可知，分区滴灌及施氮量的处理对黄瓜株高和茎粗的影响不显著；分区滴灌及施氮量的处理对黄瓜叶面积和叶片数的影响显著，P2处理叶面积最大，较C2的增大35.5%，低氮处理下，变化趋势相反，P1较C1减小18.4%，P2处理的叶片数最多，其他各处理间差异不显著。

表1　分区滴灌施氮对日光温室黄瓜形态指标的影响

处理		株高（cm）	茎粗（mm）	叶面积（m^2）	叶片数
	0	89.30 ± 4.41a	6.67 ± 0.22a	0.33 ± 0.01c	10.40 ± 0.49ab
P	1	87.33 ± 3.16a	6.63 ± 0.26a	0.31 ± 0.03c	10.20 ± 0.40b
	2	90.60 ± 9.09a	6.71 ± 0.46a	0.42 ± 0.03a	11.20 ± 0.75a
	0	89.70 ± 5.57a	6.54 ± 0.30a	0.31 ± 0.06c	10.20 ± 1.17b
C	1	94.00 ± 7.64a	6.53 ± 0.37a	0.38 ± 0.02b	11.00 ± 1.10ab
	2	89.30 ± 4.05a	6.63 ± 0.36a	0.31 ± 0.03c	10.40 ± 0.49ab

同一指标数据用不同小写字母标识表示差异显著（$P<0.05$），下表同

3.2 分区滴灌施氮对日光温室黄瓜产量及水氮利用效率的影响

由表2可知，分区滴灌施氮可显著提高黄瓜的产量，P1的产量较C1显著提高，提高的比例为12.9%，P2的产量较C2显著提高，提高的比例为11.1%，但两种滴灌方式下施氮量的增加对产量的贡献不明显。分区滴灌施氮可显著提高黄瓜的水分利用效率，P0的水分利用效率较C0显著提高，提高的比例为13.8%，P1的水分利用效率较C1显著提高，提高的比例为16.6%，P2的水分利用效率较C2显著提高，提高的比例为14.7%。P1的氮素利用效率最高，分区滴灌施氮方式可显著提高黄瓜的氮素利用效率，P1的氮素利用效率较C1显著提高，提高的比例为45.4%，P2的氮素利用效率较C2显著提高，提高的比例为215.7%；在同一种滴灌方式下，氮素利用效率随施氮量的增加而显著降低。

表 2　分区滴灌施氮对日光温室黄瓜产量及水氮利用效率的影响

处理		产量（$kg \cdot 667m^{-2}$）	灌溉量（mm）	水分利用效率（$kg \cdot m^3$）	氮素利用效率（%）
	0	13 420.09 ± 843.45bc	407.14	49.4 ± 3.10bc	—
P	1	14 505.82 ± 417.39ab	407.14	53.4 ± 1.53ab	116.0 ± 6.86a
	2	14 833.12 ± 1 406.26a	407.14	54.6 ± 5.17a	80.5 ± 21.8b
	0	12 180.19 ± 2 024.47c	419.97	43.4 ± 7.22d	—
C	1	12 847.49 ± 1 114.01c	419.97	45.8 ± 3.97cd	79.8 ± 19.1b
	2	13 351.02 ± 422.41bc	419.97	47.6 ± 1.50cd	25.5 ± 3.78c

4　讨论与结论

控制性交替灌溉多应用在果树上[9, 10]，近年在我国北方日光温室蔬菜栽培上也有报道[22]，但产量出现下降。这与本试验的结果不同，本试验中所有处理的产量都超过了预设的目标产量10 000$kg \cdot 667m^{-2}$，C0产量最低，但也超出目标产量21.8%；P2产量最高，超出目标产量48.3%，这也与其最高的叶面积和叶片数相一致；且分区滴灌较常规滴灌处理黄瓜产量显著提高，分析原因，一是试验开始前施入的有机肥贡献了部分肥力，二是分区滴灌方式有利于土壤矿化作用速率提高，释放了土壤中难以利用的肥力；三是黄瓜果实的含水率超过96%[23]，相比于叶菜类或茄果类蔬菜，水分对于黄瓜产量的贡献要远远大于氮肥，因此供给充足的水分是提高黄瓜产量的第一步，同时提供充足的肥料是第二步，这也是生产上普遍采取大水大肥的栽培模式的原因，但盲目的投入水肥不仅造成浪费还对环境造成破坏，由表2可以看到，采取分区滴灌的方式，可以在原有灌溉量和施氮量的基础上，大幅提高产量，那分区滴灌方式下能否进一步减少灌溉量以提高水分利用效率呢？

交替滴灌技术在改进灌溉方式的同时减少灌溉量，造成产量降低的现象[16]，所以交替滴灌策略如何制定才能做到在减少蒸腾耗水的同时不影响或提高产量，还缺乏证据。黄瓜为无限生长型植物，而且生产上不同于番茄、茄子、辣椒等蔬菜采取摘心的措施控制植株营养生长，黄瓜具有营养生长与生殖生长同期的特点，叶片与果实的生长同时进行，因此，盲目的采取控制性灌水策略，在降低黄瓜叶片蒸腾耗水的同时，也会降低蒸腾拉力，影响果实的膨大，造成产量的降低。本试验设计采取的分区滴灌方式与常规滴灌相比，只是滴灌方式的改变，从表2中看到两者并无灌溉量的太大差异，严格的保证滴灌方式与氮素两因素试验设计，避免出现变量增加（灌溉量不同）的情况，至于在分区滴灌施氮方式下进一步减少灌溉量，在哪个生长阶段减少灌溉量，对黄瓜生理及生长特性造成何种影响，可作为下一步研究方向。综上所述，分区滴灌减施氮处理能在显著提高黄瓜产量的同时，水氮利用效率显著提升，可作为日光温室黄瓜“节水减肥”栽培的水氮管理模式。

参考文献

[1] 史静，张乃明，包立.我国设施农业土壤质量退化特征与调控研究进展［J］.中国生态农业学报，2013，21（7）：787-794.

[2] 康绍忠，张建华，梁宗锁，等.控制性交替灌溉——一种新的农田节水调控思路［J］.干旱地区农业研究，1997，15（1）：4-9.

[3] 李平，齐学斌，樊向阳，等.分根区交替灌溉对马铃薯水氮利用效率的影响［J］.农业工程学报，2009，25（6）：92-95.

[4] 梁继华，李伏生，唐梅，等.分根区交替灌溉对盆栽甜玉米水分及氮素利用的影响［J］.农业工程学报，2006，22（10）：68-72.

[5] 刘永贤，李伏生，农梦玲.烤烟不同生育时期分根区交替灌溉的节水调质效应［J］.农业工程学报，2009，25（1）：16-20.

[6] 孙华银，康绍忠，胡笑涛，等.根系分区交替灌溉对温室甜椒不同灌水下限的响应［J］.农业工程学报，2008（6）：78-84.

[7] 孟兆江，孙景生，段爱旺，等.调亏灌溉条件下冬小麦籽粒灌浆特征及其模拟模型［J］.农业工程学报，2010，26（1）：18-23.

[8] Affi N，Fadl AE，Otmani ME，et al. Does partial rootzone drying alternation frequency enhance water stress resistance and improve water saving? ［J］. Journal of Materials and Environmental Science，2013，4（3）：468-473.

[9] 杨启良，张富仓，刘小刚，等.控制性分根区交替滴灌对苹果幼树形态特征与根系水分传导的影响［J］.应用生态学报，2012，23（5）：1 233-1 239.

[10] 周青云，康绍忠.葡萄根系分区交替滴灌的土壤水分动态模拟［J］.水利学报，2007（10）：1 245-1 252.

[11] Munitz S，Netzer Y，Schwartz A. Sustained and regulated deficit irrigation of field - grown merlot grapevines［J］. Australian Journal of Grape and Wine Research，2017，23（1）：87-94.

[12] 张强，徐飞，王荣富，等.控制性分根交替灌溉下氮形态对番茄生长、果实产量及品质的影响［J］.应用生态学报，2014，25（12）：3 547-3 555.

[13] Léllis BC，Carvalho DF，Martínez-Romero A，et al. Effective management of irrigation water for carrot under constant and optimized regulated deficit irrigation in brazil［J］. Agricultural Water Management，2017，192：294-305.

[14] 杜太生，康绍忠，胡笑涛，等.根系分区交替滴灌对棉花产量和水分利用效率的影响［J］.中国农业科学，2005，38（10）：2 061-2 068.

[15] 李培岭，张富仓.膜下分区交替滴灌和施氮对棉花干物质累积与氮肥利用的影响［J］.应用生态学报，2013，24（2）：416-422.

[16] 刘学娜，刘彬彬，崔青青，等.交替滴灌施氮对日光温室黄瓜生长、光合特性、产量及水氮利用效率的影响［J］.植物生理学报，2016，52（6）：905-916.

[17] 岳文俊，张富仓，李志军，等.日光温室甜瓜根系生长及单果重的水氮耦合效应［J］.中国农业科学，2015，48（10）：1 996-2 006.

[18] 李清明，邹志荣，郭晓冬，等.不同灌溉上限对温室黄瓜初花期生长动态、产量及品质的影响［J］.西北农林科技大学学报（自然科学版），2005，33（4）：47-51+56.

［19］张福墁.设施园艺学［M］.北京：中国农业大学出版社，2002.

［20］Cabello MJ，Castellanos MT，Romojaro F，et al. Yield and quality of melon grown under different irrigation and nitrogen rates［J］. Agricultural Water Management，2009，96（5）：866-874.

［21］Fageria NK，Baligar VC. Methodology for evaluation of lowland rice genotypes for nitrogen use efficiency［J］. Journal of Plant Nutrition，2003，26（6）：1 315-1 333.

［22］Yang H，Du TS，Qiu RJ，et al. Improved water use efficiency and fruit quality of greenhouse crops under regulated deficit irrigation in northwest china［J］. Agricultural Water Management，2017，179：193-204.

［23］李邵，薛绪掌，郭文善，等.供水吸力对温室盆栽黄瓜产量与品质的影响［J］.园艺学报，2010，37（8）：1 339-1 344.

盐胁迫下CO_2加富对黄瓜幼苗叶片光合特性及活性氧代谢的影响

厉书豪[1]，李曼[1]，张文东[1]，李仪曼[1]，艾希珍[1, 2]，刘彬彬[2]，李清明[1, 2, 3]*

（1. 山东农业大学园艺科学与工程学院，山东　泰安　271018；2. 作物生物学国家重点实验室，泰安　271018；3. 农业部黄淮海设施农业工程科学观测实验站，泰安　271018）

摘要：以'津优35号'黄瓜为试材。采用裂区设计，主区因素为CO_2浓度处理，设2个CO_2浓度水平：大气CO_2浓度（400mmol·mol^{-1}）和CO_2加富（800 ± 40mmol·mol^{-1}）。裂区因素为盐分处理，用NaCl模拟盐胁迫，设对照（0mmol·L^{-1}NaCl）、盐处理（80mmol·L^{-1}NaCl）2个盐分水平。研究了盐胁迫下CO_2加富对黄瓜幼苗生长、光合特性及活性氧代谢的影响。结果表明：CO_2加富显著提高了盐胁迫下黄瓜幼苗的株高、茎粗、叶面积及地上部鲜重，降低了叶绿素a、叶绿素b、类胡萝卜素及叶绿素（a+b）含量，但显著提高了净光合速率，同时降低了气孔导度及蒸腾速率，并且使其具有较高的表观电子传递速率及PSⅡ实际光化学效率；CO_2加富显著提高了盐胁迫下黄瓜幼苗叶片脯氨酸含量及SOD、POD、CAT活性，丙二醛、过氧化氢和超氧阴离子含量显著降低。综上所述，CO_2加富可通过提高幼苗叶片净光合速率、脯氨酸含量及抗氧化酶活性，降低蒸腾速率、减少丙二醛含量及活性氧的积累，从而缓解盐胁迫对黄瓜植株造成的伤害。

关键词：黄瓜；盐胁迫；CO_2加富；光合特性；活性氧代谢

Effects of CO_2 Enrichment on Photosynthetic Characteristics and Reactive oxygen Species Metabo lism of Cucumber Seedlings Leaves Under Salt Stress

Li Shuhao[1]，Li Man[1]，Zhang Wendong[1]，LI Yiman[1]，AI Xizhen[1, 2]，LIU Binbin[2]，LI Qingming[1, 2, 3]*

（1. *College of Horticulture Science and Engineering*，*Shandong Agricultural University*，*Tai'an*　271018；2. *State Key Laboratory of Crop Biology*，*Tai'an*　271018；3. *Scientific Observing and Experimental Station of Environment Controlled Agricultural Engineering in Huang-Huai-Hai Region*，*Ministry of Agriculture*，*Tai'an*　271018）

abstract：The effects of CO_2 enrichment on photosynthetic characteristics and reactive oxygen species metabolism of cucumber（*cucumis sativus* 'Jinyou No.35'）seedlings leaves under salt stress were investigated. Using split plot design，the main treatment had two CO_2 concentrations levels（ambient［CO_2］≈400 mmol·mol^{-1} and enrichment［CO_2］= 800 ± 40 mmol·mol^{-1}），and the subplot had two levels of salinity treatment（0 mmol·L^{-1} NaCl and 80 mmol·L^{-1} NaCl）. The results showed that Elevated［CO_2］increased the plant height，stem thickness，leaf area and shoot fresh weight significantly of cucumber seedlings under salt stress. Elevated［CO_2］

资助：国家自然科学基金项目（31471918）、山东省农业重大应用技术创新项目（鲁财农指［2016］36号）。通讯作者：gslqm@sdau.edu.cn decreased the contents of chlorophyll a, chlorophyll b, carotinoid and chlorophyll（a+b）, but markedly increased rate of photosynthesis, *ETR*, *Φ*PS Ⅱ and activities of SOD, POD, CAT, and proline content. However, Elevated ［CO_2］ reduced stomatal conductance, transpiration rate and contents of MDA, H_2O_2 and rate of $O_2^{\cdot-}$ production of cucumber seedlings leaves under salt stress. Therefore, elevated ［CO_2］ alleviated the effect of salt stress on the growth of cucumber plants through an enhanced rate of photosynthesis, proline content and the antioxidative enzyme activities, meanwhile reduced rate of transpiration, MDA content and the accumulation of reactive oxygen species.

key words: Cucumber; Salt stress; CO_2 enrichment; Photosynthetic characteristics; Reactive oxygen species metabolism

土壤盐渍化是一个全球性的问题，盐胁迫已经成为制约作物生长的主要非生物胁迫之一（Suleyman等，2000）。研究显示，全国各类盐渍土总面积约9 913.3万hm^2，约占全国土壤总面积的13.4%（李彬等，2005）。黄瓜是我国栽培面积最大的蔬菜作物之一，对盐分十分敏感。盐胁迫抑制黄瓜幼苗植株的生长、降低叶片净光合速率（王素平等，2006）。CO_2是绿色植物光合作用的原料，其浓度高低直接影响光合速率大小。目前大气CO_2浓度约为400mmol·mol^{-1}，预计2100年将上升至730～1 020mmol·mol^{-1}（杨连新等，2010）。对C_3植物来说，最适CO_2浓度约为1 000mmol·mol^{-1}，在现有条件下其光合速率受到限制，因此大气CO_2浓度升高可提高其净光合速率。高CO_2浓度提高了高温胁迫下黄瓜叶片净光合速率（潘璐等，2014），并可提高植株水分利用效率（Melgar等，2008）、促进生物量积累（Pérez-López等，2009a）及碳水化合物合成（李曼等，2017），具有显著的“CO_2施肥效应”。前人研究结果表明，高CO_2浓度可显著促进盐胁迫下花椰菜（Zaghdoud等，2016）和甜椒（Piñero等，2014）植株生长，提高小麦叶片脯氨酸含量（刘家尧等，1998）、大麦叶片抗氧化酶活性（Pérez-López等，2009b），从而对盐胁迫造成的损伤具有一定缓解作用。但盐胁迫下高CO_2浓度对黄瓜幼苗叶片光合特性及活性氧代谢的影响，鲜见报道。本文以黄瓜幼苗为研究对象，在CO_2加富和盐胁迫条件下，对其生长、光合特性和活性氧代谢进行了研究，以期为未来大气CO_2浓度升高和盐胁迫等非生物胁迫下黄瓜的优质高效生产提供理论依据和技术参数。

1 材料与方法

1.1 试验材料与试验设计

试验于2017年3—5月在山东农业大学园艺实验站进行，以“津优35号”黄瓜（*cucumis sativus*）为试材。种子经浸种催芽后，播于穴盘中育苗，基质以草炭、蛭石、珍珠岩按3：1：1比例配制。待幼苗第1片真叶完全展开时，选择整齐一致的幼苗定植于内径为37.5cm×29cm、高为12cm的涂黑的塑料盆中进行水培，每盆6株。营养液为1个剂量的山崎黄瓜营养液配方，试验期间，营养液用气泵通气，每3min通气30S，4d更换1次营养液。

试验采用裂区设计，主区因素为CO_2浓度处理，设2个CO_2浓度水平：大气CO_2浓度（400mmol·mol^{-1}）和CO_2加富（800±40mmol·mol^{-1}）。裂区因素为盐分处理，用NaCl模拟盐胁迫，设对照（0mmol·L^{-1}NaCl）、盐处理（80mmol·L^{-1}NaCl）2个盐分水平。每处理重

复10次，每个重复1盆，待幼苗长至2叶1心时进行处理。试验在自行设计的顶通风式塑料拱棚（长6m，宽6m；脊高2.6m）内进行，棚内装有环境控制系统，加富处理的CO_2由液态CO_2气体钢瓶供给。其浓度由自动控制系统控制（Auto 2000，北京奥托），当监测系统监测到CO_2浓度低于设置的下限值时，开启电磁阀补气，使其浓度保持在800 ± 40mmol·mol^{-1}。

1.2 测定指标与方法

处理第8d，取从下往上数第3片功能叶测定各项生理指标，重复3次。

株高：用直尺测定从子叶下方到生长点的距离；茎粗：用游标卡尺测定子叶下方基部的直径；叶面积：用直尺测定完全展开叶的单叶叶长（L_L），用公式$S_L=L_L^2$计算总叶面积（龚建华等，2001）。鲜干重：用去离子水冲洗干净并吸干水分，从根茎结合处剪断，分别称得地上、地下部鲜重；105℃杀青15min，80℃烘至恒重，称得干重。

采用美国LI-COR公司生产的LI-6400便携式光合测定系统测定光合参数。测定条件为：光强1 000mmol·m^{-2}·S^{-1}，CO_2测定浓度为380mmol·mol^{-1}。

采用英国Hansatech公司生产的FMS-2型调制式叶绿素荧光仪测定叶绿素荧光参数。

叶绿素含量采用80%丙酮浸提法测定、脯氨酸（Pro）含量采用茚三酮—磺基水杨酸法测定（李合生，2000）；丙二醛（MDA）含量采用硫代巴比妥酸比色法测定、超氧阴离子自由基（$O_2^{·-}$）产生速率采用羟胺氧化法测定（汤章城，1999）；过氧化氢（H_2O_2）含量采用分光光度法测定（赵世杰等，2016）；超氧化物歧化酶（SOD）活性采用氮蓝四唑光还原法测定、过氧化物酶（POD）活性采用氧化愈创木酚比色法测定（李合生，2000）；过氧化氢酶（CAT）活性采用紫外吸收法测定（Chance等，1956）。

1.3 数据处理

采用Microsoft Excel 2007和DPS软件对数据进行处理、显著性分析（Duncan’s多重极差检验，α=0.05），并采用SigmaPlot 10.0软件绘图。

2 结果与分析

2.1 盐胁迫下CO_2加富对黄瓜幼苗生长的影响

由表1可以看出，盐胁迫抑制了黄瓜幼苗的生长，差异达显著水平。在大气CO_2浓度下，80mmol·L^{-1} NaCl处理使其株高、茎粗、叶面积、地上、地下部鲜干重较对照分别降低了45.0%、24.4%、54.2%、44.1%、35.3%、42.3%、35.6%；而CO_2加富可在一定程度上缓解盐胁迫对黄瓜幼苗植株的抑制作用。具体表现在，盐胁迫下CO_2加富使其各指标较大气CO_2浓度分别增加了13.1%、5.9%、10.5%、31.5%、15.9%、36.1%、8.9%，除地上部干重和地下部鲜干重外，差异均达显著水平。

表1 盐胁迫下CO_2加富对黄瓜幼苗生长的影响

Table 1 Effects of CO_2 enrichment on the plant growth of cucumber seedlings under salt stress

CO_2浓度 (mmol·mol^{-1})	盐分处理 (mmol·L^{-1})	株高 (cm)	茎粗（mm）	叶面积 (cm^2·plant^{-1})	鲜重（g·plant^{-1}）		干重（g·plant^{-1}）	
					地上部	地下部	地上部	地下部
400	0	27.36 ± 0.6^b	7.242 ± 0.215^b	862.98 ± 4.88^b	35.01 ± 2.31^b	16.06 ± 0.98^a	3.50 ± 0.24^b	0.87 ± 0.18^{ab}
	80	15.04 ± 0.6^d	5.471 ± 0.113^d	395.21 ± 29.18^d	19.58 ± 0.61^d	10.39 ± 0.54^b	2.02 ± 0.05^c	0.56 ± 0.06^c

（续表）

CO_2浓度 (mmol·mol^{-1})	盐分处理 (mmol·L^{-1})	株高 （cm）	茎粗（mm）	叶面积 （cm^2·plant^{-1}）	鲜重（g·plant^{-1}）		干重（g·plant^{-1}）	
					地上部	地下部	地上部	地下部
800	0	30.19 ± 0.3^a	7.506 ± 0.080^a	935.13 ± 25.4^a	43.33 ± 4.32^a	17.94 ± 2.65^a	4.50 ± 0.74^a	0.92 ± 0.12^a
	80	17.01 ± 0.9^c	5.795 ± 0.144^c	436.75 ± 16.11^c	25.75 ± 0.40^c	12.04 ± 1.13^b	2.75 ± 0.12^{bc}	0.61 ± 0.03^{bc}

同列不同小写字母表示处理间差异显著（$P<0.05$）Different small letters in the same column meant significant difference at 0.05 level among treatments.下同 The same below.

2.2 盐胁迫下CO_2加富对黄瓜幼苗叶片色素含量的影响

由表2可以看出，盐胁迫下黄瓜幼苗叶片Chl a、Chl b、Car、Chl（a+b）含量降低，差异达显著水平，但对Chl a/b影响不大。在大气CO_2浓度下，80mmol·L^{-1}NaCl处理使其Chl a、Chl b、Car、Chl（a+b）含量较对照分别降低了8.7%、10.6%、12.0%、9.3%；而CO_2加富进一步降低了盐胁迫下叶片色素含量，除Chl a差异达显著水平外，其他指标均未达显著水平。

表 2　盐胁迫下CO_2加富对黄瓜幼苗叶片色素含量的影响

Table 2　Effects of CO_2 enrichment on chlorophyll content of cucumber seedlings leaves under salt stress

CO_2浓度 (mmol·mol^{-1})	盐分处理 （mmol·L^{-1}）	叶绿素a [mg·g^{-1}(FW)]	叶绿素b[mg·g^{-1} (FW)]	类胡萝卜素 [mg·g^{-1}(FW)]	叶绿素（a+b） [mg·g^{-1}(FW)]	叶绿素a/b
400	0	1.809 ± 0.078^a	0.729 ± 0.035^a	0.648 ± 0.025^a	2.539 ± 0.114^a	2.482 ± 0.014^a
	80	1.651 ± 0.028^b	0.652 ± 0.019^{bc}	0.570 ± 0.009^{bc}	2.303 ± 0.047^{bc}	2.533 ± 0.030^a
800	0	1.694 ± 0.032^b	0.687 ± 0.024^{ab}	0.593 ± 0.014^b	2.381 ± 0.055^b	2.467 ± 0.040^a
	80	1.534 ± 0.009^c	0.621 ± 0.006^c	0.543 ± 0.010^c	2.154 ± 0.013^c	2.471 ± 0.020^a

2.3 盐胁迫下CO_2加富对黄瓜幼苗叶片光合气体交换参数的影响

由图1可以看出，盐胁迫抑制了黄瓜幼苗叶片光合作用，差异达显著水平。在大气CO_2浓度下，80mmol·L^{-1}NaCl处理使其P_n、G_s、C_i、T_r较对照分别下降了13.7%、31.3%、2.8%、11.3%；而盐胁迫下CO_2加富使其P_n较对照增高了8.7%，G_s、C_i、T_r分别下降了16.1%、6.2%、13.0%，除C_i外，差异均达显著水平。说明CO_2加富可降低叶片气孔导度，从而减少水分耗散，同时增强净光合速率，进而提高光合水分利用率，以此适应盐胁迫。

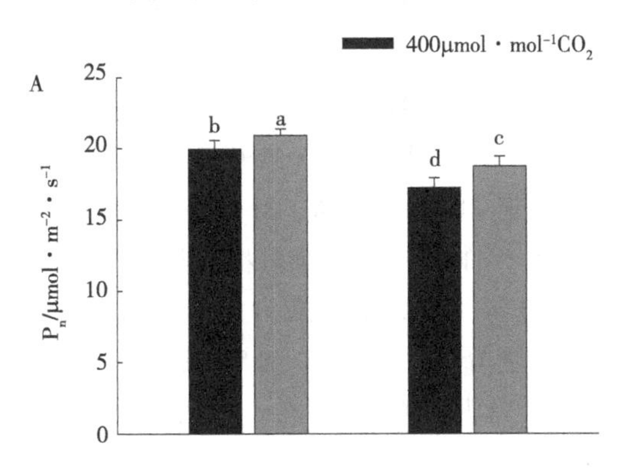

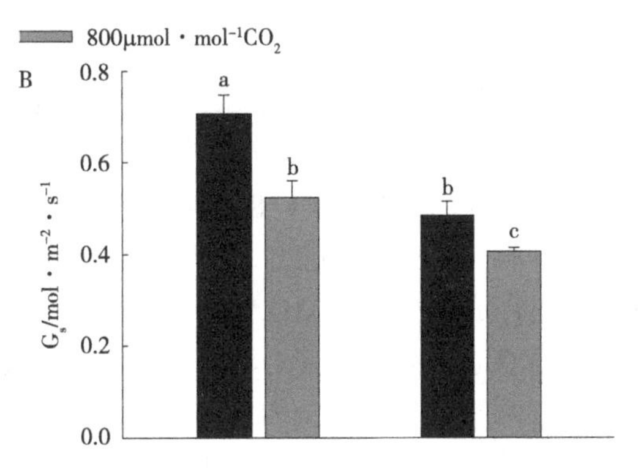

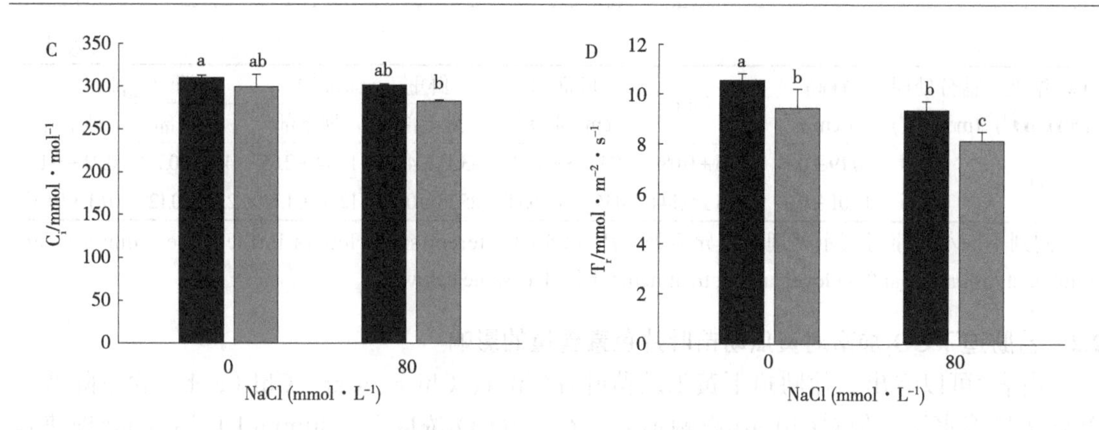

图 1　盐胁迫下CO_2加富对黄瓜幼苗叶片光合气体交换参数的影响

Figure 1　Effects of CO_2 enrichment on photosynthetic gas exchange parameters of cucumber seedlings leaves under salt stress

2.4　盐胁迫下CO_2加富对黄瓜幼苗叶片叶绿素荧光参数的影响

由表3可以看出，盐胁迫使*NPQ*上升，F_v/F_m、*Φ*PSⅡ、*ETR*、*qP*下降，除F_v/F_m外，差异均达显著水平。80mmol·L^{-1}NaCl处理时，在大气CO_2浓度下，*NPQ*较正常条件下上升了75.4%，F_v/F_m、*Φ*PSⅡ、*ETR*、*qP*则分别下降了2.7%、15.4%、15.4%、7.8%；而CO_2加富可提高盐胁迫下黄瓜叶片*Φ*PSⅡ，恢复PSⅡ光化学活性，使其*Φ*PSⅡ、*ETR*、*qP*较大气CO_2浓度分别上升7.8%、7.8%、3.3%，*NPQ*下降21.6%，除*qP*、*NPQ*外，差异均达显著水平，但对F_v/F_m影响不大。

表 3　盐胁迫下CO_2加富对黄瓜幼苗叶片叶绿素荧光参数的影响

Table 3　Effects of CO_2 enrichment on chlorophyll fluorescence parameters of cucumber seedlings leaves under salt stress

CO_2浓度（mmol·mol^{-1}）	盐分处理（mmol·L^{-1}）	F_v/F_m	*Φ*PSⅡ	*ETR*	*qP*	*NPQ*
400	0	0.829 ± 0.021^a	0.656 ± 0.019^a	220.41 ± 6.48^a	0.897 ± 0.008^a	0.481 ± 0.092^{bc}
	80	0.806 ± 0.042^a	0.555 ± 0.011^c	186.41 ± 3.85^c	0.827 ± 0.050^b	0.844 ± 0.096^a
800	0	0.831 ± 0.008^a	0.687 ± 0.002^a	230.72 ± 0.70^a	0.912 ± 0.004^a	0.412 ± 0.076^c
	80	0.818 ± 0.018^a	0.598 ± 0.024^b	201.04 ± 7.96^b	0.854 ± 0.035^{ab}	0.661 ± 0.048^{ab}

2.5　盐胁迫下CO_2加富对黄瓜幼苗叶片脯氨酸、MDA、$O_2^{\cdot-}$和H_2O_2的影响

由图2可以看出，盐胁迫下黄瓜幼苗叶片脯氨酸、MDA和ROS含量升高，差异达显著水平。在大气CO_2浓度下，80mmol·L^{-1}NaCl处理使其脯氨酸、MDA、H_2O_2含量和$O_2^{\cdot-}$生成速率较对照分别升高90.2%、37.7%、18.8%、56.7%；而CO_2加富进一步促进了盐胁迫下脯氨酸的合成，抑制了MDA和H_2O_2积累及$O_2^{\cdot-}$生成速率，使其脯氨酸含量较大气CO_2浓度升高了13.0%，MDA和H_2O_2积累量及$O_2^{\cdot-}$生成速率分别降低了19.2%、15.9%、15.7%，差异达显著水平。说明CO_2加富可通过促进黄瓜幼苗叶片脯氨酸的积累，同时降低MDA及ROS的积累，减少膜脂的过氧化，从而提高植株耐盐能力。

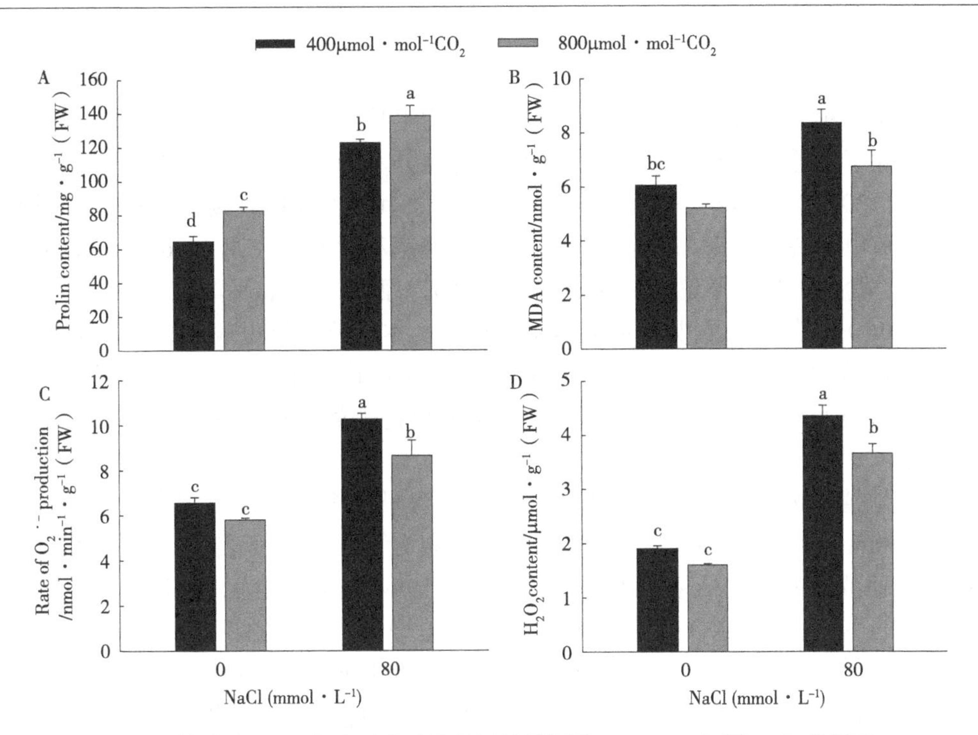

图 2　盐胁迫下CO_2加富对黄瓜幼苗叶片脯氨酸、MDA、$O_2^{\cdot-}$和H_2O_2的影响

Figure 2　Effects of CO_2 enrichment on contents of proline，MDA，H_2O_2 and rate of $O_2^{\cdot-}$ production of cucumber seedlings leaves under salt stress

2.6　盐胁迫下CO_2加富对黄瓜幼苗叶片抗氧化酶活性的影响

由图3可以看出，盐胁迫对黄瓜幼苗叶片抗氧化酶活性影响显著。在大气CO_2浓度下，80mmol·L^{-1}NaCl处理使其SOD、CAT活性较对照分别提高了10.3%、34.2%，但POD活性下降了50.7%；而CO_2加富进一步提高了盐胁迫下SOD、CAT活性，与此同时也提高了POD活性，使其SOD、POD、CAT活性较大气CO_2浓度分别提高了8.2%、62.9%、24.9%，差异达显著水平。说明CO_2加富可通过提高黄瓜幼苗叶片抗氧化酶活性，清除活性氧累积，从而缓解盐胁迫对膜造成的损伤。

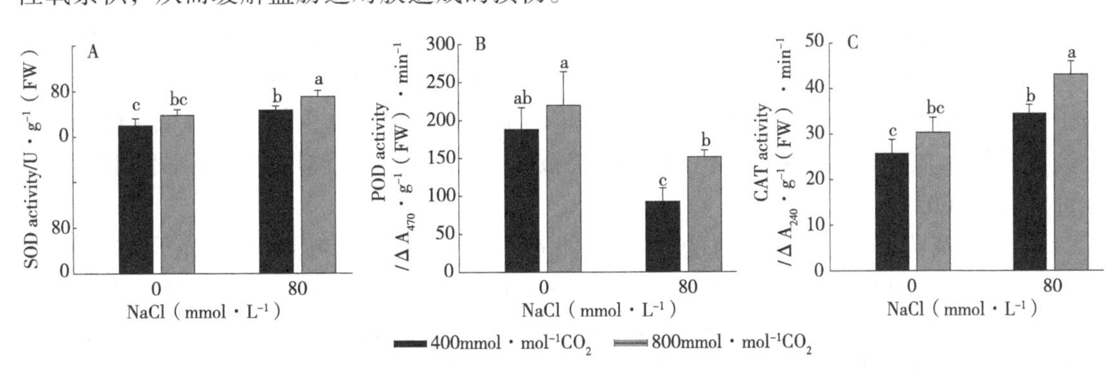

图 3　盐胁迫下CO_2加富对黄瓜幼苗叶片抗氧化酶活性的影响

Figure 3　Effects of CO_2 enrichment on antioxidative enzyme activity of cucumber seedlings leaves under salt stress

3 讨论

盐胁迫是一种重要的非生物逆境，是影响植物生长和作物产量的主要因素。盐对作物产生的毒害除了渗透胁迫和离子毒害，还会引起养分亏缺、活性氧伤害等一系列次生胁迫（张新春等，2002）。CO_2是植物光合作用的原料，其浓度高低直接影响Rubisco催化活性和方向、调节电子传递速率，进而影响植物光合速率和活性氧代谢（Bowes，1991）。前人研究表明，盐胁迫抑制了黄瓜（周珩等，2014）、西瓜（韩志平等，2008）、番茄（吴雪霞等，2006）幼苗植株的生长及干物质的积累。本试验结果也表明，80mmol·L^{-1}NaCl处理抑制了黄瓜幼苗的生长（表1），而CO_2加富缓解了盐胁迫对黄瓜幼苗植株地上部的抑制作用，这与前人在小麦（刘家尧等，1996）和花椰菜（Zaghdoud等，2013）上的研究结果类似。

叶绿体是植物进行光合作用的场所，其色素含量及组分直接影响光合作用。本试验结果表明，80mmol·L^{-1}NaCl处理降低了黄瓜幼苗叶片色素含量（表2），这与付晴晴等人（2017）在葡萄上的研究结果一致。蒋明义等（1994）研究认为，渗透胁迫降低植株叶片色素含量的主要原因是ROS的氧化作用，盐胁迫增加了ROS的积累（图3），从而促进色素的降解。有研究表明CO_2加富提高了冬小麦（张其德等，1998）和枇杷（张放等，2003）叶片色素含量，但本试验结果显示，CO_2加富降低了黄瓜幼苗叶片色素含量，这与Pérez-López（2015）和宝俐等人（2016）研究结果一致，究其原因可能是由于不同种类作物的叶绿体对CO_2加富的响应存在差异，而且CO_2加富可降低叶片蒸腾速率致使叶温过高，叶绿素分解加快，同时降低了植物体对矿质元素的吸收速率，出现缺素症状（李天来，2013）。本试验中叶片色素含量降低主要与CO_2加富下叶片快速生长引起的“稀释效应”有关。

光合作用是构成作物生产力的基础，盐胁迫抑制作物光合作用引起P_n下降，是气孔因素还是非气孔因素应同时依据G_s和C_i的变化而定（许大全，2002）。本试验中，80mmol·L^{-1}NaCl处理7d后，黄瓜幼苗叶片G_s和C_i均下降（图1），表明此时气孔限制为主要因素。另有研究揭示，逆境胁迫导致气孔关闭，在于H_2O_2的调控作用，盐胁迫使植物细胞H_2O_2水平升高（图3），促进了保卫细胞K^+外流和胞外Ca^{2+}内流，从而引起气孔关闭（简令成等，2009）。崔青青等（2017）研究发现，在干旱胁迫下，无论氮素的高低，CO_2加富均能通过降低G_s和气孔密度而提高黄瓜叶片水分利用率。本试验中，CO_2加富提高了盐胁迫下叶片P_n，从而为植物体提供更多的糖源，促进植株生长；并降低了G_s和T_r，提高光合用水效率，但G_s降低的节水效应会被叶面积增加和叶温增高的耗水作用所抵消（许大全，2002）。

叶绿素荧光作为光合作用的探针，能在一定程度反应叶绿体的光合作用能力（Maxwell等，2000）。本试验结果表明，80mmol·L^{-1}NaCl处理，显著降低了ΦPSⅡ、*ETR*和*qP*（表3），说明盐胁迫降低了PSⅡ实际光化学效率和表观光合电子传递速率，使Q_A处于较高的还原态，阻止同化力的生成，从而影响碳同化。*NPQ*显著升高说明盐胁迫下吸收的光能不能完全通过电子传递链传递给PSⅡ反应中心，热耗散的比例增加，从而调节光合量子效率及适应逆境条件（许大全，2002）。CO_2加富则提高了盐胁迫下黄瓜幼苗ΦPSⅡ、*ETR*，并在一定程度上提高了*qP*，降低了*NPQ*，从而促进光化学效率和电子传递速率，一定程度上缓解盐胁迫的负面影响，这与陈丹等（2004）在枇杷上的研究结果类似。同时，在CO_2加富和干旱胁迫下的研究结果也显示，CO_2加富可通过提高黄瓜叶片表观量子效率、光饱和的最大同化效率和光饱和的电子传递效率从而提高叶片净光合速率（李清明等，2011）。

脯氨酸作为相容性物质，主要分布于线粒体及细胞质基质中，从而维持胞质与胞液间渗透压差，使渗透胁迫下叶片叶绿体和线粒体保持良好的水分状态，保证光合和呼吸作用的正常进行（Maggio等，2000）。逆境下植物体通过钝化脯氨酸氧化降解、活化其生物合成，使其在细胞内大量累积，是植物适应逆境的重要机制（简令成等，2009）。脯氨酸可充当ROS清除剂（蒋明义等，1997），提高复合蛋白和膜系统稳定性，保护和修复叶绿体PSⅡ在逆境中伤害（Park等，2004），维持其电子传递作用。本试验中，80mmol·L^{-1}NaCl处理提高了黄瓜幼苗叶片脯氨酸含量（图2），是其适应盐胁迫的一种表现；同时CO_2加富进一步促进了其在叶片中的积累，与前人（刘家尧等，1998）在小麦上的研究结果类似。

ROS是由膜系统上的氧化酶催化产生的一类活性含氧化合物，高浓度的ROS能引起DNA、蛋白质和脂质过氧化，破坏膜结构（喻景权，2014）。MDA作为膜脂过氧化的重要产物之一，能加剧膜的损伤。在正常生理状态下，抗氧化酶系统可使ROS的产生和消除维持在一种动态平衡状态。本试验结果表明，80mmol·L^{-1}NaCl处理增加了黄瓜幼苗叶片ROS及MDA的积累量（图3），同时也提高了SOD、CAT活性（图4），说明其在一定程度上制约和调控了ROS水平，诱导植物细胞对盐胁迫做出适应性反应。CO_2加富进一步提高了盐胁迫下抗氧化酶活性，并降低了ROS和MDA积累量，从而减轻ROS的毒害作用，维持ROS产生与清除平衡，这与前人（Pérez-López等，2009b）在大麦上的研究结果一致。另外，在CO_2加富和干旱胁迫下的研究结果也表明，CO_2加富可通过提高细胞内抗氧化酶活性，减少ROS积累，减轻氧化损伤（李清明等，2010）。CO_2加富提高抗氧化能力可能是由于高CO_2浓度促进了线粒体中脯氨酸的合成，增强SOD等抗氧化酶活性，提高ROS清除能力。

综上所述，CO_2加富能通过提高P_n和降低T_r，提高水分利用效率；提高ΦPSⅡ和*qP*，降低*NPQ*，恢复PSⅡ光化学活性；提高SOD、POD、CAT活性及脯氨酸水平，抑制MDA、O_2^{-}和H_2O_2的积累量，防止膜脂过氧化，从而缓解盐胁迫对黄瓜幼苗造成的不利影响，提高其耐盐能力。这为未来大气CO_2浓度升高和盐胁迫等非生物胁迫下黄瓜的优质高效生产提供了理论依据和技术参数。

参考文献

宝俐，董金龙，段增强，等. 2016. CO_2浓度升高和氮素供应对黄瓜叶片光合色素的影响［J］. 土壤，48（4）：653-660

陈丹，张放. 2004. 水分胁迫条件下CO_2加富对枇杷叶绿素荧光及抗氧化酶的影响［J］. 浙江农业学报，16（2）：13-17

崔青青，董彦红，李清明，等. 2017. CO_2加富下水氮耦合对黄瓜叶片光合作用和超微结构的影响［J］. 应用生态学报，28（4）：1237-1245

付晴晴，孙永江，翟衡，等. 2017. 盐胁迫对葡萄种间杂交砧木F1株系光合特性的影响［J］. 植物生理学报，53（9）：1 640-1 648
龚建华，向军. 2001. 黄瓜群体叶面积无破坏性速测方法研究［J］. 中国蔬菜，（4）：7-9
韩志平，郭世荣，焦彦生，等. 2008. NaCl胁迫对西瓜幼苗生长和光合气体交换参数的影响［J］. 西北植物学报，28（4）：4 745-4 751］
简令成，王红. 2009. 逆境植物细胞生物学［M］. 北京：科学出版社
蒋明义，郭绍川，张学明（1997）. 氧化胁迫下稻苗体内积累的脯氨酸的抗氧化作用［J］. 植物生理学报，23（4）：347-352
蒋明义，杨文英，徐江，等. 1994. 渗透胁迫下水稻幼苗中叶绿素降解的活性氧损伤作用［J］. 植物学报，36（4）：289-295
李彬，王志春，孙志高，等. 2005. 中国盐碱地资源与可持续利用研究［J］. 干旱地区农业研究，23（2）：154-158
李合生. 2000. 植物生理生化实验原理和技术［M］. 北京：高等教育出版社
李曼，董彦红，崔青青，等. 2017. CO_2浓度加倍下水氮耦合对黄瓜叶片碳氮代谢及其关键酶活性的影响［J］. 植物生理学报，53（9）：1 717-1 727
李清明，刘彬彬，艾希珍. 2010. CO_2浓度倍增对干旱胁迫下黄瓜幼苗膜脂过氧化及抗氧化系统的影响［J］. 生态学报，30（22）：6 063-6 071
李清明，刘彬彬，邹志荣. 2011. CO_2浓度倍增对干旱胁迫下黄瓜幼苗光合特性的影响［J］. 中国农业科学，44（5）：963-971
李天来. 2013. 日光温室蔬菜栽培理论与实践［M］. 北京：中国农业出版社
刘家尧，衣艳君，白克智，等. 1996. CO_2盐冲击对小麦幼苗呼吸酶活性的影响［J］. 植物学报，38（8）：641-646
刘家尧，衣艳君，白克智，等. 1998. CO_2倍增环境生长的小麦幼苗对盐胁迫的生理反应［J］. 生态学报，18（4）：74-78
潘璐，刘杰才，崔世茂，等. 2014. 高温和加富CO_2温室中黄瓜Rubisco活化酶与光合作用的关系［J］. 园艺学报，41（8）：1 591-1 600.
汤章城. 1999. 现代植物生理学实验指南［M］. 北京：科学出版社
王素平，李娟，郭世荣，et al., 2006. NaCl胁迫对黄瓜幼苗植株生长和光合特性的影响［J］. 西北植物学报，26（3）：455-461
吴雪霞，朱月林，朱为民，et al., 2006. 外源一氧化氮对NaCl胁迫下番茄幼苗生长和光合作用的影响［J］. 西北植物学报，26（6）：1 206-1 211
许大全. 2002. 光合作用效率［M］. 上海：上海科学技术出版社
杨连新，王云霞，王余龙，等. 2010. 开放式空气中CO_2浓度增高（FACE）对水稻生长和发育的影响［J］. 生态学报，30（6）：1 573-1 585
喻景权. 2014. 蔬菜生长发育与品质调控：理论与实践［M］. 北京：科学出版社
张放，陈丹，张士良，等. 2003. 高浓度CO_2对不同水分条件下枇杷生理的影响［J］. 园艺学报，30（6）：647-652
张其德，朱新广，卢从明，等. 1998. 盐胁迫下CO_2浓度倍增对冬小麦叶绿体光能吸收和激发能分配的影响［J］. 生物物理学报，14（3）：157-162
张新春，庄炳昌，李自超. 2002. 植物耐盐性研究进展［J］. 玉米科学，10（1）：50-56
赵世杰，苍晶. 2016. 植物生理学实验指导［M］. 北京：中国农业出版社
周珩，郭世荣，邵慧娟，等. 2014. 等渗NaCl和$Ca(NO_3)_2$胁迫对黄瓜幼苗生长和生理特性的影响［J］. 生态学报，34（7）：1 880-1 890
Bowes G. 1991. Growth at elevated CO_2：photosynthetic responses mediated through Rubisco. Plant Cell Environ，14（8）：795-806
Chance B，Maehly AC. 1956. Assay of catalase and peroxidase. Methods Enzymol，2：764-775
Maggio A，Reddy MP，Joly RJ. 2000. Leaf gas exchange and solute accumulation in the halophyte Salvadora persica grown at moderate salinity［J］. Environ Exp Bot，44（1）：31-38
Maxwell K，Johnson GN. 2000. Chlorophyll fluorescence—a practical guide［J］. J Exp Bot，51（345）：659-668
Melgar JC，Syvertsen JP，García-Sánchez F. 2008. Can elevated CO_2 improve salt tolerance in olive trees［J］. J Plant Physiol，165（6）：631-640

Park EJ, Jeknić Z, Sakamoto A, et al. 2004. Genetic engineering of glycinebetaine synthesis in tomato protects seeds, plants, and flowers from chilling damage [J]. Plant J, 40 (4): 474

Pérez-López U, Miranda-Apodaca J, Lacuesta M, et al. 2015. Growth and nutritional quality improvement in two differently pigmented lettuce cultivars grown under elevated CO_2 and/or salinity [J]. Sci Hortic, 195: 56-66

Pérez-López U, Robredo A, Lacuesta M, et al. 2009a. The impact of salt stress on the water status of barley plants is partially mitigated by elevated CO_2 [J]. Environ Exp Bot, 66 (3): 463-470

Pérez-López U, Robredo A, Lacuesta M, et al. 2009b. The oxidative stress caused by salinity in two barley cultivars is mitigated by elevated CO_2 [J]. Physiol Plant, 135 (1): 29-42

Piñero MC, Houdusse F, Garcia-Mina JM, et al. 2014. Regulation of hormonal responses of sweet pepper as affected by salinity and elevated CO_2 concentration [J]. Physiol Plant, 151 (4): 375-389

Suleyman IA, Atsushi S, Yoshitaka N, et al. 2000. Ionic and osmotic effects of NaCl-induced inactivation of photosystems I and II in synechococcus sp [J]. Plant physiol, 123 (3): 1 047-1 056

Zaghdoud C, Carvajal M, Ferchichi A, et al. 2016. Water balance and N-metabolism in broccoli (*Brassica oleracea* L. var. Italica) plants depending on nitrogen source under salt stress and elevated CO_2 [J]. Sci Total Environ, 571: 763-771

Zaghdoud C, Motacadenas C, Carvajal M, et al. 2013. Elevated CO_2 alleviates negative effects of salinity on broccoli (*Brassica oleracea* L. var Italica) plants by modulating water balance through aquaporins abundance [J]. Environ Exp Bot, 95: 15-24

干旱胁迫对草石蚕保护酶活性和渗透调节物质的影响

班甜甜[1]，张素勤[2]，肖体菊[2]，耿广东[2*]
（1. 贵州大学生命科学学院，贵阳 550025；2. 贵州大学农学院，贵阳 550025）

摘要：研究不同干旱胁迫下草石蚕保护酶活性和渗透调节物质的变化规律，分析不同抗旱性品种在干旱胁迫下的生理变化，以期为草石蚕的抗旱性鉴定提供参考依据，并进一步为筛选和选育抗旱性品种提供理论基础。以抗旱性不同的三个草石蚕品种为试材，在设施内采用自然干旱胁迫方法，分别胁迫4d、8d、12d和16d，通过测定各生理指标的数据，研究干旱胁迫对草石蚕的影响。随干旱胁迫时间延长，3个品种的可溶性糖和Pro含量逐渐增加；“贵栽1号”可溶性蛋白含量一直升高，而“鲁引1号”和“赣引1号”则呈先升高后降低的变化趋势；SOD、POD和CAT活性逐渐升高；膜脂过氧化产物MDA含量逐渐增加。抗旱性强的“贵栽1号”可溶性糖、Pro和MDA含量增加幅度最低，SOD、POD和CAT活性升高幅度最大。草石蚕保护酶系统、渗透调节物质和膜脂过氧化产物的变化规律与干旱胁迫程度和品种抗旱性有关。

关键词：草石蚕；干旱胁迫；抗旱性；保护酶系统；渗透调节物质

Effects of Drought Stress on Physiological Property of *Stachys Sieboldii* Miq.

Ban Tiantian[1]，Zhang Suqin[2]，Xiao Tiju[2]，Geng Guangdong[2*]
（1. *College of Life Sciences*，*Guizhou University*，*Guiyang* 550025；
2. *College of Agriculture*，*Guizhou University*，*Guiyang* 550025）

Abstract：Three varieties of s*tachys Sieboldii* Miq. with different drought resistance were used to investigate the effect of drought stress on physiological property by natural drought stress in the present study.The experiment could provide theoretical basis for the production of s*tachys Sieboldii* Miq.The results showed that the content of soluble sugar，Pro and MDA rose gradually，and the activities of SOD，POD and CAT kept rising at all times under drought stress. The content of soluble sugar，Pro and MDA of the variety “Guizai NO.1” which had good drought resistance had the least increasing amplitude. While the activities of SOD，POD and CAT of “Guizai NO.1” had the highest increasing amplitude. The content of soluble protein of the varieties “Luyin No. 1” and “Ganyin No. 1” rose in the first stage，and then fell under drought stress. While the content of soluble protein of “Guizai NO.1” kept rising at all times.It could be found that protective enzyme system and osmotic adjustment substance were affected by drought press in *stachys sieboldii* Miq.，which was related to drought resistance of variety.

Key words：*Stachys sieboldii* Miq.；Drought stress；Drought resistance；Protective enzyme system；Osmotic

---

作者简介：班甜甜，女，1990年出生，贵州大学在读硕士研究生，邮箱：1574601087@qq.com
*通信作者，耿广东，教授，博士，主要从事蔬菜栽培生理与生物技术方面的教研工作。E-mail：genggd213@163.com

adjustment substance

旱害是植物生长过程中常见的危害，它对植物的影响很广泛，可以从种子的萌发期延伸到结果期（李博，2008）。干旱首先会使植株叶子出现萎蔫、凋落、黄化等现象，这些现象能直接地反映出植物受害程度的大小。根作为植物从外界吸收水分最主要的途径，干旱对其的危害很大，如阻碍毛细根的伸长和新生根的形成，降低根系活力，导致植物不能正常生长，甚至停止生长（马剑等，2011）。大量研究表明：水分缺失会导致叶绿素的合成受阻，含量降低（李得孝等，2007；冯晓敏等，2012），使叶片上的气孔关闭，阻碍植物对CO_2的吸收（夏岩石等，2011），影响光合效率（卞付萍，2014）。除此之外在逆境条件下，植物体内脯氨酸含量、可溶性糖含量和可溶性蛋白的含量都会有一定程度的积累以维持细胞的渗透压，帮助植株适应环境。植物的抗氧化酶（SOD、POD、CAT等）保护系统具有清除逆境中产生的O^{2-}、H_2O_2、OH^-等活性氧的功能（李滨胜等，2010），减轻对细胞膜透性的伤害。MDA作为膜质过氧化最重要的产物之一，对细胞有很大的毒性，MDA积累量常作为衡量植物受害程度的重要指标（李明等，2002）。除了外部形态的变化，SOD、POD、CAT活性和MDA积累量可以作为评价植物抗性的常用指标（李明等，2002）。

草石蚕（*Stachys sieboldii* Miq.）是水苏属多年生草本植物，食用部分为地下茎，营养价值丰富，具有补血、强身、清热解毒、治咳嗽等功效（赵建青，2009）。草石蚕有一定的耐旱能力，但在极端条件下，其产量和品质均会受到很大影响，因此探索草石蚕的抗旱机理，对提高其产量和品质具有重要意义。目前，对植物干旱胁迫条件下的生理生化变化的研究较多，但对草石蚕干旱胁迫方面的相关研究鲜见报道。因此，本文以3个抗旱性不同的草石蚕品种为试材，研究干旱胁迫对其保护酶活性和渗透调节物质的影响。本研究拟从生理指标的变化来探讨干旱胁迫对草石蚕的影响，以期了解草石蚕的抗旱机制，并为草石蚕生产中的抗旱问题提供理论指导。

1　材料与方法

1.1　材料

以“贵栽1号”（抗旱性强，贵州务川主栽品种）、“鲁引1号”（抗旱性中等，山东滨洲主栽品种）和“赣引1号”（抗旱性弱，江西萍乡主栽品种）3个抗旱性不同的草石蚕品种为实验材料。

1.2　方法

试验于2016年3月在贵州大学农学院教学实验基地大棚内进行，常规育苗，待幼苗长到5叶1心时，将其移栽到盆中（盆中加等量的土），每盆移栽4棵，每个处理10盆，生长15d后，进行干旱处理（先浇透水，然后自然干旱），干旱处理4d、8d、12d和16d后对其进行各种生理指标测定。以正常浇水生长的草石蚕为对照，3次重复。采样时间均为8:00—9:00，试验测定的生理指标有可溶性糖、可溶性蛋白、脯氨酸（Pro）、SOD、POD、CAT和丙二醛（MDA），试验取样材料为从上到下第3和第4片叶。

可溶性糖含量采用蒽酮比色法，可溶性蛋白采用考马斯亮兰G-250染色法，脯氨酸测定采用茚三酮比色法，MDA采用硫代巴比妥酸法（李合生，2000）；抗氧化酶SOD、POD和CAT活性的测定采用王学奎（2006）方法测定。

1.3 数据处理

采用Microsoft Excel 2013和DPS（V12.01）软件进行数据的处理和分析。

2 结果与分析

2.1 干旱胁迫对草石蚕可溶性糖含量的影响

在干旱条件下，3个品种的草石蚕随胁迫时间延长，可溶性糖含量呈一直上升的变化趋势（图1）。处理第8d时，“鲁引1号”和“赣引1号”可溶性糖含量明显上升，而“贵栽1号”可溶性糖含量上升幅度小，仅比对照提高了5.32%，但与对照达到极显著差异水平；处理第16d时，“鲁引1号”“赣引1号”和“贵栽1号”可溶性糖含量达到最大值，分别比对照提高了134.07%、80.85%和48.28%，且均极显著高于对照，其中抗旱性强的“贵栽1号”变化幅度最小，抗旱性中等的“鲁引1号”变化幅度最大。

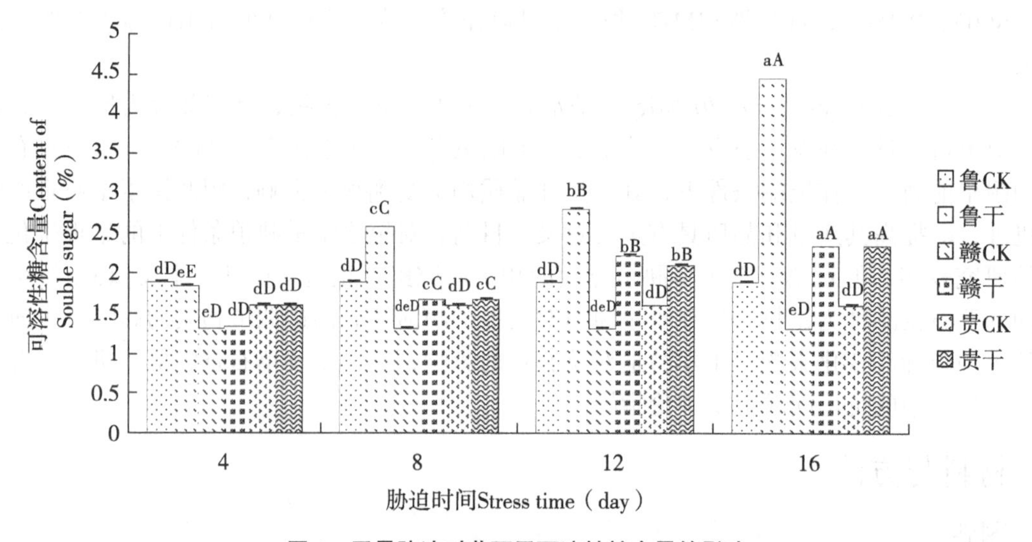

图1 干旱胁迫对草石蚕可溶性糖含量的影响

Figure 1 Effect of drought stress on soluble sugar content of *Stachys sieboldii* Miq.

2.2 干旱胁迫对草石蚕可溶性蛋白含量的影响

由图2可知，随胁迫时间延长，“鲁引1号”和“赣引1号”可溶性蛋白含量呈先升高后降低的变化趋势，而“贵栽1号”则一直升高。干旱处理第4d时，3个品种可溶性蛋白含量变化不明显，与对照未达到显著差异水平；干旱处理第8d时，“鲁引1号”和“赣引1号”比对照有明显的上升，且在第12d时达到峰值，分别比对照上升了78.15%和57.83%，极显著高于对照；处理第16d时较最大值有一定的下降，但仍高于对照。抗旱性强的“贵栽1号”一直升高，第16d时达到最大值，比对照提高了21.23%，差异达极显著性水平。

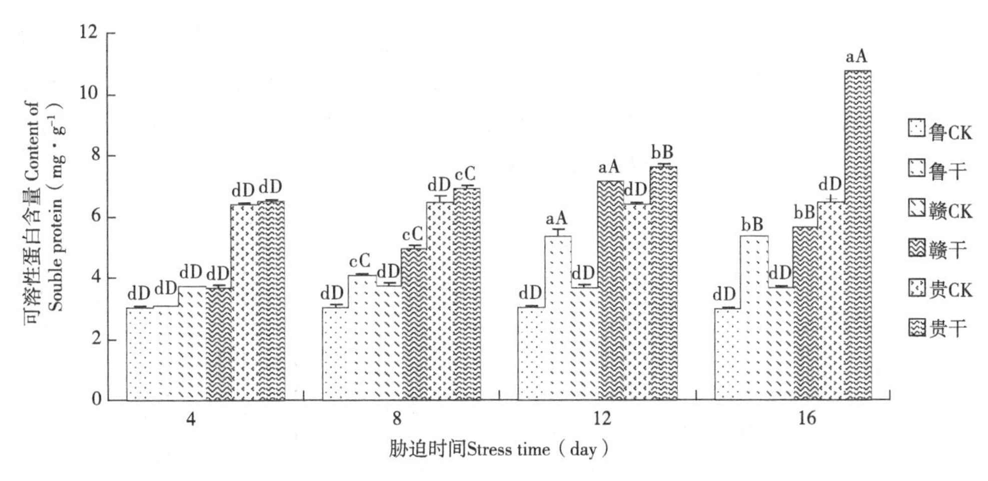

图 2　干旱胁迫对草石蚕可溶性蛋白含量的影响

Figure 2　Effect of drought stress on soluble protein content of S*tachys sieboldii* Miq.

2.3　干旱胁迫对草石蚕Pro含量的影响

随干旱胁迫时间延长，3个品种的草石蚕Pro含量呈一直增加的变化趋势（图3）。干旱胁迫第4d时，3个品种的Pro含量升高，且与对照达到极显著性差异水平；干旱胁迫第12d时，Pro含量进一步升高；干旱胁迫第16d时，3个品种的草石蚕Pro含量均达到最大值，与对照相比，“鲁引1号”“赣引1号”和“贵栽1号”分别增加了164.90%、203.29%和90.08%。由此可知，草石蚕可通过增加Pro含量来适应旱害，其中抗旱性强的“贵栽1号”Pro含量增加幅度最小，抗旱性弱的“赣引1号”的Pro含量增加幅度最大。

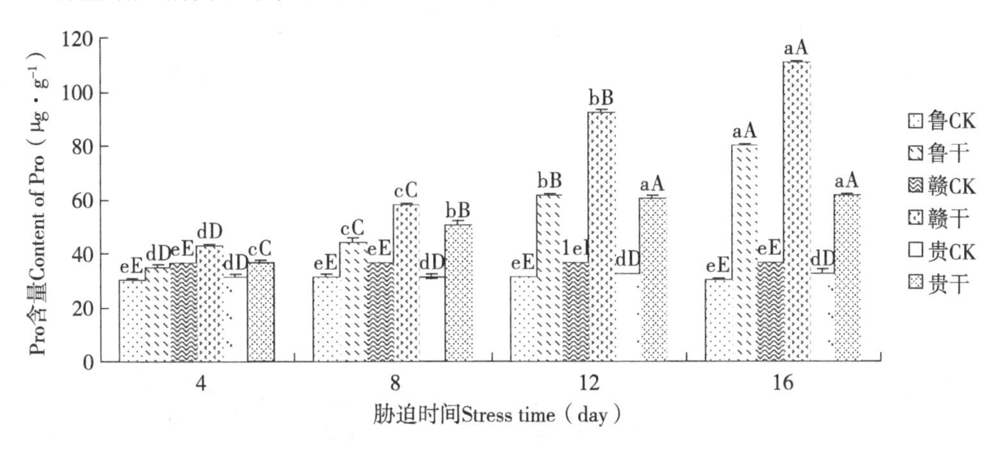

图 3　干旱胁迫对草石蚕Pro含量的影响

Figure 3　Effect of drought stress on Pro content of S*tachys sieboldii* Miq.

2.4　干旱胁迫对草石蚕SOD活性的影响

随干旱胁迫时间的延长，3种草石蚕SOD活性均呈一直升高的变化趋势（图4）。干旱胁迫第4d时，SOD的活性变化不明显；干旱胁迫第8d时，3种草石蚕SOD活性均明显升高，“鲁引1号”“赣引1号”和“贵栽1号”均比对照升高了12.27%、13.63%和13.91%，并均达到极显著性差异水平；干旱胁迫第16d时，三种草石蚕SOD活性均升高到最大值，与对照相比，“鲁引1号”“赣引1号”和“贵栽1号”分别升高了41.67%、35.43%和52.58%，均与第12d处理的达到极显著差异水平。由此可见，抗旱性强的“贵栽1号”升高幅度最大，抗旱性差的“赣引1号”升高幅度最小。

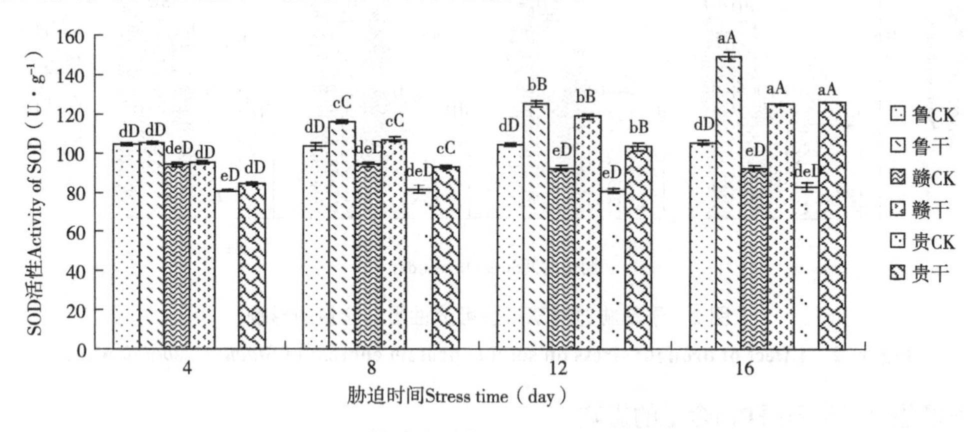

图4　干旱胁迫对草石蚕SOD活性的影响

Figure 4　Effect of drought stress on SOD activity of *Stachys sieboldii* Miq.

2.5　干旱胁迫对草石蚕POD活性的影响

随干旱胁迫时间延长，3种草石蚕的POD活性均呈一直上升的变化趋势（图5）。干旱胁迫第4d时，3种草石蚕POD活性与对照相比升高幅度不明显；干旱胁迫第8d时，3种草石蚕POD活性均有明显的升高，与对照相比均达到极显著差异水平；干旱胁迫第12d时，三种草石蚕POD活性进一步升高，与对照相比，“鲁引1号”“赣引1号”和“贵栽1号”分别升高了52.21%、82.05%和50.79%；干旱胁迫第16d时，3种草石蚕POD活性均达到最大值，“贵栽1号”的POD活性最高，“鲁引1号”次之，“赣引1号”升高幅度最小。由此可见，抗旱性强的“贵栽1号”POD活性升高幅度最大，抗旱性差的“赣引1号”升高幅度最小。

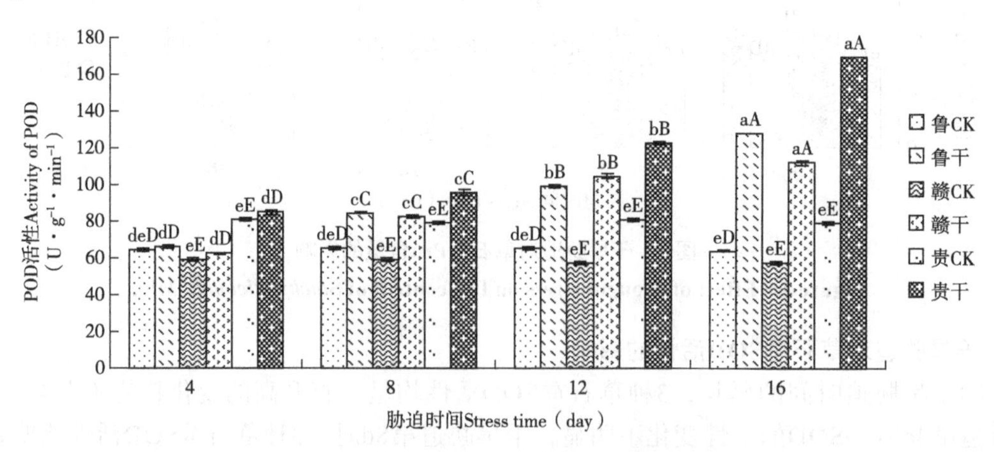

图5　干旱胁迫对草石蚕POD活性的影响

Figure 5　Effect of drought stress on POD activity of *Stachys sieboldii* Miq.

2.6 干旱胁迫对草石蚕CAT活性的影响

随干旱胁迫时间延长，3种草石蚕的CAT活性均呈一直上升的变化趋势（图6）。干旱胁迫第4d时，3种草石蚕CAT活性与对照相比变化不明显；干旱胁迫第8d时，3种草石蚕CAT活性与对照的变化明显，“贵栽1号”升高了91.90%，“赣引1号”升高了27.89%，“鲁引1号”升高了40.00%，且均达到极显著差异水平；干旱胁迫第16d时，3种草石蚕CAT活性均升到最大值，“鲁引1号”“赣引1号”和“贵栽1号”分别比对照升高了124.95%、81.37%和205.81%。由此可见，抗旱性强的“贵栽1号”SOD活性升高幅度最大，抗旱性弱的“赣引1号”升高幅度最小。

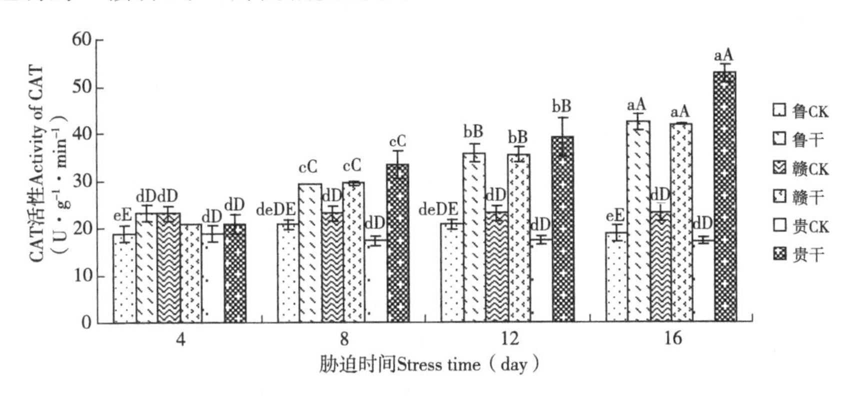

图 6　干旱胁迫对草石蚕CAT活性的影响

Figure 6　Effect of drought stress on CAT activity of *Stachys sieboldii* Miq.

2.7 干旱胁迫对草石蚕MDA积累量的影响

三种草石蚕MDA积累量随干旱胁迫时间的延长呈一直递增的变化趋势（图7）。干旱胁迫第4d时，三种草石蚕MDA积累量与对照相比变化不明显；干旱胁迫第8d时，3种草石蚕MDA积累量较对照增加幅度明显，“赣引1号”“贵栽1号”和“鲁引1号”分别增加了88.87%、49.76%和112.60%，与对照和第4d的积累量相比均达到显著性差异水平；干旱胁迫第16d时，3个品种的草石蚕MDA积累量达到最大，“赣引1号”“贵栽1号”和“鲁引1号”分别较对照增加了311.07%、150.71%和269.66%。由此可见，抗旱性强的“贵栽1号”升高幅度最小，抗旱性差的“赣引1号”升高幅度最大。

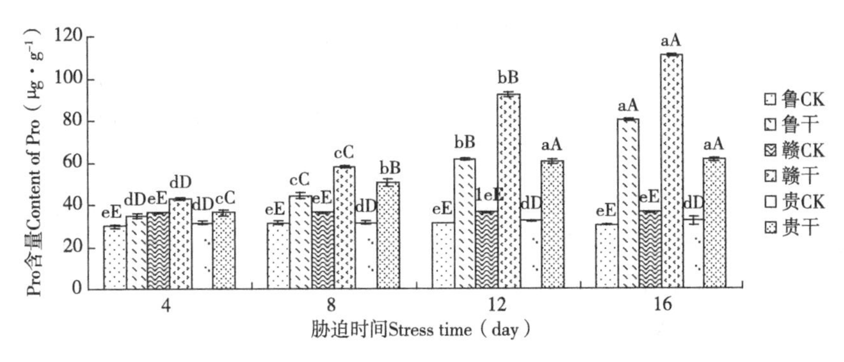

图 7　干旱胁迫对草石蚕MDA积累量的影响

Figure 7　Effect of drought stress on MDA content of S*tachys sieboldii* Miq.

3 讨论

3.1 干旱胁迫对植物渗透调节物质的影响

干旱胁迫下植物会通过积累渗透调节物质来提高细胞浓度（王晶懋等，2012），增强植物从外界吸收水分的能力，以适应干旱的环境（金不换，2009；郭春芳等，2015；李瑞智等，1989）。本试验发现，随干旱胁迫时间的延长，3种草石蚕可溶性糖含量一直升高，“赣引1号”的可溶性糖含量增加最多，“贵栽1号”增加最少。前期可溶性糖含量升高可能是因为干旱胁迫下植物光合系统的运转机能受到影响，植物体内大分子化合物的合成受到抑制，转而合成低分子量的蔗糖等化合物（Dure L.，1993）；而干旱胁迫后期，植物叶片的含水量降低，致使可溶性糖含量升高（李强等，2010），也可能是因为植物内部机能遭到破坏，生命活动降低，导致对糖的消耗降低，最终使可溶性糖含量升高。

干旱胁迫时Pro的合成和积累会急剧增多，有时甚至会增加几十倍，Pro的疏水端具有与蛋白质结合的能力，亲水端又可以和水结合，因此蛋白质就可以利用Pro来束缚大量的水，防止胁迫条件下蛋白质发生脱水现象，造成其性质改变（李瑞智等，1989），所以Pro对植物的生长有积极作用。本研究中3个品种的草石蚕随着干旱胁迫时间延长，Pro积累量一直上升，前期是因为草石蚕积极调节内部渗透物质，以适宜外部逆境，但随干旱胁迫时间延长，Pro含量的增加很大程度上是因为叶片失水造成的。

干旱胁迫时间延长时，3个品种草石蚕可溶性蛋白含量均有所增加，与张明生等（张明生等，2003）对甘薯研究得到的结论一致，是植物对逆境做出的一种适应性反应，干旱胁迫时间达到一定程度时开始下降，这可能是由2个原因造成的，一是蛋白质的合成能力降低（陈明涛等，2010），二是分解速度加快（鲁存海，2010）。抗旱性强的“贵栽1号”在试验胁迫范围内，随干旱胁迫时间延长，可溶性蛋白含量一直上升，说明其自我调节能力好。

3.2 干旱胁迫对植物抗氧化酶活性的影响

植物体内具有活性氧自由基清除剂抗氧化酶体系（SOD、POD、CAT），逆境中，植物会通过提高其活性，来减少活性氧对植株的伤害。干旱胁迫下POD活性升高可以保护植物细胞膜的结构，其活性升高的原因是活性氧的增加刺激了清除酶系统的活性（Giannopolitis C N等，1977）。SOD可以把植物体内有害的超氧自由基转化为过氧化物，再由POD和CAT转化为无危害的水（姜义宝等，2008）。CAT可以清除细胞内的有毒物质，使细胞免受毒害（何亚丽等，2002），所以3种酶的活性是鉴定植物抗逆性的重要指标，酶活性越大清除危害物的能力就越大。本试验不同草石蚕品种SOD、POD和CAT活性随干旱胁迫时间的延长一直升高，说明在试验胁迫范围内，草石蚕一直积极调动抗氧化酶活性降低细胞内有毒物质的积累，以适应外界的干旱胁迫，但其调节能力是有限的，当达到一定胁迫程度后，其活性会降低，失去调节能力，植物会受到严重伤害，从本试验结果看，试验范围内的胁迫还未达到这个程度。

3.3 干旱胁迫对植物体内MDA含量的影响

MDA是植物细胞内膜脂过氧化的产物，会引起细胞膜功能紊乱（宁建凤，2005），因此，MDA的积累量可作为评价植物在逆境下受伤害程度的重要指标。MDA的积累量越少，植株受到的危害就越轻，抗性越好，MDA的积累量越多，植物组织受到的伤害就越严重，抗性越差。本试验中，3个草石蚕品种的MDA积累量变化情况不同，但均成一直升高的变化趋势，这与在花生（贺鸿雁等，2006）、白桦（孙国荣等，2003）和红松（阎秀峰等，1999）等植物上的研究结果相似。抗旱性强的"贵栽1号"MDA积累量最少，所受伤害最小，而抗旱性差的"赣引1号"MDA积累量最多，植株受到的伤害最大。

参考文献

卞付萍. 2014. 干旱胁迫下水榆花楸生长及生理特性的研究［D］. 南京：南京林业大学，

陈明涛，赵忠，权金娥. 2010. 干旱对4种苗木根尖可溶性蛋白组分和含量的影响［J］. 西北植物学报，30（6）：1157-1165.

冯晓敏，张永清. 2012. 水分胁迫对糜子植株苗期生长和光合特性的影响［J］. 作物学报，38（8）：1 513-1 521.

郭春芳，孙云. 2015. 干旱胁迫下植物的渗透调节及脯氨酸代谢研究进展［J］. 福建教育学院学报，（1）：114-128.

贺鸿雁，孙存华，杜伟，等. 2006. PEG6000胁迫对花生幼苗渗透调节物质的影响［J］. 中国油料作物学报，28（1）：76-82.

何亚丽，刘友良，陈权，等. 2002. 水杨酸和热锻炼诱导的高羊茅幼苗的耐热性与抗氧化的关系［J］. 植物生理与分子生物学学报，28（2）：89-95.

金不换. 2009. 干旱胁迫对不同品种早熟禾形态和生理特性影响的研究［D］.哈尔滨：东北农业大学，

姜义宝，杨玉荣，郑秋红. 2008. 外源一氧化氮对干旱胁迫下苜蓿幼苗抗氧化酶活性和叶绿素荧光特性的影响［J］. 干旱地区农业研究，26（2）：65-68.

李滨胜，周玉迁，潘杰，等. 2010. 干旱胁迫下细叶景天生理生化指标的变化［J］. 北方园艺，16：105-107.

李博. 2008. 几种玉簪的水分胁迫耐受性研究［D］. 哈尔滨：东北林业大学，

李得孝，崔黎艳，陈耀锋. 2007. 渗透胁迫对玉米幼苗叶片叶绿素含量及POD活性的影响［J］. 干旱地区农业研究，25（1）：140-148.

李合生. 2000. 植物生理生化实验原理和技术［M］. 北京：高等教育出版社，

李明，王根轩. 2002. 干旱胁迫对甘草幼苗保护酶活性及脂质过氧化作用的影响［J］. 生态学报，22（4）：503-507.

李强，曹建华，余龙江，等. 2010. 干旱胁迫过程中外源钙对忍冬光合生理的影响［J］. 生态环境学报，19（10）：2291-2296.

李瑞智，黄林. 1989. 二氧化硫对小麦、绿豆幼苗游离脯氨酸的影响［J］. 重庆环境科学，11（1）：23-26.

鲁存海. 2010. 8种野生早熟禾抗旱性及草坪质量综合评价研究［D］. 兰州：甘肃农业大学.

马剑，刘桂林，颉芳芳，等. 2011. 水分胁迫对八宝景天生理特性的影响［J］. 中国农学通报，27（6）：99-102.

宁建凤. 2005. 氮对盐胁迫下库拉索芦荟生长及生理特性的影响［D］. 南京：南京农业大学.

孙国荣，彭永臻，阎秀峰，等. 2003. 干旱胁迫对白桦实生苗保护酶活性及脂质过氧化作用的影响［J］. 林业科学，39（1）：165-167.

王晶懋，张楚涵，闫庆伟，等. 2012. 植物抗旱的生理渗透调节及保护酶活性研究进展［J］. 黑龙江生态工程职业学院学报，（2）：31-33.

王学奎. 2006. 植物生理生化实验原理与技术［M］. 北京：高等教出版社，

陈健辉，李荣华，郭培国，等. 2011. 干旱胁迫对不同耐旱性大麦品种叶片超微结构的影响［J］. 植物学报，46（1）：28-36.

阎秀峰，李晶，祖元刚. 1999. 干旱胁迫对红松幼苗保护酶活性及脂质过氧化作用的影响［J］. 生态学报，19（06）：850-854.

赵建青. 2009. 草石蚕高产栽培及加工技术［J］. 农业技术与装备，04（8）：36-37.

张明生，谢波，谈锋，等. 2003. 甘薯可溶性蛋白、叶绿素及 ATP 含量变化与品种抗旱性关系的研究［J］. 中国农业科学，36（1）：13-16.

Dure L. 1993. P1ant responses to cellular dehydration during environment stress［J］. Plant Physiology，103（10）：91-93.

Giannopolitis C N，Ries S K. 1977. Superoxide dismutase II. Purification and quantitative relationship with water soluble protein in seedlings［J］. Plant Physiology，59：315-318.

基于枸杞枝条粉的复配基质对辣椒育苗效果的影响

曲继松[1, 2]，李堃[2]，高丽红[1*]，张丽娟[2]，朱倩楠[2]
（1. 中国农业大学园艺学院，北京　100093；2. 宁夏农林科学院种质资源研究所，银川　750002）

摘要：分析不同枸杞枝条复配基质对辣椒幼苗生长发育及光合参数的影响，比较复配基质的育苗效果，筛选适宜的辣椒育苗基质配比方案，为枸杞枝条复配基质的研发提供技术支撑。以枸杞枝条、珍珠岩和蛭石作为基质材料，共设10个复配处理，以“壮苗二号”育苗基质作为对照，分析不同复配基质的物理性状及其对辣椒幼苗生长发育及光合参数的影响。添加枸杞枝条可降低复配基质的容重，提高复配基质的通气孔隙和持水孔隙。T10茎粗比CK高4%，T6处理辣椒幼苗的长势最强。T6处理辣椒幼苗的地上部鲜质量达到0.657g/株，T5处理辣椒的地上部干质量较CK高11.11%。T10处理的壮苗指数比CK高41.93%。通过综合性状分析得出：V（枸杞枝条）：V（珍珠岩）：V（蛭石）=6:1:2（T10）和6:1:1（T6）为辣椒育苗的最适枸杞枝条基质配比方案，可作为园艺栽培基质进行研发和利用。

关键词：农林废弃物；园艺基质；辣椒；育苗；生长发育；光合特性

Effects of Wolfberry Branches Complexed Growing Medium on the Growth of Hot Pepper（*Capsicum annuum* L.）Seedlings

Qu Jisong[1, 2]，Li Kun[2]，Gao Lihong[*]，Zhang Lijuan[2]，Zhu Qiannan[2]
（*Institute of Germplasm Resources*，*Ningxia Academy of Agriculture and Forestry Science*，*Yinchuan*　750002）

Abstract：In order to analyze the effects of different horticultural medium of Lycium barbarum on the growth and photosynthetic parameters of pepper seedlings and compare the effect of compound substrate on the seedling growth and screen the appropriate proportioning plan of pepper seedling matrix to provide the technical support for the research and development of the matrix of Lycium barbarum. Taking wolfberry branches，perlite and vermiculite as matrix materials，a total of 10 compound treatments were taken. The seedling matrix of “Zhuang Miao 2” was used as a control. The physical characters of different compound matrix and their effects on the growth and photosynthesis of pepper seedlings and photosynthesis Effect of parameters. The addition of wolfberry branches can reduce the bulk density of the compound matrix，improve the ventilation matrix and water holding pore porosity. T10 stem 4% higher than CK，T6 treatment pepper seedlings growing the strongest，T6 fresh shoots of peppers

---

收稿日期：2018-03-08
基金项目：公益性行业（农业）科研专项 “作物秸秆基质化利用”（201503137），宁夏回族自治区重点研发计划项目（2016-2017农业科技园区项目）

作者简介：曲继松（1980—），男，吉林永吉人，主要从事设施环境调控和生物质基质化利用方面的研究。E-mail:qujs119@126.com
*通讯作者：高丽红（1967—），女，教授，博士生导师，从事设施环境调控与无土栽培方面的研究。Email：gaolh@cau.edu.cn

reached above the fresh mass of 0.657g /strain, The shoot dry weight of T5 was 11.11% higher than that of CK.T10 treatment of seedling index higher than CK 41.93%, The results showed that V（perlite）: V（vermiculite）= 6: 1: 2 and 6: 1: 1 were the optimum combination of the wolfberry branches and shoots in pepper seedling breeding. R & D and utilization as a horticultural medium.

Key words: Agriculture and forestry waste; Horticultural medium; Pepper; nursery; Growth and development; Photosynthetic parameters

草炭是现代园艺无土栽培生产中广泛使用的基质原料，在自然条件下草炭形成约需上千年时间，过度开采利用，使草炭的消耗速度加快，体现出“不可再生”资源的特点[1-2]。很多国家已经开始限制草炭的开采，导致草炭的价格不断上涨[3]，同时国内草炭资源分布不均匀，受产地所限，长途运输无疑会增加育苗成本[4]，因此，开发和利用来源广泛、性能稳定、价格低廉，又便于规模化商品生产的草炭替代基质的研究已成为热点。

目前世界上90%以上的商业性无土栽培是采用基质栽培方式，对基质的需求量连年增加，基质的研究是基质栽培的基础和关键[5-6]。世界上普遍应用的基质原料是草炭，但草炭是不可再生资源，大量开采会破坏湿地环境，加剧温室效应[7]。因此，低成本、环保型无土栽培基质成为研究的重点。国外利用蔬菜废弃物、园林废弃物、牛圈垫料、城市废弃物、修剪肥料、稻秸秆替代草炭作为园艺基质，且均获得良好效果[8-13]。国内对本地丰富廉价的资源生产替代草炭作为园艺基质做了大量的研究工作，利用麦秆、玉米秆、菇渣等材料尝试对基质进行产业化开发，效果较佳[14-16]，利用各种农产废弃物（芦苇末、竹废弃物、花生壳、醋糟等）作为原料配制有机生态型无土栽培基质栽培蔬菜，产量和品质均得到大幅度提高[17-19]。

枸杞种植是宁夏地区的农业优势主导产业，至2017年宁夏枸杞种植面积高达90万亩，占全国枸杞种植面积的50%以上。在枸杞种植管理中，每年12月至下一年的1～3月都需要对不带叶的粗硬枸杞枝条进行修剪，除此之外，在枸杞生长期还需要不断地将不结果枝条（俗称“幼条”）剪掉。目前，大部分被修剪下来的枸杞枝条仅作为燃料使用，利用价值十分低，若将每年修剪下来的枸杞枝条开发利用，作为替代草炭的育苗基质，不仅可以为基质的生产提供来源，还可以降低宁夏等西北地区的育苗成本。项目组[20-21]结合宁夏特色产业，将废弃的枸杞枝条中添加尿素、鸡粪等物质进行发酵，开发出枸杞枝条基质，不仅可为基质的生产提供来源，还可以调整基质的孔隙度、容重、pH值等性质。本研究将项目组开发的枸杞枝条基质与珍珠岩、蛭石按不同体积配比，进行辣椒育苗试验，以“壮苗二号”育苗基质作为对照，分析枸杞枝条复配基质的物理性状及其对辣椒幼苗生长的影响，筛选出适宜辣椒育苗的枸杞枝条基质复配方案，为枸杞枝条复配基质的研发和生产提供技术支撑。

1 材料与方法

1.1 供试材料

试验于2016年10月25日至2016年12月23日在宁夏农林科学院园林场试验基地的日光温室内进行，供试辣椒品种为“陇椒5号”。育苗基质材料为：枸杞枝条，珍珠岩，蛭石和“壮苗二号”育苗基质。枸杞枝条来源于宁夏枸杞种植基地，经过发酵后使用。具体发酵方法为：将修剪下来的的枸杞枝条粉碎成0.5～1.0cm的碎屑装入发酵池（1m×1m×1m），并加入3.0kg的尿素和20.0kg的消毒鸡粪，混合均匀后用塑料薄膜覆盖，保持60%～65%的相对含水量，高温密闭发酵75d。发酵枸杞枝条的化学性质为：pH值=7.80，全氮12.5g·kg^{-1}，全磷2.73g·kg^{-1}，全钾15.2g·kg^{-1}，有机碳35.8%[22]。

1.2 试验设计

试验采用随机区组设计，共设10个处理，具体如表1所示，各处理重复3次，每处理1个穴盘，CK使用“壮苗二号”育苗基质。播种前，辣椒种子温汤浸种15min后于室温中浸泡8h，然后用湿毛巾包裹放置于28℃的培养箱中催芽48h，当70%的种子露白时点播于98穴标准育苗穴盘中。试验期间日光温室环境控制为白天最高温度28℃，夜间最低温度10℃，日平均湿度为40%。不同处理的田间管理统一，由于枸杞枝条的氮、钾含量比较丰富，辣椒育苗期每天浇清水。

表1 各处理混配比例（v:v）

different processing materials mixed ratio

处理Treatment	枸杞枝条粉Wolfberry branches powder	珍珠岩Perlite	蛭石Vermiculite
CK	壮苗二号		
T1	1	—	—
T2	2	1	1
T3	3	1	1
T4	4	1	1
T5	5	1	1
T6	6	1	1
T7	3	1	2
T8	4	1	2
T9	5	1	2
T10	6	1	2

1.3 测定指标及方法

（1）基质物理性状的测定。参照郭世荣[16]的方法测算基质的干体积质量，总孔隙度，通气孔隙，持水孔隙，以及大小孔隙比。

（2）辣椒幼苗生长指标的测定。辣椒出苗40d时，分别取样测定辣椒幼苗的株高、茎粗、叶片数、最长根的根长及总根体积。辣椒幼苗三叶一心时测定地上部、地下部干鲜质量，计算壮苗指数和根冠比，计算公式为：壮苗指数=（茎粗/株高+地下部干质量/地上部干质量）×全株干质量g/株；根冠比=单株地下部干质量/单株地上部干质量。每处理取样5株，3次重复。

（3）辣椒幼苗光合及荧光指标的测定。辣椒幼苗三叶一心时，于晴天10:00测定光合指标（TPS-2便携式光合作用测定系统）和荧光指标（Handy PEA荧光仪）。测定时育苗温室内部光照强度为（1 000 ± 50）μmol·m^{-2}·s^{-1}，CO_2浓度为（400 ± 20）μmol·mol^{-1}。测定叶片为秧苗最高一片完全展开功能叶，每个处理测量3片叶片，随机选择取样。

1.4 数据处理

使用Excel 2007软件进行数据处理，使用DPS v7.05软件LSD检验法进行方差分析。

2 结果与分析

2.1 枸杞枝条复配基质的物理性状

由表2可以看出，各处理复配基质的干体积质量均在0.14～0.16g·cm^{-3}范围内，且显著低于CK，说明枸杞枝条基质与其他基质复配可以降低干体积质量。各处理复配基质的总孔隙度、通气孔隙及持水孔隙均显著高于CK，说明枸杞枝条复配基质的保水性和通气性优于壮苗二号育苗基质，育苗基质中添加枸杞枝条可以增加基质的孔隙度。由不同处理对比可知，基质配比中枸杞枝条所占比例越大，配比基质的干体积质量越小，总孔隙度越大，大小孔隙比越小；基质配比中蛭石所占比例越大，配比基质的干体积质量越大，总孔隙度越小，大小孔隙比越大。

表 2 不同复配基质的物理性状

Table 2 Physical properties of different substrates formula

处理 Treatment	干体积质量（g·cm^{-3}）Dry bulk density	总孔隙度（%）Total porosity	通气孔隙（%）Aeration porosity	持水孔隙（%）Water-holding porosity	大小孔隙比 Void ratio
CK	0.23 ± 0.02a	39.10 ± 2.44i	8.18 ± 0.07i	30.91 ± 1.38g	0.26 ± 0.01h
T1	0.14 ± 0.01c	74.97 ± 2.40ab	38.14 ± 1.73a	36.83 ± 1.69f	1.04 ± 0.01a
T2	0.14 ± 0.01c	63.52 ± 1.04f	23.63 ± 1.09e	39.89 ± 0.96e	0.59 ± 0.00e
T3	0.14 ± 0.00c	66.17 ± 0.43ef	19.90 ± 0.96f	46.27 ± 0.95b	0.43 ± 0.02f
T4	0.15 ± 0.01bc	49.54 ± 1.74h	12.48 ± 1.40h	37.07 ± 0.95f	0.34 ± 0.01g
T5	0.15 ± 0.01bc	59.92 ± 1.46g	17.66 ± 1.23g	42.26 ± 1.32cd	0.42 ± 0.00f
T6	0.15 ± 0.01bc	71.77 ± 1.41bc	31.65 ± 1.44b	40.12 ± 1.74de	0.79 ± 0.00b
T7	0.16 ± 0.01b	73.57 ± 1.10b	30.56 ± 1.42bc	43.01 ± 1.56c	0.71 ± 0.01c
T8	0.15 ± 0.01bc	69.55 ± 4.31cd	29.09 ± 1.50c	40.46 ± 1.25de	0.72 ± 0.02c
T9	0.15 ± 0.01bc	67.58 ± 1.33de	26.87 ± 1.11d	40.70 ± 1.26de	0.66 ± 0.02d
T10	0.14 ± 0.01c	76.93 ± 1.29a	19.36 ± 0.92fg	57.56 ± 1.09a	0.34 ± 0.01g

注：同列数据后标不同字母表示差异显著（P<0.05），下表同

Note：Different letters show significant difference between treatments at P<0.05 according to LSR test，the same as blow.

2.2 不同复配基质对辣椒幼苗生长指标的影响

如表3所示，T1、T2、T3和T4处理辣椒幼苗的株高、茎粗、叶片数、总根体积等生长指标较好，且优于CK；T7处理辣椒的株高、茎粗、根体积等指标最差，且低于CK。在出苗40d时，T4和T10处理的叶片数较多，与CK差异不显著。在出苗40d时，T1、T2、T3、T4和T6处理的株高分别比CK高15.72%、7.64%、10.92%、33.84%、6.55%，T1、T4、T8

和T10茎粗分别比CK高9%、16%、4.5%、4%，T4和T10的叶片数略高于CK，所有复配基质的根长均高于CK，处T9处理外，其余复配基质的总根体积均高于CK，通过上述生长指标综合分析可知，T9处理辣椒幼苗长势最弱，T6处理辣椒幼苗的长势最强。

表 3　不同复配基质对辣椒幼苗生长指标的影响

Table 3　Effects of different substrate formulas on growth index of tomato seeding

处理 Treatment	株高（cm） Plant height	茎粗（mm） Stem diameter	叶片数 Leaf number	最长根的根长（cm） Root height	总根体积（cm^3） Root volume
CK	4.58 ± 0.51cd	2.00 ± 0.07bcd	4.0 ± 0.5a	9.17 ± 0.70c	0.34 ± 0.08bc
T1	5.30 ± 0.28b	2.18 ± 0.15ab	4.0 ± 0.5a	9.97 ± 1.07bc	0.47 ± 0.04a
T2	4.93 ± 0.45bc	2.01 ± 0.12bcd	3.3 ± 0.9b	10.13 ± 1.46bc	0.43 ± 0.10ab
T3	5.08 ± 0.26b	1.94 ± 0.21cd	3.8 ± 1.0ab	10.60 ± 1.11bc	0.36 ± 0.03abc
T4	6.13 ± 0.29a	2.32 ± 0.12a	4.2 ± 0.4a	11.70 ± 2.95ab	0.40 ± 0.09abc
T5	4.32 ± 0.23de	2.00 ± 0.15bcd	3.9 ± 0.7ab	10.57 ± 1.25bc	0.36 ± 0.07abc
T6	4.88 ± 0.35bc	2.01 ± 0.16bcd	3.7 ± 0.7ab	12.17 ± 1.72ab	0.42 ± 0.05ab
T7	4.07 ± 0.216e	1.83 ± 0.06d	3.6 ± 1.0ab	10.30 ± 0.10bc	0.39 ± 0.08abc
T8	4.17 ± 0.57de	2.09 ± 0.10bc	3.9 ± 0.7ab	11.23 ± 0.76abc	0.47 ± 0.02a
T9	4.18 ± 0.19de	1.96 ± 0.14bcd	3.9 ± 0.6ab	10.87 ± 0.67abc	0.30 ± 0.02c
T10	4.20 ± 0.52de	2.08 ± 0.46bc	4.1 ± 1.0a	13.10 ± 1.81a	0.37 ± 0.05abc

2.3　不同复配基质对辣椒干鲜质量、根冠比及壮苗指数的影响

由表4可知，T6处理辣椒幼苗的地上部鲜质量最高，达到0.657g/株，其次是T4处理的地上部鲜质量较高，T6和T4处理的地上部鲜质量分别比CK高22.12%和14.31%。T5处理辣椒的地上部干质量最大，其中地上部干质量比CK高11.11%。由根冠比可以看出，各处理的根冠比与CK相比无显著性差异，其中T2处理辣椒的根冠比最大，CK和T9处理辣椒的根冠比最小。由壮苗指数可知，T10处理的壮苗指数最大，比CK高41.93%，其次为T6处理的壮苗指数较大，说明使用T10和T6处理的基质复配方案育苗时，辣椒幼苗的质量较优。

表 4　不同复配基质对辣椒干鲜质量、根冠比及壮苗指数的影响

Table 4　Effects of different substrate formulas on fresh，dry weight，root/shoot and healthy index of tomato seeding

处理 Treatment	地上部鲜质量(g) Shoot fresh weight	地上部干质量(g) Shoot dry weight	地下部鲜质量(g) Root fresh weight	地下部干质量(g) Root dry weight	根冠比 Root/Shoot	壮苗指数 Healthy index
CK	0.538 ± 0.078bc	0.045 ± 0.009a	0.380 ± 0.111de	0.022 ± 0.006cd	0.486 ± 0.078b	0.062 ± 0.019a
T1	0.580 ± 0.027ab	0.050 ± 0.004a	0.495 ± 0.039ab	0.028 ± 0.003abc	0.562 ± 0.050ab	0.075 ± 0.009a
T2	0.531 ± 0.022bc	0.044 ± 0.002a	0.482 ± 0.037abc	0.030 ± 0.002a	0.679 ± 0.072a	0.080 ± 0.008a
T3	0.541 ± 0.033bc	0.046 ± 0.008a	0.409 ± 0.014abcde	0.027 ± 0.006abcd	0.581 ± 0.054ab	0.070 ± 0.013a
T4	0.615 ± 0.116ab	0.043 ± 0.007a	0.450 ± 0.071abcd	0.025 ± 0.003abcd	0.586 ± 0.035ab	0.066 ± 0.010a
T5	0.447 ± 0.028c	0.050 ± 0.001a	0.393 ± 0.059cde	0.029 ± 0.002ab	0.575 ± 0.032ab	0.082 ± 0.005a

（续表）

处理 Treatment	地上部鲜质量(g) Shoot fresh weight	地上部干质量(g) Shoot dry weight	地下部鲜质量(g) Root fresh weight	地下部干质量(g) Root dry weight	根冠比 Root/Shoot	壮苗指数 Healthy index
T6	0.657 ± 0.030a	0.043 ± 0.006a	0.455 ± 0.042abcd	0.029 ± 0.004ab	0.574 ± 0.084ab	0.087 ± 0.013a
T7	0.546 ± 0.075b	0.045 ± 0.002a	0.413 ± 0.082abcde	0.030 ± 0.001a	0.667 ± 0.042a	0.084 ± 0.004a
T8	0.557 ± 0.057b	0.046 ± 0.003a	0.396 ± 0.054bcde	0.023 ± 0.004bcd	0.502 ± 0.061b	0.070 ± 0.017a
T9	0.446 ± 0.063c	0.041 ± 0.009a	0.334 ± 0.024e	0.021 ± 0.003d	0.483 ± 0.279b	0.068 ± 0.093a
T10	0.586 ± 0.003ab	0.045 ± 0.002a	0.498 ± 0.013a	0.030 ± 0.001a	0.667 ± 0.042a	0.088 ± 0.004a

3 讨论

复配基质需根据不同基质材料的理化性质及幼苗的生物学特性进行配比，部分研究表明，辣椒栽培中基质适宜的干体积质量范围为0.1 ~ 0.8g·cm^{-3}[24-26]，各处理枸杞枝条复配基质的干体积质量均在该范围内，说明枸杞枝条作为栽培基质是可行的。将枸杞枝条基质与其他基质复配可以降低干体积质量、增加基质的孔隙度，配比基质中枸杞枝条所占比例越大，配比基质的干体积质量越小，总孔隙度越大，大小孔隙比越小。根据蒋卫杰等[27]提出的大小孔隙比以0.25 ~ 0.5为宜的标准，T3、T4、T5和T10处理的大小孔隙比较适宜，且优于CK。根据田吉林等[28]提出的，基质总孔隙度的适宜范围为60% ~ 90%，通气孔隙度>15%、持水孔隙>45%的标准，T3和T10处理枸杞枝条复配基质的物理性状最适宜。

在育苗方面，由10种基质辣椒幼苗的株高、茎粗、叶片数、根长和根容综合分析可知，T4和T10处理的叶片数较多，T10茎粗分别比CK高4%，T9处理辣椒幼苗长势最弱，T6处理辣椒幼苗的长势最强。T6处理辣椒幼苗的地上部鲜质量最高，T5处理辣椒的地上部干质量最大，T2处理辣椒的根冠比最大，T10处理的壮苗指数最大。

将不同处理基质的物理性状与辣椒穴盘苗的指标综合分析可知，各处理中，T10处理的枸杞枝条复配基质较适于辣椒穴盘育苗，T6、T5处理次之。

4 结论

T10处理和T6处理（*V*（枸杞枝条）:*V*（珍珠岩）:*V*（蛭石）=6:1:2&6:1:1）的枸杞枝条复配基质物理性质和育苗效果较好，较适于作为辣椒穴盘育苗的基质配比方案。

参考文献

［1］ Yao H Y，Jiao X D，Wu F Z. Effect of continuous cucumber cropping and alternative rotations under protected cultivation on soil microbial community diversity［J］.Plant and Soil.2006，284：195-203.

［2］ 安忠花.我国生态农业现状与发展前景［J］.现代农业科技，2015（3）：270-271

［3］ 邓一鸣，刘聪.我国无土栽培的现状及发展趋势［J］.北京农业，2015（3）：5

［4］ Guerin V，Lemaire F，Marfa O，*et al*.Grwoth of Viburnmu tinus in Peat-based and peat-substitute growing media［J］.Seientiaortieulturae，2001，89（2）：129-142.

［5］ 李光义，李勤奋，张晶元. 木薯茎秆基质化的堆肥工艺及评价［J］.农业工程学报，2011，27（1）：320-325.

［6］ 宋成军，田宜水，罗娟，等. 厌氧发酵固体剩余物建植高羊茅草皮的生态特征［J］.农业工程学报，2015，31（17）：254-260.

［7］ 张硕，余宏军，蒋卫杰.发酵玉米芯或甘蔗渣基质的黄瓜育苗效果［J］.农业工程学报，2015，31（11）：236- 242..

［8］ Catia Santos，Piebiep Goufo，Joao Fonseca，*et al.* Effect of lignocellulosic and phenolic compounds on ammonia，nitric oxide and greenhouse gas emissions during composting［J］.Journal of Cleaner Production，2018，171：548-556

［9］ Rafaela Cáceres，Narcís Coromina，Krystyna Malin´ska，*et al.*Nitrification during extended co-composting of extreme mixtures of green waste and solid fraction of cattle slurry to obtain growing media［J］. Waste Management，2016，58：118–125

［10］ Rafaela Cáceres，Narcís Coromina，Krystyna Malinska，*et al.*Evolution of process control parameters during extended co-composting of green waste and solid fraction of cattle slurry to obtain growing media［J］. Bioresource Technology ，2015，179：398–406

［11］ J.Jara-Samaniego，M.D.Perez-Murcia，M.A.Bustamante，*et al.*Composting as sustainable strategy for municipal solid waste management in the Chimborazo Region，Ecuador：Suitability of the obtained composts for seedling production［J］. Journal of Cleaner Production，2017，141：1349-1358

［12］ A.Nietoa，G.Gascó，J.Paz-Ferreiro，*et al.*The effect of pruning waste and biochar addition on brown peat basedgrowing media properties［J］.Scientia Horticulturae ，2016，199：142–148

［13］ Jaco Emanuele Bonaguro，Lucia Coletto，Giampaolo Zanin. Environmental and agronomic performance of fresh rice hulls used as growing medium component for Cyclamen persicum L. pot plants［J］.Journal of Cleaner Production，2017，142：2 125-2 132

［14］ 李谦盛，郭世荣，李世军.利用工农业有机废弃物生产优质无土栽培基质［J］.自然资源学报.2002，17（4）：515-519.

［15］ 张唐娟，袁巧霞，陈红，等.菇渣作为有机栽培基质好氧改性的实验［J］.环境工程学报.2013，7（2）：722- 726

［16］ 刘超杰，郭世荣，王长义等.混配醋糟复合基质对辣椒幼苗生长的影响［J］.园艺学报，2010，37（4）：559-566.

［17］ 马彦霞，郁继华，张晶，等.设施蔬菜栽培茬口对生态型无土栽培基质性状变化的影响［J］.生态学报，2014，34（14）：4 071-4 079.

［18］ 辜夕容，邓雪梅，刘颖旎，等.竹废弃物的资源化利用研究进展［J］.农业工程学报，2016，32（1）：236-242.

［19］ 刘伟，余宏军，蒋卫杰.我国蔬菜无土栽培基质研究与应用进展［J］.中国生态农业学报，2006，14（3）：4-7.

［20］ 冯海萍，曲继松，杨冬艳，等.接种微生物菌剂对枸杞枝条基质化发酵品质的影响［J］.环境科学学报.2015，35（5）：1 457-1 463

［21］ 冯海萍，杨志刚，杨冬艳，等.枸杞枝条基质化发酵工艺及参数优化［J］.农业工程学报，2015，31（5）：252- 260.

［22］ 鲍士旦.土壤农化分析（第三版）［M］.北京：中国农业出版社，2000.

［23］ 高俊凤.植物生理学实验指导［M］.西安：世界图书出版西安公司，2000.

［24］ Raviv M，Oka Y，Katsn J，*et al.*High-Nitrogen Compost as a Medium for Organic Container-Grown Crops［J］.Bioresource Technology，2005，96（4）：419-427.

［25］ 徐文俊，程智慧，孟焕文，等.农业废弃有机物基质配方对辣椒生长及产量的影响［J］.西北农林科技大学学报（自然科学版），2012，40（4）：127-133.

［26］ De BOODT M，VERDONDK O. The physical properties of the substrates in hort［J］. Acta Horticulturae，1972，26：37-44.

［27］ 蒋卫杰，杨其常.小康之路·无栽培特选项目与技术［M］. 北京：科学普及出版社，2008.

［28］ 田吉林，奚振邦，陈春宏，等.无土栽培基质的质量参数（孔隙性）研究［J］.上海农业学报，2003，19（1）：46-49.

水氮耦合对泥炭培番茄氮素吸收与分配的影响

张延平，温祥珍，李亚灵
（山西农业大学园艺学院，太谷　030801）

摘要：为了了解水肥对番茄氮素吸收与分配规律，以进口泥炭为盆栽基质对“红尊贵”番茄进行水氮耦合试验。研究了在不同氮素（每株每次施氮分别为N1：0.05g、N2：0.2g）和水分（每株每次精确灌水分别为：W1：600ml、W2：750ml、W3：900ml）条件下番茄的氮素吸收利用效果。结果表明：施氮水平显著影响了植株氮素含量，与N1相比，N2水平下提高了植株各器官的氮素含量，并且N2植株最终氮素含量（27g/kg）比N1（18g/kg）增加了50%，植株累积吸氮量比N1增加了90%。累积吸氮量随着灌水量的增加而增加，W2、W3比W1分别增加了7%和17%。施氮水平和灌水量对氮素的物质分配也有一定的影响，N2的茎和叶氮素所占比例比N1多4%和2%，根和果氮素所占比例比N1少1%和5%；W3处理下茎叶中氮素所占比例（42%）比W1和W2（38%）多4%，果实所占比例（56%）比W1和W2（60%）少4%。结论认为，本试验条件下，增加施氮量可以提高植株氮素含量和累积吸氮量，增加灌水量可以提高累积吸氮量和氮的利用效率，水肥对氮素的吸收具有相互促进的关系；植株各器官中，茎中氮素含量变化最敏感，可根据茎中氮素含量来诊断植株是否缺氮；低氮低水可促进氮素往果实中分配，高氮高水有利于氮素往茎叶中分配。

关键词：番茄；水氮耦合；氮利用效率

Effect of Water and Nitrogen Coupling on Nitrogen Absorption and Allocation of Tomato by Peat Cultivation

Zhang Yanping，Wen Xiangzhen，Li Yaling
（*College of Horticulture*，*Shanxi Agricultural University*，*Taigu*　030801）

Abstract：To understand the regularity of nitrogen absorption and allocation of tomato by water and fertilizer, the experiment of water and nitrogen potted “*Hongzungui*” tomato by substrate of imported peat. Researched nitrogen absorption and utilization effect of tomato under different nitrogen（nitrogen application of per plant of per time were：N1:0.05g，N2:0.2g）and water（the exact irrigation of per plant of per time were: W1:600mL，W2:750mL，W3:900mL）conditions. The results showed that nitrogen application significantly affected the nitrogen content of the plant. Compared with N1，the N2 level increased the nitrogen content of the plant organs，the final nitrogen content of N2 plants（27g/kg）increased by 50% than that of N1（18g/kg），and the accumulation of N uptake in plant increased by 90% over N1. Accumulation of N uptake increased with the

---

基金项目：国家自然科学基金重点项目（61233006）；山西省煤基重点科技攻关项目（FT201402-05）
作者简介：张延平（1991—），女，河北邯郸人，在读硕士研究生，主要从事蔬菜栽培生理研究
通讯作者：李亚灵（1962—），女，山西灵石人，教授，博士，博士生导师，主要从事蔬菜栽培生理研究

increase of irrigation water, compared with W1, W2 and W3 increased by 7% and 17% respectively. Nitrogen application level and irrigation water also have a certain impact on the allocation of nitrogen, compared with N1, the percentage of nitrogen in stem and leaf of N2 increased by 4% and 2%, but in root and fruit reduced 1% and 5%. The percentage of nitrogen in stem and leave of W3（42%）was 4% more than W1 and W2（38%）, but in fruit（56%）, it was 4% less than W1 and W2（60%）. The conclusion showed, under the experimental conditions, increasing the amount of nitrogen can increase the plant nitrogen content and cumulative nitrogen uptake, increase the irrigation water can increase the cumulative nitrogen uptake and nitrogen use efficiency. Water and fertilizer on absorption of nitrogen has mutual promotion. The changes of nitrogen content in stem is the most sensitive in all organs of plant, and the nitrogen content in stem can be used to diagnose if the nitrogen is deficiency in plant. Combination of low nitrogen and low water can promote the allocation of nitrogen in fruit, and combination of high nitrogen and high water is beneficial to the allocation of nitrogen in stem and leave.

Key words: Tomato; Water and nitrogen coupling; Nitrogen use efficiency

1 引言

水资源紧缺和肥料利用效率低是制约作物产量和农业发展的重要因素[1]。探讨新的节水灌溉技术，提高农业生产中的水分利用效率，是当前农业发展中的首要问题[2]。N素是获得作物高产必不可少的因素[3]，并且适宜的施氮量可提高番茄产量和品质[4]，但在土壤和品种间的最佳施氮量有很大的不同[5]，国内对不同作物的氮素营养及氮肥施用量的研究表明，氮肥的利用率低至28%~41%[6]，而番茄在种植中，具有需水量大，根系发达，吸水能力强等特点[7]，当在供水充足条件下，给予足够的施氮量，番茄的产量可大幅提升[8]。但过量的水氮对环境有一定负作用[9]。目前国内外关于水氮耦合的研究[10-12]多是在土壤栽培条件下对番茄生长、干物质生产、水肥利用效率等方面的探讨，而对于氮素吸收与分配上缺乏量化研究。基质栽培能够充分发挥作物增产的潜力[13]，还具有避免土传病害、减轻连作障碍，成本低、易于标准化管理和产品品质好等优点[14]。因此，本试验将探索泥炭栽培条件下水氮耦合对番茄氮素吸收与分配的影响，以更好地指导生产实践。

2 材料方法

2.1 试验材料及场地

试验材料为番茄“红尊贵”品种，是由国外引进的杂交一代新品种（产自西安华番农业发展有限公司），属于无限生长型早熟品种。

试验于山西农业大学设施农业工程研究所进行，番茄种植在用阳光板修建的长12m、宽11.5m、顶高2.5m、墙高2m的生长室内。在2017年4月17日至8月3日109d生长期内，平均温度为26℃，相对湿度为40%。

2.2 试验方法

试验设置两个氮素水平：即低氮（N1）：0.05g/次/株、高氮（N2）：0.2g/次/株，每个氮素水平下设置三个水分处理：即每次精确灌水分别为低水（W1）：600mL/株、中水（W2）：750mL/株、高水（W3）：900ml/株（苗期W1、W2、W3用量分别为400ml/

株、500ml/株、600ml/株），共6个处理。每个处理种植40株番茄幼苗，重复4次（其中1次重复为计产区），共计960株，随机区组排列。

试验于4月17日定植，将生长一致的4片叶的幼苗定植在放有防漏袋的黑色营养钵中（型号为30×28cm），营养钵中有7.5L的苔藓泥炭（产自德国福洛伽公司）基质。从定植开始定期浇灌相应浓度的营养液（表1）。全生长期N1、N2总施氮量为1.55g/株和6.2g/株（图1-A），W1、W2和W3总灌水量分别为26.6L/株、32.5L/株和38.4L/株（图1-B）。

表 1　试验期间浇灌营养液的配方（mg/株/次）

Table 1　The formula of nutrient solution for irrigation during the experiment period

处理 Treatment	$Ca(NO_3)_2$	KNO_3	NH_4NO_3	KH_2PO_4	$MgSO_4.7H_20$	EC（mS/cm）
N1	177	202	2.857	45.56	123	1.08～1.26
N2	177	202	431.429	45.56	123	1.55～1.92

注：EC值表示电导率值，由便携式电导率仪（HI8733）在给植株浇灌营养液之前进行测定。

Note：The EC value indicates the conductivity value，it measured by a portable conductivity meter（HI8733）before watering the nutrient solution to the plant.

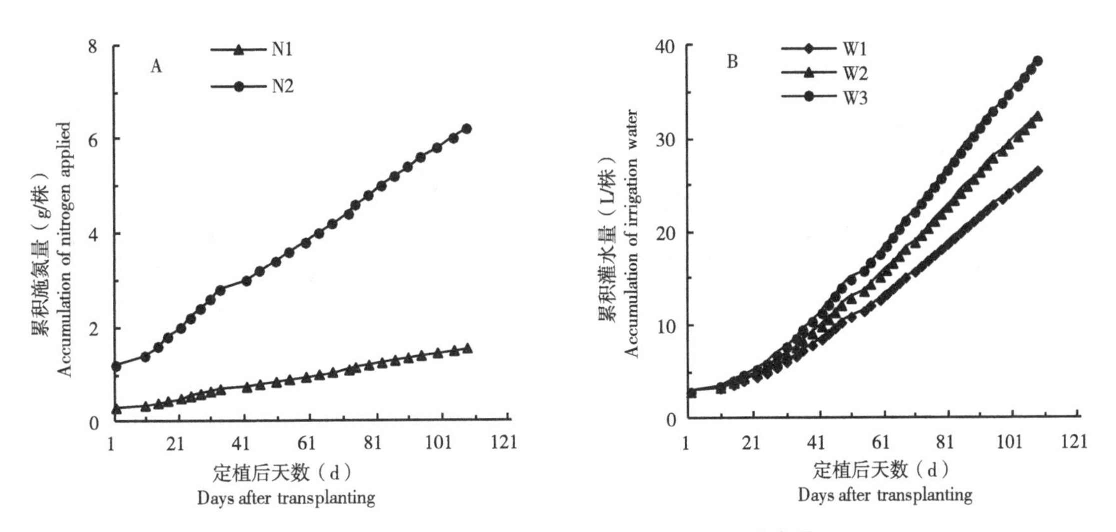

图 1　试验期间不同处理累积施氮量和累积灌水量

Figure 1　Accumulation of nitrogen applied and irrigation water under different treatments during the experiment period

2.3　测定内容及方法

番茄植株干鲜重的测定：定植后，每10d（最后两次为15d）每个处理选取3株长势一致的植株，6个处理3次重复共54株，将植株从营养钵中连同基质和完整的根系取出，用水冲洗干净，并将植株分解，分别测定根、茎、叶、果的鲜重，再将它们在105℃下杀青20min，然后在80℃下烘干至恒重，称其干重。

植株各器官含氮量测定：将植株各器官干样用粉碎机粉碎，利用H_2SO_4-H_2O_2消煮，然后用凯氏定氮法测定各器官的含氮量，各器官累积吸氮量（g）=器官含氮量（g/kg）×器官干物质量（kg/株），根、茎、叶、果累积吸氮量相加即为植株氮累积吸收量。植株含氮量（g/kg）=植株累积吸氮量（g/株）/植株干物质量（kg/株）。

3 结果与分析

3.1 水氮耦合对番茄根、茎、叶和植株中氮素含量的影响

水氮耦合处理会直接影响番茄植株根、茎、叶、果实中氮素含量，测定结果如图2（N1W1、N1W2、N1W3、N2W1、N2W2和N2W3的果实氮素含量为23g/kg、18g/kg、21g/kg、28g/kg、28g/kg、26g/kg）所示。同一水分处理下，N2水平下番茄各器官的氮素含量明显高于N1，并且随着试验的进行，茎中氮素含量差异逐渐变大；同一氮素水平下，水分对各器官氮素含量影响相对较小。N2水平下，灌水量越大，植株氮素含量越小；N1水平下，W2植株氮素含量最低，W3次之，W1最高。

整个生长期中，根系中氮素含量相对稳定，茎、叶中氮素含量在缓苗后有所上升，开花坐果后茎、叶中氮素含量逐渐降低，且茎中氮素含量下降幅度较大，采收初期相对稳定。在番茄整个生长期中，叶片中氮含量始终高于根和茎，这主要可能是植物体内氮主要存在于蛋白质和叶绿素中，而叶绿素又主要集中在叶片中。

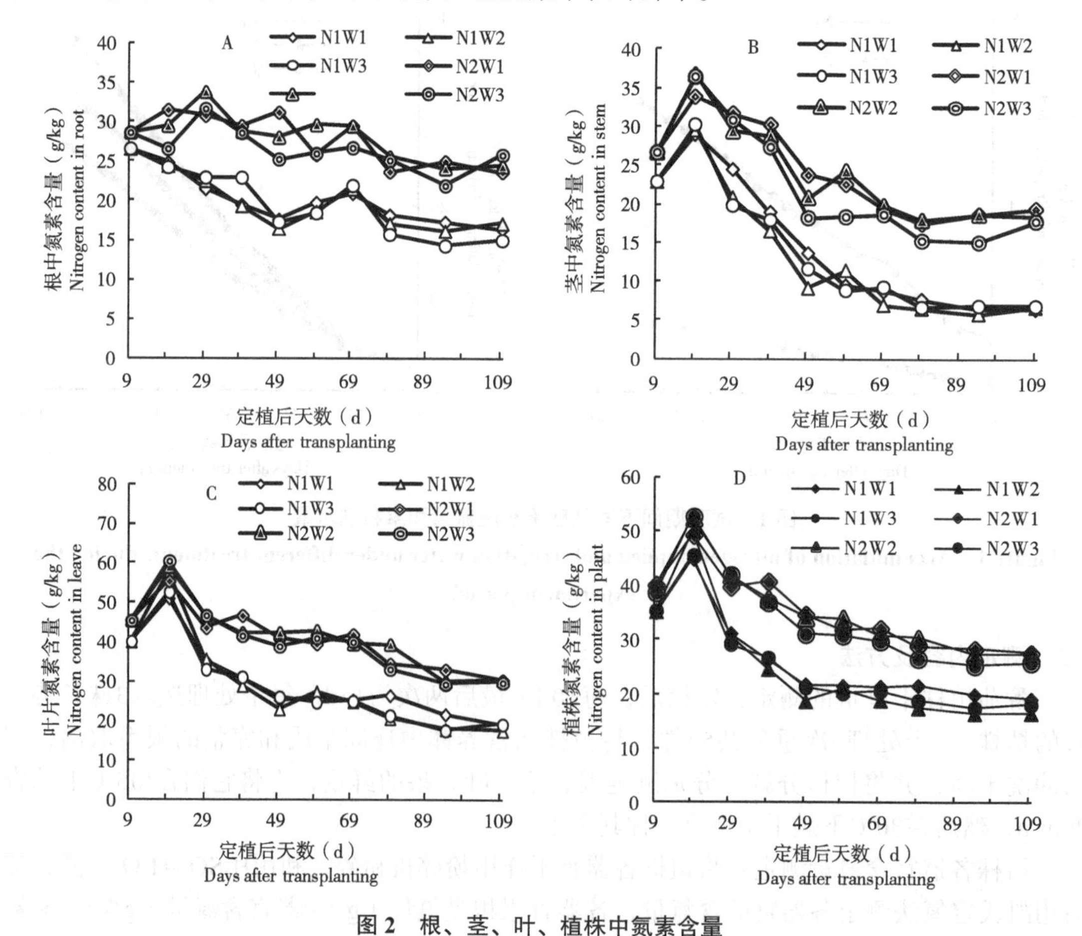

图2 根、茎、叶、植株中氮素含量

Figure 2 Nitrogen content in root，stem，leave，and plant

具体来说，定植后19d时，茎、叶中氮素含量最高，此时，茎中N1、N2氮素含量分别为30g/kg和37g/kg，叶中氮素含量分别为54g/kg和58g/kg，N2水平下茎、叶氮素含量分别比N1增加了24%和8%。根、茎、叶氮素含量趋于稳定时，N1氮素含量分别为16g/kg、7g/kg和19g/kg，N2根、茎、叶的氮素含量分别为24g/kg、18g/kg、31g/kg，分别比N1增加了50%、157%和63%。

植株中氮素含量和根、茎、叶的变化规律类似，氮素含量第19d达到最大，N1和N2氮素含量最大分别为47g/kg和51g/kg，最终稳定在18g/kg和27g/kg，N2比N1增加了50%。

3.2 水氮耦合对番茄根、茎、叶、果及植株累积吸氮量的影响

不同施氮水平和水分处理影响了番茄植株各器官氮素含量，从而影响植株累吸氮量，根、茎、叶、果和植株累积吸氮量，其测定结果如图3所示。

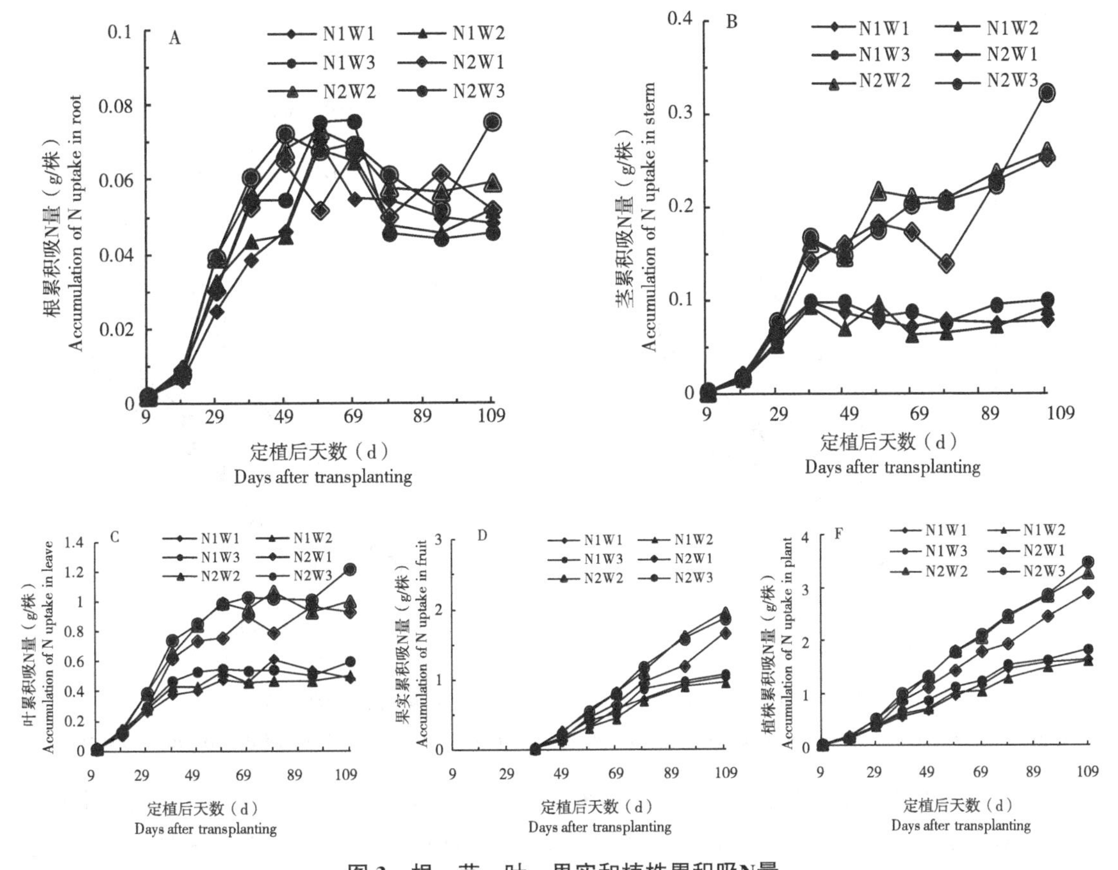

图3 根、茎、叶、果实和植株累积吸N量

Figure 3 Accumulation of N uptake in root，stem，leave，fruit，and plant

由图3可知，番茄从定植到坐果前期（定植后30d），根系所吸收的氮素主要分配给茎和叶；从坐果到实验结束，茎、叶中氮素增加幅度减小，果实吸N量逐渐增加，且增幅并未减慢，这与加工番茄的生长重心转移有关。因此根系吸收的氮素逐渐转移到果实中，在收获末期果实中的氮吸收量占其总吸收量的54%～63%。

同一水分处理下，N2水平下植株的茎、叶、果实以及植株的累积吸N量远高于N1，且N2水平下植株累积吸氮量平均比N1高90%，主要是因为N2植株的氮素含量和植株总干物质量均高于N1。N2水平下，随着灌水量的增加，植株累积吸氮量逐渐增加；N1水平下，植株累积吸氮量反而是W2处理下的最低。

各处理下，各器官最大累积吸氮量均表现为果>叶>茎>根，N1W1、N1W2、N1W3、N2W1、N2W2、N2W3果实累积吸氮量占植株总吸氮量的比例分别为62.80%、60.00%、59.34%、59.29%、59.63%和53.45%，N2水平下，随着灌水量的增加，果实累积吸氮量所占比例越小，N1水平下，W2和W1的果实累积氮素差异不大，W3果实累积吸氮量所占比例明显低于W1和W2。

进一步研究氮素和水分对植株氮素吸收的影响，分别以定植后生长天数为横坐标，植株累积吸氮量为纵坐标，在Excel中作散点图（图4），添加趋势线得出相应的函数式（表2）。

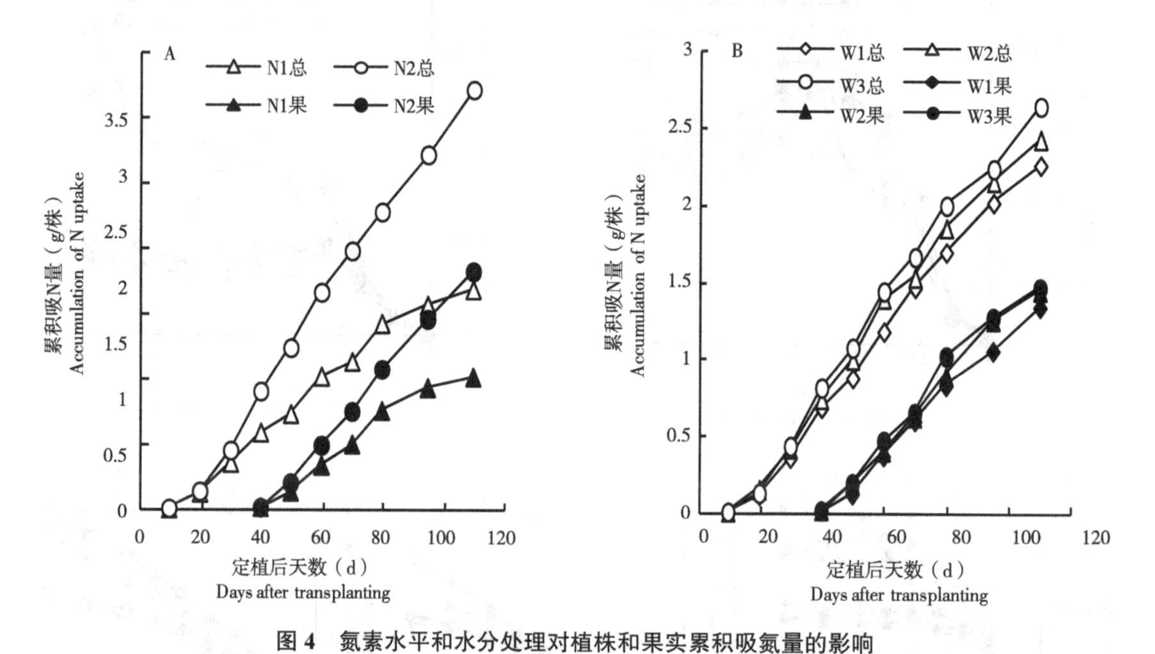

图4　氮素水平和水分处理对植株和果实累积吸氮量的影响

Figure 4　Effects of nitrogen level and water treatment on accumulation of N uptake in plant and fruit

由图4-A可知，前30d内不同氮素水平对植株累积吸氮量的影响不大，30d以后出现差异，并逐渐变大。由表2累积吸氮量函数式可知，单株植株累积吸氮量Y_{PN1}和Y_{PN2}的吸收速率为0.018g/d和0.033 4g/d，N2比N1高86%，虽然N2水平下氮素利用效率只有N1的1/2，但试验结束时N2的植株累积吸氮量比N1高90%，差异显著。就果实的生长来看，定植后约40d开始有果实出现，前期果实生长差异不大，60d后差异逐渐变大，由表2生长量的函数式可知单株果实累积吸氮量Y_{FN1}和Y_{FN2}的吸收速率分别为0.015 3g/d和0.026 5g/d，N2比N1高73%，N2的最终果实累积吸氮量比N1高78%，差异显著。

表 2　氮素水平和水分处理对植株和果实累积吸氮量的影响

Table 2　Effects of nitrogen level and water treatment on accumulation of N uptake in plant and fruit

处理 Treatment	植株累积吸氮量的函数式 The functional formula of accumulation of N uptake in plant	果实累积吸氮量的函数式 The functional formula of accumulation of N uptake in fruit	总吸氮量（g）Total N uptake	果实吸氮量（g）N uptake in fruit	果实N素占比（%）Proportion of N in fruit	氮利用效率（g/g）Nitrogen use efficiency
N1	Y_{PN1}=0.018X−0.13（R^2=0.982 4）	Y_{FN1}=0.015 3X−0.553 7（R^2=0.971 1）	1.69	1.02	60	109%
N2	Y_{PN2}=0.033 4X−0.391 5（R^2=0.995 4）	Y_{FN2}=0.026 5X−1.046 1（R^2=0.997 8）	3.21	1.82	57	52%
W1	Y_{PW1}=0.024X−0.260 5（R^2=0.993 9）	Y_{FW1}=0.019 5X−0.761 6（R^2=0.993）	2.27	1.35	59	59%
W2	Y_{PW2}=0.025 6X−0.244 9（R^2=0.992 2）	Y_{FW2}=0.021 4X−0.824 4（R^2=0.993 3）	2.43	1.45	60	63%
W3	Y_{PW3}=0.027 6X−0.276 9（R^2=0.992 6）	Y_{FW3}=002 18X−0.813 6（R^2=0.984）	2.65	1.47	55	68%

注：Y_P代表植株累积吸氮量，Y_F代表果实累积吸氮量。

Note：Y_P represents cumulative N uptake in plant，Y_F represents cumulative N uptake in fruit.

从图4-B可知，前40d植株累积吸氮量差异不大，40d以后差异逐渐变大。由表2累积吸氮量函数式可知，单株植株累积吸氮量的吸收速率随着灌水量的增加而增大，Y_{PW2}和Y_{PW3}的吸收速率分别比Y_{PW1}高7%和15%，试验结束时，W2、W3的植株累积吸氮量分别比W1高7%和17%，并且氮的利用效率随着灌水凉的增加而增大。从生长60d到试验结束，果实累积吸N量一直保持$Y_{FW3}>Y_{FW2}>Y_{FW1}$。果实氮素累积吸收速率也随着灌水量的增加而增大，W2、W3的吸收速率分别比W1高10%和12%。试验结束时，W2、W3果实累积吸氮量比W1分别高7%和9%。

3.3　施氮水平对对根、茎、叶、果氮素分配的影响

为了研究氮素水平和水分处理对植株氮素分配的影响，分别将同一氮素水平（水分处理）的植株合并计算，以定植后天数为横坐标，各器官的氮素比例为纵坐标，在Excel中作散点图（图5）。

由图5可知，植株前期（定植后生长30d）氮素主要分布在根、茎、叶中，各处理叶片中氮素所占比例都在75%以上，随着植株长大并开始坐果，氮素分配在根、茎、叶中的分配比例逐渐减小，而向果实的分配增加，后期分配比例趋于稳定。

由图5-A施氮水平对氮素分配的影响来说，在分配比例趋于稳定阶段，N1和N2植株氮素最终分配在根、茎、叶、果的比例分别为3%、5%、32%、60%和2%、9%、34%、55%。就这两组数据来看，N1的根、果所占比例分别比N2的多1%、5%，N2的茎、叶所占比例比N1多4%和2%，所以在N1水平下植株氮素往根、果实分配的更多，在N2水平下往茎、叶分配的更多。表明当氮肥充足时，有利于干物质往果实的分配，当氮肥供应不足

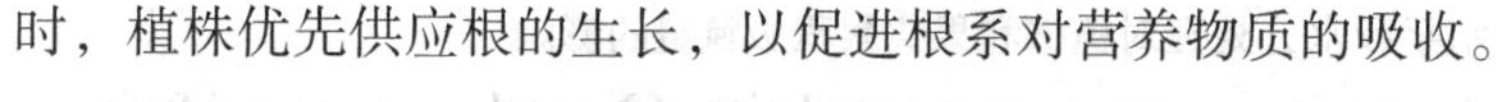

时，植株优先供应根的生长，以促进根系对营养物质的吸收。

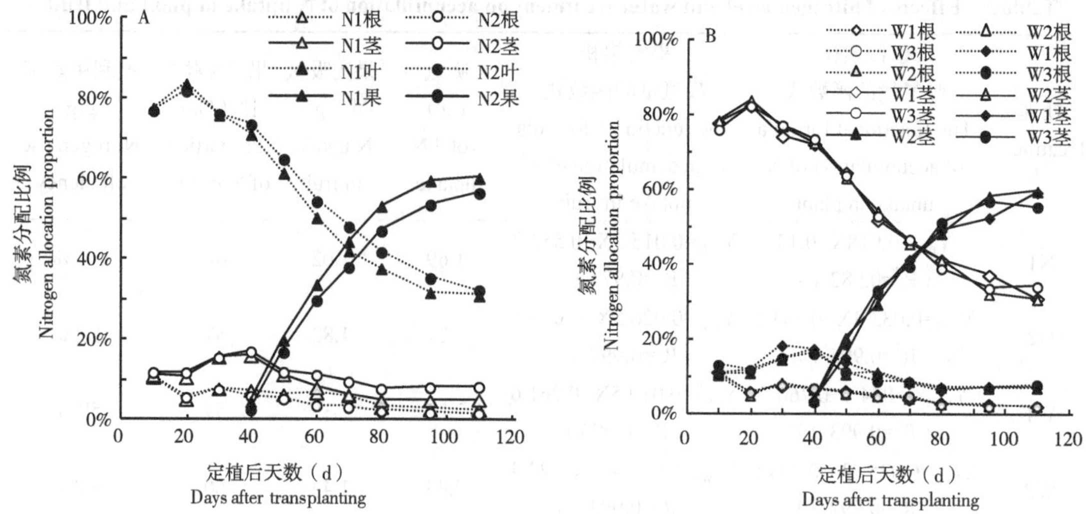

图5　氮素水平和水分处理对根、茎、叶、果分配的影响

Figure 5　Effects of nitrogen levels and water treatment on allocation of root，stem，leaf and fruit

由图5-B可知，从定植到果实收获初期，水分处理对氮素分配的影响差异不大，在拉秧期，各处理根中氮素所占比例均为2%，W1、W2和W3的茎叶所占比例为38%、38%和43%，果所占比例分别为60%、60%、56%。因此，W3条件下，即水分充足时，氮素更倾向往茎叶分配，即更有利于营养生长；当水分较少时（W2和W3），氮素更倾向往果实分配，即抑制营养生长，促进生殖生长。

4　讨论

4.1　水氮耦合对氮素含量和累积吸氮量的影响

养分是构成植物体的组成成分，也是植株进行生理活动不可或缺的参与者。Badr[15]研究表明较低的氮肥供应减少了作物氮的吸收，进而导致了茎叶氮浓度的降低，有限的氮抑制了全冠生长和作物产量，本研究表明，高氮水平下番茄根、茎、叶、果的氮素含量明显高于低氮的，因此在一定范围内，增施氮肥可增加植株各器官氮素含量，从而促进植株的生长和作物产量。柳美玉[16]研究表明随着营养液浓度的提高，植株各元素吸收逐渐增加。本研也究表明，高氮水平下，植株累积吸氮量增加了90%，因此，在一定范围内提高营养液中氮的浓度可促进植株对氮素吸收。

本试验结束时，低氮水平的植株老叶存在失绿变黄情况，最终叶片氮素含量18～19g/kg，仍高于杨明凤[17]、薛琳[18]等拉秧期叶片氮素含量（17.8g/kg和17.9g/kg），而薛琳试验表明，拉秧期植株还未达到自然衰老，因此虽然试验中N1水平施氮量略显不足，但也能满足植株基本生长需求。

从开花坐果后，低氮水平植株的茎中N素含量呈降低趋势，从果实收获初期茎中N素含量就降低为6～7g/kg（低于高方胜[19]、杨明凤[17]等人的研究结果），显著低于高氮水平（17～20g/kg）。氮肥稍有不足时，茎中氮素含量降低较快，因此，可以根据茎中氮素含量来诊断植株的营养状况。

4.2　水氮耦合对氮素利用效率影响

试验中施氮水平对氮素利用效率有极显著的影响，高氮水平下氮的利用效率约为低氮水平下氮的利用效率的1/2，即施氮量多反而氮肥利用效率低，这种结果在许多的田间作物[20]及温室蔬菜作物氮肥试验[21]中都类似，这可能是氮肥施用后，不能完全被植株吸收，氮肥的转化效率不完全等所致。而表2中，低氮水平的氮素利用效率高于100%，即植株累积吸氮量高于施氮量，可能是施氮量太少，不能满足植株生长需要，迫使植株吸收基质中的氮素，这也是N1水平下老叶在拉秧时期失绿变黄的原因。

低水、中水、高水处理下的氮素利用效率分别为59%、63%、68%，即氮肥的利用效率随着灌水量的增加而增加。低氮水平下，高水比低水的氮素利用效率增加了11%，高氮水平下，高水比低水氮素利用效率增加19%（图3），因此，在一定的范围内增加灌水量，可提高氮素利用效率，尤其在高氮条件下，增加灌水量氮的利用效率提高的更多。并且高氮水平下的植株和果实累积吸氮量显著高于低氮（表2），因此在一定范围内，增施氮肥可促进番茄根系的发育、改善根系的营养状况[22]，从而增加根系对氮素的吸收，提高氮素的利用效率。

4.3　施氮水平对对根、茎、叶、果氮素分配的影响

植株前期（定植后生长30d）氮素仅分布在根、茎、叶中，各处理叶中的氮素所占比例都在75%以上，随着植株长大并开始坐果，氮素分配在根、茎、叶中的分配比例逐渐减小，而向果实的分配增加，果实收获期，各处理果实中的氮素占总吸氮量比例都在55%以上。汤明尧[6]研究也表明，番茄生长前期氮素累积主要分配在叶中，比例达75%，成熟期，氮素由营养器官茎叶向生殖器官花果中转移，茎叶中的氮占总吸氮量的比例继续减小，果实增加，果实中氮素所占比例达50%。

本试验研究表明，高氮水平下，茎、叶所占比例比低氮多4%和2%，根和果氮素所占比例比低氮低1%和5%，Badr等研究表明较低的氮肥供应导致了茎叶氮浓度的降低[15]，从而导致了茎叶累积吸氮量的减少，降低茎叶中的氮素所占植株中的比例，因此，低氮条件下，氮素往果实分配的更多。高水处理下，氮素往茎叶分配比例（42%）高于低水和中水（38%），氮素往果实分配比例（56%）低于低水和中水（60%），因此，灌水量越大，更容易促进植株的营养生长，不利于往果实分配，这与高方胜[19]研究结果一致。

5　结论

施氮水平显著影响了植株氮素含量，与低氮相比，高氮水平下最终根含氮量24g/kg、茎含氮量18g/kg、叶含氮量31g/kg、果含氮量27g/kg和植株含氮量27g/kg增幅分别为50%、157%、63%、32和50%，同时植株累积吸N量3.21g/株也增加了90%；增加灌水量可以提高植株累积吸氮量，中水和高水比低水分别增加了7%和17%，从而增加了氮的利用效率，因此水肥对氮素的吸收具有相互促进的关系，在氮素吸收上施氮水平起主要作用。

施氮水平和灌水量对氮素的物质分配也有一定的影响，高氮水平下茎和叶氮素所占比例比N1多4%和2%，根和果氮素所占比例比低氮少1%和5%；高水处理下茎叶中氮素所占比例（42%）比低水和中水（38%）多4%，果实所占比例（56%）比低水和中水（60%）少4%，因此低氮低水可促进氮素往果实中分配，高氮高水有益于氮素往茎叶中分配。

在植株整个生长周期中，茎中氮素含量变化最敏感，因此可根据茎中氮素含量来诊断植株是否缺氮。

参考文献

[1] 邢英英，张富仓，张燕，等. 滴灌施肥水肥耦合对温室番茄产量、品质和水氮利用的影响［J］.中国农业科学，2015，48（4）：713-726.

[2] 潘丽萍，李彦，唐立松.局部根区灌溉对棉花主要生理生态特性的影响［J］.中国农业科学，2009，42（8）：2 982-2 986.

[3] Badr M A，El-Tohamy W A，Zaghloul A M. Yield and water use efficiency of potato grown under different irrigation and nitrogen levels in an arid region［J］. Agricultural Water Management，2012，110（3）：9-15.

[4] 王新，马富裕，刁明，等. 滴灌番茄临界氮浓度、氮素吸收和氮营养指数模拟［J］.农业工程学报，2013，29（18）：99-108.

[5] Danb J，Thomasc K，Toms C. Economically Optimal Nitrogen Rates of Corn：Management Zones Delineated from Soil and Terrain Attributes［J］. Agronomy Journal，2011，103（4）：1 026-1 035.

[6] 汤明尧，张炎，胡伟，等. 不同施氮水平对加工番茄养分吸收、分配及产量的影响［J］.植物营养与肥料学报，2010，16（5）：1 238-1 245.

[7] 赵雪雁，赵海莉.干旱区内陆河流域产业结构效益分析——以黑河流域中游张掖市为例［J］.西北师范大学学报（自然科学版），2007，43（1）：91-94.

[8] Scholberg J，Mcneal B L，Boote K J，et al. M. Nitrogen stress effects on growth and nitrogen accumulation by field-grown tomato.［J］. Agronomy Journal，2000，92（1）：159-167.

[9] Benincasa P，Guiducci M，Tei F. Nitrogen Use Efficiency：Meaning and Sources of Variation--Case Studies on Three Vegetable Crops in Central Italy［J］. Horttechnology，2011，21（3）：266-273.

[10] 张军，李建明，张中典，等. 水肥对番茄产量、品质和水分利用率的影响及综合评价［J］.西北农林科技大学（自然科学版），2016，44（7）：215-221.

[11] 罗勤，陈竹君，闫波，等. 水肥减量对日光温室土壤水分状况及番茄产量和品质的影响［J］.植物营养与肥料学报，2015，21（2）：449-457.

[12] Farneselli M，Benincasa P，Tosti G，et al. Tei F. High fertigation frequency improves nitrogen uptake and crop performance in processing tomato grown with high nitrogen and water supply［J］. Agricultural Water Management，2015，154：52-58.

[13] 张广楠. 无土栽培技术研究的现状与发展前景［J］.甘肃农业科技，2004（2）：6-8.

[14] 刘升学，于贤昌，刘伟，等. 有机基质配方对袋培番茄生长及产量的影响［J］.西北农业学报，2009，18（3）：184-188.

[15] Badr M A，Abou-Hussein S D，El-Tohamy W A. Tomato yield，nitrogen uptake and water use efficiency as affected by planting geometry and level of nitrogen in an arid region［J］.Agricultural Water Management，2016，169：90-97.

[16] 柳美玉，曹红霞，杜贞其，等. 营养液浓度对番茄营养生长期干物质累积及养分吸收的影响［J］.西北农林科技大学（自然科学版），2017，45（4）：119-126.

[17] 杨明凤，陈荣毅，朱惠芝，等. 覆膜滴灌加工番茄肥料的吸收与分配规律研究［J］.新疆农垦科技，2014，37（12）：37-40.

[18] 薛琳，田丽萍，王进.滴灌覆膜条件下加工番茄的养分吸收与分配规律的研究［J］.石河子大学学报（自然科学版），2004（5）：389-392.

［19］高方胜，徐坤.土壤水分对不同季节番茄产量及氮磷钾吸收分配特性的影响［J］.核农学报，2014，28（9）：1 722-1 727.

［20］李升东，张卫峰，王法宏，等.施氮量对小麦氮素利用的影响［J］.麦类作物学报，2016，36（2）：223-230.

［21］姜慧敏，张建峰，杨俊诚，等.不同氮肥用量对设施番茄产量、品质和土壤硝态氮累积的影响［J］.农业环境科学学报，2010，29（12）：2 338-2 345.

［22］刘长庆.不同氮肥用量对保护地番茄产量、品质及土壤微生物的影响［J］.山东农业科学，2012，44（12）：75-77.

不同灌水量对温室秋茬茄子蒸腾速率及水分利用效率的影响

陶虹蓉[1, 2]，杨宜[1, 2]，李银坤[1, 3*]，郭文忠[1, 3]，李海平[2]，李灵芝[2]

（1. 北京农业智能装备技术研究中心，北京　100097；2. 山西农业大学园艺学院，太谷　030801；3. 北京市农业物联网工程技术研究中心，北京　100097）

摘要：本试验以茄子为材料，借助称重式蒸渗仪试验平台，以直径20cm标准蒸发皿蒸发量为灌溉依据，设置了I1（K_{cp1}:0.6）、I2（K_{cp2}:0.8）和I3（K_{cp3}:1.0）3种灌水水平，研究了不同灌水量下温室秋茬茄子的蒸腾规律、产量及其水分利用效率。结果表明：不同灌水处理在各生育期的日蒸腾速率均呈单峰曲线变化，峰值在12:00～13:00出现，蒸腾速率峰值随灌水量的增加而升高。温室茄子全生育期的总蒸腾量为84.11～128.85mm，其中开花结果期蒸腾量最高，可占总蒸腾量的38.74%～41.98%。与处理I1相比，处理I2和I3的总蒸腾量分别增加了24.86%和53.20%。增加灌水量能够提高温室茄子产量，其中处理I2的产量与处理I3相比无显著差异，但比处理I1显著增加了44.67%。水分利用效率以处理I2最高，为25.49kg·m^{-3}，相比处理I1与I3分别增加了15.97%与13.26%。综合考虑温室茄子蒸腾量、产量及水分利用效率，处理I2（K_{cp2}：0.8）在比处理I3减少20%灌水量的条件下，仍具有较高的产量与水分利用效率，为供试条件下较优灌水处理。

关键词：灌水量；茄子；蒸腾速率；水分利用效率

Effect of Different Irrigation Amount on Transpiration Rate and Water Use Efficiency of Autumn Eggplant in Greenhouse

Tao Hongrong[1, 2], Yang Yi[1, 2], Li Yinkun[2, 3*], GuoWenzhong[2, 3], Li Haiping[1], Li Lingzhi[1]

（1. *National Research Center of Intelligent Equipment for Agriculture*, *Beijing*　100097；2. *College of Horticulture*, *Shanxi Agricultural University*, *Taigu*　030801；3. *Beijing Engineering Technology Research Center of Agricultural Internet of Things*, *Beijing*　100097）

Abstract: In this experiment, eggplant was used as the material, and with the help of weighing lysimeter, the evaporation of the evaporator pan in diameter 20 cm was used as the basis of irrigation.I1（K_{cp1}:0.6）, I2（K_{cp2}:0.8）and I3（K_{cp3}:1.0）were set used to study the transpiration, yield and water use efficiency of greenhouse eggplant under different irrigation levels.The results showed that the daily transpiration rate of different

---

基金项目：国家重点研发计划（2017YFD0201503）、国家自然科学基金（41501312）

作者简介：陶虹蓉（1993—），女，山西翼城人，硕士研究生，研究方向为蔬菜栽培与生理。E-mail：thr93520@163.com

通讯作者：李银坤（1982—），男，博士，高级工程师，研究方向为作物水肥高效利用。E-mail：lykunl218@163.com irrigation treatments showed a single peak curve, and the peak appeared at 12:00 ~ 13:00.The peak transpiration rate increased with the increasing of irrigation volume.The total transpiration of greenhouse eggplant during the whole growing period was 84.11 ~ 128.85 mm, the highest transpiration was in flowering stage, accounting for 38.74% ~ 41.98% of total transpiration.Compared with treatment I1, the total transpiration of treatment I2 and I3 increased by 24.86% and 53.20% respectively.Increasing the amount of irrigation could increase the yield of greenhouse eggplant.The yield of treatment I2 was not significantly different from that of treatment I3, but increased significantly by 44.67% compared with that of treatment I1.The highest water use efficiency of treatment I2 was 25.49 kg·m^{-3}, compared with the treatment of I1 and I3 increased by 15.97% and 13.26%, respectively. Considering the transpiration rate, yield and water use efficiency of greenhouse eggplant, I2 (Kcp2: 0.8) still had higher yield and water use efficiency under the condition of 20% less irrigation than that of treatment I3, so it was the better irrigation treatment.

Key words: Irrigation volume; Eggplant; Transpiration rate; Water use efficiency

1 前言

水是影响蔬菜生长最为关键的生态因子之一。水分的过多与不足，均会对蔬菜的产量和品质产生很大的影响。温室不能直接利用天然降水，需要依靠灌溉来补充水分，而在我国蔬菜生产中仍主要依靠传统经验灌水，灌溉量大，灌溉水有效利用系数低，严重浪费了水源，造成设施蔬菜发病率增高，产量和效益低下[1-2]。

蒸腾是植物生理特征的重要指标，而蒸腾速率大小反映了植物潜在耗水能力的强弱[3]。现有研究表明，蔬菜作物的蒸腾耗水在其生长过程中具有类似的变化规律，一般随着生育期的推进呈先增大后降低的变化趋势[4]，其中结果盛期的蒸腾耗水量最大，可占全生育期的69.6%左右[5]。作物的蒸腾耗水除受其自身生理特性影响外，还受到温度、湿度和光照强度等环境因子的影响[6-7]，牛勇等[5]研究认为温室黄瓜的蒸腾量与太阳净辐射的相关性最大，其次是温室内的相对湿度。但是在土壤供水亏缺或过饱和时，土壤含水量则成为影响蒸腾的主导因子[8]。张婷华等[9]研究结果表明，灌水量不足将严重抑制植株蒸腾，影响植株生长，其中40% ~ 50%田间持水量（T3）相比75% ~ 80%田间持水量（CK），温室番茄蒸腾量降低了81.1%。可见，适宜的土壤水分条件是减少水分蒸腾，提高水分利用效率的关键[10]。而探讨蔬菜作物蒸腾速率及生育期蒸腾量对不同灌水量的响应规律，对解释温室蔬菜节水机理及制定高产高效的灌溉制度具有重要的理论与实际意义。

国内外有关植物蒸腾耗水量的测定方法很多，而称重式蒸渗仪被认为是测量作物腾发量和渗漏量的标准仪器[11]。当前基于蒸渗仪开展温室茄子蒸腾速率动态变化的相关研究也较为少见。本试验以茄子为材料，以称重式蒸渗仪为试验平台，以20cm标准蒸发皿蒸发量为灌溉依据，研究不同灌溉量下温室茄子的日蒸腾速率以及茄子生育期内蒸腾量的动态变化，为温室茄子制定合理的灌溉制度，实现节水生产提供科学理论依据。

2 材料与方法

2.1 试验区概况

试验于2017年8月至12月在北京市农林科学院试验温室内进行。供试温室长38m，宽

11m，南北走向，土质为砂壤土。试验前0～20cm土壤理化性质为：土壤容重1.40g·cm^{-3}，田间体积持水量28.0%，有机质含量15.89g·kg^{-1}，全氮质量分数0.60g·kg^{-1}，速效钾含量0.15g·kg^{-1}。温室中央安装有自动气象监测系统，实时采集温室内部气象数据，包括：气温、相对湿度和太阳辐射等。

2.2 试验设计

试验根据冠层水面蒸发量设置3个灌水量水平：I1，累计水面蒸发量的60%（K_{cp1}：0.6）；I2，累计水面蒸发量的80%（K_{cp2}：0.8）；I3，累计水面蒸发量的100%（K_{cp3}：1.0）。

供试茄子品种为“京茄黑宝”，于2017年8月8日定植，定植时选取4叶1心的秧苗。采用双行错位定植法，行距50cm，株距45cm，其中蒸渗仪定植茄子4株，覆盖地膜。灌溉方式为膜下滴灌，滴头间距10cm。在茄子定植前基施商品有机肥30 000kg·hm^{-2}（N、P_2O_5、K_2O含量分别为7.14、0.06、3.68g·kg^{-1}），定植后23d、42d和65d进行追肥，肥料类型为大量元素水溶性肥料，施肥时先将肥料在水中充分溶解，然后随滴灌施入。本试验将茄子生育期划分为4个阶段：苗期（定植29d内），结果初期（定植30～56d），结果盛期（定植57～101d）和结果末期（定植102d至拉秧）。

2.3 测定指标与方法

2.3.1 蒸发皿蒸发量及灌水量

温室内设置有直径为20cm的标准蒸发皿，其高度始终与茄子冠层保持一致。每日下午18:00测量蒸发皿水面蒸发量，蒸发皿剩余水量与最初倒入的10mm的差值即为水面蒸发量（E_p），水面蒸发量的累积值乘以分别乘以灌溉系数，即为实际灌水量，其计算公式为：

$$I=K_{cp}\times\sum E_p \tag{1}$$

式中，I为灌水量，mm；K_{cp}为灌溉系数，E_p为累积水面蒸发量，mm。

2.3.2 茄子蒸腾量

温室安装有称重式蒸渗仪，长100cm，宽60cm，土体深90cm。称重式蒸渗仪每1h记录一次土柱重量（精度为0.01mm），蒸腾量的计算原理根据水量平衡方程，即：

$$T\times S=(W_{t-1}-W_t)/\rho+I \tag{2}$$

式中，T为时间段内蒸腾量，mm；S为蒸渗仪箱体的表面积，0.6m^2；W_{t-1}、W_t为$t-1$时刻和t时刻蒸渗仪箱体内土、水的质量，g；ρ为水的密度，1g·cm^{-3}。

2.3.3 茄子产量Y（kg·hm^{-2}）

进入结果期后，对每次采收的茄子进行称重计产，并折算成公顷产量。

2.3.4 水分利用效率（WUE，kg·m^{-3}）

$$WUE=Y/ET \tag{3}$$

式中：Y为产量，kg·hm^{-2}，ET为总蒸腾量，mm。

2.4 统计与方法

试验数据由Excel 2003进行整理作图，用SPSS 23.0进行统计分析。

3 结果分析

3.1 蒸发皿水面蒸发量及温度变化规律

由图1可知，蒸发皿水面蒸发量与温度的变化相似，均呈波动下降趋势。二者具有极显著的相关性（相关系数r=0.767 7）。试验期间温度呈逐渐下降趋势，平均温度19.78℃。水面蒸发量在定植22d内波动上升趋势，最大日蒸发量为3.90mm·d^{-1}，然后随茄子生育期的推进呈下降趋势，波动幅度较大，定植99d后水面蒸发量在0.50mm·d^{-1}上下浮动。试验期间的水面蒸发总量为174.35mm。

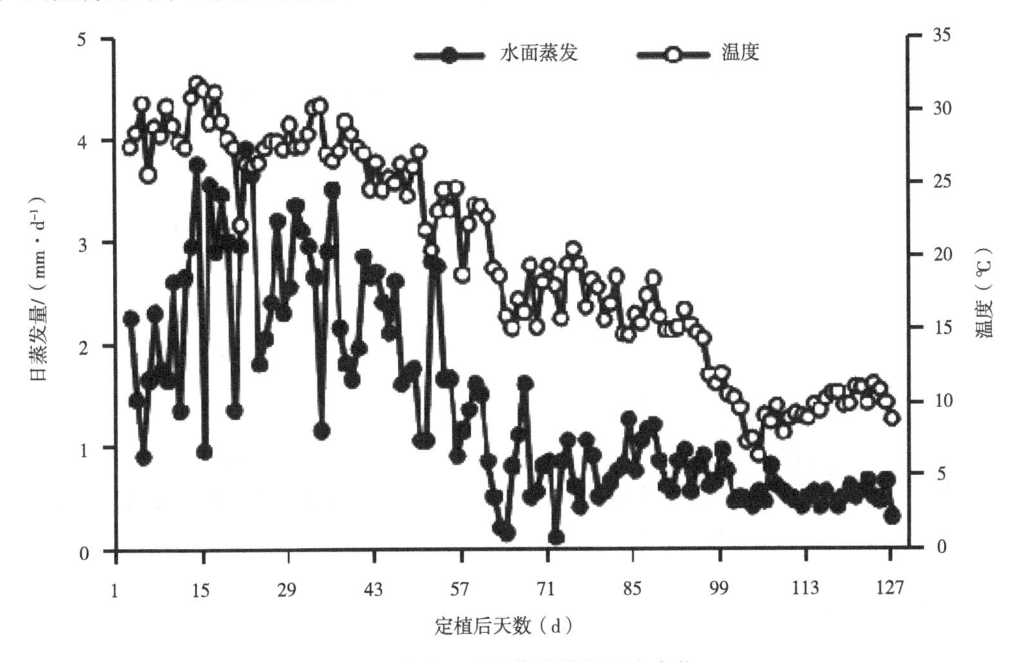

图1 蒸发皿水面蒸发量及温度变化

3.2 温室茄子各生育期蒸腾速率日变化

茄子各生育期内蒸腾速率的日变化过程如图2所示。从图2中可以看出，不同灌水处理下茄子各生育期的蒸腾速率和温度的变化趋势一致。蒸腾作用主要发生在白天，夜间也有发生，但相对较小且较为稳定。各生育期的蒸腾速率日变化均呈单峰曲线，其中苗期、结果初期和结果盛期的蒸腾速率均在8:00随着温度的升高而逐渐增大，在13:00温度达到最高，蒸腾速率也出现峰值，峰值之后蒸腾速率随着温度的降低迅速减小，并在19:00趋于稳定。而结果末期的蒸腾速率在9:00时才开始上升，其中温度在13:00时最高，在18:00时降至稳定，而蒸腾速率则在12:00时达到峰值，在15:00逐渐稳定。

茄子的蒸腾速率峰值在不同生育期差异很大。其中结果初期的蒸腾速率峰值最大，其次为结果盛期，结果末期的蒸腾速率峰值最小。不同灌水处理在结果初期的蒸腾速率峰值分别0.20、0.28和0.31mm·h^{-1}，与处理I1相比，处理I2和I3在结果初期蒸腾速率峰值可分别增加40.00%和55.00%，而处理I2和I3的差异较小。

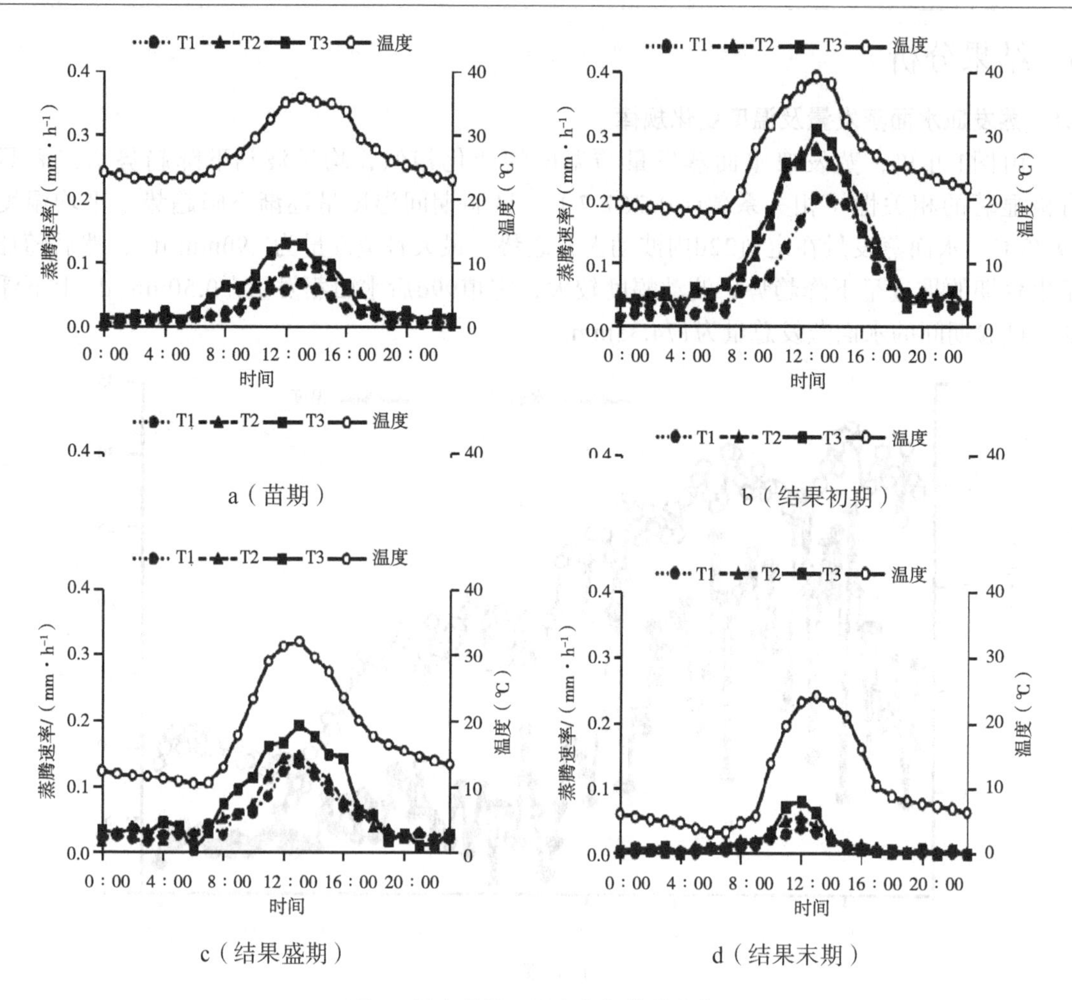

a（苗期） b（结果初期）

c（结果盛期） d（结果末期）

图2　温室茄子不同生育期蒸腾速率

3.3　不同灌水处理下茄子日蒸腾量和累积蒸腾量的动态变化

温室茄子日蒸腾量和累积蒸腾量的动态变化如图3所示。可以明显看出，各处理的日蒸腾量随生育期的变化均呈先升高后降低的变化趋势。日蒸腾量在茄子苗期呈逐渐增大趋势，在结果初期达到峰值2.45～2.93mm·d^{-1}，结果盛期日蒸腾量呈波动下降趋势，到结果末期日蒸腾量明显减小，且波动幅度较小，该阶段平均日蒸腾量仅为0.25～0.33mm·d^{-1}。

各处理累积蒸腾量变化趋势均为“S”形曲线（图3），苗期增长缓慢，结果初期增长速率加快，到结果盛期和结果末期增长速率又逐渐减小。温室茄子的总蒸腾量为84.11～128.85mm，其中结果初期的阶段累积蒸腾量最高，可占全生育期累积蒸腾量的38.74%～41.98%。其次为结果盛期，结果末期的蒸腾量最小，仅占总蒸腾量的6.07%～8.11%。灌水量越高其累积蒸腾量也越大，其中结果初期处理I2和I3的阶段累积蒸腾量分别比处理I1增加了20.36%和41.38%，结果盛期处理I2和I3分别比处理I1增加了41.34%和91.32%，而苗期和结果末期各处理间蒸腾量差异较小。处理I2和I3的总蒸腾量与处理I1相比则分别增加了24.86%和53.20%。

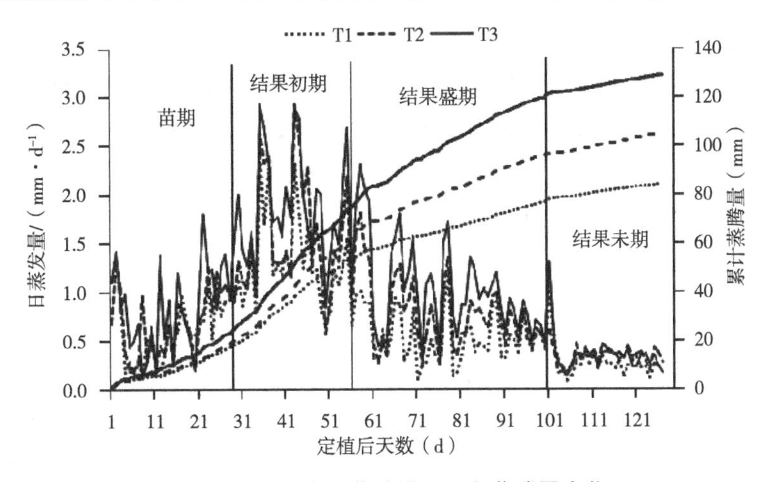

图3 生育期内日蒸腾量及累积蒸腾量变化

3.4 不同灌水处理下茄子的产量及水分利用效率

由图4可知，茄子的产量随着灌水量的增大而增加，处理I2和I3的产量无显著差异，但与处理I1相比，分别增加了44.67%（$P<0.05$）和56.86%（$P<0.05$）。各处理的水分利用效率无显著性差异，其中处理I2的水分利用效率最高，为25.49kg·m^{-3}，相比处理I1和I3分别增加了15.97%和13.26%。

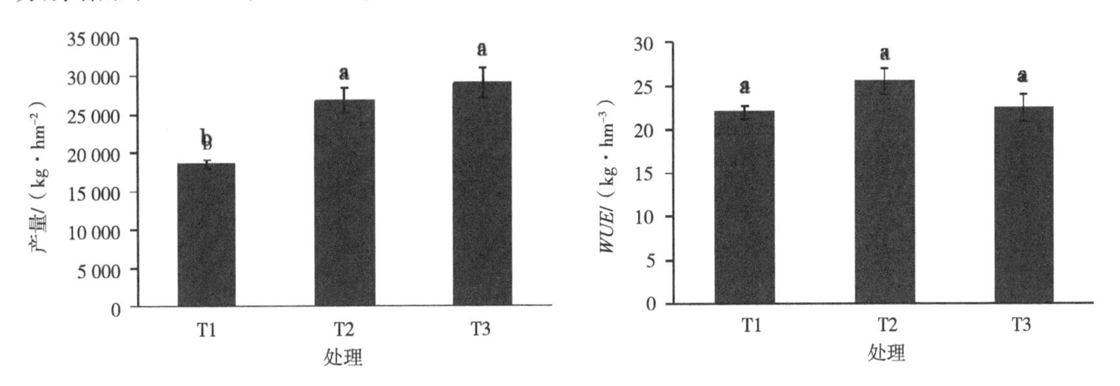

图4 茄子产量及水分利用效率

注：图中不同小写字母表示差异达显著水平（$P<0.05$）

4 结论

（1）不同处理下各生育期茄子的日蒸腾速率均呈单峰曲线，峰值在12:00～13:00出现。结果初期蒸腾速率最大，为0.20～0.31mm·h^{-1}，结果末期蒸腾速率最小，仅为0.04～0.08mm·h^{-1}。各生育阶段的蒸腾速率均随着灌水量的增大而增大，与处理I1相比，处理I2和I3在结果初期的蒸腾速率峰值分别增加了40.00%和55.00%。

（2）茄子累积蒸腾量为84.11～128.85mm，灌水量大的处理累积蒸腾量也较大，与处理I1相比，处理I2和I3的总蒸腾量分别增加了24.86%和53.20%。结果初期的阶段累积蒸腾量最高，可占总蒸腾量的38.74%～41.98%。其次为结果盛期，占总蒸腾量的

28.90%～36.10%。结果末期的蒸腾量仅占总蒸腾量的6.07%～8.11%。

（3）增加灌水量能够提高温室茄子产量，与处理I1相比，处理I2和I3的产量分别增加了44.67%（$P<0.05$）和56.86%（$P<0.05$），但处理I2和I3无显著差异。水分利用效率以处理I2最高，为25.49kg·m^{-3}，相比处理I1与I3分别增加了15.97%与13.26%。处理I2（K_{cp2}：0.8）在比处理I3减少20%灌水量的条件下，仍具有较高的产量与水分利用效率，为供试条件下较优灌水处理。

参考文献

［1］胡占阳.中国水资源贫乏与水资源浪费的矛盾分析［J］.辽宁工程技术大学学报（社会科学版），2013（1）：74-77

［2］郭全忠.不同灌水量对设施番茄产量和品质的影响研究［J］.陕西农业科学，2012，58（6）：34-36.

［3］陈倩倩，范阳阳，郝影宾，等.不同土壤水分含量对玉米气孔发育过程和蒸腾耗水量的影响［J］.干旱地区农业研究，2011，29（3）：75-79.

［4］刘浩，孙景生，王聪聪，等.温室番茄需水特性及影响因素分析［J］.节水灌溉，2011（4）：11-14.

［5］牛勇，刘洪禄，吴文勇，等.基于大型称重式蒸渗仪的日光温室黄瓜蒸腾规律研究［J］.农业工程学报，2011，27（1）：52-56.

［6］周继华，毛思帅，薛绪掌，等.负水头灌溉系统供营养液番茄生产及耗水研究［J］.节水灌溉，2014（11）：1-5.

［7］刘浩，孙景生，王聪聪，等.温室番茄需水特性及影响因素分析［J］.节水灌溉，2011（4）：11-14.

［8］张大龙，常毅博，李建明，等.大棚甜瓜蒸腾规律及其影响因子［J］.生态学报，2014，34（4）：953-962.

［9］张婷华.土壤水分胁迫对温室番茄蒸腾的影响及模拟研究［D］.南京：南京信息工程大学，2014.

［10］杨晓婷，张恒嘉，张明，等.灌水频率与灌水量对南瓜耗水特征与水生产力的影响［J］.华北农学报，2016，31（4）：192-198.

［11］姜峻，都全胜，赵军，等.称重式蒸渗仪系统改进及在农田蒸散研究中的应用［J］.水土保持通报，2008，28（6）：67-72.

基于光辐射灌溉量的温室番茄产量、品质和水肥利用研究

魏晓然，杨其长，程瑞锋*

（中国农业科学院农业环境与可持续发展研究所，北京　100081）

摘要：为了探讨基于光辐射的不同灌溉量在番茄产量、品质和水肥利用效能上的差异，在起垄内嵌式栽培条件下，采用滴灌施肥技术，灌溉模式分为常规时间间隔灌溉（以下称“常规灌溉”，CK）和基于光辐射的灌溉两种，其中基于光辐射的灌溉模式分为处理T1、T2和T3。结果表明：基质水分含量处理CK>T3>T2>T1，灌溉量越多，基质水分含量越高。相比于处理CK，处理T1、T2和T3的灌溉水分利用效率分别提高了34.9%、37.9%和30.1%，处理T3的产量增加了18.3%，达到67.0t/hm^2，处理T2的产量增加不明显，处理T1的产量减少了20.8%，表明基于光辐射的适宜灌溉量显著提高了灌溉水分利用效率和果实产量，过少的灌溉则抑制了产量的增加。处理T1和T2的可溶性固形物、可溶性糖、糖酸比和维生素C含量显著高于其它处理，处理T1因过度亏缺灌溉干物质含量最高，而处理T3的维生素C含量和糖酸比也显著高于处理CK，表明基于光辐射的适宜灌溉量不但提高了产量，还保证了品质。相比于处理CK，在晴天和阴天时，处理T3的废液排出率分别减少了40.0%和52.4%，处理T1和T2无废液排出，表明基于光辐射的灌溉模式有效节约了水肥，避免了水肥的大量浪费。综合考虑番茄产量、品质和水肥利用情况，本研究推荐基于光辐射的处理T3灌溉量58.9L/株作为番茄整个生育期的适宜灌溉量。

关键词：光辐射；灌溉量；番茄产量；品质；水肥利用

Research on the Tomato Yield，Quality，Water and Fertilizer Utilization Based on Solar Radiation Irrigation in Greenhouse

Wei Xiaoran，Yang Qichang，Cheng Ruifeng

（*Institute of Environment and Sustainable Development in Agriculture*，*Chinese Academy of Agricultural Sciences*，*Beijing*　100081）

Abstract：In order to discover the difference of tomato yield，quality，water and fertilizer utilization based on the different solar radiation irrigation amount. Adopting drip irrigation technique under the condition of soil ridge substrate embedded cultivation，irrigation mode was divided into regular time interval（hereinafter referred to as “normal irrigation”，CK）and based on solar radiation irrigation，which was divided into treatment T1，T2 and T3. The results showed that substrate moisture content was treatment CK > T3 > T2 > T1，The greater irrigation amount was，the higher substrate moisture content was. Compared with treatment CK，the irrigation water utilization efficiency of T1，T2 and T3 had increased by 34.9%，37.9% and 30.1% respectively. The yield of treatment T3 had increased by 18.3%，reached 67.0t/hm^2. The yield of treatment T2 hadn’t significantly

作者简介：魏晓然，山东滨州人，主要从事设施栽培水肥灌溉制度的研究。

项目资助：科技部科技伙伴计划资（KY201702008）

Email：weixiaoran199216@163.com

increased, and the yield of T1 had reduced by 20.8%. It' s indicated that the suitable irrigation amount based on solar radiation had significantly improved the irrigation water utilization efficiency and fruit yield, and too little irrigation had inhibited the increase of yield. The soluble solids, soluble sugar, sugar acid ratio and Vc content of treatment T1 and T2 were significantly higher than those of other treatments. Due to excessive deficit irrigation, the dry matter content of treatment T1 is the highest, while Vtamin C content and sugar acid ratio were also significantly higher than treatment CK. It' s indicated that the suitable irrigation amount based on solar radiation hadn' t only increased the yield, but also guaranteed the quality. compared with the treatment CK, the drainage rate of treatment T3 had reduced by 40.0% and 52.4% respectively on sunny and cloudy day and treatment T1 and T2 had no drainage. It' s indicated that based on solar radiation irrigation mode saved water and fertilizer efficiently and avoided the waste of water and fertilizer. Based on the comprehensive consideration of tomato yield, quality, water and fertilizer utilization, this study recommended treatment T3 58.9L/plant as the suitable irrigation amount based on solar radiation irrigation mode for the whole growth period of tomato.

Key words: Solar radiation; Irrigation amount; Tomato yield; Quality; Water and fertilizer utilization

1 引言

在国内外设施栽培过程中，灌溉模式和灌溉量对番茄产量、品质和水肥利用的影响已有大量研究。传统的时间间隔灌溉模式多采用过量的水肥管理，造成了水肥的浪费；而蒸发皿水面蒸发量Epan（Ep）为参数估算需求量的模式，在温室中已经广泛应用，但该模式不能实现灌溉的自动化，不能满足实时天气状况下番茄对水肥的需求。Baselga等研究表明灌溉量增加了产量却降低了番茄果实内可溶性糖和有机酸的含量；陈秀香等研究发现，在亏缺灌溉时果实中可溶性糖、有机酸、可溶性固形物含量增加，果实营养和风味品质改善。

2 材料及方法

2.1 试验地点及材料

试验在北京市顺义区大孙各庄镇的节能型日光温室（纬度40° 13 ' N，经度116° 65 ' E）中进行，温室东西走向，长60m，跨度8m，脊高3.8m，靠近后墙有2m走道。试验于2017年8月20日进行番茄育苗，9月7日当番茄处于“两叶一心”时开始定植，双行栽培，试验区两侧分别有两垄保护行（图1）。

2.2 灌溉模式

常规灌溉采用传统的滴灌设备，按照时间间隔定时进行灌溉；基于光辐射的灌溉采用水肥一体机进行，水肥一体机连接有太阳辐射传感器，太阳辐射传感器距离温室地面2.0m，距离后墙4.5m，试验根据文献[25]和温室光照情况设定光辐射累积阈值1 600kJ/m^2，达到阈值水肥一体机开始启动。

试验各处理均采用营养液滴灌，每垄布置一条滴灌管，滴灌管直径20mm，滴箭间距20cm，常规灌溉滴箭流速1.6L/h，水肥一体机灌溉滴箭流速2.2L/h，滴箭都位于番茄植株根区附近，具体试验布置如图2所示。

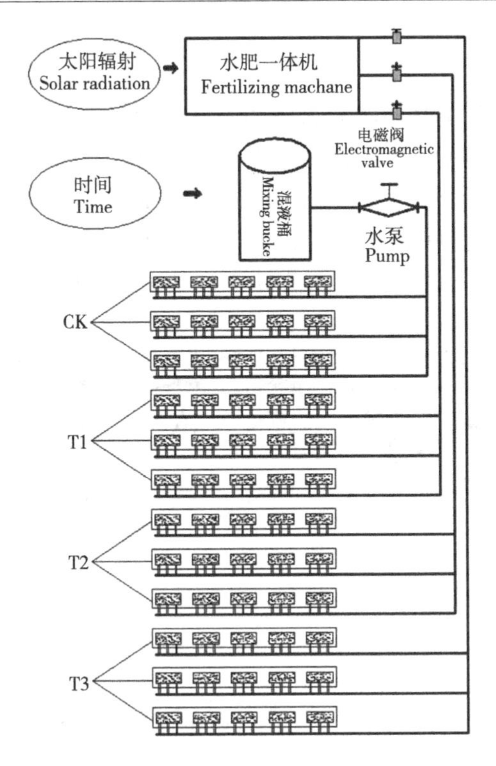

图1　试验布置示意

Figure 1　Schematic diagram of experiment arrangement

2.3　测定项目及方法

2.3.1　环境参数测定

基质水分含量：采用国产基质专用高精度湿度传感器（TM-JZ01）测量基质水分含量，插在根区附近，距离滴箭10cm。

2.3.2　番茄产量和灌溉水分利用效率测定

番茄产量：果实成熟后每隔一周采摘1次，直至番茄拉秧，最终对各个处理产量统计汇总。

番茄灌溉量：每个处理滴灌灌溉的管路上都连接有旋翼式水表（LXS-25E），用于记录每个处理的灌溉量。

灌溉水分利用效率：灌溉水分利用效率是指作物利用单位灌水量生产的经济作物产量，单位为kg/m^3，计算公式为：

$$IWUE=Y/I$$

式中：IWUE（irrigation water use efficiency）为灌溉水分利用效率；Y为产量（kg）；I为灌溉量（m^3）。

2.4　数据处理及统计分析

方差分析采用SPSS 18数据处理软件，不同处理间的多重比较采用Duncan法，图表中不同小写字母表示在a=0.05水平下差异显著（p=0.05），数据统计和作图采用Excel 2010。

3 结果

3.1 不同灌溉模式和灌溉量下基质水分含量和温度的变化

图3为温室不同处理基质水分含量变化情况。由图2可知，不同的灌溉模式和灌溉量下基质水分含量都有显著性差异，基质水分含量处理CK>T3>T2>T1，灌溉量越多，基质水分含量越高。在一天内，处理CK、T3、T2、T1基质水分含量分别在94.5～99.6%、89.3～93.8%、76.2～89.4%、63.1～78.6%区间波动，基于光辐射的灌溉模式变化幅度更大。

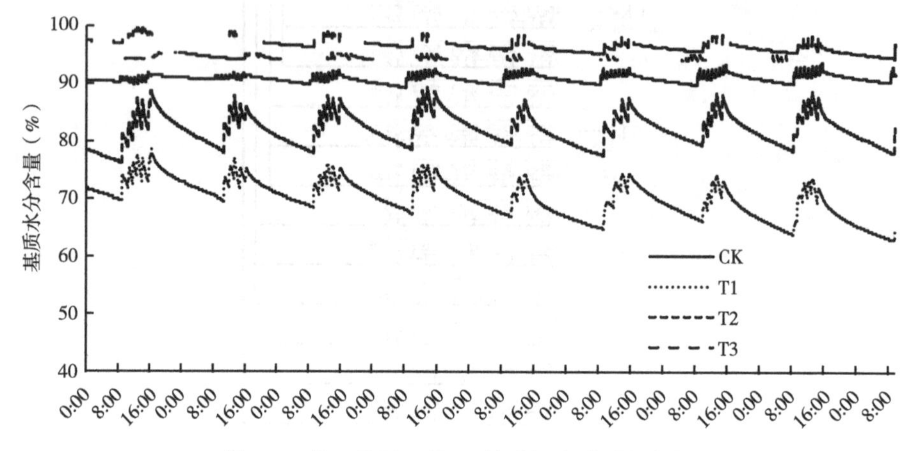

图2　11月28日到12月6日基质温度随时间变化

3.2 灌溉模式和灌溉量对果实产量和灌溉水分利用效率的影响

由图3可知，不同灌溉模式和灌溉量对果实产量和灌溉水分利用效率有显著影响。与处理CK相比，处理T3的产量增加了18.28%，达到67.03t/hm²，处理T2的产量增加不明显，处理T1产量减少了20.84%。可见，基于光辐射的适宜灌溉量能显著提高果实产量，灌溉量减少则抑制了果实产量的提高。从灌溉水分利用效率的角度分析，相比于处理CK，处理T1、T2、T3分别提高了34.90%、37.94%、30.10%。可见，相比于常规灌溉，基于光辐射的灌溉模式显著提高了灌溉水分的利用效率。

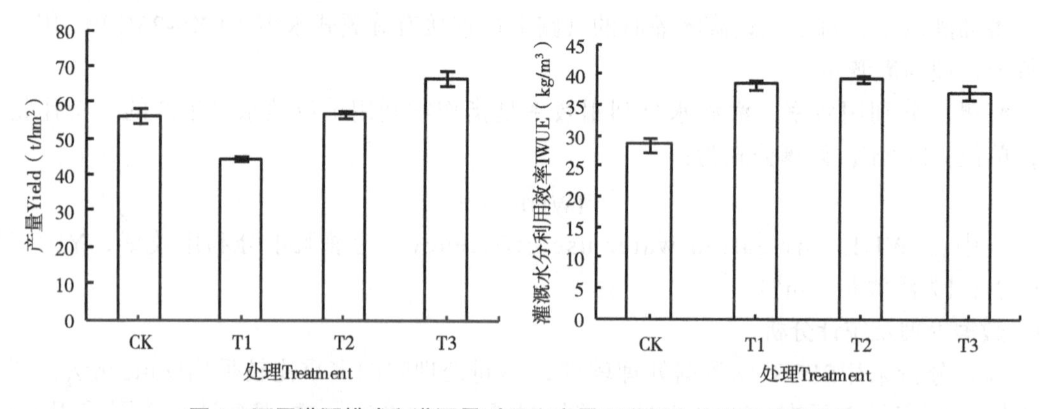

图3　不同灌溉模式和灌溉量对果实产量和灌溉水分利用效率统计

3.3 灌溉模式和灌溉量对果实品质的影响

由表1可知，不同灌溉模式和灌溉量对果实品质有显著影响。处理T1含水量显著小于其他处理，干物质含量较高；对于番茄风味品质方面，相比于处理CK，处理T1和T2可溶性固形物含量、糖酸比显著提高，而处理T3差异不明显。对于营养品质和口感方面，处理T1和T2可溶性糖和维生素C含量显著高于处理CK，处理T3 维生素C含量和糖酸比也显著高于处理CK。可见，相比于常规灌溉，基于光辐射的灌溉模式能显著提高番茄的风味和营养品质，灌溉量越少，干物质含量越高。

表1 不同灌溉模式和灌溉量对果实品质的影响

处理 Treatment	含水量 Water content（%）	可溶性固形物 Soluble solid（%）	维生素C mg/100g	可溶性糖 Soluble sugar（mg/g）	糖酸比 sugar-acid ratio
CK	94.79 ± 0.19a	4.17 ± 0.09c	10.78 ± 0.19c	24.54 ± 0.93c	6.23 ± 0.16c
T1	93.88 ± 0.13b	5.00 ± 0.12a	14.40 ± 0.33ab	30.87 ± 0.55a	7.54 ± 0.33ab
T2	94.85 ± 0.06a	4.67 ± 0.03b	15.62 ± 0.59a	28.17 ± 1.05b	8.45 ± 0.33a
T3	95.09 ± 0.26a	4.23 ± 0.03c	13.46 ± 0.58b	24.75 ± 0.23c	7.35 ± 0.36b

3.4 灌溉量对水肥利用的影响（表2）

表2为连续6d不同处理水肥灌溉量和废液排出量情况，由表2可知，不同天气基于光辐射的灌溉量有显著差异，晴天的灌溉量显著多于阴天，而常规灌溉无显著差异。常规灌溉废液排出量显著多于基于光辐射的灌溉，在晴天和阴天时，处理CK的废液排出率平均值分别为20.8%、33.0%，处理T3的废液排出率平均值分别为12.5%、15.7%，处理T1和T2无废液排出；相比于处理CK，在晴天和阴天时，处理T3的废液排出率分别减少了40.0%、52.4%。可见，相比于常规灌溉，基于光辐射的灌溉模式有利于水肥的节约，减少了水肥的大量浪费。

表2 不同灌溉量对水肥利用的影响

日期 date	天气 weather	灌溉量 irrigation amount（L）				排出量 drainage amount（L）		排出率 drainage ratio（%）	
		CK	T1	T2	T3	CK	T3	CK	T3
1月9日	晴	33.2	17.6	30.0	39.1	7.2	5.4	21.7	13.8
1月10日	晴	32.9	21.1	33.9	47.1	7.3	6.6	22.2	14.0
1月11日	晴	36.4	21.5	34.9	46.9	6.7	4.5	18.4	9.6
1月12日	阴	34.9	14.1	23.3	30.3	11.6	4.5	33.2	14.9

（续表）

日期 date	天气 weather	灌溉量 irrigation amount（L）				排出量 drainage amount（L）		排出率 drainage ratio（%）	
		CK	T1	T2	T3	CK	T3	CK	T3
1月13日	阴	33.4	13.1	23.3	32.1	10.2	5.4	30.5	16.8
1月14日	阴	32.6	14.0	23.2	32.1	11.5	4.9	35.3	15.3

4 结论与讨论

灌溉模式和灌溉量对番茄产量、品质和水肥利用情况有显著影响，本研究表明，在起垄内嵌式栽培条件下，采用滴灌施肥技术，与常规灌溉相比，基于光辐射的灌溉模式更容易满足番茄对水肥的需求，处理T3显著提高了果实产量和灌溉水分利用效率，同时保证了果实品质，节约了水肥，避免了水肥的大量浪费。综合考虑番茄产量、品质和水肥利用情况，推荐基于光辐射的处理T3灌溉量58.9L/株作为番茄整个生育期的适宜灌溉量。

参考文献

［1］王鹏勃，李建明，丁娟娟，等. 水肥耦合对温室袋培番茄品质、产量及水分利用效率的影响［J］.中国农业科学，2015，48（2）：314 -323.

［2］邢英英，张富仓，吴立峰，等. 基于番茄产量品质水肥利用效率确定适宜滴灌灌水施肥量. 农业工程学报，2015，31（增刊1）：110-121.

［3］张燕，张富仓，袁宇霞，等. 灌水和施肥对温室滴灌施肥番茄生长和品质的影响［J］. 干旱地区农业研究，2014，32（2）：206-212.

［4］李邵，薛绪掌，郭文善，等. 负水头灌溉对温室番茄生长、产量及品质的影响［J］. 农业工程学报，2008，24（Supp.2）：225-229.

［5］刘燕，邹志荣，陈立宇，等. 不同水肥处理对有机基质型砂培番茄产量和品质的影响［J］. 西北农业学报，2012，20（9）：89-94.

［6］蔡东升，李建明，李 惠，等. 营养液的供应量对番茄产量、品质和挥发性物质的影响. 应用生态学报，2018，29（3）

［7］王秀康，邢英英，张富仓. 膜下滴灌施肥番茄水肥供应量的优化研究. 农业机械学报，2016，47（1）：141-150.

［8］Mahajan G，Singh K G. Response of greenhouse tomato to irrigation and fertigation. Agricultural Water Management，2006，84：202-206.

［9］李银坤，郭文忠，薛绪掌，等. 不同灌溉施肥模式对温室番茄产量、品质及水肥利用的影响［J］. 中国农业科学，2017，50（19）：3 757-3 765.

［10］Zotarelli L，Dukes M D，Scholberg J M S，et al. Tomato nitrogen accumulation and fertilizer use efficiency on a sandy soil，as affected by nitrogen rate and irrigation scheduling. Agricultural Water Management，2009，96（8）：1 247-1 258.

［11］Zotarelli L，Scholberg J M，Dukes M D，et al. J. Tomato yield，Biomass accumulation，root distribution and irrigation water use efficiency on a sandy soil，as affected by nitrogen rate and irrigation scheduling. Agricultural Water Management，2009，96（1）：23-34.

［12］Uttam K S，Athanasior P P，Xiu M H. Irrigation Strategies for Greenhouse Tomato Production on Rockwool，Hortscience .2008，43（2）：484-493

小白菜N素动态积累及N素需求分析研究

熊鑫[1]，常丽英[1]，章竞瑾[1]，黄丹枫[1，2*]

（1. 上海交通大学农业与生物学院，2. 农业部都市农业重点实验室（南方中心），上海 200240）

摘要： 掌握作物全生长周期氮素需求规律是实现氮素精量供给的关键。本研究以小白菜（*Brassica chinensis* L.）为研究对象，在室内土壤盆栽试验条件下，测定了4个氮水平（0，0.05，0.1和0.2g N/kg土，分别由N1，N2，N3和N4表示）处理，5个生长期（11d，18d，31d，38d和45d，分别由G1，G2，G3，G4和G5表示）的植株含N量及相应的生长、品质指标。利用主成分分析法对小白菜生长和品质指标做多目标综合评价，根据综合指标评判结果分析了小白菜各生长期的N素需求量，同时分析了小白菜各生长期的生长和品质指标与植株含氮量的相关性。结果表明：①G1-G5期，品质最佳处理分别为N2、N3、N4、N4和N4，小白菜地上部干重的积累量分别为0.019，0.075，0.374，0.587和0.937g/株，对应的N素积累量为1.32，4.54，20.99，24.85和29.56mg/株；②N1-N4处理下，G3期小白菜的氮素积累比例均较高，分别为27.08%，34.26%，39.24%和51.71%；③鸡毛菜收获期（G2），硝态氮含量与植株N浓度呈极显著正相关（r=0.95）；青菜收获期（G5），叶面积、地上鲜重和硝态氮含量均与植株N浓度呈极显著正相关（r=0.904，0.959，0.943）。本研究表明，G1-G5生长期小白菜N素需求量分别占整个生长期N素需求量的4.62%，11.93%，53.46%，13.51%和16.48%，可以为小白菜氮素的精准管理提供理论指导。

关键词： 小白菜；精准施肥；N需求；N积累；植株氮浓度

Dynamic Accumulation and Demand Analysis of Nitrogen in Pak Choi Plant

Xiong Xin[1] Chang Liying[1] Zhang Jingjin[1] Huang Danfeng[1，2*]

（1. *College of Agriculture and Biology*，*Shanghai Jiaotong University*，*Shanghai* 200240；2. *Key Laboratory of Urban Agriculture*（*South center*），*Ministry of Agriculture*，*Shanghai* 200240）

Abstract： It is the key to achieve precisely nitrogen supply by grasping the characteristics of nitrogen demand during the crop growth cycle. To clarify the nitrogen demand of pak choi（*Brassica chinensis* L.），the nitrogen accumulation and indices associated with growth and quality were determined by pot experiment in this study. Pak choi were planted under 4 different nitrogen levels（N1，N2，N3 and N4 represent 0，0.05，0.1 and 0.2 g N / kg soil，respectively），and the samples were collected at different growth stages（G1，G2，G3，G4 and G5 represent 11，18，31，38 and 45 d，respectively）. The comprehensive quality was evaluated by method of principal component analysis（PCA），and the nitrogen demand was analyzed based on the comprehensive evaluation result. The results showed that：（1）From G1 to G5，the optimal nitrogen treatments were N2，N3，

作者简介：熊鑫（1990—），女，陕西咸阳人，博士生，研究方向：蔬菜栽培生理，E-mail：xiongxin 1989@126.com

*通讯作者：黄丹枫（1956—），女，上海人，教授，博士生导师，研究方向：设施园艺技术，E-mail：hdf@sjtu.edu.cn

N4, N4 and N4, and the accumulation of dry weight in pakchoi was 0, 0.019, 0.075, 0.374, 0.587 and 0.937 g /plant , respectively. Correspondingly the amounts of nitrogen accumulation were 1.32, 4.54, 20.99, 24.85 and 29.56 mg / plant, respectively. (2) Compared to other stages, the proportion of nitrogen accumulation in G3 was higher; they were 27.08%, 34.26%, 39.24% and 51.71% under the treatment of N1-N4, respectively. (3) At G2, there was a significant positive correlation (r=0.95) between NO_3^--N and pant nitrogen concentration. At G5, leaf area, fresh weight and NO_3^--N were significantly correlated with plant nitrogen concentration and r was 0.904, 0.959 and 0.943, respectively. The result indicated that the proportion of nitrogen demand for pak choi was 4.62%, 11.93%, 53.46%, 13.51% and 16.48% from G1 to G5, respectively, which could promote the precise nitrogen management for pak choi.

Key words: Pak Choi; Precision fertilization; Nitrogen demand; Nitrogen accumulation; Plant nitrogen concentration

1 前言

小白菜（*Brassica chinensis* L.）又名不结球白菜、青菜，十字花科芸薹属，具有很高的营养价值，是我国分布最广、栽培面积最大、最为大众化的蔬菜之一[1，2]。施用氮肥是小白菜生产中的重要措施，合理施氮能提高小白菜产量和品质[3-6]。目前小白菜的N肥施用策略相关研究报道较多，严瑾[7]研究表明当小白菜纯氮施用量为167kg·hm^{-2}时产量最高，王荣萍[8]研究表明尿素为氮源，底肥70%、追肥30%能够兼顾产量和品质，周亮[9]研究表明有机无机肥配施能够降低小白菜硝酸盐含量。然而，由于土壤类型、气候条件、品种等的差异，这些研究结果并没有很好的普适性，生产实践中依然存在着盲目施氮、过量施氮的现象[10]。

过量或不合理施氮不仅不能达到高产优质的目的，还会降低氮肥利用率，增加氮肥损失，并且造成土壤质量退化、地表水和地下水体硝酸盐含量超标等一系列环境问题，严重影响到农田的可持续生产[11，12]。以作物需求为导向，根据作物的需N规律，将N素定量，定时直接提供给作物，不仅可以保证优质高产，还能使土壤养分收支平衡，减少土壤氮素残留，降低对环境的污染，保证土壤的可持续生产。

因此，本研究以小白菜全生长期的N素需求规律为目标，在不同N水平条件下测定了小白菜五个生长点的植株N含量及相应的生长和品质指标，利用主成分分析法对小白菜生长和品质做多目标综合评价，根据综合指标评判结果计算了小白菜各生长期的N素需求量，同时分析了小白菜各生长期的生长和品质指标与植株含氮量的相关性，为小白菜的合理施氮和评价小白菜N营养状况的指标选择提供了理论依据。

2 材料与方法

2.1 供试材料

小白菜品种为“华王”。供试土壤为潮土，主要理化性质为：有机质39.73g/kg，全氮1.43g/kg，速效磷23.31g/kg，速效钾0.136g/kg，硝态氮13.3mg/kg，铵态氮8.5mg/kg，pH值为8.29。试验选用肥料为尿素（AR）和KH_2PO_4（AR）。

2.2 实验设计

试验在上海交通大学农业工程训练中心温室进行。土壤盆栽，采用规格为13.5cm×11cm×8.9cm的试验盆，每盆装土（自然风干，过60目筛）500g。试验共设置4个N素水平的处理：0、0.05、0.1、0.2g/kg土（尿素添加量为0、0.053 6、0.107、0.214g/盆，分别以CK、T1、T2、T3表示）。各处理P和K用量相同，P添加量为$0.15g \cdot kg^{-1}$，K添加量为$0.189g \cdot kg^{-1}$（KH_2PO_4添加量为0.329 1g/盆）。每处理3次重复，每重复15盆，共180盆。于2016年9月2日将尿素、KH_2PO_4和土壤混合均匀装盆，加水至田间最大持水量的75%±5%，待第2d土壤稳定后播种，每盆10粒。待小白菜现两片真叶时，每盆留苗3株。小白菜生长期间不追施任何肥料，但进行定量浇水和除虫工作。分别于播种后11d、18d、31d、38d和45d采集小白菜地上部分。

2.3 测定项目与方法

于每个采样期，各处理随机选取9盆，挑选20株长势较为一致的小白菜用于测定植株生长生理性状。其中，15株用于考察其生长指标，测量其株高、茎粗、叶面积、单株干鲜重及植株N含量。每处理3次重复，每重复5株；另外5株保存于−20℃冰箱中，待整体采样结束后用于测定生理指标，包括叶绿素、硝态氮、可溶性蛋白、可溶性糖和维生素C含量。每处理重复3次。

2.3.1 小白菜生长指标的测定

株高（小白菜叶基部到生长点的距离）的测定采用钢尺，精度为0.01cm；茎粗（第一叶基部）的测定采样游标卡尺，精度为0.02mm；叶面积（整株所有叶片面积之和）测定采用扫描仪，利用image J软件提取数据；单株干鲜重的测定采用电子秤称量，精确到0.000 1g。每处理重复3次，每重复5株。

2.3.2 小白菜植株N含量的测定

测完小白菜的生长指标后，在105℃下迅速杀青30min，然后在80℃下烘干至恒重，研磨过60目筛。称取粉末5.00mg左右，采用Vario ELⅢ元素分析仪（德国Elementar）测定植株含N量。小白菜植株氮素积累量（mg/株）=小白菜地上干重（g/株）×小白菜植株含氮量（%）×1 000

2.3.3 小白菜生理指标的测定

可溶性蛋白含量测定参照考马斯亮蓝—G250法，可溶性糖含量测定采用蒽酮法，维生素C含量测定采用钼蓝比色法，硝态氮和叶绿素含量测定参照高俊凤[13]《植物生理学实验指导》。

2.3.4 数据分析

采用Microsoft Excel 2007和SPSS16.0软件对数据进行统计分析，采用Duncan's新复极差法（0.05）进行差异显著性检测，采用Bivariate Correlation方法进行指标间的相关性分析，采用Factor过程的Principal components方法进行主成分分析。

3 结果与分析

3.1 小白菜植株的N素累积

3.1.1 植株含氮量的动态变化

由表1可知，整个生长期，小白菜地上部干物质氮的浓度在1.6%~7%变化。对于同1个取样日，小白菜植株地上部氮浓度值随施氮量的增加而提高。第11、第18、第31、第38d和45d时，各施氮处理组小白菜地上部干物质氮浓度分别显著高于空白对照5.83%~6.75%，0.18%~11.74%、32.76%~142.24%、28.81%~139.55%和1.88%~96.88%。对于同一施氮量的处理下，小白菜地上部氮浓度值随小白菜生长期的延长而降低。整个观察期内，空白对照组小白菜地上部氮浓度值由6.52%下降到1.60%，T1、T2和T3处理组分别由6.93%下降到1.63%，6.96%下降到2.33%，6.90%下降到3.15%（表2）。

表 1 不同氮水平条件下小白菜地上干物质氮浓度（%）的动态变化

Table 1 Dynamic change of nitrogen concentration in shoot dry matter of pak choiunder different nitrogen levels

处理	生长期Growth periods（d）				
Treatment	11	18	31	38	45
CK	6.52 ± 0.04b	5.45 ± 0.05b	2.32 ± 0.04d	1.77 ± 0.03d	1.60 ± 0.03c
T1	6.93 ± 0.01a	5.46 ± 0.04b	3.08 ± 0.03c	2.28 ± 0.03c	1.60 ± 0.01c
T2	6.96 ± 0.03a	6.06 ± 0.03a	4.10 ± 0.03b	3.07 ± 0.05b	2.33 ± 0.04b
T3	6.90 ± 0.04a	6.09 ± 0.05a	5.62 ± 0.07a	4.24 ± 0.02a	3.15 ± 0.015a

注：表中数据为平均值 ± 标准差，同一列不同字母表示不同处理的差异显著性达0.05水平，下同。Data of the table represent average value ± standard deviation and those with the different letters in the same column are significantly different（$p< 0.05$）.The same below.

表 2 不同氮水平条件下小白菜地上干重（g/plant）的动态累积

Table 2 Dynamic accumulation of pak choi shoot dry mater under different nitrogen levels

处理	生长期Growth periods（d）				
Treatment	11	18	31	38	45
CK	0.019 ± 0.003	0.076 ± 0.001ab	0.276 ± 0.002b	0.421 ± 0.077b	0.490 ± 0.11b
T1	0.019 ± 0.001	0.082 ± 0.013ab	0.327 ± 0.035ab	0.423 ± 0.055b	0.499 ± 0.079b
T2	0.016 ± 0.003	0.075 ± 0.012b	0.297 ± 0.017ab	0.397 ± 0.03b	0.619 ± 0.077b
T3	0.016 ± 0.001	0.094 ± 0.005a	0.374 ± 0.076a	0.587 ± 0.102a	0.937 ± 0.103a

注：表中数据为平均值 ± 标准差，同一列不同字母表示不同处理的差异显著性达0.05水平，下同。Data of the table represent average value ± standard deviation and those with the different letters in the same column are significantly different（$p< 0.05$）.The same below.

表 3 不同生长阶段小白菜植株氮素积累量

Table 3 The amount of nitrogen accumulation in pak choi plant at different growth period

处理 Treat-ment	0 ~ 11d		11 ~ 18d		18 ~ 31d		31 ~ 38d		38 ~ 45d		收获期
	积累量 NAA (mg/株)	积累比例 PNA (%)	积累量 NAA (mg/株)	积累比例 PNA (%)	积累量 NAA (mg/株)	积累比例 PNA (%)	积累量 NAA (mg/株)	积累比例 PNA (%)	积累量 NAA (mg/株)	积累比例 PNA (%)	积累量 NAA (mg/株)
CK	1.21 ± 0.11ab	14.56	2.94 ± 0.16b	35.38	2.25 ± 0.21c	27.08	1.08 ± 0.51b	13.00	0.83 ± 0.15b	9.98	8.31 ± 0.34d
T1	1.32 ± 0.08a	12.26	3.13 ± 0.28b	29.06	3.69 ± 0.63c	34.26	1.49 ± 0.51b	13.83	1.13 ± 0.12b	10.59	10.77 ± 1.15c

（续表）

处理 Treatment	0～11d 积累量 NAA (mg/株)	0～11d 积累比例 PNA (%)	11～18d 积累量 NAA (mg/株)	11～18d 积累比例 PNA (%)	18～31d 积累量 NAA (mg/株)	18～31d 积累比例 PNA (%)	31～38d 积累量 NAA (mg/株)	31～38d 积累比例 PNA (%)	38～45d 积累量 NAA (mg/株)	38～45d 积累比例 PNA (%)	收获期 积累量 NAA (mg/株)
T2	1.13 ± 0.08b	7.85	3.41 ± 0.35b	23.68	5.65 ± 0.90b	39.24	1.99 ± 0.21b	13.82	2.23 ± 1.36b	15.41	14.40 ± 1.07b
T3	1.11 ± 0.09b	3.76	4.60 ± 0.30a	15.57	15.28 ± 1.28a	51.71	3.86 ± 1.03a	13.06	4.71 ± 1.76a	15.90	29.55 ± 1.93a

注：表中数据为平均值 ± 标准差，同一列不同字母表示不同处理的差异显著性达0.05水平，下同。Data of the table represent average value ± standard deviation and those with the different letters in the same column are significantly different（$p< 0.05$）.The same below.

NAA = amount of nitrogen accumulation；PNA = proportion of nitrogrn accumulation

表 4 不同生长期小白菜植株的生长和生理指标变化

Table 4 The change of physiological indices in pak choi at different growth period

样本编号 Sample number		株高 (cm) plant height	茎粗 (mm) Stem diameter	叶面积 (cm^2) Leaf area	地上鲜重(g/plant) Fresh weight	地上干重(g/plant) Dry weight	叶绿素 (mg/g) chlorophyll	可溶性蛋白 (mg/g) Soluble protein	可溶性糖 (mg/g) Soluble sugar	维生素C ($mg·100g^{-1}$) Vitamin C	硝态氮 (μg/g) Nitrate-N	植株氮浓度% Nitrogen concentration
11d	CK	3.05b	1.36	7.26b	0.25b	0.02	0.80a	5.53a	3.09c	38.34b	—	6.52b
	T1	3.44a	1.45	9.10a	0.33a	0.02	0.66bc	3.38c	4.29b	48.37b	—	6.93a
	T2	3.55a	1.41	7.35b	0.25b	0.02	0.56c	5.36ab	6.79a	74.44a	—	6.96a
	T3	3.48a	1.33	7.70b	0.26b	0.02	0.71ab	4.50b	2.88c	63.34a	—	6.90a
18d	CK	3.91ab	2.28ab	35.11b	1.03b	0.08ab	0.44	7.93b	10.46b	214.72a	6.67c	5.45b
	T1	4.08ab	2.46a	37.19b	1.15ab	0.08ab	0.48	17.15a	15.46a	194.83ab	8.92c	5.46b
	T2	4.29a	2.52a	41.71a	1.32a	0.07b	0.57	19.64a	10.42b	157.56b	43.88b	6.06a
	T3	3.79b	1.98b	32.85b	1.00b	0.09a	0.42	19.79a	6.17c	222.78a	67.17a	6.09a
31d	CK	4.39b	3.11b	147.57c	3.32c	0.28b	0.81b	3.62b	14.63ab	148.61	1.88b	2.32d
	T1	5.58a	3.46a	182.68bc	4.17c	0.33ab	1.03ab	3.66b	16.38a	139.72	1.67b	3.08c
	T2	5.43a	3.36ab	222.84b	5.34bc	0.30ab	1.13a	4.97b	12.50bc	123.61	2.04b	4.10b
	T3	5.59a	3.43a	296.40a	7.61a	0.37a	1.11a	10.98a	8.96c	120.56	25.17a	5.62a
38d	CK	4.94b	3.98	400.47c	4.78b	0.42b	0.75b	2.22b	20.92a	130.00a	0.92c	1.77d
	T1	5.65a	4.09	507.33b	5.01b	0.42b	0.99a	3.44b	22.46a	132.16a	1.04bc	2.28c
	T2	5.49ab	3.83	521.06b	4.85b	0.40b	1.01a	5.55a	17.92b	113.61a	1.46b	3.07b
	T3	5.58a	4.40	901.08a	9.07a	0.59a	1.05a	5.21a	12.33c	86.39b	8.29a	4.24a
45d	CK	4.84b	4.21c	452.36b	4.56c	0.49b	0.53b	4.26	25.50a	131.67	4.21c	1.60c
	T1	5.72a	4.38bc	472.70b	4.96c	0.50b	0.97a	4.25	23.17a	113.05	3.67c	1.63c
	T2	5.73a	4.93a	587.64b	6.56b	0.62b	0.58b	4.98	25.46a	124.72	7.13b	2.33b
	T3	5.64a	6.26a	1 117.56a	11.15a	0.94a	0.92a	5.38	19.29b	114.44	10.71a	3.15a

注：—，表示未采集到样品。不同字母表示同一采样日不同处理间差异显著性达0.05水平。

—，no samples.Different letters in the same column and sampling date are significantly different（$p< 0.05$）.

3.1.2 地上部干重的动态累积

表3可知，小白菜植株地上部干物质的量随施氮量的增加而提高。第11d，各施氮处理间小白菜地上部干物质量无显著差异；第18，第31，第38和45d，仅T3处理组小白菜地上部干物质量显著升高，分别高于空白对照23.68%、35.51%、39.43%和91.22%，而CK，T1和T2施氮水平间的小白菜地上部干物质量为统计意义上的相等。

3.1.3 植株N素的动态累积

随施氮量增加，收获期小白菜植株N素积累量增加。供试氮水平（0～0.2mgN/g土）范围内，不同处理收获期植株N素积累量差异显著，T1，T2和T3处理分别较空白对照升高29.60%，73.29%和255.60%。施氮量对小白菜植株氮素积累的影响在不同生长阶段也存在差异。生长至11d时，T1处理的N素积累量大于其他处理；生长至18d后，随施氮量增加，阶段N素积累量也增加，以T3处理的N素积累量最大，T1，T2处理和空白对照之间无显著差异。小白菜从播种到生长至4片真叶（0～18d），两个调查时间点植株氮素积累量占全生长期氮素积累量的比例随施氮量增加而减小，而4片真叶（18d）以后则反之。说明本试验条件下，小白菜生长前期（4片真叶之前）不需要大量的施氮即可满足植株对氮素的吸收，而小白菜生长后期只有高氮处理（T2和T3）才能促进植株对氮素的吸收，低氮处理（T1）则无显著促进效应。

3.2 小白菜的生长和品质指标

3.2.1 不同N水平处理下小白菜生长和品质指标的变化

不同生长期，小白菜的形态和生理指标测定结果见表4。同一取样日，不同氮处理条件下小白菜的各生长和生理指标存在差异。对于小白菜的形态，施氮处理组较空白对照组，株高升高、茎加粗、叶片面积之和显著增加；对于小白菜的生长，前期低氮处理（T1）小白菜鲜重显著升高，干重和叶绿素含量无显著差异；后期高氮处理（T3）小白菜干鲜重和叶绿素含量显著升高；对于小白菜的品质，前期（0～18d）可溶性蛋白，可溶性糖和维生素C含量随施氮量的变化规律不明显；后期（18～45d），随着施氮量的升高，可溶性蛋白含量升高，可溶性糖含量降低，维生素C含量无显著变化，硝态氮含量显著升高。

3.2.2 小白菜多项生长和品质指标的综合评价

在评价小白菜时，单项生长或品质指标很难判断小白菜的综合品质，因而采用多目标综合评价方法来评价小白菜的综合品质。本文选取株高（X_1）、茎粗（X_2）、叶面积（X_3）、地上部鲜重（X_4）、地上部干重（X_5）、硝态氮（X_6）、可溶性蛋白（X_7）、可溶性糖（X_8）、维生素C（X_9）、叶绿素（X_{10}）和植株N浓度（X_{11}）这11个生长和品质因素作为评价因子，首先，对这11个指标进行同趋化（低优指标）和标准化，然后进行主成分分析，并得到相关矩阵的特征根及特征向量累积贡献率（表5），前三项特征根的累积贡献率达到了92.33%，大于90%，所以可用第一、第二和第三主成分作为评价的综合指标，且评价的可信度为92.33%。通过计算各指标与前三个主成分的关系为：

$$Z_1=0.592X_1+0.884X_2+0.922X_3+0.57X_4+0.912X_5-0.048X_6-0.264X_7+0.401X_8-0.605X_9+0.25X_{10}-0.309X_{11}$$

$Z_2=-0.335X_1-0.414X_2-0.209X_3-0.085X_4-0.334X_5+0.827X_6+0.795X_7-0.855X_8+0.335X_9-0.094X_{10}+0.901X_{11}$

$Z_3=0.65X_1+0.184X_2+0.208X_3+0.467X_4+0.201X_5-0.392X_6-0.437X_7-0.147X_8-0.634X_9+0.945X_{10}-0.091X_{11}$

综合评值：$Z=0.669\ 07Z_1+0.153\ 15Z_2+0.101\ 08Z_3$，由第一、第二和第三主成分与客观权重之积，得到不同N水平处理下不同生长期小白菜品质综合评判结果（表6），综合评价值越高，小白菜品质越好。

表5　主要主成分的特征值、贡献率和累积贡献率

Table 5　Eigenvalues（E），contribution proportions（CR）andcumulative contribution proportions（CCR）of mainprinciple component

主成分 Principle component	特征值 E	贡献率 CR（%）	累积贡献率 CCR（%）
1	7.360	66.907	66.907
2	1.685	15.315	82.222
3	1.112	10.108	92.330
4	0.368	3.342	95.673
5	0.210	1.906	97.578
6	0.131	1.189	98.767
7	0.066	0.604	99.371
8	0.050	0.452	99.823
9	0.013	0.121	99.944
10	0.005	0.048	99.991
11	0.001	0.009	100.000

Extraction Method：Principal Component Analysis.

表6　不同N水平处理小白菜品质综合评判结果

Table 6　Comprehensive evaluation results of different treatments

处理Treatment		Z1	Z2	Z3	Z	排序Sort
11d	CK	−15.30	17.27	−23.15	−9.93	2
	T1	−18.40	37.54	−28.23	−9.42	1
	T2	−35.38	26.27	−46.42	−24.34	4
	T3	−29.73	25.13	−38.27	−19.91	3
18d	CK	−92.04	69.97	−133.07	−64.31	3
	T1	−78.26	67.65	−125.44	−54.67	2
	T2	−55.75	89.85	−114.66	−35.13	1
	T3	−107.31	137.02	−167.14	−67.71	4
31d	CK	58.91	9.75	−62.41	34.60	4
	T1	98.96	−2.18	−48.28	60.99	3
	T2	144.33	−10.39	−29.43	91.99	2
	T3	211.57	1.22	−22.34	139.47	1

（续表）

处理Treatment		Z1	Z2	Z3	Z	排序Sort
38d	CK	308.90	−57.83	3.28	198.14	4
	T1	407.03	−79.56	24.10	262.56	3
	T2	478.92	−81.84	38.01	311.71	2
	T3	796.16	−159.87	134.71	521.78	1
45d	CK	−67.89	−67.89	9.86	−54.81	4
	T1	387.13	−77.26	27.66	249.96	3
	T2	488.26	−95.63	161.31	328.31	2
	T3	985.50	161.31	161.31	700.27	1

表7　生长、生理指标与植株N浓度在小白菜不同生长期的相关性

Table 7　Correlation analysis between nitrogen concentration of pakchoi plant and physiological indices at different growth period

生长期 Growth Periods	株高 Plant height	茎粗 Stem diameter	叶面积Leaf area	地上鲜重Fresh weight	地上干重Dry weight	叶绿素 Chlorophyll	可溶性蛋白 Soluble protein	可溶性糖 Soluble sugar	维生素C VitaminC	硝态氮 Nitrate-N
11d	0.85**	0.243	0.397	0.352	−0.185	−0.664*	0.437	0.510	0.690*	–
18d	0.420	−0.264	0.133	0.186	0.305	0.231	0.726**	−0.655*	0.224	0.95**
31d	0.666*	0.460	0.919**	0.934**	0.593*	0.605*	0.916**	−0.786**	−0.524	0.86**
38d	0.486	0.339	0.923**	0.815**	0.628*	0.688*	0.764**	−0.902**	−0.835**	0.893**
45d	0.441	0.922**	0.904**	0.959**	0.895**	0.299	0.520	−0.640*	−0.261	0.943**

注：*，**分别表示在0.05和0.01水平上的相关性，下同。*，**，Correlation is significant at the 0.05 and 0.01 level，the same below.

3.2.3　不同生长期小白菜生长和品质指标与植株N浓度的相关性

同时，我们分析了不同生长期小白菜的生长和品质指标与植株N浓度的相关性。结果（表7）表明，小白菜的生长、品质指标与植株N浓度的相关性在不同生长期不同。第11d，株高与植株N浓度之间呈强正线性关系，维生素C含量与植株N浓度之间呈正线性关系，叶绿素含量与植株N浓度之间呈负线性关系；第18d，可溶性蛋白含量、硝态氮含量均与植株N浓度之间呈强正线性关系，而可溶性糖含量与植株N浓度之间呈负线性关系；第31d，除了可溶性蛋白含量和硝态氮含量外，叶面积和地上鲜重也与植株N浓度之间呈强正线性关系，株高、地上干重和叶绿素含量与植株N浓度之间呈正线性关系，可溶性糖含量与植株N浓度之间呈强负线性关系；第38d，叶面积、地上鲜重、可溶性蛋白和硝态氮含量均与植株N浓度之间呈强正线性关系，地上干重和叶绿素含量与植株N浓度之间呈正线性关系，可溶性糖和维生素C含量与植株N浓度之间呈强负线性关系；第45d，茎粗，叶面积、地上鲜重、地上干重和硝态氮含量均与植株N浓度之间呈强正线性关系，而可溶性糖含量与植株N浓度之间呈负线性关系。因此，在不同生长阶段通过选择不同的生长生理指标即可反映小白菜植株的N浓度，如鸡毛菜收获时期，可以选择可溶性蛋白或硝态氮；小青菜收获时期，可以选择叶面积或地上鲜重等。

4 讨论

4.1 小白菜不同生长期的N素需求量分析

本研究结果（表1和表3）表明，小白菜植株N浓度值和收获期植株N素积累量随施氮量的增加而提高，说明在供试N水平范围内，增施N肥可以促进小白菜对N素的吸收和积累。同时，图1和表1还表明，小白菜植株氮浓度值随小白菜地上部生物量的增加而下降，这主要是由于小白菜生长期内干物质的累积速率大于N的吸收速率，该研究结果符合“作物临界N浓度稀释曲线”的模型[14]。

小白菜品质指标综合评判结果（表6）表明，11d时，T1处理的小白菜品质最佳，故T1处理的N素积累量可看作是0～11d这一时期小白菜生长最佳的N素需求量，为1.32mg/株；同理，11～18d，18～31d，31～38d，38～45d这4个生长期小白菜的N素需求量分别为3.41mg/株，15.28mg/株，3.86mg/株和4.71mg/株，则小白菜整个生长期的N素需求总量为28.58mg/株。

然而，本实验提供的N水平范围有限，在后期（31～45d）不能确定小白菜的生长是否受到N素营养亏缺的制约，所以计算出的N素需求总量有可能偏小。因此，有必要继续扩大N水平范围，直至小白菜成熟期地上部生物量不再随施N量的增加而提高，再根据品质指标综合评判标准确定各生长期小白菜的N素需求量，为小白菜生产的精准施N提供理论指导。

4.2 小白菜N素吸收的关键时期分析

本研究结果（表3）表明，不同生长阶段小白菜植株氮素积累存在差异。不同N肥施用量条件下，18～31d这一阶段小白菜植株N素积累的比例在整个生长期均最高，达27%～52%，说明小白菜生长至4～8片真叶的时期（18～31d）是N素吸收的关键时期。该研究结果正好为生产实践“四叶一心时期追施N肥有利于提高小白菜的产量”提供了理论依据。此外，本研究还发现施氮量对阶段小白菜植株氮素积累的比例也有影响。在小白菜生长至4片真叶以前（0～18d），不施氮处理组（CK）小白菜N素的积累比例显著高于施氮处理（T1，T2和T3），4片真叶以后，则相反。因此，在小白菜工厂化育苗中，少施N肥更有利于壮苗，在小白菜栽培实践中，应该轻施基肥，四片真叶以后适量追肥。然而要更精确地了解小白菜各生长期的N素需求量及N素吸收速率，还应该缩短取样间隔，扩大N水平范围，进行深入研究。

4.3 小白菜不同生长阶段N营养状况的评价指标分析

植株N浓度能够诊断小白菜的营养状况，然而植株氮浓度的测定相对于形态和其他生理指标的测定比较费时费力，因此本文分析了而各生长、品质指标在不同生长期与小白菜植株氮浓度的相关性。研究表明，4片真叶以前（0～18d），硝态氮含量最能反映植株N浓度，与氮浓度极显著正相关，r值为0.95。4片真叶以后，形态、生长和生理指标均能很好的反映N浓度。其中，叶面积、地上鲜重和硝态氮与N浓度呈极显著正相关。因此，鸡毛菜生产实践中，推荐将硝态氮含量作为评判N营养状况的指标，青菜收获期，推荐叶面积、地上鲜重和硝态氮含量作为评判N营养状况的指标。

参考文献

[1] 俞晓琴. 小白菜食用价值及栽培技术 [J]. 吉林蔬菜，2011（6）：19-20

[2] 刘朝阳，李贞霞，于丹. 小白菜营养成分测定分析 [J]. 中国园艺文摘，2014（4）：29-31.

[3] 严瑾，黄丹枫，张屹东，等. 菜地土壤基础供氮量对小白菜产量及氮素利用率的相关关系研究 [J]. 安徽农业科学，2014（16）：5071-5073.

[4] 曹小闯，吴良欢，陈贤友，等. 氨基酸部分替代硝态氮对小白菜产量、品质及根际分泌物的影响 [J]. 植物营养与肥料学报，2012（3）：699-705.

[5] 朱伟锋，林咸永，金崇伟，等. 氮肥对不同白菜品种抗氧化物质含量和抗氧化活性的影响 [J]. 浙江大学学报（农业与生命科学版），2009，（3）：299-306.

[6] 张彩峰，李珍珍，陆奕，等. 氮肥浓度及形态对青菜产量及品质的影响 [J]. 上海：上海农业学报，2014（1）：75-78.

[7] 严瑾. 土壤基础供氮能力与小白菜氮素利用率的相关关系研究 [D]. 上海交通大学，2014.

[8] 王荣萍，蓝佩玲，李淑仪，等. 氮肥品种及施肥方式对小白菜产量与品质的影响 [J]. 生态环境，2007（3）：1 040-1 043.

[9] 周亮，荣湘民，郭春铭，等. 有机无机肥配施对小白菜产量及品质的影响 [J]. 湖南农业科学，2012（23）：52-55.

[10] 熊鑫，沈海滨，王辉. 上海市小白菜高效安全生产现状与建议 [J]. 长江蔬菜，2016（22）：12

[11] Albornoz F. Crop responses to nitrogen overfertilization：A review [J]. Scientia Horticulturae，2016（205）：79-83.

[12] 韦高玲，卓慕宁，廖义善，等. 不同施肥水平下菜地耕层土壤中氮磷淋溶损失特征 [J]. 生态环境学报，2016（6）：1023-1031.

[13] 高俊凤. 植物生理学实验指导 [M]. 北京：高等教育出版社，2000.

[14] Greenwood D J，Lemaire G，Gosse G，et al. Decline inpercentage N of C3 and C4 crops with increasing plant mass [J]. Annals of Botany，1990，66（4）：425-436.

不同氮水平下增施CO_2对番茄植株叶片养分的影响[1]

荀志丽[2]，张玲，温祥珍，李亚灵[3]

（山西农业大学园艺学院，太谷　030801）

摘要： 为探究不同氮水平对番茄植株叶片养分的影响，以及增施CO_2的效果。以番茄“鸿途”为材料，在2个自然光照人工气候室内分别设置CO_2浓度为300（C1）、600（C2）μl·L^{-1}，采用苗钵基质栽培，定期定量浇营养液的氮素水平分别是50（N1）、150（N2）、250（N3）、350（N4）、450（N5）mg·L^{-1}，每个处理重复3次，分别在植株开花期、坐果期、果实膨大期测定叶片中硝态氮和矿质元素的含量。结果表明，在番茄开花期，适宜氮（N3-N4）水平下，增施CO_2，最能促进植株叶片中矿质元素的含量；在低氮（N1）水平下，增施CO_2，对植株生长影响不大；在高氮（N5）水平下，增施CO_2，不仅有利于植株对养分的吸收，还能缓解氮肥过高对植株的不利影响。但在番茄坐果期和果实膨大期，不论何种N水平下增施CO_2，其叶片中养分含量都几乎没有变化。综合本试验番茄生长各项指标及经济效益，在营养液N浓度为250mg/L·L^{-1}（N3）−350mg/L·L^{-1}（N4）时，增施CO_2浓度到600μl/L·L^{-1}（C2）最能促进开花期番茄植株叶片中养分含量的增加。

关键词： N水平；增施CO_2；总干重；矿质元素

The Effect of Elevated CO_2 on Leaf Nutrient of Tomato Plant Under Different Nitrogen Levels Supply

Xun Zhili，Zhang Ling，Wen Xiangzhen，Li Yaling

（*College of Horticulture*，*Shanxi Agricultural University*，*Taigu*　030801）

Abstract: To explore the effect of the nutrient of tomato plant leaves on different nitrogen levels and the effect of elevated CO_2. The material was Tomato variety of “HongTu”，in two natural light artificial climate room，CO_2 concentration were set up for 300（C1），600（C2）μl·L^{-1}，the pot culture and irrigated with nutrient solution termly and quantify was adopted with 5 levels，50（N1），150（N2），250（N3），350（N4），450（N5）mg·L^{-1}，each sample repeat 3 times，respectively，in the stage of plant flowering，fruit，fruit enlargement，the contents of nitrate nitrogen and mineral elements in the leaves were determined. The results showed in tomato flowering period，On the preference N level（N3-N4），elevated CO_2，the contents of mineral elements in the plant leaf were the maximum value；On the lower N level（N1），elevated CO_2，little influence to plant growth；On the higher N level（N5），elevated CO_2，was not only beneficial to the plant to

---

基金项目：国家自然科学基金重点项目“温室环境作物生长模型与环境优化调控”（编号：61233006），山西省煤基重点科技攻关项目“设施蔬菜高效固碳技术研究与示范”（项目编号：FT201402-05）

作者简介：荀志丽（1986—），山西霍州人，在读博士研究生

通讯作者：李亚灵（1962—），山西灵石人，教授，博士生导师，联系方式E-mail：yalingli1988@163.com absorb nutrients，also would alleviate the negative impact from the high nitrogen fertilizer. However，the nutrient content in the leaves showed little change at the period of tomato fruit and fruit expansion，regardless of the N levels. Comprehensive index and economic benefit，the nutrient contents of tomato leaves during flowering period were increased rapidly，in the condition of nutrient solution N concentration of 250 mg·L^{-1}（N3）－350 mg·L^{-1}（N4），and elevated CO_2 concentration to 600μl·L^{-1}（C2）.

Key words：N levels；Elevated CO_2；The total dry weight；Mineral element

CO_2是植物光合作用的必要原料，而温室是一个相对封闭的空间，空气流动性差，使得白天温室内CO_2浓度低，很大程度上限制了蔬菜产量，因此，设施栽培中过低的CO_2已成为光合作用的主要限制因素，它制约了植株的生长发育，降低了蔬菜的产量和品质[1]。国内外的学者在这方面也进行了许多详细的研究，早在1995刘保才等用通径分析法证明，影响日光温室蔬菜栽培的光照、有效积温、CO_2浓度三大环境要素中，CO_2浓度对产量的贡献最大[2]。多数研究也表明，CO_2浓度升高对作物生长发育、生理指标以及养分吸收等产生正向促进作用[3]。

植物对CO_2施肥的响应依赖于养分的供应强弱，高养分条件促进植物生长，低养分条件抑制植物生，且CO_2施肥有利于植物组织中碳水化合物的积累，降低了含氮量。蔬菜栽培生产过程中，对氮肥的需求量最大，氮素是植物必需的大量营养元素之一，是植物生长的重要物质基础，对植物的器官建造、物质代谢、生化过程等都有不可替代的作用，通常被称为“生命元素”，直接或间接影响着植物的生长发育。因此，本研究以番茄“鸿途”为材料，在自然光照人工气候室内，设置2个CO_2浓度，5个不同氮水平的营养液，以期探究不同氮水平下番茄植株叶片养分的变化情况，以及不同氮水平下增施CO_2对植株叶片养分的影响。

1　材料与方法

1.1　试验地点和材料：

试验在山西太谷（北纬37° 25′，东经112° 25′）山西农业大学设施农业工程中心的非对称三连栋温室内进行。在其内搭建了2个自然光照人工气候室，其结构坐北朝南，长3m、宽2.4m、前屋面高1.5m，后屋面高2.3m。

试验所用番茄品种为“鸿途”，无限生长型。

1.2　试验处理

试验在自然光照人工气候室中进行，各人工气候室中都安装有温度、湿度、CO_2传感器和小吊扇，通过CO_2+温湿度一体式控制器智能控制。其CO_2由CO_2钢瓶供应，温湿度由立式空调自动调节。

试验均设计2个CO_2浓度处理300μl·L^{-1}、600μl·L^{-1}，分别通过2个自然光照人工气候室控制，以C1、C2表示，C1处理的浓度最为接近大气CO_2浓度，作为对照处理。CO_2浓度处理（阴雨天气除外）时间从9:00～16:00，共7h，该时段正是太阳光照强，温室番茄光合作用旺盛的时期。期间打开小吊扇以促进空气流通，使其内部CO_2浓度均匀。

试验设计5种氮素水平，氮素含量分别是50、150、250、350、450mg·L^{-1}，以N1、N2、N3、N4、N5表示。采用苗钵基质（椰壳）栽培，将番茄植株定植在苗钵中，然后定期定量浇不同氮素水平的营养液，所用苗钵内均套有黑色塑料袋，以避免营养液外漏。每隔3d浇1次营养液，每次浇400ml。营养液中的氮素梯度是在日本山崎配方（表1）的基础上进行改进得到的，微量元素采用通用配方。

表1　两次试验的番茄营养液配方（mg·L^{-1}）

Table 1　Nutrient solution composition for tomato growing

N浓度（mg·L^{-1}）	50	150	250	350	450
$Ca(NO_3)_2$	292.9	292.9	292.9	292.9	292.9
KCl	248.1	248.1	248.1	248.1	248.1
NH_4NO_3	0.0	285.7	571.4	857.1	1 142.9
KH_2PO_4	91.1	91.1	91.1	91.1	91.1
$MgSO_4·7H_2O$	245.9	245.9	245.9	245.9	245.9
EC值	2.28	2.37	2.94	3.30	3.75

注：表中数值是由mmol·L^{-1}转换为mg·L^{-1}的营养液中的各药品的质量。表中EC表示电导率值，由便携式电导率仪（HI 8733）在试验开始时进行测定，单位为mS·cm^{-1}

试验于2015年8月20日定植，定植时的苗龄为60d，定植后缓苗8d。每个N水平有8株，重复3次，随机选取3株长势均匀的植株进行定株观察。番茄植株在9月21日开始坐果，10月5日试验结束，共计47d。

1.3　测定项目与方法

环境指标：自然光照人工气候室中的气温、空气相对湿度和CO_2浓度均由北京烁光盛业科技发展有限公司提供的SG605二氧化碳+温湿度一体式控制器监控，每10min电脑自动记录1次数据，为瞬时值。

（1）干重。用分析天平分别测定植株叶片、茎、根的干重。

（2）硝态氮含量。硝基水杨酸比色法。

（3）矿质元素。分别在定植后第19天（番茄开花期）、第33天（坐果期）、第47天（果实膨大期）测定了番茄叶片中矿质元素含量。全氮含量的测定用H_2SO_4-H_2O_2蒸馏法；P、K、Ca、Mg含量用火焰原子吸收法。

2　结果与分析

2.1　试验期间自然光照人工气候室中的环境状况

由图1A和B可以看出，试验期间C1和C2处理下自然光照人工气候室中温度和空气相对湿度基本一致，分别约为20.5℃和78.8%。图1C以2015年9月2日为例，呈现了当天C1和C2两个人工气候室内的CO_2浓度控制状况，各人工气候室中的CO_2浓度在夜间均较高，在6:00即日出之后CO_2浓度直线下降，最低降至40μl·L^{-1}，在9:00开始通入CO_2气体之后，浓度迅速上升，在CO_2施用时间段内，各人工气候室内CO_2浓度基本达到控制目标，实际测定的CO_2浓度均值分别约为360μl·L^{-1}和630μl·L^{-1}。

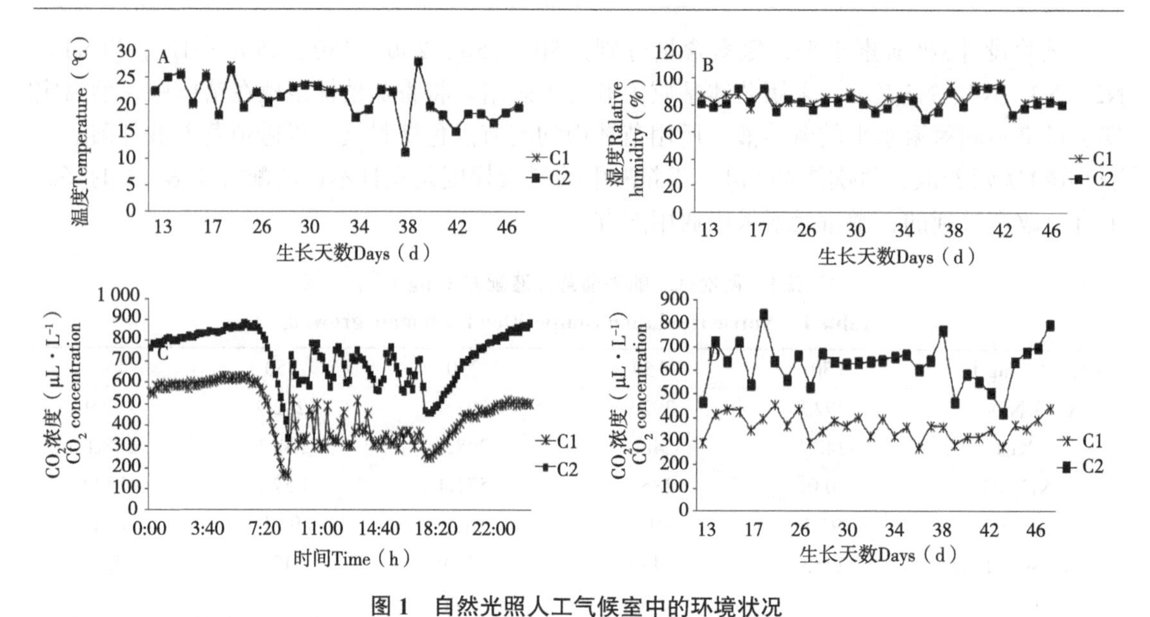

图1　自然光照人工气候室中的环境状况

Figure 1　The temperature situation in natural light artificial climate rooms

注：图1中数值是两次试验期间每天24h的温度均值（00:00到次日00:00）共144个数据的均值

2.2　增施CO_2对番茄植株生物量的影响

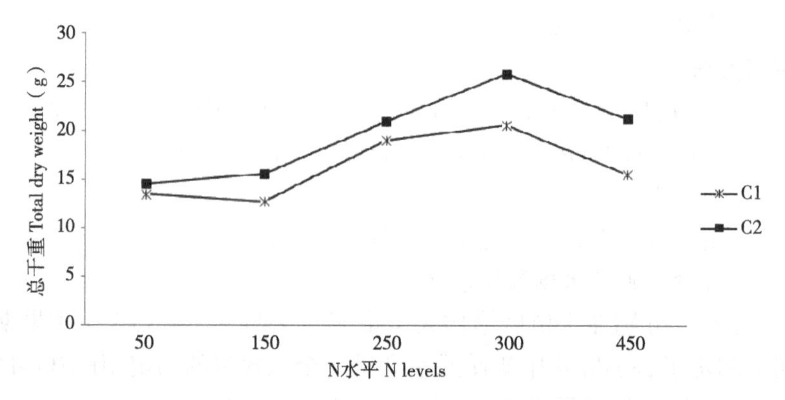

图2　CO_2浓度对番茄植株生物量的影响

Figure 2　The influence of enriched CO_2 on biomass

注：图中各点为试验期间每次测定的各CO_2处理下3次重复的单株总干重均值

图2是不同氮水平下施用CO_2以后植株总干重的含量。从图2中可知，在低氮的N1水平下，植株总干重较低，较N3水平低5.55g，随营养液氮素水平的增加，总干重逐渐增加，在营养液N4水平下达到了最高值，当营养液氮素水平继续增加到高氮的N5水平时，番茄植株的总干重不再增加反而降低，较N3水平低3.43g，降低了18.03%。

在适宜的N3水平下，增施CO_2，植株的总干重较C1浓度高1.94g，增加了10.19%；在低氮N1水平下，增施CO_2，植株的总干重几乎没有变化；高氮的N5水平下，增施CO_2，植株的总干重较C1浓度高5.60g，即增加了35.92%。

2.3 增施CO_2对叶片硝态氮含量的影响

植物体内硝态氮含量可以反映土壤氮素供应情况，常作为施肥的指标。图3是番茄生长不同时期不同氮水平下施用CO_2以后叶片硝态氮含量的结果。由图3可知，植株不同生长时期，不同氮水平下，各CO_2浓度下，叶片中硝态氮含量变化趋势几乎相同。在低氮的N1水平下，番茄生长的3个时期，叶片中硝态氮含量都较低，分别较N3水平低487.57$\mu g \cdot g^{-1}$、845.77$\mu g \cdot g^{-1}$、386.75$\mu g \cdot g^{-1}$，降低了58.13%、73.09%、63.4%；在高氮N5水平下，叶片中硝态氮含量较高，分别较N3水平高959.19$\mu g \cdot g^{-1}$、666.67$\mu g \cdot g^{-1}$、378.11$\mu g \cdot g^{-1}$，增加了111.44%、57.61%、61.99%。说明叶片中硝态氮含量与营养液中氮水平呈正相关。

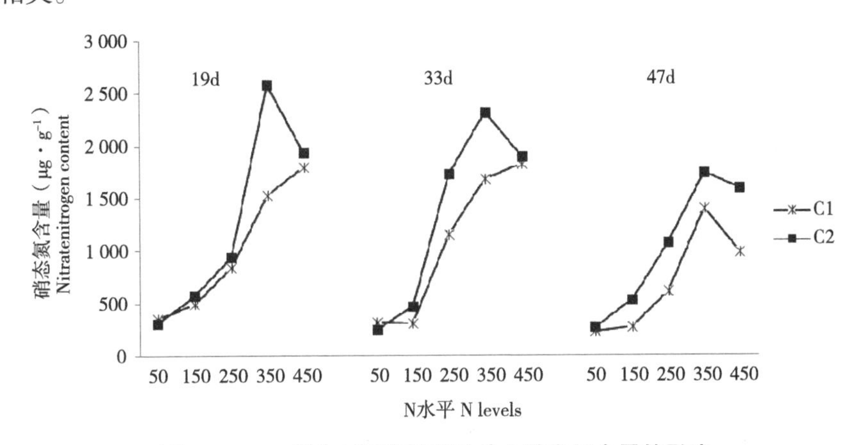

图 3 CO_2-N耦合对番茄植株叶片中硝态氮含量的影响

Figure 3 The influence of CO_2-N treats on nitrate nitrogen content of tomato leaves

注：图3中各处理所用数据是试验期间5次测定的叶片硝态氮含量均值，每次测定每个处理有3次重复

在对照N3水平下，增施CO_2，番茄叶片中硝态氮含量在坐果期较C1浓度高572.14$\mu g \cdot g^{-1}$，增加了49.44%；在低氮的N1水平下，增施CO_2，叶片中硝态氮含量几乎没有变化；在高氮的N5水平下，增施CO_2，叶片中硝态氮含量几乎没有变化；在较高氮的N4水平下，增施CO_2，硝态氮含量在开花期和坐果期较C1浓度分别高1 049.65$\mu g \cdot g^{-1}$、626.86$\mu g \cdot g^{-1}$，即增加了68.59%，37.21%。叶片中硝态氮含量在果实膨大期几乎没有变化。

2.4 对矿质元素含量的影响

2.4.1 叶片中全氮含量

图4是在番茄生长的不同时期，不同N水平营养液条件下增施CO_2后，其叶片含氮量的结果。从图4可知，番茄开花期、坐果期、果实膨大期这3个时期中，在适宜的N3水平下，其叶片中的含氮量均较高分别为36.97$mg \cdot g^{-1}$、35.1$mg \cdot g^{-1}$、34.79$mg \cdot g^{-1}$；在低氮的N1水平下，这3个时期中番茄叶片含氮量都较低，分别较N3水平低6.53$mg \cdot g^{-1}$、13.99$mg \cdot g^{-1}$、12.11$mg \cdot g^{-1}$，分别降低了17.66%、39.86%、34.8%；在高氮N5水平下，叶片含氮量与N3接近几乎没有变化。

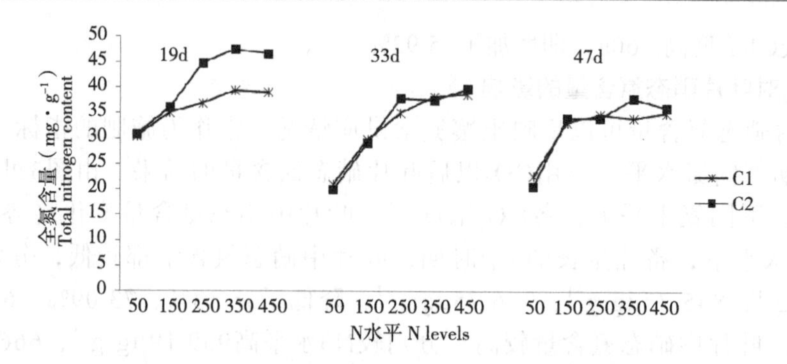

图 4　CO_2-N耦合对番茄植株叶片中全氮含量的影响

Figure 4　The influence of CO_2-N treats on total nitrogen content of tomato leaves

注：图中各处理所用数据是试验期间测定的叶片全氮含量共计3次的均值，每次测定每个处理有3次重复

番茄开花期，在对照N3水平下，其叶片中的含氮量较高，较C1浓度高7.82mg·g^{-1}，增加了21.15%，在低氮N1水平下，增施CO_2，3个时期中叶片含氮量无变化，在高氮的N5水平下，增施CO_2，叶片中含氮量同样较高，较C1浓度高7.46mg·g^{-1}，即增加了19.06%；番茄坐果期和果实膨大期，不同氮水平下增施CO_2效果不明显。

2.4.2　叶片中全磷含量

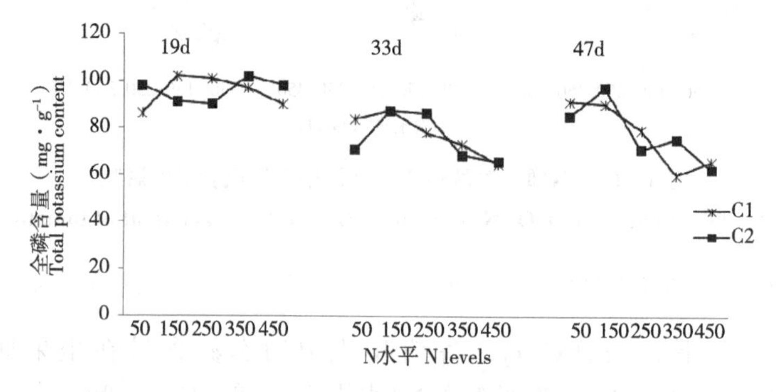

图 5　CO_2-N耦合对番茄植株叶片中全磷含量的影响

Figure 5　The influence of CO_2-N treats on total phosphorus content of tomato leaves

注：图中各处理所用数据是试验期间测定的叶片全磷含量共计3次的均值，每次测定每个处理有3次重复

图5是营养液不同氮水平下施用CO_2以后叶片含磷量的结果。由图5可知，适宜的N3水平下，番茄在开花期、坐果期、果实膨大期其叶片中的含磷量分别为100.84mg·g^{-1}、77.99mg·g^{-1}、79.19mg·g^{-1}；在低氮的N1水平下，叶片中含磷量在开花期较N3水平低14.43mg·g^{-1}，降低了14.3%，在坐果期和果实膨大期，较N3水平分别高9.62mg·g^{-1}、10.82mg·g^{-1}，即增加了12.33%、13.66%；在高氮的N5水平下，叶片中含磷量在3个时期中都较低，分别较N3水平低10.83mg·g^{-1}、13.23mg·g^{-1}、13.23mg·g^{-1}即降低了10.74%、16.96%、16.71%。

在番茄生长的3个时期中，不论何种N水平下，增施CO_2，叶片中磷含量都几乎没有变化。

2.4.3 叶片中全钾含量

图6是不同氮水平营养液供应条件下，增施CO_2以后，番茄在3个不同生长时期中其叶片含钾量的结果。由图6可知，在对照N3水平下，3个时期中含钾量分别为13.39mg·g^{-1}、12.68mg·g^{-1}、11.19mg·g^{-1}；在低氮的N1水平下，番茄在开花期、坐果期和果实膨大期叶片含钾量都较低，较N3水平分别低1.99mg·g^{-1}、1.29mg·g^{-1}、1.56mg·g^{-1}，即降低了14.86%、10.17%、13.94%；在高氮的N5水平下，在番茄开花期、坐果期，含钾量较低，较N3水平分别低1.42mg·g^{-1}、1.84mg·g^{-1}，分别降低了10.6%、14.51%，而在果实膨大期含钾量较高，较N3水平高2.41mg·g^{-1}，增加了21.54%。

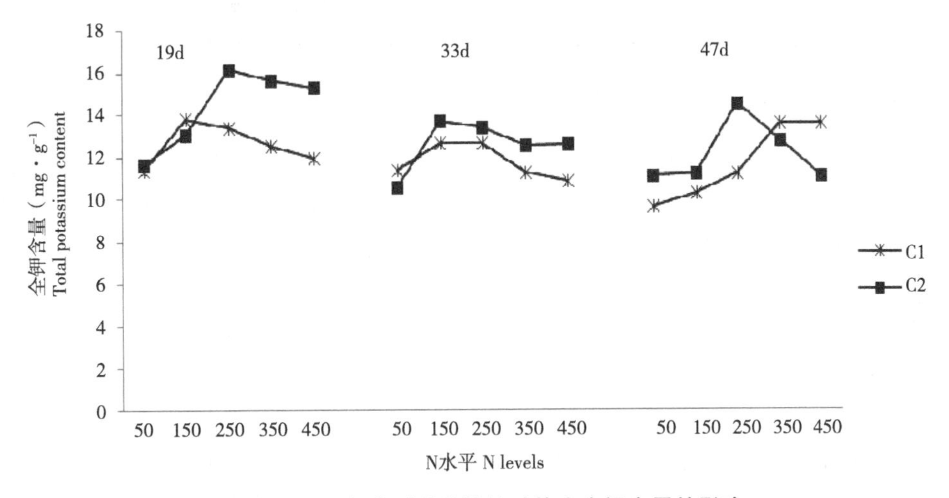

图 6　CO_2-N耦合对番茄植株叶片中全钾含量的影响

Figure 6　The influence of CO_2-N treats on total potassium content of tomato leaves

注：图中各处理所用数据是试验期间测定的叶片全钾含量共计3次的均值，每次测定每个处理有3次重复

在对照N3水平下，增施CO_2，番茄开花期和果实膨大期叶片中含钾量较高，分别较C1浓度高2.77mg·g^{-1}、3.33mg·g^{-1}，即增加了20.69%，29.76%；在低氮的N1水平下，增施CO_2处理，对番茄3个生长时期叶片中的含钾量影响都较小；在高氮的N5水平下，增施CO_2，叶片中含钾量在番茄开花期和坐果期较高，较C1浓度分别高3.33mg·g^{-1}、1.77mg·g^{-1}，即增加了27.82%、16.33%，但在果实膨大期叶片中含钾量较C1浓度低2.48mg·g^{-1}，即降低了18.24%。

2.4.4 叶片中Ca含量

图7是营养液不同氮水平下施用CO_2以后叶片含钙量的结果，由图7可知，适宜的N3水平下，番茄在开花期、坐果期、果实膨大期其叶片中的含钙量分别为18.39mg·g^{-1}、17.0mg·g^{-1}、31.02mg·g^{-1}；在低氮的N1水平下，叶片中含钙量在番茄生长3个时期都较低，分别比N3水平低7.39mg·g^{-1}、2.74mg·g^{-1}、12.96mg·g^{-1}，即降低了40.18%、16.12%、41.8%；在高氮的N5水平下，叶片中含钙量在番茄开花期和坐果期与N3水平接近都较高，但在番茄果实膨大期分别较N3水平低4.32mg·g^{-1}，即降低了13.9%。

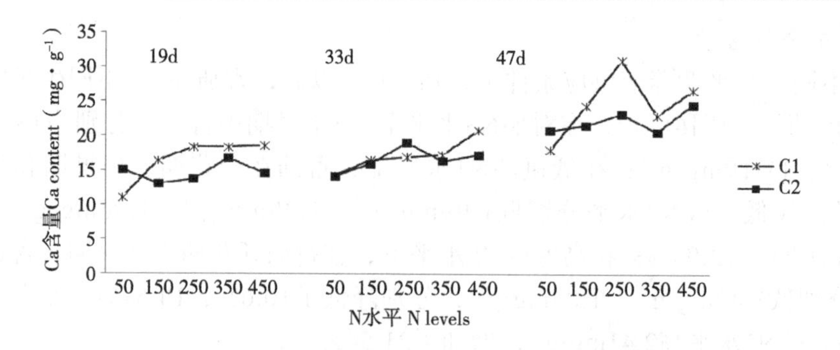

图7 CO_2-N耦合对番茄植株叶片中Ca含量的影响

Figure 7 The influence of CO_2-N treats on Ca content of tomato leaves

注：图中各处理所用数据是试验期间3次测定的叶片Ca含量的均值，每次测定每个处理有3次重复

在对照N3水平下，增施CO_2，叶片中钙含量在番茄开花期和果实膨大期降低，较C1浓度分别低4.72mg·g^{-1}、7.6mg·g^{-1}，即降低了25.67%、24.5%；在低氮的N1水平下，增施CO_2，叶片中钙含量在番茄开花期和果实膨大期增加，较C1水平下分别高4.15mg·g^{-1}、2.79mg·g^{-1}，增加了37.72%、15.45%；在高氮的N5水平下，增施CO_2，叶片中钙含量在番茄开花期较C1浓度低3.96mg·g^{-1}，降低了21.37%。不论在何种N水平下，增施CO_2，番茄含钙量在坐果期几乎没有影响。

2.2.5 Mg

图8是营养液不同氮水平下施用CO_2以后番茄不同生长时期叶片含镁量的结果，由图8可知，适宜的N3水平下，番茄在开花期、坐果期、果实膨大期其叶片中的含镁量分别为2.46mg·g^{-1}、2.29mg·g^{-1}、4.37mg·g^{-1}；在低氮N1水平下，在番茄开花期和果实膨大期，叶片中含镁量较N3水平分别低0.75mg·g^{-1}、0.83mg·g^{-1}，降低了43.77%、18.99%；在高氮N5水平下，叶片中含镁量在果实膨大期较N3水平低0.78mg·g^{-1}，降低了17.85%，开花期和坐果期叶片中含镁量与N3水平接近；番茄坐果期，不同氮水平下，叶片中镁含量没有明显变化。

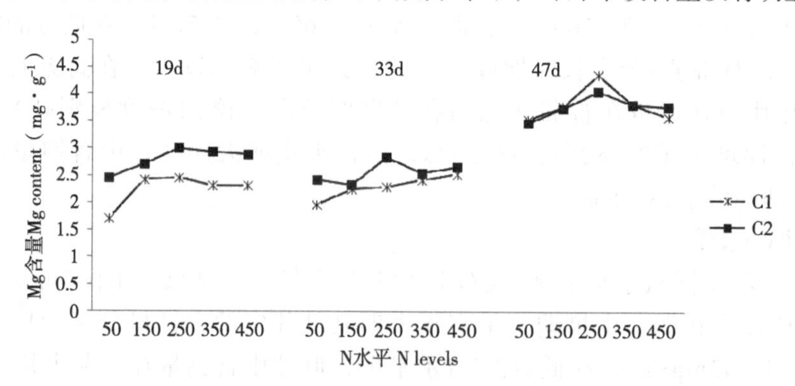

图8 CO_2-N互作对番茄植株叶片中Mg含量的影响

Figure 8 The influence of CO_2-N treats on Mg content of tomato leaves

注：图中各处理所用数据是试验期间3次测定的叶片Mg含量的均值，每次测定每个处理有3次重复

对照N3水平下，增施CO_2，叶片中含镁量在开花期较C1浓度下高0.55$mg \cdot g^{-1}$，增加22.36%；在低氮的N1水平下，增施CO_2，叶片中含镁量在开花期较高，比C1浓度下高0.75$mg \cdot g^{-1}$，增加了43.86%；高氮N5水平下，增施CO_2，叶片中含镁量在开花期较C1浓度下高0.58$mg \cdot g^{-1}$，增加了24.87%。不论何种N水平下，增施CO_2，叶片中含镁量在番茄坐果期和果实膨大期几乎都没有影响。

3 讨论

3.1 营养液不同氮水平对叶片中养分含量的影响

本研究中，叶片中硝态氮和N含量在营养液中氮浓度达到N4水平时达到最大值，在过高的N5水平时，其含量降低。植物对氮的吸收情况，国内外已做了不少研究，早在1967年Legg和Standford就观测到施氮有助于植物对氮的吸收，并指出与激发效应有关，提出氮肥的激发效应有随土壤供氮力值增长而加大的趋势。综合本研究结果和有关研究报道，说明植物生长过程中N浓度不宜过低，适当的高N浓度有利于植物对氮和其他养分的吸收。

此外，在番茄开花期，不同N水平对叶片Ca和Mg含量影响不同。有研究表明，由于蔬菜栽培设施内相对封闭的生长环境，土壤缺少雨水的淋溶，随着营养液氮素含量的增加[4]，H^+离子在$NH4^+$和NO_2^-氧化形成NO^{3-}积累的过程中得到释放，基质PH值下降，由此根系吸收如Ca、Mg等矿质元素过程受到土壤酸化的制约，Schulzeetal等在1989年提出NH^{4+}存在会抑制植物对Ca^{2+}、Mg^{2+}的吸收，本研究中Mg的含量变化与上述结论一致，Ca不一致，可能原因是钙离子在植物体内属于不易移动的营养元素。

3.2 增施CO_2对番茄开花期叶片中养分含量的影响

3.2.1 适宜氮水平下增施CO_2对叶片中养分含量的影响

李娟[5]试验表明，营养液浓度较高时，植株对CO_2的响应程度也会较高。同样，也有研究表明植物在N供应充足条件下对CO_2浓度升高的响应比N供应不足时强烈[6]。CO_2浓度升高会加速植物对矿质养分的吸收，提高植株矿质养分的含量，从而满足自身生长的需要。本研究中，营养液中氮浓度在适宜的N3水平时，增施CO_2处于番茄开花期的叶片中N、K的含量显著增加，与上述研究相符。

前人研究表明，CO_2浓度升高通过提高Ca^{2+}和Mg^{2+}向光合器官叶片的运输能力，促进四季竹生长[7]，本研究中，增施CO_2，叶片中Mg含量增加，与上述结论一致，但Ca含量下降，Ca元素在植株体内与植株的抗逆性有关，增施CO_2后使生长环境更适宜植株的生长，因此Ca含量也减少，来减弱这种抑制作用。

3.2.2 低氮水平下增施CO_2对叶片中养分含量的影响

在本研究中，低氮水平下，增施CO_2，植株干重、硝态氮含量以及N、P、K含量几乎都没有变化。可能原因是营养液中氮供应严重不足，植株蛋白质合成有限，即使人为增施CO_2，植株也没有能力利用这些CO_2。于佳等[8]利用顶式气室，通过盆栽试验表明，当CO_2浓度为760μl/L时，作物光合作用会因N肥含量低而受到限制，而在N肥含量好的条件下则能使作物的光合能力提高。因此，在低氮条件下，增施CO_2几乎不影响植株对养分的吸收。但，低氮水平下，增施CO_2，促进叶片中Ca含量增加，一定程度上提高了植株对低肥逆境的抗性，从而使番茄植株能够正常生长。

3.2.3　高氮水平下增施CO_2对叶片中养分含量的影响

前面已讨论到，植株生长需要充足的氮源，植株对CO_2的响应依赖于营养液的养分含量，N供应充足条件下对CO_2浓度升高的响应强，有研究表明，CO_2施肥能增加番茄对养分的利用率，也能减少由养分过高而带来的负面影响[9]。也有研究表明CO_2浓度升高可以促进植物对N的吸收[10]。本研究中，在高氮的N5浓度下，增施CO_2，番茄植株叶片中矿质元素的含量变化趋势与适宜的N3浓度下一致，同样是N、K、Mg的含量增加，Ca含量降低，另外，增施CO_2还缓解了氮过高导致叶片中矿质元素含量下降的现象，与上述结论相符。

4　结论

综上所述，叶片中各种养分的含量不仅仅与增施CO_2这一个因素相关，同时还受N供应水平和植株生长时期的影响。在番茄开花期，适宜氮（N3-N4）水平下，增施CO_2，最能促进植株叶片中矿质元素的含量，在低氮（N1）水平下，增施CO_2，对植株生长影响不大，在高氮的（N5）水平下，增施CO_2不仅有利于植株对养分的吸收，还能缓解氮肥过高对植株的不利影响。但在番茄坐果期和果实膨大期，不论何种N水平下增施CO_2，其叶片中养分含量都几乎没有变化。因此，番茄温室栽培中，增施CO_2同时需要综合考虑各种因素。综合本试验番茄生长各项指标及经济效益，在营养液N浓度为$250mg·L^{-1}$（N3）$-350mg·L^{-1}$（N4）时，增施CO_2浓度到$600\mu l·L^{-1}$（C2）最能促进开花期番茄植株叶片中养分含量的增加。

参考文献

[1] Tongbai P，Kozai T，Ohyama K. CO_2 and air circulation effects on photosynthesis and transpiration of tomato seedings［J］. Scientia Horticulturae，2010，126（3）：338-344.

[2] 刘保才，赛富昌. 影响日光温室蔬菜产量的三大要素分析［J］. 河南农业科学，1995（11）：34-35.

[3] Leakey A D B，Ainsworth E A，Bernacchi C J，*et al*. Elevated CO_2 effects on plant carbon，nitrogen，and water relations：six important lesions from FACE［J］. Journal of Experimental Botany，2009，60（10）：2859-2876

[4] 李廷轩，张锡洲，王昌全，等. 保护地土壤次生盐渍化的研究进展［J］. 西南农业学报，2001（S1）：103-107.

[5] 李娟，周建明. CO_2与NH_4^+/NO_3^-比互作对番茄幼苗培养介质pH、根系生长及根系活力的影响［J］. 植物营养与肥料学报，2007，13（5）：865-878.

[6] Rogers G，Miiham P，Thibaud M，et al. Interactions between rising CO_2 concentration and nitrogen supply in conon growth and leaf nitrogen concentration［J］. Functional Plant Biology，1996，23：119-1 255

[7] 庄明浩，李迎春，郭子武，等. CO_2浓度升高对毛竹和四季竹叶片主要养分化学计量特征的影响［J］. 植物营养与肥料学报，2013，19（1）：239-245.

[8] 于佳，于显枫，郭天文，等. 施氮和大气CO_2浓度升高对春小麦拔节期光合作用的影响［J］. 麦类作物学报，2010，30（4）：651-655.

[9] 李娟，周健民，段增强，等. 养分与CO_2交互作用对番茄幼苗生长及一些生理指标的影响［J］. 西北农业学报，2005（4）：10-13

[10] Kogawara S，Norisada M，Tange T，et al. Elevated atmospheric CO_2 concentration alters the effect of phosphate supply on growth of Japanese red pine（Pinus densiflora）seedlings［J］. Tree Physiology，2006，26：25-33.

基于称重式蒸渗仪的秋茬礼品西瓜耗水特征分析

杨宜[1]，陶虹蓉[1]，李海平[1]，郭文忠[2, 3]，李银坤[2, 3*]，李灵芝[1]

［1. 山西农业大学 园艺学院，太谷 030801；2. 北京农业智能装备技术研究中心，北京 100097；3. 农业部都市农业（华北）重点实验室，北京 100097］

摘要： 以“京秀”西瓜为试材，以称重式蒸渗仪为技术手段，以水面蒸发量为灌溉标准，研究了秋茬礼品西瓜耗水特征及与环境因素的相关性，以期为礼品西瓜精准灌溉提供依据。结果表明：①西瓜的耗水规律表现为前期小、中期大、后期小的变化规律，全生育期蒸散量为114.79mm，日均蒸散量为1.11mm。西瓜苗期、伸蔓期、开花坐果期、果实膨大期和成熟期的蒸散强度分别为：1.04mm·d^{-1}、1.20mm·d^{-1}、1.34mm·d^{-1}、1.08mm·d^{-1}、0.81mm·d^{-1}。②西瓜日蒸散量与光合有效辐射、太阳辐射、最高温度、相对湿度、平均温度、最低温度均呈极显著正相关，相关系数依次为：0.797、0.764、0.599、0.419、0.274。日蒸散量与相对湿度呈极显著负相关，相关系数为−0.485。③西瓜生育后期日蒸散量与水面蒸发量呈极显著线性相关（p<0.01），相关系数为0.801 1。

关键词： 蒸渗仪；温室；礼品西瓜；蒸散量；环境因素

Water Consumption Analysis of Autumn Mini-watermelon on the Basis of Weighing Lysimemer

Yang Yi[1]，Tao Hongrong[1]，Li Haiping[1]，Guo Wenzhong[2, 3]，Li Yinkun[2, 3]，Li Lingzhi[1]

［1. *College of Horticulture*，*Shanxi Agricultural University*，*Taigu* 030801；2. *National Research Center of intelligent Equipment for Agriculture*，*Beijing* 100097；3. *Key Laboratory of Urban agriculture*（*North China*），*Ministry of Agriculture*，*Beijing* 100097］

Abstract：Taking watermelon of “Jingxiu” as test material，weighing lysimeter as technical measure and water evaporation as irrigation standard，the water consumption characteristics of autumn mini-watermelon and its relationship with environmental factors were studied in order to Watermelon provides a basis for precise irrigation. The results showed as follows：①The water consumption law of watermelon showed small，middle and late changes. The total evapotranspiration during the whole growth period was 114.79 mm and the average daily evapotranspiration was 1.11 mm. The evapotranspiration intensity of watermelon at seedling stage，spreading vine stage，flowering and fruitsetting，fruit enlargement and maturity were 1.04mm · d-1，1.20mm · d-1，1.34mm · d-1 and 1.08mm · d- 1，0.81 mm · d-1.②The daily evapotranspiration of watermelon had significant

基金项目：国家重点研发计划（2017YFD0201503）、国家自然科学基金（41501312）

作者简介：杨宜（1994—），女，河北张家口人，硕士研究生，研究方向为蔬菜栽培与生理。E-mail：Yangyi9401@163.com

通讯作者：李银坤（1982—），男，博士，高级工程师，研究方向为作物水肥高效利用。E-mail：lykunl218@163.com.

positive correlation with photosynthetically active radiation, solar radiation, maximum temperature, relative humidity, average temperature and minimum temperature, the correlation coefficients were 0.797, 0.764, 0.599, 0.419, 0.274.Daily evapotranspiration and relative humidity showed a very significant negative correlation with a correlation coefficient of −0.485.③There was a significant linear correlation between daily evapotranspiration and water surface evaporation（$p<0.01$）in the late growth stage of watermelon, with a correlation coefficient of 0.801 1.

Key words: lysimeter; Greenhouse; Mini-watermelon; Evapotranspiration; Environmental factors

1 引言

近年来，礼品西瓜作为广受生产者和消费者喜爱的高档果品种类，在设施中栽种面积不断扩大，但由于秋冬温度过低及灌水施肥不合理，导致秋茬礼品西瓜出现裂瓜、畸形和延迟上市等问题[1]。研究作物耗水规律是建立合理灌溉制度的前提，对于发展节水农业具有十分重要的意义[2]。目前国内针对设施作物耗水特征主要针对番茄与黄瓜等大宗蔬菜，对礼品西瓜的研究较少。任自力等（2011）[3]采用膜下滴灌栽培模式研究得出，西瓜蒸散量呈缓慢升高趋势，全生育期平均需水强度为4.45mm/d。郑健等（2009）[4]利用盆栽试验研究得出，温室小型西瓜全生育期最佳需水量为107mm，需水高峰出现在开花结果期与果实膨大期。称重式蒸渗仪（Lysimeter）作为一种高精度的获取蒸散量的技术手段，常被作为标准校核其他方法。目前国内外基于蒸渗仪实测值的研究以露天作物为主[5-7]，与温室作物蒸散规律有很大区别。为此，本研究针对目前设施生产中灌水超标、水分利用效率低等问题，以礼品西瓜为研究对象，以直径20cm的蒸发皿水面蒸发量为灌水依据，基于称重式蒸渗仪实测值研究其耗水特征及影响因素，对温室礼品西瓜科学高效灌溉具有现实意义。

2 材料与方法

2.1 研究区概况

试验在北京市农林科学院玻璃温室内进行。该地区位于东经116.29°，北纬39.94°，供试温室长38m，宽11m，南北走向。土质为砂壤土，覆盖无滴聚乙稀薄膜。试验前0～20cm土壤理化性质为：土壤容重1.40$g \cdot cm^{-3}$，田间体积持水量28.0%，有机质含量15.89$g \cdot kg^{-1}$，全氮质量分数0.60$g \cdot kg^{-1}$，速效钾含量0.15$g \cdot kg^{-1}$。试验用称重式蒸渗仪的长宽高为：1m×0.6m×0.9m，每小时自动记录数据，分辨率为0.01mm水深。

2.2 试验设计

供试小型礼品西瓜品种为"京秀"。2017年8月12日选取约3叶1心的西瓜秧苗，移栽4株在称重式蒸渗仪上。定植前底施商品有机肥30 000$kg \cdot hm^{-2}$，有机肥养分含量为：有机质212.2$g \cdot kg^{-1}$，全氮7.14$g \cdot kg^{-1}$，有效磷0.06$g \cdot kg^{-1}$，速效钾3.68$g \cdot kg^{-1}$。蒸渗仪灌定植水15mm，栽培畦灌定植水30mm。温室内布置2个直径为20cm的蒸发皿，以每日17:30测定的水面蒸发量累积值乘以系数0.8作为灌水依据，灌水周期为7～10d。选取2017年9月22日（阴天）和9月23日（晴天）2个典型天气分析西瓜的日蒸散规律。

2.3 测定项目与方法

2.3.1 蒸散量

称重式蒸渗仪每h记录1次土柱重量（分辨率0.01mm），蒸散量的计算原理为：

$$T \times A=（W_{t-1}-W_t）/\rho+I \quad （2）$$

式中，T为时间段内蒸散量，mm；A为蒸渗仪箱体的表面积，0.6m2；W_{t-1}、W_t为t−1时刻和t时刻蒸渗仪箱体内土、水的质量，g；ρ为水的密度，$1g \cdot cm^{-3}$；I为时段内的灌水量，mm。

2.3.2 环境因子

利用小型气象站对温室内平均温度、相对湿度和太阳辐射等环境因子进行监测，采集间隔为1d及1h。

2.3.3 统计与方法

采用Microsoft 2010进行试验数据的处理及相关图表的制作，利用SPSS19.0进行统计分析。

3 结果与分析

3.1 温室内环境因子特征分析

由图1可知温室内平均温度在1～33d（苗期和伸蔓期）呈平缓波动趋势。开花结果期（34d）开始后温度持续下降，平均温度为22.28℃，最低温达6.32℃。苗期与伸蔓期温室内湿度较低，平均湿度范围为53.71%～88.10%。开花坐果期开始后由于室外温度过低通风时间减少，相对湿度逐渐上升，果实膨大期与成熟期相对湿度范围在70.3%～93.3%。整体表现为苗期和伸蔓期温度高、湿度低；开花坐果期、果实膨大期和成熟期温度低、湿度高。温度与相对湿度呈相反的变化趋势。

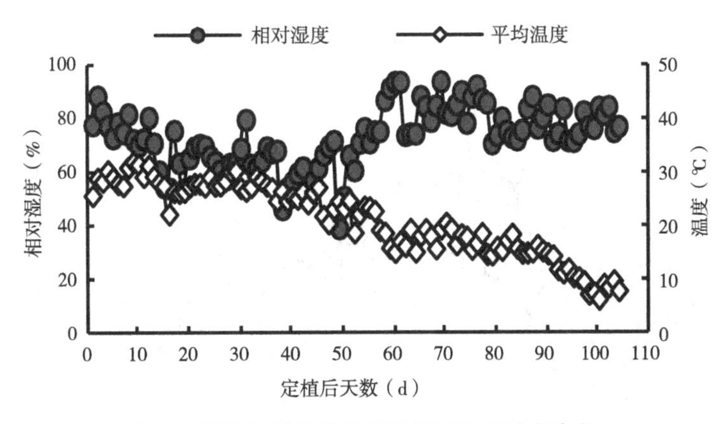

图1 温室西瓜平均温度及相对湿度动态变化

Figure 1 Greenhouse watermelon average temperature and relative humidity dynamic changes

3.2 西瓜典型日蒸散规律分析

图2显示晴天西瓜日蒸散量变化趋势为双峰曲线，9:00开始缓慢增加至11:00出现第一个峰值0.23mm。随着太阳辐射的增加，蒸散速率在11:00～13:00呈先减小后增加的趋势，

到13:00左右出现第二个峰值0.27mm，随后蒸散量缓慢下降至平缓。阴天日蒸散量变化呈单峰曲线，9:00开始增加至2:00出现峰值0.20mm，随后下降至平缓。太阳辐射与蒸散量具有相同的变化趋势，晴天时太阳辐射在11:00到达峰值747.08W/m^2，随后迅速下降。阴天辐射强度与变化幅度较小，太阳辐射于13:00出现峰值258.54W/m^{2}后缓慢降低。

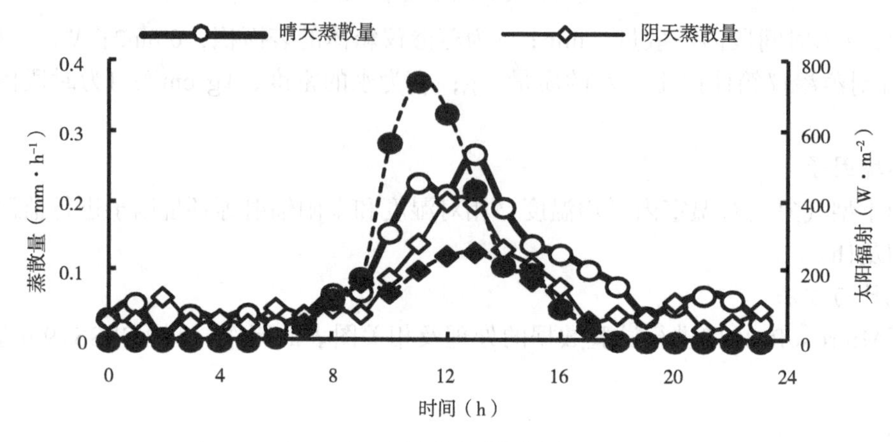

图 2　西瓜典型日蒸散规律分析

Figure 2　Evapotranspiration analysis of a typical watermelon weather

3.3　西瓜全生育期蒸散规律分析

由图3可知西瓜全生育期累积蒸散量为114.79mm。西瓜的累积蒸散量变化趋势总体表现为慢—快—慢的S形曲线。在定植后1～31d缓慢增加，变化幅度较小，定植后32～80d累积蒸散量迅速增加，变化幅度较大，定植后81d累积蒸散量缓慢增加。西瓜日均蒸散量为1.11mm，变化趋势呈单峰曲线，前期缓慢上升，在39d达到峰值2.35mm，随后日蒸散量逐渐降低至平缓。

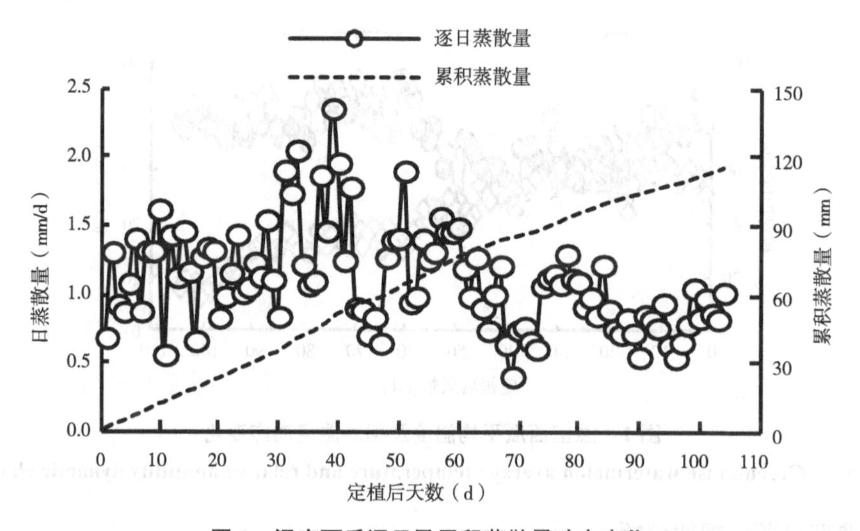

图 3　温室西瓜逐日及累积蒸散量动态变化

Figure 3　Daily and cumulative evapotranspiration of greenhouse watermelon dynamics

3.4 西瓜各生育阶段蒸散规律分析

由表1可以看出西瓜各生育阶段蒸散量的大小依次为果实膨大期>伸蔓期>开花坐果期>成熟期>苗期。苗期植株较小，蒸腾作用最弱，蒸散量占全生育期蒸散量的9.94%，蒸散强度为1.04mm/d。伸蔓期蒸散强度升高至1.20mm/d，蒸散量为26.34mm，占全生育期的22.95%。开花结果期温度升高，蒸散强度增加至1.34mm/d。进入果实膨大期后蒸散量为43.25mm，占全生育期的37.68%。此时平均温度仅有18.29℃，蒸散强度为1.08mm/d，略低于开花坐果期。进入成熟期后蒸散强度降低为0.81mm/d，蒸散量为13.69mm，占全生育期的11.92%。

3.5 西瓜日蒸散量与环境因素的相关性分析

如表2所示，除相对湿度与其他各项呈极显著负相关外（$P<0.01$），其他各因素及与蒸散量之间均呈极显著正相关（$P<0.01$），光合有效辐射、太阳辐射、最高温度、平均温度、最低温度和相对湿度与日蒸散量的相关系数分别为0.797、0.764、0.599、0.419、0.274、−0.485。西瓜日蒸散量与气象因素之间密切相关。

表1　黄瓜各生育阶段蒸腾规律

Table 1　Cucumber transpiration rate of each growth stage

生育期	苗期 1 ~ 11d	伸蔓期 12 ~ 33d	开花坐果期 34 ~ 48d	果实膨大期 49 ~ 88d	成熟期 89 ~ 104d	合计 104d
蒸散量	11.41	26.34	20.10	43.65	13.69	114.79
蒸散强度	1.04	1.20	1.34	1.08	0.81	1.11
阶段蒸散量占比	9.94	22.95	17.51	37.68	11.92	100.00
平均温度	29.12	27.68	25.53	18.29	10.70	21.12

表2　各气象因子与西瓜蒸散量的相关系数

Table 2　Correlation coefficients of meteorological factors and watermelon evapotranspiration

变量	最低温度	最高温度	平均温度	相对湿度	光合有效辐射	太阳辐射	日蒸散量
最低温度	1						
最高温度	0.678**	1					
平均温度	0.963**	0.828**	1				
相对湿度	−0.385**	−0.432**	−0.474**	1			
光合有效辐射	0.382**	0.797**	0.568**	−0.456**	1		
太阳辐射	0.371**	0.792**	0.557**	−0.426**	0.988**	1	
日蒸散量	0.274**	0.599**	0.419**	−0.485**	0.797**	0.764**	1

注：**表示在0.01水平上显著相关

3.6 蒸散量与水面蒸发量的相关性分析

如图4a所示，西瓜全生育期水面蒸发量与日蒸散量的变化趋势基本相同，尤其在定植30d后水面蒸发量与蒸散量一致性较好。通过对西瓜定植后31 ~ 90d水面蒸发与日蒸散量统计分析表明：二者呈极显著的相关性，R^2=0.801 1（$p<0.01$）（图4b）。温室西瓜从开花结果期到果实膨大期（31 ~ 90d）的水面蒸发量与蒸散量具有更好的一致性。

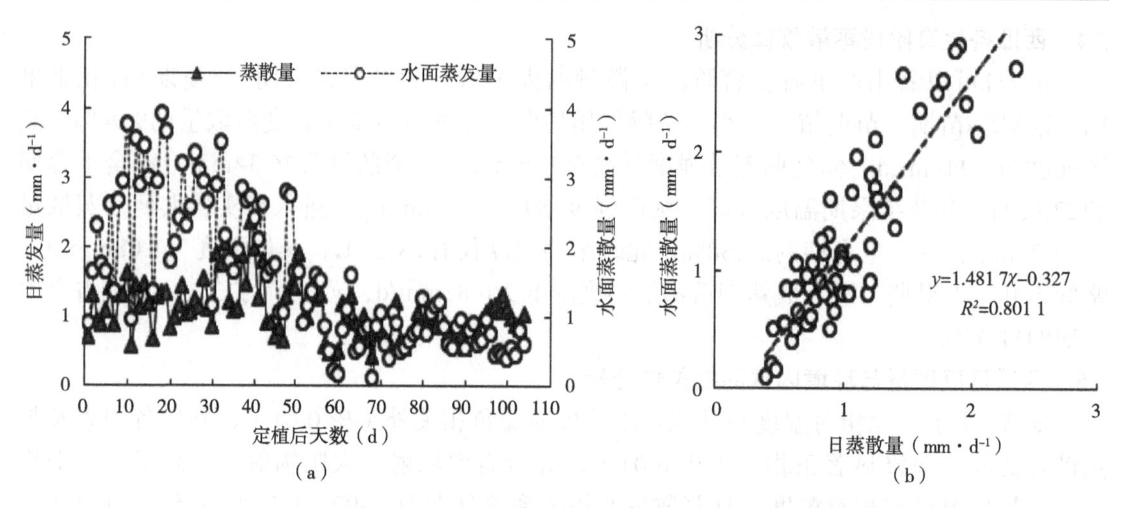

图 4　西瓜日蒸散量与水面蒸发量的动态关系（a）及统计分析（b）

Figure 4　Dynamic relationship between watermelon evapotranspiration and water surface evaporation（a）and statistical analysis（b）

4　结论与讨论

4.1　讨论

许金香[8]等研究认为，秋冬茬番茄的蒸散量呈现逐渐降低的趋势，开花结果期番茄蒸散量最高，本试验与许金香等[8]结论一致。造成蒸散量降低的原因是西瓜进入果实膨大期后温度逐渐降低，无法满足西瓜旺盛生长所需的温度。因此，在进行温室管理时可优先保障西瓜伸蔓期和开花坐果期的水分和温度需求。气象因素中太阳辐射、光合有效辐射与蒸散量均有较好的相关性，而温度与蒸散量相关性稍差，这与牛勇等[9]和刘浩[10]等研究结论一致。蒸散量与水面蒸发量在生育前期相关性较差，主要是由于前期温度较高且植株叶面积较小，此时地表蒸发以土壤蒸发为主，水面蒸发量高于实际蒸散量。进入开花结果期后土壤蒸发减少，水面蒸发量与蒸散量相关性较好（R^2=0.801 1）。

4.2　结论

（1）温室西瓜全生育期内（104d）蒸散总量为114.79mm，平均日蒸散量为1.11mm。温室西瓜的需水规律表现为前期小—中期大—后期小，需水高峰出现在开花坐果期。其中苗期、伸蔓期、开花坐果期、果实膨大期和成熟期的蒸散强度分别为：1.04mm·d^{-1}、1.20mm·d^{-1}、1.34mm·d^{-1}、1.08mm·d^{-1}和0.81mm·d^{-1}。

（2）温室西瓜日蒸散量与光合有效辐射、太阳辐射、最高温度、相对湿度、平均温度、最低温度均呈极显著正相关，相关系数依次为：0.797、0.764、0.599、0.419和0.274。蒸散量与相对湿度呈极显著负相关，相关系数为-0.485。

（3）西瓜开花坐果期至果实膨大期，蒸散量与水面蒸发量呈良好的线性正相关，R^2为0.801 1。

参考文献

［1］赵春江，郭文忠. 中国水肥一体化装备的分类及发展方向［J］. 农业工程技术，2017，37（7）：10-15.

［2］倪宏正，尤春，倪玮.设施蔬菜水肥一体化技术应用［J］.中国园艺文摘，2013，29（4）：140-141+192.［2017-08-15］.

［3］任自力，张显，朱盼盼，等. 大棚西瓜需水强度与环境因子的关系［J］. 北方园艺，2013（9）：42-46.

［4］郑健，蔡焕杰，王健，等. 日光温室西瓜产量影响因素通径分析及水分生产函数［J］. 农业工程学报，2009，25（10）：30-34.

［5］López-Urrea R，Martín De Santa Olalla F，Fabeiro C，et al. Testing evapotranspiration equations using lysimeter observations in a semiarid climate［J］. Agricultural Water Management. 2006，85（1-2）：15-26.

［6］Piouceau J，Panfili F，Bois G，et al. Actual evapotranspiration and crop coefficients for five species of three-year-old bamboo plants under a tropical climate［J］. Agricultural Water Management. 2014，137：15-22.

［7］张建君. 农田日蒸散量估算方法研究［D］.北京：中国农业科学院，2009.

［8］许金香. 日光温室番茄栽培需水规律的研究［D］. 北京：中国农业大学，2004.

［9］牛勇，刘洪禄，吴文勇，等. 基于大型称重式蒸渗仪的日光温室黄瓜蒸腾规律研究［J］. 农业工程学报，2011，27（1）：52-56.

［10］刘浩，孙景生，王聪聪，等. 温室番茄需水特性及影响因素分析［J］. 节水灌溉，2011（4）：11-14.

夜间日光温室黄瓜顶部叶片结露时长模拟

贺威威，马沙一，王蕊，须晖，李天来

（沈阳农业大学园艺学院　设施园艺省部共建教育部重点实验室　辽宁省设施园艺重点实验室　辽宁省设施蔬菜工程实验室，沈阳　110161）

摘要：黄瓜作为冬季日光温室主要栽培作物之一，研究发现长期生长在高湿（相对湿度≥85%）环境下，不仅其物理特性改变，如表皮透性增加、气孔敏感度下降和根系吸水能力下降等，而且增加真菌病害发生的几率。研究表明高湿环境下，叶片凝结的露水是真菌病害发生的直接原因。为准确模拟夜间日光温室顶部黄瓜叶片结露时长，以沈阳农业大学北山试验基地一栋日光温室的实时采集数据为例，提出了使用ROC曲线分析日光温室内的相对湿度阈值模型和露点温差阈值模型模拟夜间黄瓜顶部叶片结露时长，并将模拟结果与叶片湿度传感器测量值拟合分析比较。试验结果表明：相对湿度阈值（91.3%）模型模拟黄瓜顶部叶片结露时长比较准确，模拟值与观察值结果一致（=0.959），结露时长模拟值误差范围在2h以内；露点温差阈值（1.02℃）模型模拟黄瓜顶部叶片结露时长更加准确，模拟值与观察值结果一致（=0.969），结露时长模拟值误差范围在1h以内。针对东北地区日光温室环境，露点温差阈值（1.02℃）模型虽然模拟结露时长误差较小，但考虑到实用性和可推广性，应选择相对湿度阈值（91.3%）模型模拟黄瓜顶部叶片结露时长，并且此ROC曲线分析黄瓜叶片结露阈值的方法可为其他作物建立结露时长模型提供参考依据。

关键词：相对湿度；ROC曲线；露点温差；结露时长

Simulation of Nocturnal Cucumber Top Leaf Dew Duration in Solar Greenhouse

He Weiwei，Ma Shayi ，Wang Rui ，Xu Hui，Li Tianlai

（*College of Horticulture/ Key Lab of Protected Horticulture，Ministry of Education/Key Lab of Protected Horticulture of Liaoning Province/Key Lab of Protected Vegetables Engineering of Liaoning Province，Shenyang Agricultural University，Shenyang*　110161）

Abstract：Cucumber is one of the main crops cultivated in solar greenhouse in winter. The dew duration is long and the dew yield is large in solar greenhouse winter nocturnal time. The study found that long-term growth in high humidity（relative humidity≥85%）environment，which not only the physical properties change，such as epidermal permeability increased，stomatal sensitivity decreased and root water absorption capacity decreased and increase the probability of fungal diseases occurred. The research shows that the dew of the leaves in the high humidity environment is the direct cause of the occurrence of fungal disease. In order to accurately simulate the cucumber leaf dew duration on the top of the solar greenhouse，the real time collection data of a solar greenhouse in the North Mountain test base of shenyang Agricultural University is taken as an example. The ROC curve is applied to analyze the relative humidity threshold and dew point temperature deviation threshold method for simulating the nocturnal cucumber top leaves dew duration，and the simulation results are simulated and compared with the measured values of the leaf humidity sensor.

作者简介：贺威威（1991—）男 河南商丘人，沈阳农业大学园艺学院设施环境方向在读研究生。E-mail：heweiwei@syau.edu.cn

The experimental results show that the relative humidity threshold（91.3%）was accurate in simulating the top leaves of cucumber. The simulated values were consistent with the observed values（=0.959），and the simulation error of dew duration was within 2 hours. The dew point temperature deviation threshold（1.02℃）model simulates the dew duration of nocturnal cucumber top leaves is more accurate，and the simulated value is consistent with the observed value（=0.969），the simulation error of dew duration is less than 1 hours. For the solar greenhouse environment in Northeast China，the simulated nocturnal cucumber top leaves dew duration of the dew point temperature deviation threshold（1.02℃）model is small，but considering the practicality and promotion，should choose the relative humidity threshold（91.3%）model. And the ROC curve analysis method of cucumber leaf dew threshold can provide a reference for other crops simulating dew duration.

Key words：Relative humidity；ROC curve；Dew point temperature deviation；Dew duration

1 引言

日光温室是我国北方越冬蔬菜生产的主要栽培方式，具有生产效率高和经济效益好等特点，但同样存在部分问题，特别是冬季日光温室内湿度过高所引发的病害[6]。日光温室内由湿度过高引起的作物结露不仅雾化温室棚膜、减弱透光率，而且减弱植物光合效率，影响作物的生长[8]。同时，植物叶片结露时长是植物病害预警的重要指标，其中以相对湿度阈值建立的结露时长模型广泛应用于植物病害防治领域[9]。

模拟植物叶片结露时长的方法主要可以归纳为两大类：经验模型如相对湿度阈值和露点温差模型等[13]；物理模型如能量平衡模型和神经网络模型等[2]。郑海山等[4]根据叶湿频率和相对湿度的关系，得出温室大棚黄瓜叶片结露的临界相对湿度值93%，并估算一天内结露时长，误差在2h左右。Kruit等[11]发现草地上使用相对湿度阈值87%时误差最小，并对70%～87%间的相对湿度下的结露情况进行了进一步阐述。Kim等[10]人选用空气温度与露点温度差小于3.7℃，建立了模糊逻辑模型比较准确的模拟了叶片结露时长。陆佩玲等[3]研究发现91.3%的相对湿度阈值与叶片表面露水凝结出现不一致，并建立了叶温能量平衡模型模拟了玉米顶部冠层的凝露情况，得出叶片夜间结露量呈倒U型，经历快—慢—快三个时段的变化。Sentelhas等[15]人通过叶片湿度传感器设计了不同倾角和高度下的草地结露时长试验方案，发现0°～150°倾角下结露时长显著高于30°～450°倾角，在距冠层顶部30厘米处结露量最大。Bassimba等[6]利用ROC曲线分析了柑橘冠层结露时长的相对湿度阈值，发现相对湿度阈值为87.45%，此阈值下结露时长模拟结果最好。

虽然植物冠层结露时长可以通过叶片湿度传感器测量[5]，但价格昂贵，而且需要和电脑配套使用，不方便携带。本试验通过研究日光温室内气象因子的变化，以ROC曲线更加快速和准确的找出叶片结露初始时所对应的结露时长模型阈值，并比较了单环境因素下的相对湿度阈值模型与多环境因子下的露点温差阈值模型的差异性，以期得出更加准确高效的日光温室黄瓜顶部叶片结露时长测量方法。

2 材料与方法

2.1 材料

试验于2016年10月8日至2016年12月30日，沈阳农业大学园艺学院试验基地（北纬41.82°、东经123.56°）日光温室内进行。温室长60m，脊高4m，跨度8.5m，北墙高2.8m，后坡长1.5m，覆盖材料使用日本明净化高保温PO膜，厚度0.1mm，后墙为砖墙，温室8:30左右揭开保温被，16:00左右覆盖保温被，夜间不加温。试验温室内栽培越冬黄瓜，黄瓜于9月5日定植，试验时黄瓜平均高度1.5m左右，选择日光温室中部黄瓜顶部叶片距地面约1.5m处，超过2m进行落蔓处理，重新下降到1.5m处，地面覆盖黑色地膜，地膜下面黄瓜根部附近有滴灌带，试验期间灌溉1周1次。试验数据利用数据采集器，供试EM50数据采集器（美国Decagon公司），适配空气温度（测量范围-40～80℃，精度0.1℃，准确度0.5℃）和相对湿度传感器（测量范围0～100%，精度0.1%，准确度2%），叶面湿度传感器；FLUKE 2638A（美国FLUKE公司）数据采集器，叶面温度（T型热电偶线，测量范围-270～400℃，精度0.01℃，准确度0.62℃）；时间间隔5min。

2.2 方法

试验期间测定了日光温室内空气温度、叶片湿度、空气相对湿度和叶片温度。其中温室中部黄瓜顶部冠层（距地面1.5m处）放置了3个叶片湿度传感器，3个测量叶片温度T型热电偶线，3个空气温度和空气相对湿度的传感器，计算结露时长采用3个测点数据的平均值，日光温室内测点分布如图1所示。

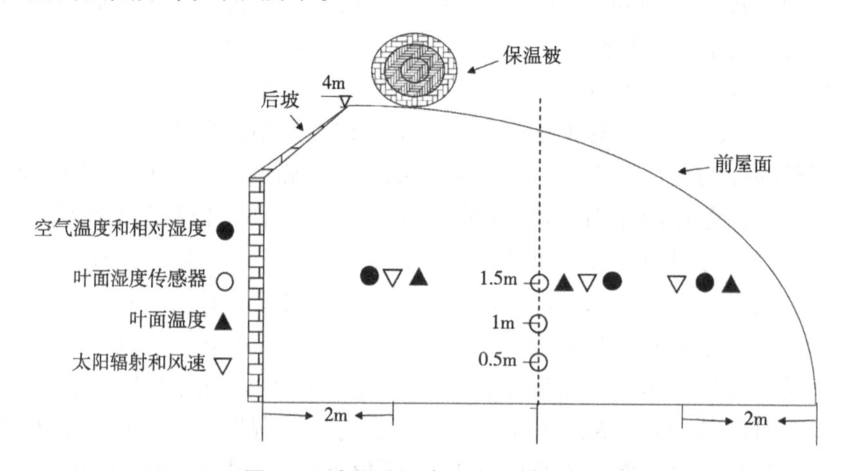

图1 日光温室测点分布及试验图片

Figure 1 Schematic diagram and photo of the solar greenhouse monitoring point and main sensor

结露临界点阈值确定多以肉眼观察为主，随机性和误差较大，适用性较低，而ROC曲线使用二进制分类统计的方法可以准确、高效地得出诊断阈值，此方法适用性强。通过ROC曲线分析评估传感器和模型的性能，基于二进制分类（叶片结露和叶片干燥）输入结果为（1和0），可以更加高效得出此模型的最佳阈值。ROC曲线x轴显示错误观测值（100—特异性），y轴显示正确观测值（灵敏度），ROC曲线下面积（AUC）表示此分类方法的正确率，AUC在1.0和0.5之间。在AUC>0.5的情况下，AUC越接近于1，说明判断效果越好。AUC在0.5～0.7时有较低准确性，AUC在0.7～0.9时有一定准确性，AUC在0.9以上时有较高准确性。AUC≤0.5时，说明判断方法完全不起作用[14]。最佳阈值的计算则依据Youden J（表1）统计[16]：

$$J=\frac{ad-bc}{(a+b)(c+d)} \tag{1}$$

试验随机选取10月、11月和12月各3d用于试验数据采集和分析，MedCalc软件作ROC曲线分析模型阈值，得到空气相对湿度模型结露阈值S和露点温差模型结露阈值M，并依据模型结露阈值，计算相对应模型模拟的结露时长，并与叶片湿度传感器的测量结果对比分析，采用情形分析表2比较两模型的优缺点和适用性。结露阈值即为最佳结露临界点，此临界值满足敏感性和特异性最优。以此结露临界点阈值建立结露时长模型，更准确地模拟黄瓜叶片结露时长。

表1　黄瓜叶片结露情形分析

Table 1　Cucumber leaf dew contingency table

叶片传感器 \ 模型模拟	1	0
1	a	b
0	c	d

$$sensitivity=\frac{a}{a+b} \tag{2}$$

$$specificity=\frac{d}{c+d} \tag{3}$$

归纳可知

$$J=sensitivity+specificity-100 \tag{4}$$

最佳阈值为J函数的最大值。设置95%的置信区间计算AUC面积[7]。

情形分析表[14]比较两种结露时长模型的模拟结果。

表2　黄瓜叶片结露时长情形分析

Table 2　Cucumber leaf dew duration contingency table

观测结露时长（h）\ 模拟结露时长（h）	叶片湿润	叶片干燥
叶片湿润	A	B
叶片干燥	C	D

说明：模拟和观测值都发生结露的时间输入A；模拟值不发生结露，观测值发生结露的时间输入B；模拟值发生结露，观测值不发生结露的时间输入C；模拟值和观测值均不发生结露的时间输入D

$$F_C=\frac{A+D}{A+B+C+D} \tag{5}$$

$$B_S=\frac{A+C}{A+B} \tag{6}$$

$$F_{AR}=\frac{C}{A+C} \tag{7}$$

公式（5）F_C表示模型精确度，介于0～1之间，越接近1，越准确；

公式（6）B_S小于1，预测值小于观察值，B_S大于1，预测值大于观察值；

公式（7）F_{AR}预测的错误率，介于0～1之间，越小，越准确。

2.2.1　基于相对湿度阈值模拟黄瓜顶部叶片结露时长

相对湿度值在结露阈值S之上，则认为此叶片处于结露状态，结露时长即计算黄瓜叶片从18:00到第二天9:00之间，空气相对湿度值大于S的累积时长数（h）。即相对湿度（RH）超过某一阈值S（单位%）的总累积时间来估测叶片结露时长。具体公式如下：

$$RH \geqslant S\ (0 \leqslant S \leqslant 100\%) \tag{8}$$

参考李明等基于冠层相对湿度的日光温室黄瓜叶片温差时间估计模型[1]，本试验中规定1h内有30min以上（包含30min）相对湿度值大于S，结露时长记作1h；低于30min记作0h。

2.2.2　基于露点温差阈值模拟黄瓜顶部叶片结露时长

叶片温度低于空气露点温度时，叶片表面将会凝结水珠。因此，本试验通过计算叶片温度与露点温度差值建立一种新的模拟叶片结露时长的方法。具体公式如下：

$$T_f - T_d \geqslant \mathrm{M} \tag{9}$$

其中，叶片温度（℃），露点温度（℃），阈值（℃）

露点温度可依据Lawrence推导公式[12]近似计算：

$$T_d = \mathrm{T}_a - \left(\frac{100 - RH}{5}\right) \tag{10}$$

同上文1.2.1，本试验规定1h内有30min以上（包含30min）露点温差值大于M，结露时长记作1h；低于30min记作0h。

3　结果与分析

3.1　相对湿度阈值模拟夜间顶部黄瓜叶片结露时长结果

ROC曲线分析可知（图2），最佳结露阈值91.3%，此时ROC曲线下面积（AUC）为0.854，标准误差0.004 17，AUC在95%的置信区间0.85～0.858，Youden指数J=0.616，敏感性83.3，特异性78.3。说明将相对湿度阈值设置为91.3%，具有一定的准确性。此阈值下，夜间顶部黄瓜叶片结露时长模拟值和观察值较为一致，x=0.916 7系数接近于1，且线性关系较大R^2=0.959（图4）。试验期间模拟值和观察值的平均结露时长分别为10.3h和11.8h，模拟值与观察值差异较小，且两者最大误差不超过2h（图3）。

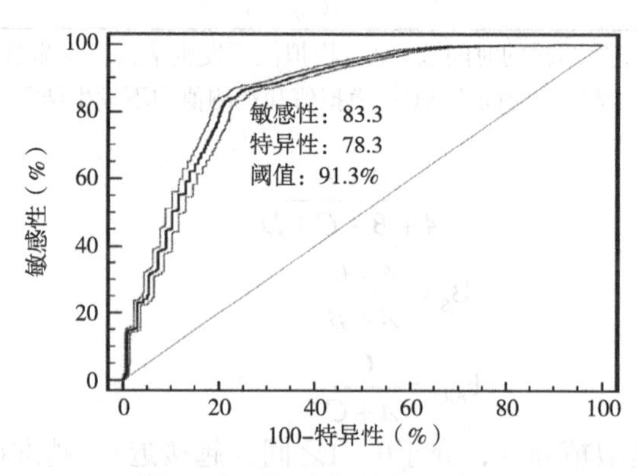

图 2　相对湿度阈值ROC分析

Figure 2　The relative humid threshold analysis of receiver operating characteristic

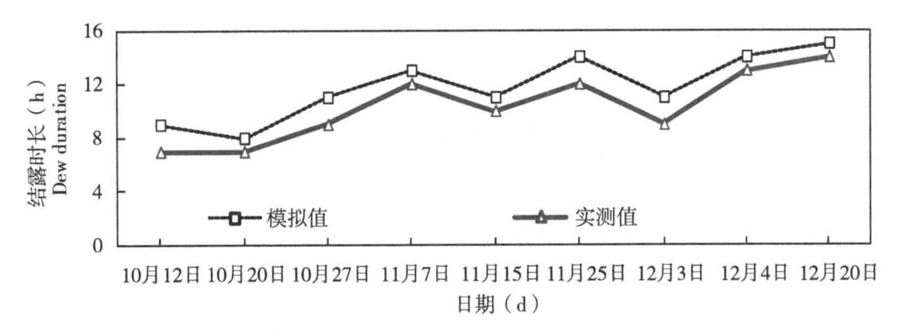

图3　黄瓜叶片结露时长实测值与预测值比较

Figure 3　Comparison of measured and predicted values of the Cucumber leaf dew duration

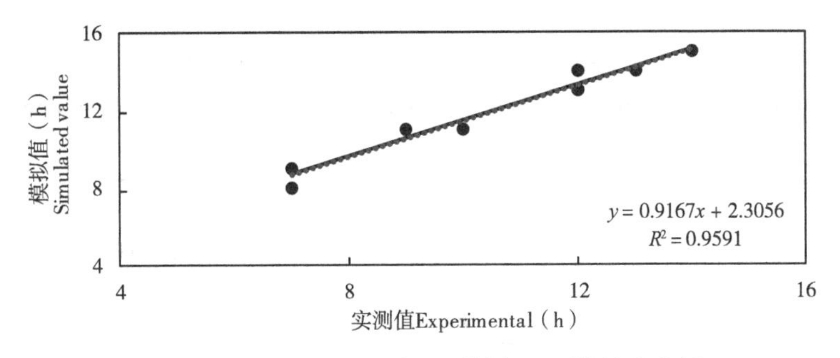

图4　黄瓜叶片结露时长实测值与预测值拟合分析

Figure 4　Fitting analysis of measured and predicted values of the Cucumber leaf dew duration

3.2　露点温差阈值模拟顶部黄瓜叶片结露时长结果

ROC曲线分析可知（图5），露点温差模型最佳结露阈值1.02℃，此时ROC曲线下面积（AUC）为0.966，标准误差0.001 13，AUC在95%的置信区间0.964～0.968，Youden指数J=0.856 7，敏感性96.01，特异性89.66。说明将露点温差阈值设置为1.02℃，具有较高准确性。此阈值下，夜间顶部黄瓜叶片结露时长模拟值和观察值较为一致，x=0.885系数接近于1，且线性关系较大=0.969（图7）。试验期间模拟值和观察值的平均结露时长分别为11h和11.8h，模拟值与观察值差异较小，且两者最大误差不超过1h（图6）。

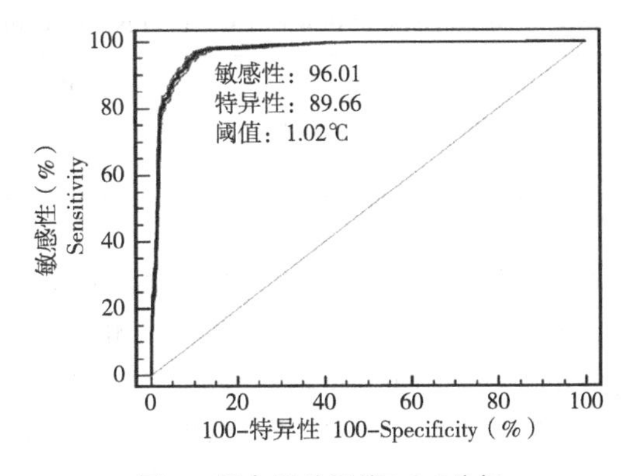

图5　露点温差阈值ROC分析

Figure 5　The dew point deviation threshold analysis of receiver operating characteristic

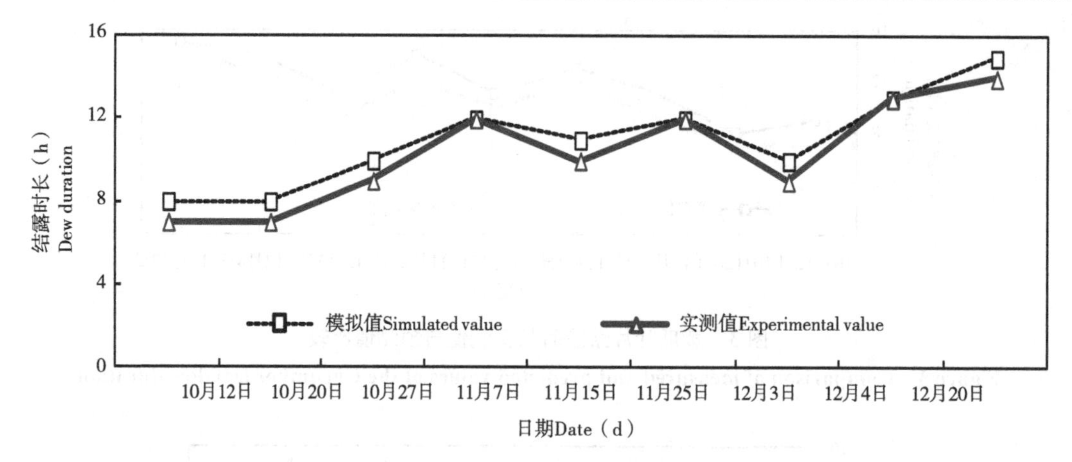

图 6　黄瓜叶片结露时长实测值与预测值比较

Figure 6　Comparison of measured and predicted values of the Cucumber leaf dew duration

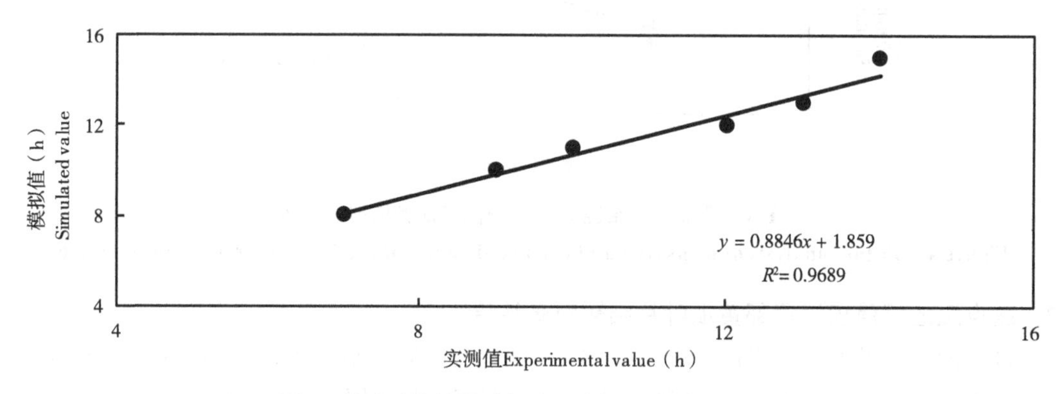

图 7　黄瓜叶片结露时长实测值与预测值拟合分析

Figure 7　Fitting analysis of measured and predicted values of the Cucumber leaf dew duration

3.3　结露时长阈值模型比较分析

表 3　结露时长阈值模型分析

Table 3　Comparison of dew duration threshold value

阈值模型	F_C	F_{AR}	B_S	R^2
相对湿度阈值（91.3%）	0.83	0.07	1.08	0.959
露点温差阈值（1.02℃）	0.87	0.06	0.91	0.969

表3分析可知，两种模型均可以模拟夜间日光温室黄瓜顶部叶片结露时长，两种结露时长模型之间比较可知，露点温差阈值（1.02℃）模型与相对湿度阈值（91.3%）模型相比，F_C值更接近于1，F_{AR}值也更接近于0，R^2值更接近于1，B_S值也同样接近于1；说明露点温差阈值（1.02℃）模型在模拟夜间日光温室黄瓜顶部叶片结露时长中，模拟结果准确性更高，错误率更低，拟合结果也更好，B_S低于1，说明结露时长模拟值低于实际结露时长观察值，差异较小。而相对湿度阈值（91.3%）模型B_S高于1，说明高估了夜间黄瓜顶

部叶片结露时长，且综合表现均低于露点温差阈值（1.02℃）模型。但黄瓜叶片温度不易测量，且对测量过程中对黄瓜叶片有损伤，因此采用ROC曲线分析得到的相对湿度阈值（91.3%），在结露时长误差不超过2h内，更有利于推广研究，露点温差模型则耗费较高更多地局限于试验研究。

4 讨论与结论

温室通风，灌溉和夜间空气温度均影响日光温室内黄瓜顶部叶片结露时长，白天通风不仅可以降低温室内空气温度，还可以将温室内湿空气排出，降低温室内空气含水量，较低的空气含水量不利于露水凝结。黄瓜叶片结露与黄瓜叶片气孔吐水很难区分，由于吐水量较小因此试验没有考虑气孔吐水因素，但仍需进一步探究吐水是否发生在露水凝结前，以及是否对结露时长产生显著影响。

相对湿度阈值模型中，将相对湿度作为唯一环境参数模拟夜间黄瓜顶部叶片结露时长，ROC曲线分析得到最佳结露阈值（91.3%），同时满足了较优的特异性和敏感性，但发现仍有较大误差，平均结露时长误差达到1.5h，说明夜间黄瓜叶片结露并非受相对湿度单因素影响，模拟结果的误差不能通过改变相对湿度阈值大小改进，因此还需进一步考虑结露形成的机理，并进一步完善相对湿度阈值模型，但相对湿度指标仍可作为判定结露时长的主要因素，进行日常温室内环境监测和判定。

露点温差阈值（1.02℃）模型，则规避了上述问题，ROC曲线分析得到露点温差阈值（1.02℃）AUC面积更大，露点温差阈值判断结露与否更加准确，同时考虑了空气温度、相对湿度和叶片温度的影响。计算得到的露点温差模型阈值并非是理论值0℃，而是ROC分析得到的1.02℃，除了T型热电偶线自身测量误差外，还受能量平衡方程模拟黄瓜顶部叶片温度结果误差的影响，能量平衡方程高估了黄瓜顶部叶片温度，T型热电偶线叶片温度测量误差在0.62℃，最终导致露点温差阈值升高了1.02℃，因此下一步需进一步提高黄瓜叶片温度测量值以及黄瓜叶片温度的模拟值。

试验通过ROC曲线分析确定了相对湿度阈值（91.3%）和露点温差阈值（1.02℃），进一步由情形分析表比较了两模型的准确性和适用性，结果表明，在夜间日光温室黄瓜顶部叶片结露时长的模拟中，以露点温差阈值（1.02℃）模型模拟结果明显优于相对湿度阈值（91.3%）模型，露点温差阈值（1.02℃）模型模拟夜间黄瓜顶部叶片结露时长误差在1h以内，而相对湿度阈值（91.3%）模型模拟结果误差达到了2h。因此将黄瓜叶片结露时长模拟局限于夜间黄瓜叶片结露时长的模拟。而相对湿度阈值模型则相对单一，但误差来源也更少，且利于日光温室内直接监测，相对湿度传感器比较常用，因此在结露时长误差不超过2h内，应优先选择阈值为91.3%的相对湿度进行估算，此方法更加简便。本试验中所选取的两种结露时长阈值在不同地区或不同作物间可能存在适用性较低的情况，但ROC曲线判断结露阈值的方法，适用于不同地区和不同作物之间的结露时长阈值模型的筛选，并可为其他作物建立结露时长模型提供参考依据。

参考文献

[1] 李明，赵春江，乔淑，等.基于冠层相对湿度的日光温室黄瓜叶片湿润时间估计模型［J］.农业工程学报，2010，26（9）：286-291.

[2] 刘淑梅，薛庆禹，黎贞发，等.基于BP神经网络的日光温室气温预报模型［J］.中国农业大学学报，2015，20（1）：176-184.

[3] 陆佩玲，任保华，于强.玉米冠层凝露的观测与模拟［J］.生态学报，1998（6）：53-58.

[4] 郑海山，陈端生，刘汉中.塑料大棚内黄瓜叶片结露时长的计算方法［J］.农业工程学报，1991，3（2）：43-49.

[5] Acharya B S，Stebler E，Zou C B. Monitoring litter interception of rainfall using leaf wetness sensor under controlled and field conditions［J］. Hydrological Processes，2017，31（1）：240-249.

[6] Bassimba D D M，Intrigliolo D S，Dalla Marta A，et al. Leaf wetness duration in irrigated citrus orchards in the Mediterranean climate conditions［J］. Agricultural and Forest Meteorology，2017，234：182-195.

[7] DeLong E R，DeLong D M，Clarke-Pearson D L. Comparing the areas under two or more correlated receiver operating characteristic curves：a nonparametric approach［J］. Biometrics，1988：837-845.

[8] Fanourakis D，Carvalho S M P，Almeida D P F，et al. Postharvest water relations in cut rose cultivars with contrasting sensitivity to high relative air humidity during growth［J］. Postharvest Biology and Technology，2012，64（1）：64-73.

[9] Huber L，Gillespie T J. Modeling leaf wetness in relation to plant disease epidemiology［J］. Annual review of phytopathology，1992，30（1）：553-577.

[10] Kidron G J，Herrnstadt I，Barzilay E. The role of dew as a moisture source for sand microbiotic crusts in the Negev Desert，Israel［J］. Journal of Arid Environments，2002，52（4）：517-533.

[11] Kruit R J W，Van Pul W A J，Jacobs A F G，et al. Comparison between four Methods to estimate Leaf Wetness Duration caused by Dew on Grassland［C］//26th Conference on Agricultural and Forest Meteorology. 2004.

[12] Kuroyanagi T. Current usage of air circulators in greenhouses in Japan［J］. Japan Agricultural Research Quarterly：JARQ，2016，50（1）：7-12.

[13] Lawrence M G. The relationship between relative humidity and the dew point temperature in moist air：A simple conversion and applications［J］. Bulletin of the American Meteorological Society，2005，86（2）：225-233.

[14] Metz C E. Basic principles of ROC analysis［C］//Seminars in nuclear medicine. WB Saunders，1978，8（4）：283-298.

[15] Sentelhas P C，Gillespie T J，Gleason M L，et al. Evaluation of a Penman–Monteith approach to provide “reference” and crop canopy leaf wetness duration estimates［J］. Agricultural and Forest Meteorology，2006，141（2）：105-117.

[16] Youden W J. Index for rating diagnostic tests［J］. Cancer，1950，3（1）：32-35.

营养液高度对生菜生长及元素吸收的影响

张伟娟[1, 2]，郭文忠[1]，王晓晶[1, 2]，李灵芝[2]，李海平[2]，陈晓丽[1, 1*]
（1. 北京农业智能装备技术研究中心，北京 100097；2. 山西农业大学园艺学院，太谷 030801）

摘要：在植物工厂水培条件下，探究了霍格兰配方营养液不同的供液高度（2cm、3cm、4cm、5cm、6cm）对生菜10种矿质元素（K、P、Ca、Mg、Na、Fe、Mn、Zn、Cu、S）吸收及积累的影响。采用OPTIMA 3300DV型电感耦合等离子体原子发射光谱仪（ICP-AES，PE，USA）测定矿质元素含量。结果表明，生菜地上部生物量以及光合色素含量均在4cm供液高度处理下最大，而生菜地下部生物量以及根长均随着营养液供液高度的增加而升高；生菜地上部10种矿质元素的含量均在供液高度6cm处理下最大，而矿质元素在生菜中的单株累积量则表现为Fe、Mn元素在6cm供液高度下最大，Ca、Mg、Na、Zn、Cu、S元素在4cm供液高度下最高，K、P元素在5cm供液高度下最大。实际生产中可根据不同的生产目的对水培作物营养液的供液高度进行调节，该试验研究结果为植物工厂功能性蔬菜的生产提供了理论依据。

关键词：生菜；水培；营养液；矿质元素；植物工厂

Effects of Nutrient Solution Levels on the Growth and Mineral Element Absorption of Lettuce

Zhang Weijuan[1, 2], Guo Wenzhong[1], Wang Xiaojing[1, 2], Li Lingzhi[2], Li Haiping[2], Chen Xiaoli[1, *]
（1. *Beijing Research Center of Intelligent Equipment for Agriculture*, *Beijing* 100097；
2. *College of horticulture*, *Shanxi Agricultural University*, *Taigu* 030801）

Abstract: The effects of different nutrient levels（2 cm, 3 cm, 4 cm, 5 cm, 6 cm）of Hoagland nutrient solution on the contents and accumulation of ten mineral elements（K, P, Ca, Mg, Na, Fe, Mn, Zn, Cu, S）in lettuce hydroponically cultured in a plant factory were investigated in the present study. ICP-AES was used to determine the mineral element content. The results indicated that the biomass and photosynthetic pigment contents in the lettuce shoot were the highest under the 4 cm treatment, while the root length and root biomass increased with the enhancement of nutrient solution level. The contents of all mineral elements were the highest under the treatment of 6 cm, while the accumulation of Fe and Mn reached the largest at 6 cm, the accumulation of Ca, Mg, Na, Zn, Cu and S were the highest at 4 cm, the K and P were the largest at 5 cm. Different production purposes can be achieved by adjusting during the actual production the level of nutrient solution. The results of this study provided a theoretical basis for the production of functional vegetables in plant factories.

作者简介：张伟娟，女，硕士研究生，研究方向为蔬菜栽培与生理。E-mail：zhwj3466@163.com
基金项目：国家重点研发计划（2017YFD0201503）

Key words：Lettuce；Hydroponics；Nutrient solution；Mineral element；Plant factory

1 引言

生菜（*Lactuca sativa* L.）是品种丰富的设施主栽蔬菜，含有较多的糖类、蛋白质、矿物质等营养成分，具有助消化、降低胆固醇等功效，深受消费者喜爱[1-2]。矿质元素含量不仅是其重要的品质指标也是无土栽培营养液配方管理的重要依据[3]。

水培是无土栽培的重要技术之一，水培作物根部所需的氧气可以来源于营养液中的溶解氧，也可以依靠裸露于空气中的根系部分直接从空气中获得，后者的根部吸氧量与裸露于空气中的根系长度及数量密切相关[4]。当水培容器高度一定时，营养液的供液高度直接影响根系在营养液及空气中的分布。营养液供液高度低，裸露于空气中的根系吸氧量增加，但营养液太浅可能无法满足植物生长所需的养分，且根系环境较不稳定。营养液供液高度高，根系吸氧量减少，深层营养液中的植物根系容易缺氧，严重时会造成根系腐烂[5-6]。因此，研究不同营养液供液高度对生菜生物量及矿质元素吸收的影响在实际生产中有着重要意义。付利波等[4]研究发现，营养液深度一定时，适当的悬根长度有利于提高生菜的产量和品质。李胜利等[5]探究了不同营养液深度对生菜生长的影响，结果表明增加营养液深度有利于根长和地下部生物量的提高，而适当的营养液深度有利于地上部生物量的提高。但目前有关营养液深度对生菜矿质元素吸收影响的研究报道甚少。由此，本试验在植物工厂中进行，通过调节水培槽内营养液的供液高度，探究了不同营养液供液高度下生菜的生长、生物量积累以及矿质元素的吸收情况，以期为水培蔬菜生产中营养液的供应提供理论依据。

2 材料与方法

2.1 材料

试验在北京农业智能装备技术研究中心的全人工光型植物工厂中进行，生菜品种为大速生（购自国家蔬菜工程技术研究中心）。将经过4℃催芽的生菜种子播于育苗海绵块中培育15天，育苗结束后选取长势一致的幼苗定植于水培槽中。

2.2 试验设计

以LED白光为生菜生长光源，光照强度为130 ± 5μmol/（$m^2 \cdot s$），光周期为16h（光）/8h（暗），环境温度设定为24℃/20℃（光期/暗期），相对湿度（RH）60%，CO_2浓度400μmol/mol。营养液采用Hoagland全营养液配方（pH5.8 ~ 6.0；EC1.2 ~ 1.3ms/cm）。水培槽高度统一为6cm，各处理中营养液供液高度（自槽底部至液位）分别设定为2cm、3cm、4cm、5cm、6cm，每2d补充一次营养液至原高度，每6d更换一次新液。自播种45天后收获生菜样品，进行各项指标的测定。

2.3 测定项目及方法

2.3.1 生长参数的测定

采用电子天平测定地上部和地下部鲜重，然后将植株地上部和地下部置于105℃烘箱中杀青0.5h，后于85℃烘干至恒重，测定干重；根长采用钢卷尺测量。

S/R=地上部干重/地下部干重

2.3.2 叶绿素含量的测定

称取剪碎的新鲜生菜叶片0.2g于研钵中，加少量$CaCO_3$和石英砂及95%乙醇2～3ml研成匀浆，再加入95%乙醇10ml继续研磨至组织变白。然后倒入漏斗中，滤入25ml容量瓶中，期间用95%乙醇多次冲洗滤纸和残渣，直至无色为止，定容到25ml。用分光光度计分别在波长470nm、649nm、665nm下测定吸光度，以95%乙醇为空白对照[7]。

2.3.3 矿质元素含量的测定

精确称取0.50g生菜地上部干粉末样品于消煮管中，加入浓硝酸和高氯酸的混合酸15ml（体积比为4:1）于180℃消煮6～8h至溶液接近无色时冷却，过滤后滤液用去离子水定容至50ml，以同样的方法制备空白对照。采用OPTIMA 3300DV型电感耦合等离子体原子发射光谱仪（ICP-AES，PE，USA）测定[3]。

2.4 数据处理

采用Excel以及SPSS 22.0软件进行数据处理以及显著性差异分析。

3 结果与分析

3.1 营养液供液高度对生菜生长的影响

表1 营养液供液高度对生菜生长的影响

Table 1 Effects of nutrient solution levels on lettuce growth

处理 Treatment	地上部Overground（g）		地下部Underground（g）		根长 Root length（cm）	S/R
	鲜重 Fresh weight	干重 Dry weight	鲜重 Fresh weight	干重 Dry weight		
2cm	19.33c	1.54b	2.85c	0.61b	30.58b	2.52d
3cm	25.67b	2.11ab	6.08ab	0.68ab	32.25ab	3.13c
4cm	30.67a	2.94a	7.50a	0.72ab	33.00ab	4.10a
5cm	30.67a	2.46a	5.27b	0.73ab	33.23a	3.37b
6cm	30.33a	2.15ab	7.56a	0.89a	34.67a	2.43d

注：小写字母表示处理间在0.05水平上的差异显著性。下同

Note：Lowercase indicate significant difference among treatments at 0.05 level. The same as below.

由表1可知，不同处理间，地上部生物量随营养液供液高度的增加呈先升高后降低的趋势。地上部鲜重在4cm和5cm处理下达到最大，显著高于3cm和2cm处理，但与6cm处理差异不显著。地上部干重在4cm处理下最大，较3cm、5cm、6cm处理分别增加了39.34%、19.51%、36.74%。不同处理间，根长和地下部生物量的变化趋势相同，均随营养液供液高度的增加而升高，其中6cm处理下的地下部干重较其他处理增加了21.92%～45.90%。

3.2　营养液供液高度对生菜叶片光合色素含量的影响

表2　营养液供液高度对生菜叶片光合色素含量的影响（$mg \cdot kg^{-1}$ FW）

Table 2　Effects of nutrient solution levels on photosynthetic pigment contents in lettuce leaves

处理 Treatment	叶绿素a Chlorophyll a	叶绿素b Chlorophyll b	类胡萝卜素 Carotenoid	叶绿素（a+b） Chlorophyll（a+b）
2cm	256d	74b	67c	330d
3cm	315c	94ab	86b	409c
4cm	482a	129a	123a	610a
5cm	376b	104ab	100b	480b
6cm	359b	98ab	95b	457bc

由表2可见，不同处理间生菜叶片叶绿素a、叶绿素b以及类胡萝卜素的含量均随营养液供液高度的增加呈先升高后降低的趋势，均在营养液供液高度2cm处理下最低，而在4cm处理下达到最大。其中，营养液供液高度为4cm处理下的生菜叶片总叶绿素含量较其他处理显著的增加了27.08%～84.85%，表明适当的营养液供液高度可能促进了生菜叶片光合色素的形成。

3.3　营养液供液高度对生菜矿质元素吸收的影响

3.3.1　生菜大量元素含量

表3　营养液供液高度对生菜大量元素含量的影响（$mg \cdot g^{-1}$ DW）

Table 3　Effects of nutrient solution levels on the contents of macroelements in lettuce

处理 Treatment	K	P	Ca	Mg	Na	S
2cm	15.93e	2.42d	6.29d	3.20c	12.56d	1.10b
3cm	18.86c	2.81c	7.94b	3.40b	14.35b	1.29a
4cm	17.71d	2.85c	7.17c	3.24c	12.93c	1.16b
5cm	22.25b	3.62b	6.68cd	3.15c	14.40b	1.35a
6cm	23.73a	3.71a	8.97a	3.99a	15.45a	1.36a

如表3所示，除4cm处理之外，K、P、Na、S 4种元素含量在其他处理间均随营养液供液高度的增加而升高，在6cm处理下达到最大，其中6cm处理下的生菜K、P、Na含量较其他处理的增加值达到显著水平（$P<0.05$）。Ca、Mg元素含量随营养液供液高度的变化波动较大，但均表现为6cm处理值最大，且含量均显著高于其他处理。

3.3.2　生菜微量元素含量

表4　营养液供液高度对生菜微量元素含量的影响（$\mu g \cdot g^{-1}$ DW）

Table 4　Effects of nutrient levels on the contents of microelements in lettuce

处理 Treatment	Fe	Mn	Zn	Cu
2cm	38.74c	39.37e	18.24a	2.85c

（续表）

处理 Treatment	Fe	Mn	Zn	Cu
3cm	35.40d	52.09b	12.41c	1.41e
4cm	37.31c	44.00d	16.42b	4.87b
5cm	51.77b	49.74c	15.88b	1.72d
6cm	105.82a	63.58a	18.62a	5.14a

由表4可知，在各处理间，微量元素Fe、Mn的含量与前述大量元素K、P、Na、S呈现相似的变化趋势，即元素含量随着营养液供液高度的增加而升高，在6cm处理下达到最大，且显著高于其他处理。Zn、Cu元素含量随营养液供液高度的变化波动较大，但均表现为6cm处理下值最大，其中Cu元素含量显著高于其他处理。

3.4 营养液供液高度对生菜矿质元素单株累积量的影响

3.4.1 生菜大量元素的单株累积量

表 5 营养液供液高度对生菜大量元素单株累积量的影响（$mg·株^{-1}$）

Table 5 Effects of nutrient solution levels on accumulation of macroelements in lettuce

处理 Treatment	K	P	Ca	Mg	Na	S
2cm	24.53d	3.73d	9.69d	4.92c	19.35e	1.69c
3cm	39.79c	5.93c	16.75c	7.17b	30.29d	2.72b
4cm	52.07b	8.39ab	21.07a	9.52a	38.03a	3.41a
5cm	54.74a	8.91a	16.43c	7.74b	35.42b	3.33a
6cm	51.01b	7.98b	19.28b	8.58ab	33.21c	2.93b

如表5所示，生菜单株Ca、Mg、Na、S元素累积量均在4cm处理下达到最大，其中Ca、Na元素单株累积量较其他处理的增加值达到显著水平（$P<0.05$）。K、P元素的单株累积量在5cm处理下达到最大，其中，K元素单株累积量较其他处理的增加值达到显著水平（$P<0.05$），而P元素单株累积量与4cm处理间差异不显著。

3.4.2 生菜微量元素的单株累积量

表 6 营养液供液高度对生菜微量元素单株累积量的影响（$\mu g·株^{-1}$）

Table 6 Effects of nutrient solution levels on accumulation of microelements in lettuce

处理 Treatment	Fe	Mn	Zn	Cu
2cm	59.66e	60.63e	28.09d	4.40c
3cm	74.70d	109.91d	26.18c	2.97d
4cm	109.70c	129.37b	48.28a	14.33a
5cm	127.35b	122.35c	39.05b	4.22c
6cm	227.50a	136.70a	40.04b	11.06b

由表6可见，Fe、Mn元素的单株累积量随营养液供液高度的增加而升高，均在6cm处理下达到最大，6cm处理下的Fe、Mn元素单株累积量较其他处理分别显著升高了78.64%～281.33%，5.67%～125.47%。Zn、Cu元素单株累积量随营养液供液高度的变化波动较大，但均表现为4cm处理下值最大，且均显著高于其他处理。

4 结论与讨论

矿质元素以离子的形式被植物根系吸收和运输，根吸收矿质元素的方式主要有主动吸收和被动吸收。当膜内离子浓度低于外界浓度时，细胞内外产生的势能差使离子从外界进入膜内，不消耗能量；反之，植物细胞需利用呼吸作用所产生的能量，并通过膜上载体将离子从外界运输至膜内，即为主动吸收[8]。因此，根际矿质元素浓度以及含氧量是影响离子向根内运输的重要因素。同时，根的生理活动如根呼吸也会影响矿质元素向植物地上部的转运进而影响植物地上部的生长及物质转化[9-10]。

随着根对元素的吸收，营养液中的离子浓度呈现一定的动态变化，营养液液位低的处理中根际含氧量虽高，但营养液离子浓度在动态变化中下降较快，不利于元素的吸收。本试验中，所有矿质元素含量都表现为在2cm处理下最低。营养液液位高的处理中，营养液中离子浓度下降缓慢，有利于根对大部分离子的吸收，本试验中，所有矿质元素含量都表现为在6cm处理下达到最高。同时，根长和地下部生物量随着营养液深度的增加而升高，这与李胜利等[5]的研究结果一致，说明营养液深度的增加有利于生菜根系的生长。但根际含氧量低不利于根的呼吸作用，矿质元素向植物地上部分的运输受到影响，进而影响了植物地上部的物质转化和积累，这可能是本试验中6cm处理下生菜地上部生物量没有相应地达到最大的原因之一。而Ca、Mg、Na、Zn、Cu、S元素的单株累积量以及生菜地上部生物量均表现为4cm处理下最大，可能是由于4cm处理下的营养液液位高度在营养液离子动态浓度和根际含氧量这两个因素上达到一定的平衡。综上，在水培槽高度一定时，营养液供液高度通过根际含氧量以及元素动态浓度进一步影响植物地上部的生长和物质积累，实际生产中，可根据生产目的对营养液供液高度进行选择和调整。

参考文献

［1］ 李润儒，朱月林，高垣美智子，等.根区温度对水培生菜生长和矿质元素含量的影响［J］.上海农业学报，2015，31（3）：48-52.

［2］ Fu W，Li P，Wu Y. Effects of different light intensities on chlorophyll fluorescence characteristics and yield in lettuce［J］. Scientia Horticulturae，2012，135：45-51.

［3］ 陈晓丽，郭文忠，薛绪掌，等.LED组合光谱对水培生菜矿物质吸收的影响［J］.光谱学与光谱分析，2014，34（5）：1 394-1 397.

［4］ 付利波，贾非，杨志新，等.悬根长度对水培生菜产量质量和水溶氧的影响［J］.西南农业学报，2015，28（4）：1775-1 779.

［5］ 李胜利，王建辉，孙治强.营养液深度对水培生菜生长的影响［J］.中国农学通报，2007，23（8）：343-345.

［6］ 徐志豪，张德威，P.Adams，等.改善水培作物根际氧气供给的原理和实践［J］.浙江农业学报，1994，（1）：44-48.

［7］ 王学奎. 植物生理生化实验原理和技术（第二版）［M］.北京：高等教育出版社，2006.
［8］ 赵素娥. 植物对矿质元素的吸收、运输和分配［J］.生物学通报，1985（8）：1-3.
［9］ Tesi R，Lenzi A，Lombardi P.Effect of different O_2 levels on spinach（Spinacia oleracea L.）grown in a floating system［J］. Acta Horticulturae，2003，641：631-637.
［10］ Armstrong W.The Use of Polarography in the Assay of Oxygen Diffusing from Roots in Anaerobic Media［J］.Physiologia Plantarum，1967，20（3）：540-553.

Overhead Supplemental Far-red Light Stimulates Greenhouse Tomato Growth under Intra-canopy Lighting

Zhang Yating, Zhang Yuqi, Yang Qichang and Li Tao*

(1. *Institute of Environment and Sustainable Development in Agriculture*, *Chinese Academy of Agriculture Sciences*, *Beijing* 100081; 2. *Key Laboratory of Energy Conservation and Waste Management of Agricultural Structures*, *Ministry of Agriculture*, *Beijing* 100081)

Abstract: Intra-canopy lighting with LEDs are widely applied for high-wire fruit vegetable production and it shows a substantial production improvement. However, further exploring the production potential under intra-canopy lighting system is still stringent in order to save energy. In this study, high-wire tomato plants were grown under intra-canopy lighting thatconstituted of red (peak wavelength at 640 nm) and blue (peak wavelength at 450 nm) LEDs, and combined with overhead supplemental Far-red light (peak wavelength at 735 nm) . Plants were exposed to three photoperiods of Far-red light: 6:00—18:00 (FR12) , 18:00—19:30 (EOD-FR1.5) , 18:00—18:30 (EOD-FR0.5) , and control without supplemental Far-red light. Our data showed that supplemental Far-red light significantly stimulated stem elongation which resulted in a greater plant height. Moreover, Far-red light modified the leaf morphology with a greater leaf length/width ratioand leaf area. The plant architecture modifications that induced by Far-red light resulted in a more homogeneously canopy light distribution. Furthermore, plant total biomass production was increased by 9%~16% under supplemental Far-red light in comparison with control, which led to 7%~12% increases in ripe fruit yield. Interestingly, soluble sugar contents of the ripe tomato fruit slightly decreased with the increasing dosage of Far-red light. Dry matter partitioning to different plant organs were not substantially affected by the Far-red light treatments in general. There were no significant differences observed between the three Far-red light treatments in plant morphology as well as yield and biomass production. This indicates that plants may already reached to the saturation dosage of Far-red light in all treatments. In this context, from the economical point of view, supplementary of low dosage Far-red light at end of day is more favorable, asit uses less electricity but induces similar effects on plant morphology modification and production.

Key words: Far-red light, Intra-canopy lighting, Plant morphology, Light interception, Tomato, *Lycopersivone-sculentum*.

Abbreviation: FR, Far-red light; Red light; EOD; endofday; PPFD; photosynthetic photo flux density; GVA; Graphical Vector Analysis

1 Introduction

Light is the most important factor in determining plant growth and production. To achieve a year-round supply of high quality horticultural product, supplementary assimilation light is widely applied in greenhouses in northern countries to compensate for the low natural radiation intensities

*Corresponding Author: litao06@caas.cn, +86 (10) 82105983

in winter season. Compared with supplementary light on the top, plants are known to benefit from evenly light distribution throughout the crop canopy (Li et al., 2014). In the high-wire tomato cultivation system with high plant density, most of the supplemental light can only be intercepted by the upper part of the plant canopy. Intra-canopy lighting is a recently developed technique that can overcome this problem by arranging lamps within the crop canopy to give lights to the middle or lower part of high-wire plants (Trouwborstet al., 2010). Light–Emitting Diodes (LEDs) are considered as suitable light source for intra-canopy lighting, which have narrow light spectrum and the emittance of Near-infrared (NIR) can be absent. Recently, intra-canopy lighting with LEDs are widely applied for high-wire fruit vegetable production in greenhouses. Yield can be enhanced by 10%~20% under intra-canopy lighting condition for sweet pepper and cucumber (Guo et al., 2016; Kumar and Hao, 2016).

It is highly recognized that intra-canopy lighting can substantially improve the light distribution in plant canopy (Gómez and Mitchell, 2016). Light distribution is largely affected by plant architecture, which depends on the arrangement of plant organs(e.g. leaf, stem)(Sarlikioti et al., 2011). Due to the complexity of plant architecture system, it is difficult to realize an optimally uniform light distribution in high-wire plant canopy even under intra-canopy lighting. As leaf photosynthesis shows a saturating response to light intensity, improving light distribution in the plant canopy may contributes to crop photosynthesis enhancement (Li et al., 2014). Study with functional-structural plant model showed that an ideal-type tomato plant with more spacious canopy architecture due to long internodes and long and narrow leaves, which can substantially increase light interception and crop photosynthesis (Sarlikioti et al., 2011).

Studies have reported that plant architecture (or morphology) is highly affected by light quality (Hogewoninget al., 2010; Wang et al., 2016). Red/Far-red ratio (R/FR ratio) is considered as one of the most important light signal that induces adaptive biochemical and morphological response known as the shade avoidance syndrome (Franklin and Whitelam, 2005; Franklin, 2008). It is well known that increasing the fraction of Far-red light increases internode length and plant height (Kasperbauer and Peaslee, 1973; Blomet al., 1995; Rubertiet al., 2012), which could facilitate light distribution in plant canopy and consequently stimulates plant growth and yield of fruit vegetables such as greenhouse tomato. Yang et al. (2012) also showed that supplemental Far-red light increased leaf area and, consequently, improved light interception and promoted the growth of lettuce. Stem and petiole elongation by a higher fraction of Far-red light either during daytime or end of day (EOD) are widely studied in dicotyledonous and ornamental species (Demotes-Mainardet al., 2016). In grasses like *Loliummultiflorum*, *Paspalumdilatatum* or sunflower, leaf length can be increased by low R/FR ratio during daytime or by applying Far-red light at end of day (EOD-FR) (Casal and Sadras, 1987). Tomato seedlings under EOD-FR light treatments had a longer leaf length than applying red light at end of day (Decoteauet al., 1988).

Although the effect of Far-red light on leaf photosynthesis, physiological and morphological properties have been reported substantially, most of these studies have been limited to plant seedlings or ornamental plants (Li and Kubota 2009; Islam et al., 2014). To our knowledge, there are limited studies carried out on the main greenhouse fruiting vegetables such as tomato, which has a deep plant canopy. This study focuses on the responses of fruiting tomato plant growth with overhead supplemental Far-red light combined with intra-canopy lighting cultivation system. The Far-red light is supplied either during the daytime or the EOD period. The study aims at investigating the potential of tomato plant growth by overhead supplemental Far-red light with the intra-canopy lighting system. Plant production, morphology, fruit quality as well as light distribution in the canopy were investigated.

2 Materials and Methods

Plant material and growth condition. Tomato seeds (*Lycopersicon esculentum* Mill.) cv. Ruifen 882 were sow into 72-cell plug trays (2.7cm× 2.7cm; 12.0 ml volume) at a Chinese Solar Greenhouse on 20 Oct. 2016. After four weekstomato seedlings were transported to an experimental glasshouse at Chinese Academy of Agricultural Sciences (CAAS, Beijing). Thereafter, seedlings were transplanted into the customized substrate bags. The nutrient solution (110.4N, 32.3P, 83.6K, 122.0Ca, 48.4Mg, 2.8Fe, 0.5Mn, 0.05 Zn, 0.02 Cu, 0.5B, and 0.01 Mo, in $mg \cdot L^{-1}$) was irrigated twice per day with drip irrigation system. The mean pH of the nutrient solution was 5.5 and mean EC (Electrical conductivity) was 2.0 $dS \cdot m^{-1}$. Growth gutters were in the north to south orientation with 8 meters long. The distancebetween growth gutters was 150 cm. Plant soneach gutter were alternatively trained totwohighwireswhichwere30cm to the rightandleftofthegrowthgutter. All plants were grown with single shoot. Plant density was 4.36 $plants \cdot m^{-2}$ in the greenhouse. Environmental conditions in the greenhouse were monitored and logged with Eco-Watch Data Acquire System (Eco-Watch, ZealquestScientific Technology Co., Ltd. China) every 10 s and averages over 10 min were recorded. During the experiment the average day/night temperature were23.4℃ /16.0℃, relative humidity were 50.4%/57.6% and CO_2 was ambient. Frequency distribution of the incident photosynthetic photon flux density (PPFD) was provided in supplementary data (S.1).The experiment was finished on 20 Mar. 2017, which resulted in 17 weeks for plant growth.

Table 1 Photoperiod and daily integral of Far-red light in the treatments

Treatment	Photoperiod of Far-red light	Daily integral of Far-red light ($mol \cdot m^{-2}$)
FR12	12h (6:00—18:00)	1.848
EOD-FR1.5	1.5h (18:00—19:30)	0.232
EOD-FR0.5	0.5h (18:00—18:30)	0.077
Control	\	\

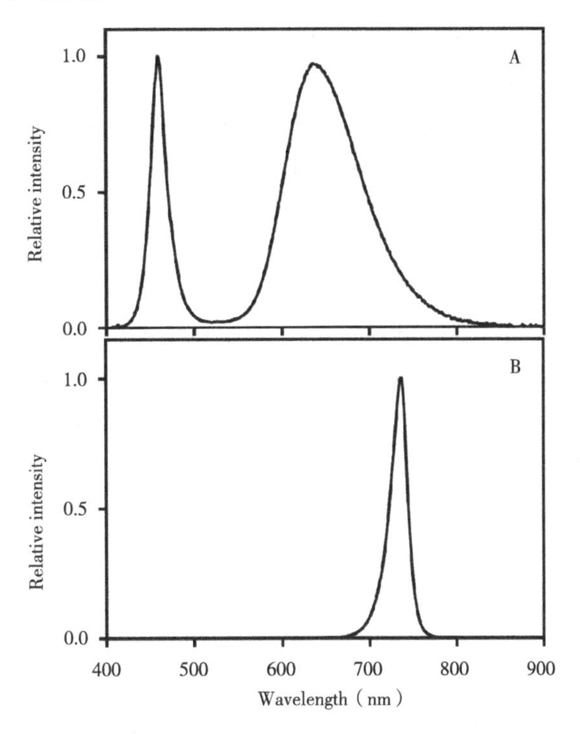

Figure 1　Spectral distribution of two supplemental lighting source between 400 and 900 nm from light-emitting diodes（LEDs）.（A）Intra-canopy lighting LED lamps delivering75% red light（600 to 700 nm, peak wavelength of 640nm）and 25% blue light（400 to 500 nm, peak wavelength of 450nm）.（B）Top lighting with Far-red LED lamps with peak wavelength of 735nm（700 to 800 nm）. The spectral distribution was measured by a spectrometer（AVANTES 2 500, The Netherlands）.

Light treatments. All plants were subjected to the intra-canopy lighting condition when they reached the high-wire（plant height >1 m）. The intra-canopy lighting system was supplied by two layers LEDs（24W, 220V, Shanghai Jing fei Company, China）which were installed above the growth gutter. The distance between the two LED layers was 70 cm. The light spectrum was a mixture of 75% red（peak wavelength at 640 nm）and 25% blue（peak wavelength at 450 nm）（Figure 1A）. Irradiance was emitted from both side of the lamps with a PPFD of 144 $\mu mol \cdot m^{-2} \cdot s^{-1}$ at 10 cm away from the lamps.The photoperiod was 12 h from 6:00 to 18:00. Positions of the LEDs were adjusted regularly to maximize the canopy light interception according to the plant heights.

The greenhouse was equally divided into four sections and each section included two growth gutters. Plants on one growth gutter that between each section were considered as boarder plants. Far-red LED lines were installed 30 cm above the plant canopy in each section. Light intensity of the Far-red LEDs was 43 $\mu mol \cdot m^{-2} \cdot s^{-1}$ at 20 cm below the lamps with peak wavelength at 735nm. Photoperiod of the Far-red LED lamps differed from each other between sections, which resulted in 4 treatments (details see Table 1) .White curtains were applied between each treatment to avoid the interaction of different treatments.

Destructive harvest. At the end of the experiment, five plants from each treatment were randomly selected and destructively harvested. Fresh and dry weight of leaves, fruitsand stems were determined. Plant organs were dried for at least 48 h at 105℃ in a ventilated oven to determine their dry weights. Leaf area was measured with a leaf area meter (LI-3100C, LI-CORInc., Lincoln, NE, USA) .The regularly removed leaves and harvested fruits were weighted and dried, their dry weights were added to obtain the cumulative dry weights per plant. Area of the removed leaves was also determined and added to the total leaf area. Stem length, leaf and truss number were also recorded. Dry matter partitioning of plant organs wascalculated with the dry weight of plant organ divided by total plant dry weight.

Yield determination. Ripe fruits were harvested three times during the experiment. Fresh weight was recorded after each harvest, the accumulated fresh weight of ripe fruits was considered as ripe fruit yield. To determine the fruit dry matter content, five ripe tomato trusses from each treatment were selected at different harvesting periods and their fresh and dry weight wererecorded.

Leaf morphology. To determine the effect of Far-red light on leaf morphology, four plants from each treatment were randomly selected andall their leaf length and width were measured.

Photosynthesis rate measurement. Net photosynthesis rates of the 10th leaf from top of canopy were measured with a portable gas exchange device equipped with a leaf chamber fluorometer (LI-6400; LI-COR) . In the measurement chamber, CO_2 concentration was 400 $\mu mol \cdot mol^{-1}$, block temperature was 25℃, relative humidity was 65%, and the airflow was 500 $\mu mol \cdot s^{-1}$. Photosynthesis rate was measured under the PPFD (10% blue, 90% red) of 100 and 400 $\mu mol \cdot m^{-2} \cdot s^{-1}$, respectively.Three leaves from different plants in each treatment were measured, which were considered as three replicates.

Light distribution measurements. Canopy light distribution measurements were carried out on a fully cloudy day. During the measurement, all supplemental LEDs were switched off. The distribution of PPFD within the canopy was measured with a line sensor (LI-191R, LI-COR, USA) in relation to a reference sensor (LI-190R, LI-COR, USA) above the canopy, both sensors were connected with a data logger (LI-1500, LI-COR, USA) . The line sensor was positioned perpendicularly to the row. PPFDs were measured from the top to the bottom of the canopy at 50 cm intervals, resulted in five height levels (i.e.0, 0.5, 1.0, 1.5 and 2.0 m) , and four measurements were taken at equal distance (about 15 cm) from each other at each height level, they were averaged as one replicate, six replicates were taken in each treatment.

Fruit soluble sugar content measurement. To determine the sugar content ofripe tomato fruit, six ripe fruits from each treatment were selected in the second harvest as six replicates. The samples were freeze-dried using a freeze dryer (STELLAR® Laboratory Freeze Dryer, Millrock Technology, Kingston) . 20 mg dry samples were extracted with 80% ethanol, and thesupernatants were dried with Pressure Blowing Concentrator（MTN-2800D，Autoscience，Tianjin）and dissolved with distilled water. Soluble sugar in the solution was determined by Ultra Performance Liquid Chromatography（Waters，AquirtyH-class UPLC，USA），which equipped with refractive index detector（RID）as described by Yin et al.（2010）with modifications.

Data analysis. Treatment effects on plant growth parameters were evaluated by one-way ANOVA（analysis of variance）using R software（R 3.3.3 for Mac OS），followed by Fisher's protected least significant difference test（LSD）at 95% confidence.

3 Results

Plant production. Overhead supplementalFar-red light significantly increased ripefruit yield and plant total biomass（Figure 2）. Specifically，ripe fruit yield was increased by 7.6%，12% and 12.6% in FR12，EOD-FR1.5，and EOD-FR0.5 treatments，respectively，compared with that in control（Figure 2A）. Plant total biomass was increased by 9%~16% in three Far-red light treatments（Figure 2B），which resulted from a higher dry weight of fruits，leaves as well as stems（Table 2）. No significant difference could be observed between the three Far-red light treatments in fruit yield and total biomass production.

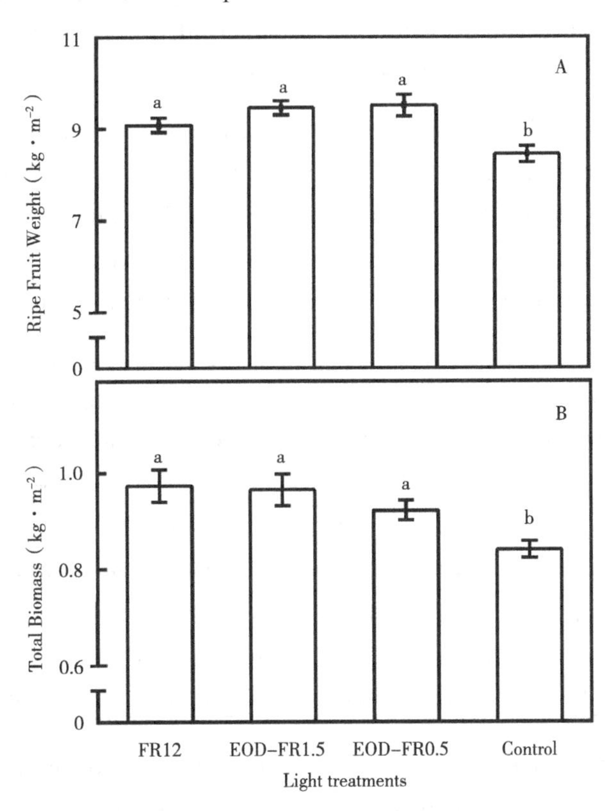

Figure 2 The effects of different Far-redlight treatments on tomato plant internode length（A）andstem length（B）. Different letters indicate statistically significant differences between treatments（$P \leqslant 0.05$）. Error bars show ± s.e.（n=5）.

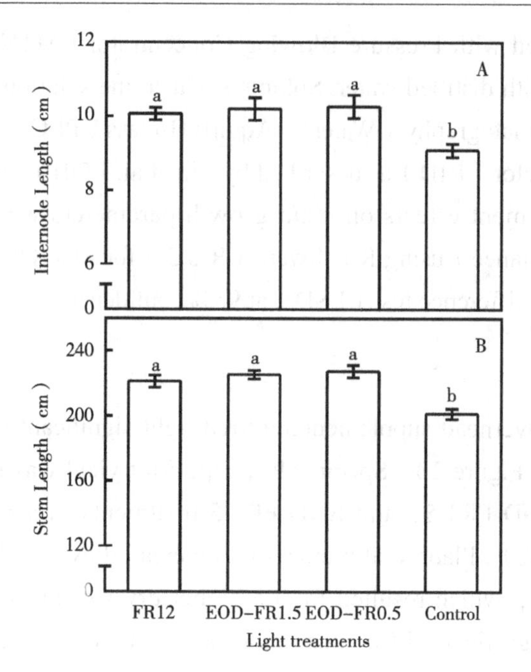

Figure 3 The effects of different Far-red light treatments on ripe fruit yield（A）and plant total dry biomass（B）. Different letters indicate statistically significant differences between treatments（$P\leq0.05$）. Error bars show ± s.e.（n=5）.

Dry matter partitioning.Regarding the dry matter partitioningto plant organs, no clear trends could be observed among the treatments. The lowest dry matter partitioning to fruit and highest partitioning to leaf were observed in FR12 treatment. EOD-FR treatments did not affect dry matter partitioning to both fruits and leaves. Treatments did not significantly affect dry matter partitioning to stems although a higher value was observed in the FR12 treatment（Table 2）.

Table 2 The effects of different Far-red lighttreatments on dry weight of plant organs as well as plant dry matter partitioning to different organs. Presented values are means ± standard error.Different letters in the same column indicate significant difference（$P\leq0.05$, n=5）

Treatments	Dry Weight（$g\cdot m^{-2}$）			Dry Matter Partitioning（%）		
	Fruit	Leaf	Stem	Fruit	Leaf	Stem
FR12	581.8 ± 10.2a	237.8 ± 20.5a	153.21 ± 16.1a	60.1 ± 2.4b	24.28 ± 1.4 a	15.61 ± 1.6a
EOD-FR1.5	605.6 ± 9.9 a	224.5 ± 11.3a	148.57 ± 5.2 a	63.0 ± 1.5ab	23.24 ± 0.6ab	13.77 ± 1.2a
EOD-FR0.5	608.8 ± 14.9a	198.6 ± 8.1ab	127.64 ± 6.1ab	66.1 ± 1.2 a	21.58 ± 1.0 b	12.35 ± 1.5a
Control	540.8 ± 11.5b	184.3 ± 8.5 b	115.45 ± 2.2 b	64.3 ± 0.5 a	21.88 ± 0.7ab	13.77 ± 0.5a

Fruit soluble sugar content.Soluble sugar contents of the ripe tomato fruit decreased with the increasing of Far-red light dose（Figure 4）. For glucose and fructose contents, control treatment showed the highest value, FR12 treatment was the lowest, and both EOD-FR treatments were in between.

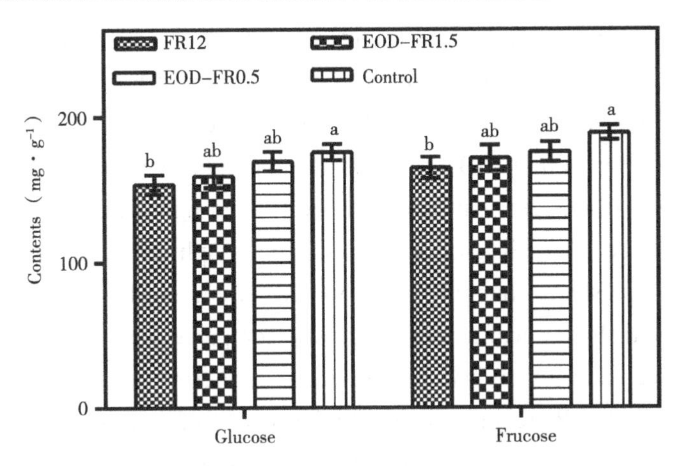

Figure 4　The effects of different Far-redlight treatments on glucose and fructose content in ripe tomato fruit. Different letters indicate statistically significant differences between treatments（$P \leqslant 0.05$）. Error bars show ± s.e.（n=5）

Leaf photosynthesis rate. Leaf photosynthesis rate in the upper part of the canopy was not affected by the Far-red light treatment as indicated by the similar net photosynthesis rates under 100 as well as 400 $\mu mol \cdot m^{-2} \cdot s^{-1}$ PPFD in all the treatments（Figure 5）.

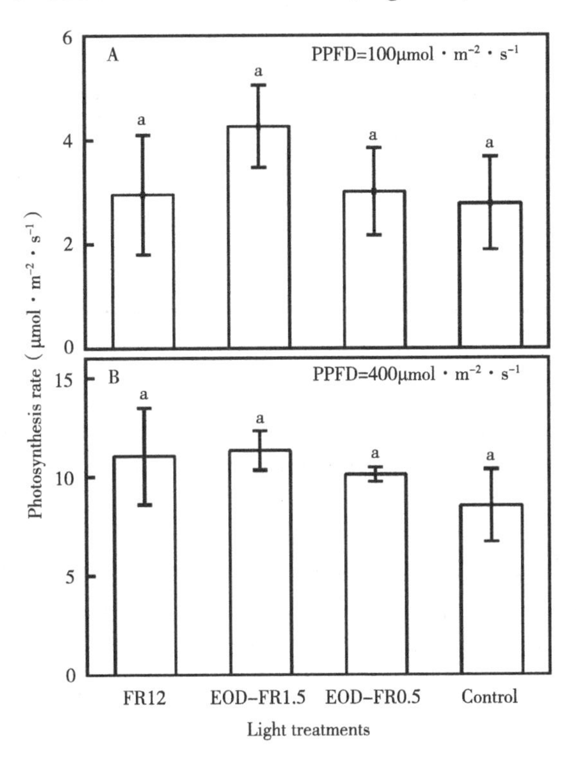

Figure 5　The effects of different Far-red light treatments on leaf photosynthesis rate under PPFD of 100$\mu mol \cdot m^{-2} \cdot s^{-1}$（A）and 400$\mu mol \cdot m^{-2} \cdot s^{-1}$（B）.Error bars show ± s.e.（n=3）. No significant difference was observed between treatments as indicated by the same letter（$P>0.05$）

Plant morphology.Tomato plants grown underoverhead supplemental Far-red light treatment had significantly longer stem in comparison with that incontrol (Figure 3) . Specifically, stem length was increased by 9.7%, 11.8% and 12.8% in the treatments of FR12, EOD-FR1.5 and EOD-FR0.5, respectively (Figure 3B) . As internode number was not significantly affected by the treatments (S.2) , the longer stem length in the Far-red light treatments resulted in a greater internode length (Figure 3A) . Regarding leaf shape, graphical vector analysis (GVA) was used to visualize the leaf shape changing between leaf length and leaf widthdirectly. All Far-red light treatments showed a higher leaf length/width ratio compared with that in control, particularly for plants in FR12 treatment which had a significantly higher ratio than the others (Figure 7) . Plant leaf area in the EOD-FR treatments were higher than leaf area of FR12 and control, while significantly difference occurred only in the EOD-FR1.5 treatment (Figure 6) .

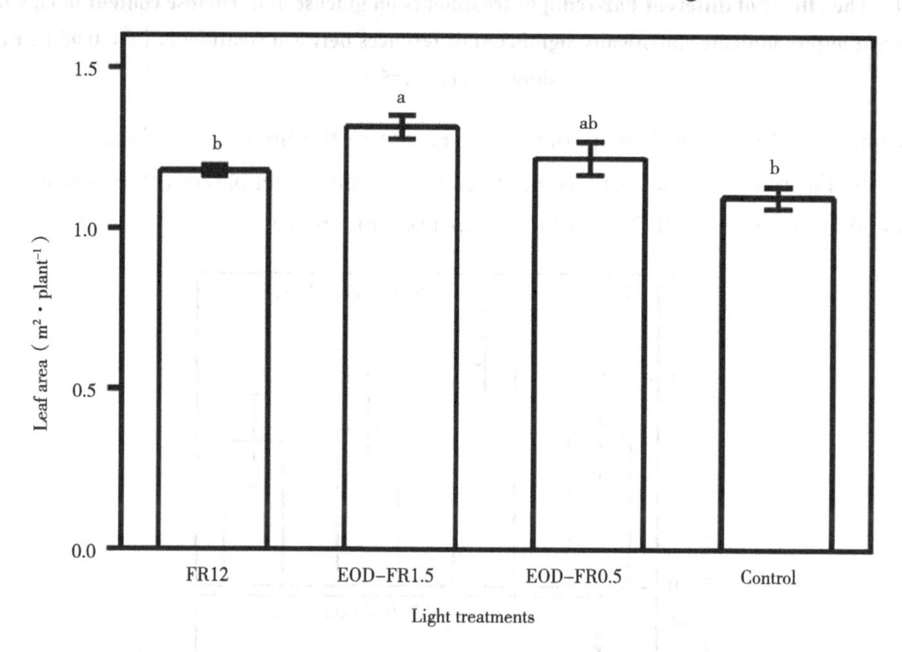

Figure 6 The effects of different Far-red light treatments on leaf area of single tomato plant. Different letters indicate statistically significant differences between treatments ($P \leq 0.05$) . Error bars show ± s.e. (n=4)

Vertical light distribution. Light intensity in the plant canopy decreases with canopy depth (Figure 8) .All Far-red light treatments showed a slower attenuation of light intensity in the canopy compared with control as indicated by a relatively higher light intensity in the same height level of the lower canopy. Of all the treatments, light intensities in the lower canopy was highest in the FR12 treatment, while it was lowest in the control, and both EOD-FR treatments were in between (Figure 8) .

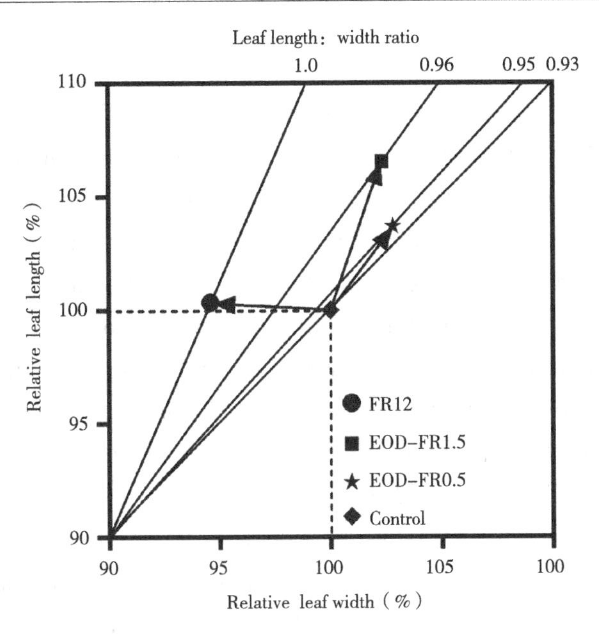

Figure 7 Graphical vector analysis (GVA) of the effect of different Far-red light treatments on average leaf width and length of tomato plant. The leaf shape change was expressed as relative leaf length (y-axes), relative leaf width (x-axes) and leaf length: width ratio (the top makers). Data from control treatment were considered as reference.

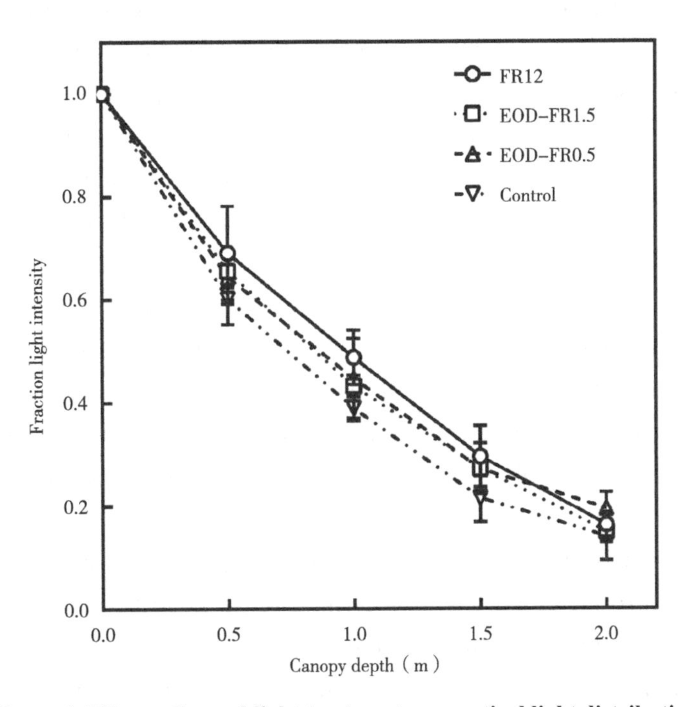

Figure 8 The effects of different Far-red light treatments on vertical light distribution in tomato plant canopy. Light intensity was measured from top to bottom of the canopy. Error bars show ± s.e. (n=6).

4 Discussion

Light is the driving force for crop photosynthesis, which often limiting crop production in greenhouses (Li, 2015) . It has been well recognized that intra-canopy lighting remarkably increases production of high-wire fruit vegetables in greenhouse (Pettersen et al., 2010; Kumar et al., 2016) . Here we showed that tomato production under intra-canopy lighting can be further stimulated by overhead supplementary of small dosage of Far-red light (Figure 2A) . Moreover, the effect of supplemental Far-red light at end of day on plant production is comparable with the supplemental Far-red light during day time which had 8 or even 24 times higher Far-red light dosage than Far-red light treatments at end of day. Such stimulating effects on production occurs are most likely resulted from the changed plant morphology which may correlates with higher canopy light interception (Hogewoning et al., 2012) .

The amount of intercepted radiation is a decisive factor for crop growth and biomass production, and depends mainly on leaf area and canopy structure (ie. plant height, petiole) (Trouwborst et al., 2010; Sarlikioti et al., 2011) . In this study, supplemental Far-red light during daytime or end of day significantly stimulated stem elongation (Figure 3) by increasing the internode length rather than increasing the internode number (S.2) . The promoted stem elongation resulted in a more uniformly light distribution in the vertical profile of the canopy as indicated by the higher fraction light intensity in the middle crop canopy (Figure 8) . Although presented data reflected light distribution only under natural light condition (Figure 8) , a more homogeneously light distribution can be inferred under intra-canopy lighting system particularly in the horizontal profile of the canopy as plants had a more spacious architecture. Plants are known to benefit from even light distribution throughout the crop canopy, as photosynthetic rate of a single leaf shows a saturation response to the light intensity (Li et al., 2014) . Sarlikiotiet al. (2011) tested with a functional structural plant model to investigate to what extent plant architecture affects light absorption and crop photosynthesis in tomato, and showed that increasing internode length leads to an increase in canopy light absorption and crop photosynthesis rate. Gouet al. (2016) also reported that supplemental Far-red light with LEDs promoted stem elongation and plant height, which increased the light interception and improved plant growth and fruit production. These studies all support our finding that supplemental Far-red light enhanced stem elongation and consequently a higher production.

A higher leaf area is highly relevant for crop photosynthesis, as long as the fraction of light interception is also increased (Wunsche and Lakso, 2000) . Supplemental Far-red light slightly increased leaf area although significant difference occurred only in the EOD-1.5 treatment (Figure 6) . This might not contribute to canopy light interception enhancement, as crop intercepts more than 90% of the incident light when leaf area index (LAI) higher than 3 (Reffye et al., 2009) . Furthermore, we observed that Far-red light increased the leaf length/width ratio, this may also positively contribute to canopy light interception, as Sarlikiotiet al. (2011) showed that increasing the length/width ratio of tomato leaves increased the light absorption and crop photosynthesis. But to what extent the Far-red light induced leaf morphology modification could contribute to canopy light absorption need to be further explored with functional-structural plant model.

The observed plant architectural modifications under increased Far-red light are not surprising as many plants display a remarkable increase in the elongation of stems and petioles, which is termed as the typical response of shade-avoidance syndrome (Fankhauser and Batschauer, 2016). Such phenomenon is often at the expense of leaf and storage organ development and growth (Pierik and de Wit, 2014). This may explain the significantly lower dry matter partitioning to fruits under FR12 treatment (Table 2). Although dry matter partitioning to plant organs are slightly influenced by the treatments, supplemental Far-red light enhanced dry weight of fruit, leaf as well as stem compared with control (Table 2), which resulted from a higher total plant dry weight (Figure 2B). Interestingly, glucose and fructose content of ripe tomato fruits were negatively correlated with the Far-red light dosage (Figure 4), indicating that the increased ripe tomato fruit production in supplementary of Far-red light is at the expense of fruit quality. The underlying mechanisms of this phenomenon need to be further explored.

Photosynthesis is the ultimate basis for plant growth. Although leaf photosynthesis is highly responsive to Far-red light (Demotes-Mainard et al., 2016), we showed that the photosynthesis rates of single leaves were similar among treatments (Figure 5). This is probably because in greenhouse the supplementary of low dosage of Far-red light could be of minor importance for leaf photosynthesis due to the presence of natural irradiance. We measured photosynthesis rates only under PPFD of 100 and 400 $\mu mol \cdot m^{-2} \cdot s^{-1}$, because the PPFD between 100 and 400 $\mu mol \cdot m^{-2} \cdot s^{-1}$ accounted for 0 ~ 60% of the incident irradiance during the plant growth period (S.1) and (S.2).

Supplementary data

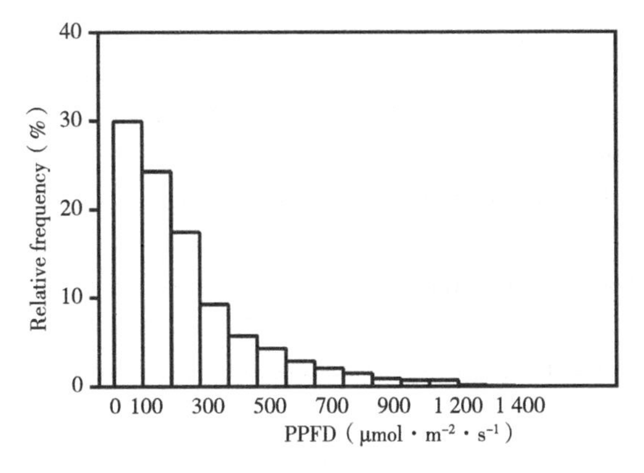

S.1 Frequency distribution of the incident photosynthetic photon flux density in the greenhouse during the plant growth period.

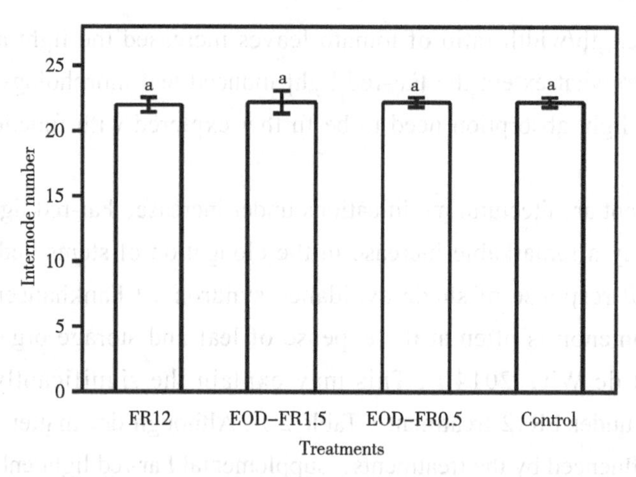

S. 2 The effects of different Far-red light treatments on internode number Error bars show ± s.e.（n=5）. No significant difference was observed between treatments as indicated by the same letter.

Plant shows a saturation response to Far-red light at end of day（Cao et al., 2016）. In this study, supplemental Far-red light at end of day for 0.5 hour（EOD-FR0.5）showed similar effect on plant growth as EOD-FR1.5 did. This indicates that EOD-FR 0.5 treatment may already reached the saturation point, but what is exactly the saturation dosage of Far-red light for tomato is still unclear, which need to be further explored. From the economical point of view, supplementary of low dosage Far-red light at end of day is more favorable, which use less electricity and have similar effects on plant morphology modification and production.

5 Acknowledgement

This work was supported by the National Key Research and Development Program of China（2017YFB0403902）and the Young Elite Scientists Sponsorship Program by CAST（2016QNRC001）.

Reference

Blom, T.J., M.J. Tsujita and G.L. Roberts. 1995. Far-red at End of Day and Reduced Irradiance Affect Plant Height of Easter and Asiatic Hybrid Lilies. Hortscience 30: 1 009-1 012.

Cao, K., J. Yu, L. Ye, H. et al. 2016. Optimal LED far-red light intensity in end-of-day promoting tomato growth and development in greenhouse.T. Chin. Soc. Arg. Eng. 32: 171-176.

Casal, J.J. and V.O. Sadras. 1987. Effects of end-of-day red/far-red ratio on growth and orientation of sunflower leaves. Botanical Gazette 148: 463-467.

Decoteau, D.R., M.J. Kasperbauer, D.D. Daniels and et al. 1988. Plastic mulch color effects on reflected light and tomato plant growth. Scientia Hort. 34: 169-175.

Demotes-Mainard, S., T. Péron, A. Corot, J. et al. 2016. Plant responses to red and far-red lights, applications in horticulture. Environ. Exp. Bot. 121: 4-21.

Fankhauser, C. and A. Batschauer. 2016. Shadow on the Plant: A Strategy to Exit. Cell 164: 15-17.

Franklin, K.A. 2008. Shade avoidance. New Phytol. 179: 930-944.

Franklin，K.A. and G.C. Whitelam. 2005. Phytochromes and shade-avoidance responses in plants. Ann. Bot. 96：169-175.

Gómez，C. and C.A. Mitchell. 2016. In search of an optimized supplemental lighting spectrum for greenhouse tomato production with intracanopy lighting.Acta Hort. 1134：57-62.

Guo，X.，X. Hao，S. Khosla，et al. 2016. Effect of LED inter-lighting combined with overhead HPS light on fruit yield and quality of year-round sweet pepper in commercial greenhouse.Acta Hort. 1134：71-78.

Hogewoning，S.W.，P. Douwstra，G. Trouwborst，et al. 2010. An artificial solar spectrum substantially alters plant development compared with usual climate room irradiance spectra. J. Expt.Bot. 61：1 267-1 276.

Hogewoning，S.W.，G. Trouwborst，E. Meinen et al. 2012. Finding the Optimal Growth-Light Spectrum for Greenhouse Crops. VII International Symposium on Light in Horticultural Systems. 956：357-363.

Islam，M.A.，D. Tarkowská，J.L. Clarke，et al. 2014. Impact of end-of-day red and far-red light on plant morphology and hormone physiology of poinsettia. Sci. Hort. 174：77-86.

Kasperbauer，M.J. and D.E. Peaslee. 1973. Morphology and Photosynthetic Efficiency of Tobacco Leaves That Received End-of-Day Red and Far Red Light during Development. Plant Physiol. 52：440-442.

Kumar，K.G.S.，X. Hao，S. et al. 2016. Comparison of HPS lighting and hybrid lighting with top HPS and intra-canopy LED lighting for high-wire mini-cucumber production. Acta Hort.1134：111-118.

Li，Q. and C. Kubota. 2009. Effects of supplemental light quality on growth and phytochemicals of baby leaf lettuce. Environ. Expt. Bot. 67：59-64.

Li，T.，E. Heuvelink，T.A. Dueck，et al. 2014. Enhancement of crop photosynthesis by diffuse light：quantifying the contributing factors. Ann. Bot. 114：145-156.

Li，T.，2015. Improving Radiation Use Efficiency in greenhouse production systems. PhD thesis，Wageningnen University，Wageningen，the Netherlands.

Pettersen，R.I.，S. Torre and H.R. Gislerød. 2010. Effects of intracanopy lighting on photosynthetic characteristics in cucumber. Sci. Hort. 125：77-81.

Pierik，R. and M. de Wit. 2014. Shade avoidance：phytochromesignalling and other aboveground neighbour detection cues. J. Expt. Bot. 65：2 815-2 824.

Reffye，d.P.，E. Heuvelink，Y.Guo，et al. 2009. Coupling process-based models and plant architectural models：A key issue for simulating crop production. Crop Modeling and Descision Support，Springer.pp. 130-147.

Ruberti，I.，G. Sessa，A. Ciolfi，et al. 2012. Plant adaptation to dynamically changing environment：The shade avoidance response. Biotechnol. Adv. 30：1 047-1 058.

Sarlikioti，V.，P.H. De Visser，G.H. Buck-Sorlin et al. 2011. How plant architecture affects light absorption and photosynthesis in tomato：towards an ideotype for plant architecture using a functional-structural plant model. Ann. Bot. 108：1 065-1 073.

Trouwborst，G.，J. Oosterkamp，S.W. Hogewoning，et al. 2010. The responses of light interception，photosynthesis and fruit yield of cucumber to LED-lighting within the canopy. Physiol. Plant. 138：289-300.

Wang，J.，W. Lu，Y. Tong et al. 2016. Leaf Morphology，Photosynthetic Performance，Chlorophyll Fluorescence，Stomatal Development of Lettuce（*Lactuca sativa* L.）Exposed to Different Ratios of Red Light to Blue Light. Front. Plant Sci. 7：250.

Wünsche，J.N. and A.N. Lakso. 2000. Apple tree physiology - implications for orchard and tree management. Compact Fruit Tree 33：82-88.

Yang，Z.C.，C. Kubota，P.L. Chia et al. 2012. Effect of end-of-day far-red light from a movable LED fixture on squash rootstock hypocotyl elongation. Scientia Hort. 136：81-86.

Yin，Y.G.，T. Tominaga，Y. Iijima，K. Aoki，et al. 2010. Metabolic alterations in organic acids and gamma-aminobutyric acid in developing tomato（*Solanum lycopersicum* L.）fruits. Plant Cell Physiol. 51：1300-1314.

提供肾脏病患食用的三低水耕萵苣之栽培

钟兴颖，方炜

（台湾大学生物产业机电工程学系）

摘要：末期肾脏疾病（End-Stage Renal Diseases，ESRD）之发生率与盛行率随经济发展逐年增加，ESRD病人常会伴随高血钾与高血压，又已知人体摄取硝酸盐后会转化为具致癌风险的亚硝胺，所以在饮食上对于钾、钠离子与硝酸盐的摄取都需有所限制。本研究选定每百克萵苣鲜重之钾、钠与硝酸盐含量上限分别为80mg，30mg与250mg；三项数值均低于上限值者以三低萵苣称之。试验在台大植物工厂内使用湛水式水耕系统（Deep Flow Technique，DFT）进行，主要栽培香波绿萵苣（frill-ice lettuce），使用LED为人工光源，平均光量为200μmol·m^{-2}·s^{-1}、光期16h、日夜温25/20℃、二氧化碳浓度1 200μmol·mol^{-1}。对照组为栽培全程（由播种至采收42d）使用1.5倍强度（EC 1.2mS.cm^{-1}）的山崎养液萵苣配方，其他4个处理组采用相同养液配方，其中3个处理组分别为采收前3、前2与前1周更换为无钾配方（以硝酸铵取代山崎养液配方中的硝酸钾），另一个处理组为采收前2周将养液更换为清水。结果显示，对照组鲜重为108.67g，每百克鲜重之钾离子含量为365.52mg。清水处理组鲜重损失高达69%，显然此处理方式在商业上为不可行，更何况钾含量也仅降低约48%。本研究开发出2种栽培流程，分别可生产生食用与熟食用的三低萵苣，前者可降低钾含量约90%，但鲜重约降低48%；后者鲜重降低约20%，钾含量降低约46%，经煮食后应可降低超过90%。本研究的各项处理组以1周为间隔，若改以更细致的区隔应可得出符合三低要求且鲜重降低较少的更理想的栽培流程，此有待后续持续研究。

关键词：末期肾脏病；高血钾症；水耕；硝酸盐；钾离子；钠离子

Cultivating Hydroponically Grown Lettuce with Three Low Features for ESRD Patients

Zhong xingying，Fang Wei

（*Department of Bio-Industrial Mechatronics Engineering*，*Taiwan University*）

Abstract：Patients with end stage renal disease（ESRD）often experience hyperkalemia，a condition that requires limiting their dietary intake of potassium. Fresh vegetables are the largest source of potassium intake for people，it is vital to provide vegetables with low potassium concentration for ESRD patents. Prior study was conducted in Japan for low-potassium vegetable production，however，some products are low in potassium but high in sodium. The objective of this study was to develop an operating procedure enable to grow low potassium，low sodium and low nitrate lettuces without scarifying too much of the fresh weight. The frill ice lettuces were grown in a plant factory with artificial lighting（PFAL）using hydroponic system with deep flow technique（DFT）. The lettuces were grown at environment with average light intensity of 200 μmol·m^{-2}·s^{-1}，light period of 16 hours，day/night temperatures of 25/20℃ and carbon dioxide concentration of 1 200 μmol·mol^{-1}. The Yamasaki solution for lettuce with 1.5 strength（EC 1.2 mS·cm^{-1}）was used from sowing to harvest（totally 42 days）as the control group.

Other three treatments were to replace potassium nitrate with ammonium nitrate three, two and one week before harvest and the fifth treatment was to replace nutrient solution with tap water two weeks before harvest. Results shown that the potassium concentration and fresh weight for the control group is 365.52 mg per 100 g fresh weight and 108.67 g per plant, respectively. The tap water treatment was not suggested for the loss of fresh weight was too big and the reduction of potassium concentration were not accepted. Two cultural procedures were recommended. They are for "ready to eat" and "eat after cook" lettuce. This study shows the potential of producing low potassium (< 80 mg per 100 g fresh weight) , low sodium (< 30 mg per 100 g fresh weight) and low nitrate (< 2 500 × 10^{-6}) frill-ice lettuce in a PFAL. More study need to be done to find the better cultural practices.

Key words: End Stage Renal Disease (ESRD) ; Hyperkalemia; Hydroponics; Nitrate; Sodium; Potassium.

1 前言与研究目的

全人工光型植物工厂不仅能“量产”民生相关种苗、蔬菜、香草与小型蔬果类作物，更能对作物“加值”，譬如透过栽培参数的调整来创造逆境刺激植物体内二次代谢物的生成（Lee and Kim，2014），以提供特定病患或养生相关的食材。量产高质量、无农药、无重金属、无虫卵、无尘土的蔬菜只是基本需求，更进一步地针对小众市场的需求去调整植物内的成分譬如高硒、高钙、高铁、高抗氧化、高维生素A、维生素C、维生素E或低钾、低硝酸盐等，从而创造价值与利润，已是植物工厂产业的现在进行式。

慢性肾脏病（Chronic Kidney Diseases，CKD）患者在每一个阶段皆有不同的营养照护要求，也强调对食物中矿物质含量摄取的控管，其中，钾离子含量是一个重要的指标。第1至第4期的肾衰竭患者，其对钾的需求需依血钾检验值来做限制，第5期（末期）且接受血液透析的患者，钾离子限制在每天2 000 ~ 3 000mg；腹膜透析者的钾需求则每天可增加到3 000 ~ 4 000mg（Beto and Bansal，2004；Tritt，2004）。若摄取的食物中钾离子含量过高，会造成高血钾症进而引起心律不整。肾脏病人需减少高钾蔬果之摄取，卫生福利部食品药物管理署定义之高钾蔬菜为每100g蔬菜大于300mg的钾，对于CKD患者，特别是末期肾脏病（End Stage Renal Diseases，ESRD）患者，一般建议蔬菜（包含生菜）必须经过川烫或油炒来降低钾离子含量方可适合食用，然而如果生菜经过这样处理，不仅口感不佳，许多营养成分及抗氧化物质也都被破坏。Wu et al.（2011）提到中国台湾居民的饮食习惯中，钾的摄取主要来自新鲜蔬菜，约占30%。如果能将蔬菜中钾含量降低，对于需要控制钾含量饮食的病患可以多一个选择。

欧、美、日等国家和地区ESRD之发生率与盛行率均随着经济发展而年年增加，患者往往必须要接受血液透析、腹膜透析与肾脏移植等替代性方式治疗。ESRD的盛行率以中国台湾、日本与美国为全球前三名，每百万人口中依序分别有3 219位、2 505位和2 076位患者（Saran et al.，2017）。低钾蔬菜作为商品选项，有其地域性背景，在日本与中国台湾都已有相关商品贩卖，皆以生菜为主。蔬菜中的植化素被认为对人类健康有帮助的物质也是优良的抗氧化剂，但蔬菜经过烹煮10min后莴苣的抗坏血酸、类胡萝卜素会下降40%，总酚损失了30%，抗氧化能力也显著地降低（Vinha et al.，2015）。Gutiérrez et al.（2014）研究发现饮食习惯中食用较多的生鲜蔬果有助于降低ESRD患者之死亡率。

CKD患者伴随高血压的发生率很高，Agarwal et al.（2003）发现有86%的美国洗肾患者合并有高血压问题，高血压患者对于钠离子的摄取是需要特别留意的。另外，硝酸盐对人体并无特别危险，但是其衍生物如亚硝胺就可能造成癌症的发生（Archer，1989），针对日常饮食的蔬菜订定硝酸盐含量的上限是应该的。

钾离子是植物生长所需的巨量元素之一，主要与酶活化、光合作用和渗透压平衡有关（Schachtman and Liu，1999）。Ogawa et al.（2007；2012）指出在栽培前期植物缺钾，会强烈抑制生长，采收前两周使用硝酸钠取代硝酸钾，可以有效降低菠菜、莴苣和西红柿的钾含量，但会显著增加钠离子含量。Yoshida et al.（2014）的低钾莴苣专利内容显示其养液是使用氯化钠（NaCl）中的钠离子去取代钾离子，这虽然能让莴苣中钾离子降低，但钠离子也显著上升，摄取此种低钾蔬菜反而会增加肾脏与心脏的负担，对某些特定病患存在相当的风险。

本研究首创在山崎养液中使用硝酸铵置换硝酸钾的配方设计，目标在开发出适合肾脏病患的莴苣栽培流程，不仅降低莴苣中钾离子含量并同时维持低钠离子与低硝酸盐含量。

2 材料与方法

2.1 试验材料与栽培方法

莴苣种子购于德城种子公司，不结球莴苣（*Lactuca sativa* L.），商业品种名为”香波绿”。本试验使用两种养液配方：①山崎养液配方（N1）；②无钾养液配方（N2），两种养液配方皆使用药品如下：KNO_3、$NH_4H_2PO_4$、$MgSO_4 \cdot 7H_2O$、$C_{10}H_{12}N_2NaFeO_8$、$Ca(NO_3)_2 \cdot 4H_2O$、$MnSO_4 \cdot H_2O$、H_3BO_3、$CuSO_4 \cdot 5H_2O$、$ZnSO_4 \cdot 7H_2O$和$Na_2MoO_4 \cdot 2H_2O$。N2养液配方以N1为基础，但以硝酸铵取代硝酸钾为无钾养液配方。两种配方中各离子浓度如表1所示。栽培过程中使用0.1 N NaOH调整pH值至6.0 ± 0.2，养液之电导度值维持在$1.2 \pm 0.2 mS \cdot cm^{-1}$。

莴苣栽培于校内的完全人工光型植物工厂，栽培期间依栽培密度分为育苗期（播种后天数，Days After Sowing，DAS 1-7）、成长期（DAS 8-21）与调质期（DAS 22-42）等3个阶段，莴苣种子于气温20℃播于培养皿中催芽24小时后（DAS 0），将发芽之莴苣种子移至海绵（DAS 1）开始栽培，育苗期24h光照，成长期开始光周期调整为16/8h，日/夜温25/20℃。光源使用LED 6500K冷白灯管（新世纪股份有限公司，台湾），光量为$200 \mu mol \cdot m^{-2} \cdot s^{-1}$，共栽培6周。

本研究依照不同养液与使用周数共有1个对照组与四个处理组：①栽培全期（DAS 1～42）皆使用N1养液栽培作为对照组（代号为6K）；②育苗期与成长期皆使用N1养液，采收前1（DAS 36～42）、2（DAS 29～42）与3周（DAS 22～42）更换为N2养液，代号分别为5K、4K与3K；③育苗与成长期皆使用N1养液，采收前两周更换为清水（N3）进行栽培，代号为t.w.（tap water）。详细栽培环境参数与流程如表2所示。栽培过程中，DAS 21、28、35、42取样测量莴苣地上部鲜重、植物叶片与养液中阴、阳离子浓度。

表 1　不同养液处理配方

Table 1　Different nutrient solution formula

养液配方	NO_3-N（$mg \cdot L^{-1}$）	NH_4-N（$mg \cdot L^{-1}$）	P（$mg \cdot L^{-1}$）	K（$mg \cdot L^{-1}$）	Ca（$mg \cdot L^{-1}$）	Mg（$mg \cdot L^{-1}$）	S（$mg \cdot L^{-1}$）
N1	84.0	7.0	15.4	156.0	40.0	12.2	16.2
N2		63.0		0.0			

注：微量元素：Fe（3.2$mg \cdot L^{-1}$），Mn（0.5$mg \cdot L^{-1}$），Cu（0.02$mg \cdot L^{-1}$），Zn（0.05$mg \cdot L^{-1}$），Mo（0.01$mg \cdot L^{-1}$），Na（1.3$mg \cdot L^{-1}$）。

N1：山崎养液配方（莴苣），1.5 倍强度

N2：无钾养液配方

表 2　低钾莴苣栽培参数代码

Table 2　Operating codes during the treatment stages for low potassium lettuce production

处理组	育苗期 DAS 1 ~ 7	成长期 DAS 8 ~ 21	调质期 1 DAS 22 ~ 28	调质期 2 DAS 29 ~ 35	调质期 3 DAS 36 ~ 42
6K（CK）	N1 × 6周				
5K	N1 × 5周				N2 × 一周
4K	N1 × 4周			N2 × 两周	
3K	N1 × 3周		N2 × 3周		
t.w.	N1 × 4周			N3 × 两周	

注：DAS：播种后天数（Days After Sowing）

育苗期（DAS：1-7）：N1_E1.2_L200_d910_H24_A25_C1200

成长期（DAS：8-21）：N1_E1.2_L200_d45_H16_A25/20_C1200

调质期（DAS：22-42）：Nx_E1.2_L200_d26_H16_A25/20_C1200

操作代码：

Nx：N，养液配方

N1：山崎养液（100%钾离子）

N2：无钾养液（0%钾离子）

N3：清水（EC=0.1$mS \cdot cm^{-1}$）

Ex：E，养液的电导度（EC）

x，电导度值，单位：$mS \cdot cm^{-1}$.

lx：L，LED灯管（冷白，色温6500K）

x：光量的值，单位：$\mu mol \cdot m^{-2} \cdot s^{-1}$

dx：d，作物栽培密度

x：栽培密度之值，单位：$plts \cdot m^{-2}$

Hx：H，光照时间

x：每日照光时数，单位：$h \cdot day^{-1}$

AdT/nT：A，日/夜温，单位：℃

Cx：C，二氧化碳浓度

x，二氧化碳浓度之值，单位：μmol·mol^{-1}

2.2　分析方法

2.2.1　*鲜重与产量*

莴苣采收后，将地下部与海绵切除并以电子天秤量测鲜重。产量计算为平均鲜重乘上调质期栽培密度（d，26株/m^2）即为产量。

2.2.2　*植物叶片与养液中矿物质浓度*

参考He et al.（1998）植物取叶片在−20℃下48h后取得叶片萃取液。样本使用离子分析仪（IA-300，DKK-TOA corporation，Japan）与使用阴阳离子管柱（Cationic column：PCI-2051，Anion column：PCI-322）分析其阴、阳离子成分。

2.3　统计方法

本试验分析使用SAS软件包，在重复数方面，地上与地下部鲜重皆是10重复，矿物质含量皆为5重复。使用邓肯式多变域分析（Duncan's multivariate analysis）分析（$P<0.05$），绘图使用SigmaPlot 10.0软件。

2.4　结果

如图1所示，6K表示6周都栽培在山崎莴苣养液中，4K表示栽培期间的前四周使用山崎养液，采收前两周改用无钾养液栽培。无钾养液会对植物地上部与地下造成抑制，使用周数越多抑制情况越严重。t.w.处理也会使生长受到抑制。

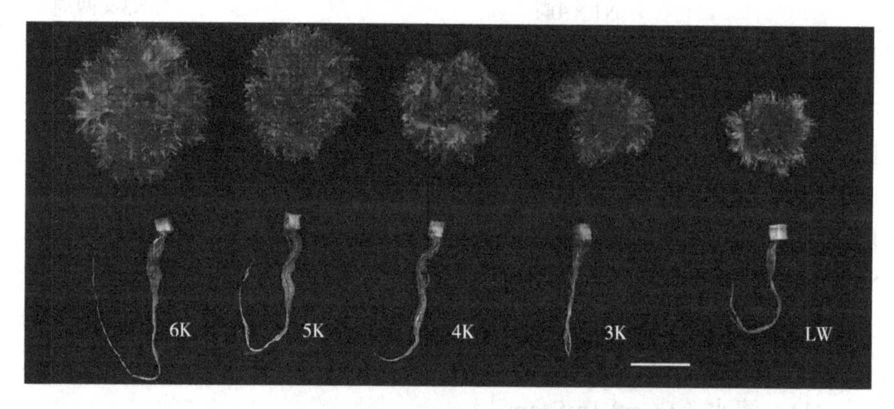

图1　不同周数搭配无钾养液对莴苣生长之影响

Figure 1　Effects of three week treatments（DAS 21 ~ 42）with potassium-free solution on the growth of frill-ice lettuce harvest on DAS 42（Label 6K means 6 weeks of regular nutrient solution with full strength on potassium（100% K），4K means 4 weeks with 100% K and 2 weeks prior to harvest with potassium-free（0% K）solution，t.w. means 4 weeks with 100% K and 2 weeks prior to harvest with tap water.）. Bar = 10cm

如图2所示，更换无钾养液与清水1周后植物生长皆会受到抑制，随着无钾养液与清水的栽培时间增加，莴苣鲜重与产量也显著降低。4K与t.w.处理组皆是处理2周，但前者为无钾养液，后者为清水，与t.w.相比4K处理组显著提高莴苣鲜重。t.w.处理组与无钾养液处理3周3K处理组在鲜重上没有显著差异。

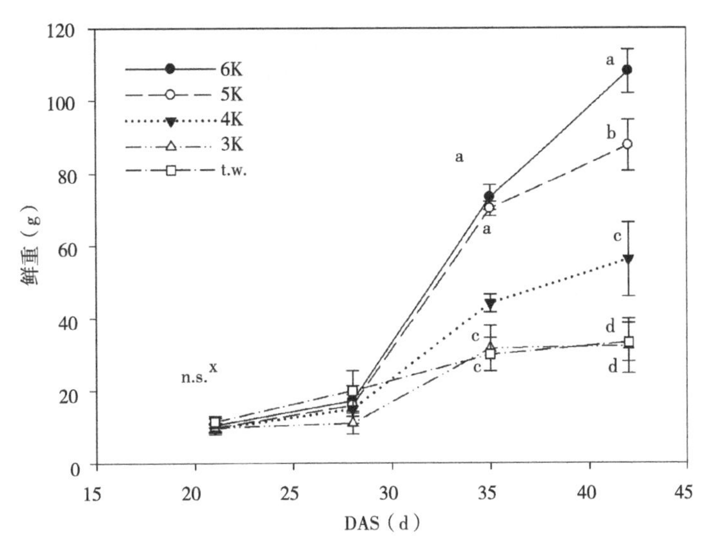

图 2　DAS 21—42 中不同栽培养液（山崎、无钾配方与清水）对莴苣鲜重之影响

Figure 2　Effect of three week treatments（DAS 21 ~ 42）of potassium-free nutrient solution on the fresh weight of frill-ice lettuce. Error bar is the standard error of mean（n=10）.（Label 6K means 6 weeks of regular nutrient solution with full strength on potassium（100% K），4K means 4 weeks with 100% K and 2 weeks prior to harvest with potassium-free（0% K）solution，t.w. means 4 weeks with 100% K and 2 weeks prior to harvest with tap water.）[x]Means with the same letter are not significantly different at 5% level by Duncan's multiple range test. n.s. means non-significant

由表3与图3观察到在DAS 42d时，5K与t.w.处理组的钾离子浓度无显著差别，但前者的单位面积总产量为后者的2.62倍；4K与3K处理组的钾离子浓度没有显著差异，但前者的产量为后者的1.74（1416/839）倍。无钾养液在收获前处理1、2、3周（5K、4K、3K）与对照组相比，其单位面积产量降为对照组的80%、52%与30%，但钾离子浓度也分别降为对照组的54%、10%与7%。随着处理周数增加地上部鲜重与产量显著降低，t.w.处理组会显著抑制地下部鲜重，其他处理根部没有显著差异，根茎比也会显著增加（表3）。

表 3　不同处理对莴苣地上部、地下部鲜重与产量之影响

Table 3　Effect of three week treatments（DAS 21 ~ 42）of potassium-free nutrient solution on the fresh weight of frill ice lettuce harvested on DAS 42

处理	地上部鲜重（$g·plt^{-1}$）	收获产量（$g·m^{-2}$）	产量相对值	钾离子浓度（$mg·100g^{-1}$）	钾离子相对值	根部鲜重（$g·plt^{-1}$）	根/茎
6K	108.67 ± 6.8 a[x]	2 825.42 ± 176.8 a	1.00	365.52 ± 70.6 a	1.00	11.16 ± 3.6 a	0.10 c
5K	87.40 ± 7.5 b	2 272.40 ± 195.0 b	0.80	197.68 ± 31.8 b	0.54	10.76 ± 2.2 a	0.12 c
4K	56.20 ± 10.8 c	1 461.20 ± 280.8 c	0.52	36.96 ± 9.3 c	0.10	9.35 ± 2.3 a	0.17 b
3K	32.30 ± 7.5 d	839.80 ± 195.0 d	0.30	26.90 ± 2.5 c	0.07	9.54 ± 1.6 a	0.30 a
t.w	33.31 ± 5.3 d	866.06 ± 137.8 d	0.31	189.87 ± 55.2 b	0.52	5.28 ± 0.9 b	0.16 b

[x]Means followed by the different letters in each column are significantly different at 5% level by Duncan's Multiple Range Test. n=10（shoot and root fresh weight），n=5（Potassium ion concentration）

如图3所示，6K处理组以正常养液栽培6周的莴苣钾离子呈现缓慢累积的情形，在DAS 35d达到高峰而后增长持平。无钾养液使用1周就会降低钾离子含量，约降低50%，而使用超过2周的无钾养液降低的比例会趋缓。

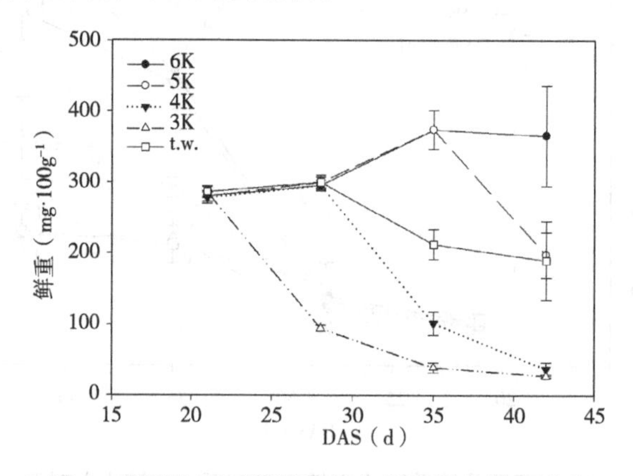

图3　不同处理对莴苣叶片中钾离子含量之影响

Figure 3　Effect of three week treatments（DAS 21 ~ 42）of potassium-free nutrient solution on the potassium ion concentration of frill-ice lettuce. Error bar is the standard error of mean（n=5）.（Label 6K means 6 weeks of regular nutrient solution with full strength on potassium（100% K），4K means 4 weeks with 100% K and 2 weeks prior to harvest with potassium-free（0% K）solution，t.w. means 4 weeks with 100% K and 2 weeks prior to harvest with tap water.）[x]Means each DAS with the same letter are not significantly different at 5% level by Duncan’s Multiple Range Test. n.s. means non-significant

图4所示为两种养液在DAS 21 ~ 42栽培过程中之变化，图4A为使用山崎养液，可以明显看到在DAS28起钾离子与硝酸根离子有明显下降趋势，镁、铵与钠离子则无明显变化。图4B为使用无钾养液，硝酸根离子缓步下降，但最后1周（DAS35至42）却出现前者提高，后者持续下降的状况，可见得此阶段莴苣对后者有较大的吸收量。在此6周的栽培期间内钠离子与镁离子没有明显变化（图4B）。

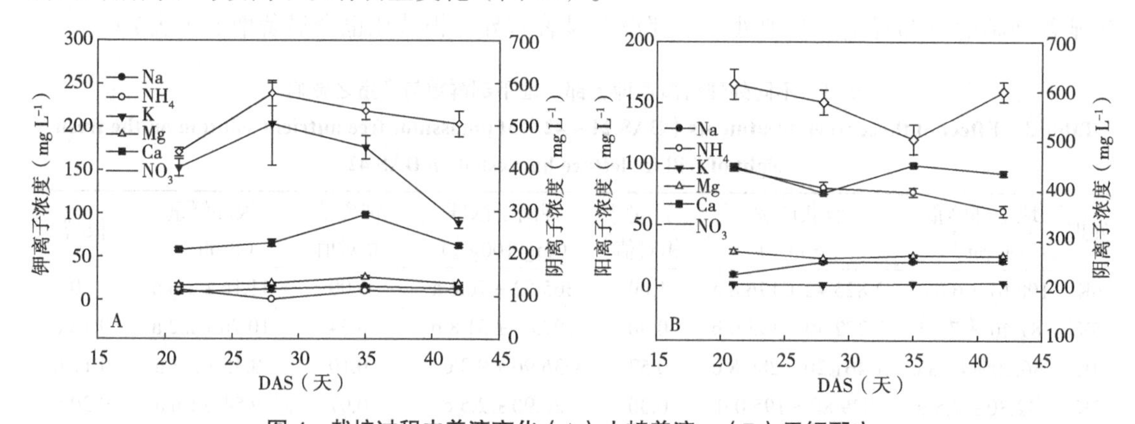

图4　栽培过程中养液变化（A）山崎养液，（B）无钾配方

Figure 4　Change of various ion concentration during DAS 21 to 42.（A）Yamazaki formula（B）Potassium free modified Yamazaki formula（n=5）

由图5A与B数据可知经过无钾养液处理莴苣叶片中铵与钠离子会显著累积，对照组的铵离子随着栽培天数则是逐步降低。由图5B与图3所示，使用无钾养液配方之处理（5K、4K与3K），会显著降低叶片中钾离子浓度，同时也显著提高叶片钠离子浓度，使用无钾配方的周数提高，钾离子逐渐降低而钠离子逐渐提高。t.w处理与对照组处理钠离子没有提高之现象。镁离子则是使用无钾养液会显著提高植物叶片中镁离子含量，对照组的镁离子则是随着栽培天数逐步降低（图5C）。无钾离子栽培莴苣会显著降低叶片中钙离子含量（图5D）。对照组会随栽培天数增加硝酸根离子含量，无钾离子处理组会因使用天数增加而减少叶片中硝酸盐含量（图5E）。

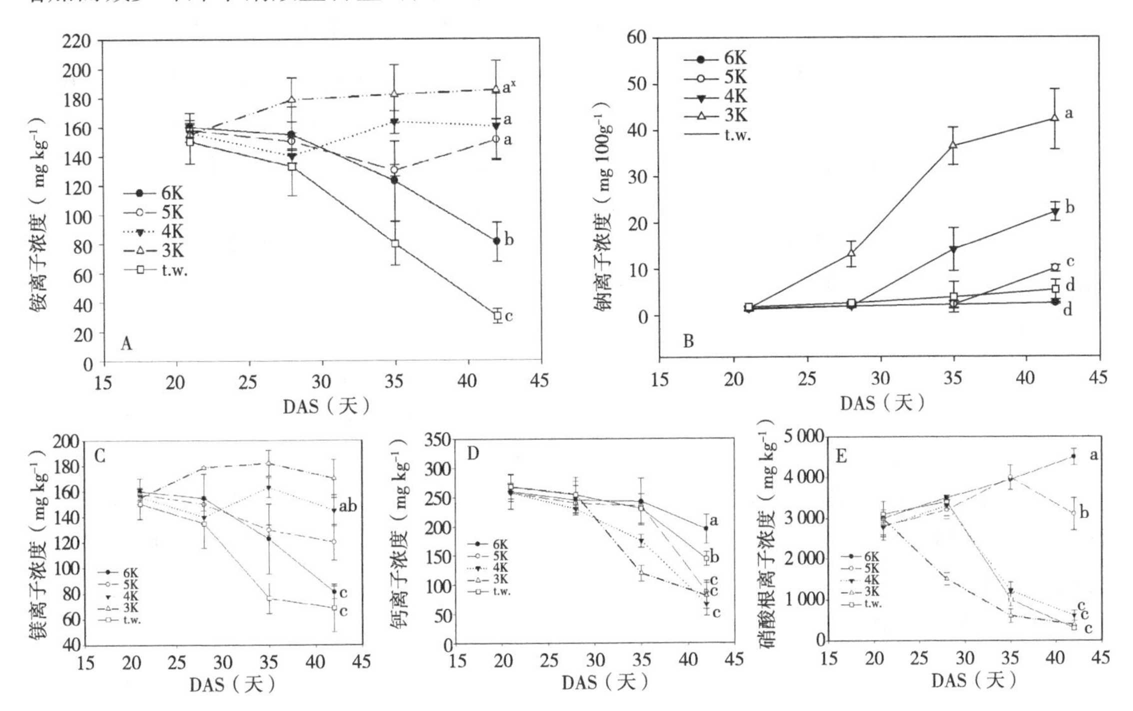

图5　不同养液处理对莴苣叶片不同离子之变化。（A）铵离子（B）钠离子（C）镁离子（D）钙离子（E）硝酸根离子

Figure 5　Effect of different nutrient solution on various ion concentrations in lettuce.（A）ammonium（B）sodium（C）magnesium（D）calcium（E）nitrate（Label 6K means 6 weeks of regular nutrient solution with full strength on potassium（100% K），4K means 4 weeks with 100% K and 2 weeks prior to harvest with potassium-free（0% K）solution，t.w. means 4 weeks with 100% K and 2 weeks prior to harvest with tap water.）[x]Means with the same letter are not significantly different at 5% level by Duncan's Multiple Range Test（n=5）

3　讨论

3.1　三低莴苣之目标值设定

慢性肾脏病（CKD）的病人常会伴有高血钾症，营养治疗低钾饮食是临床一种重要的方法（Saxena，2012）。台湾老年男性（大于65岁）的饮食中钾的平均摄取量为每日2 798mg，其中蔬菜约占30%（Wu et al.，2011），CKD患者每日建议钾的摄取量为1 500～2 500mg·d^{-1}（Cano et al.，2009），低钾莴苣若低于80mg/100g，假设病患1d生食摄取200g之生菜，其他饮食不调整也可符合建议量。蔬菜经过烹煮会大量降低钾离子含量约22～90%（Martinez-Pineda et al.，2016）以符合低钾饮食的标准，但蔬菜中的植化素与纤维素会大量减少。蔬菜经过烹煮10min后莴苣的抗坏血酸、类胡萝卜素会下降40%，总酚损失了30%，抗氧化能力也显著地降低（Vinha et al.，2015）。若能生食低钾莴苣也能保持植物纤维素与营养同时达到低钾饮食需求。

每日摄入硝酸盐有80%来自蔬菜水果（Hord et al.，2009），硝酸盐进入肠胃道容易转变成亚硝酸盐与亚硝胺，存在形成胃癌的高风险，印度与中国的研究也发现水和蔬菜中硝酸盐浓度提高与癌症风险较高有关（Taneja et al.，2017；Zhang et al.，2018），欧盟也规范了莴苣硝酸盐浓度的上限大致为2 500～4 500mg·kg^{-1}（依照季节与栽培环境不同）（European Food Safety，2008）。植物工厂生产的蔬菜应该以较严格的标准来规范，所以本研究之莴苣硝酸盐上限设定为2 500mg·kg^{-1}。前人研究低钾莴苣时，常会使用不同比例氯化钠或者硝酸钠等含钠离子的营养盐来抑制植物钾离子之含量，都会使每百克鲜重的钠离子显著增加达98～103mg（Ogawa et al.，2012；Yoshida et al.，2014）。美国莴苣平均每百克鲜重的钠含量大约在5～30mg（Kim et al.，2016），为了降低钾离子含量却增加了钠离子含量有可能反而增加病患的风险，所以本研究将钠离子的标准设置在每百克鲜重的浓度应小于30mg。本研究针对香波绿莴苣的目标值分别设定为每百克叶片鲜重中的钾、钠与硝酸盐浓度上限分别为80mg、30mg与250mg，硝酸盐浓度的单位一般使用$\times 10^{-6}$或mg·kg^{-1}，前述的每百克鲜重250mg相当于2 500mg·kg^{-1}。本试验4K与3K处理组皆符合此目标值（图3与图5），但4K处理组的鲜重比3K多了74%（表3），所以4K处理组是较佳的栽培模式。

3.2 缺钾对植物生长之影响

栽培42d之莴苣，鲜重与根系发展明显随着无钾养液的处理周数增加而有下降的趋势。随栽培周数的增加，鲜重显著降低，而地下部鲜重则是没有显著差异，硝酸盐含量则是无钾处理周数增加会显著降低植物中硝酸盐之含量。钾元素为植物生长中的必要大量元素之一，钾影响着植物的酵素活动、气孔开闭、光合作用、蛋白质合成与作物质量等（Pettigrew，2008），钾以离子态存在植物体内，钾的移动性大，若缺乏时老叶先出现黄化症状进而影响植物生长。本研究无钾处理一周显著降低莴苣鲜重，表示植物受到低钾逆境与前人研究相符。Hermans et al.（2006）研究发现，缺乏钾离子的拟南芥地下部会受到抑制，其推测是植物缺钾时将醣类运移至地上部进行光合作用之需求而抑制地下部发展，此发现与本研究相似（表3）。

3.3 铵离子作为钾离子取代剂对植物之影响

低钾莴苣的研究中常常会研究以某个阳离子来取代钾离子作为渗透调节的功能，希望可以达到植物降钾与维持植物生长，前人有使用钠离子（$NaNO_3$）作为取代（Ogawa et al.，2012；Ogawa et al.，2007），但这会伴随钠离子提高到正常莴苣之1.3～1.6倍。本研究利用硝酸铵作为取代钾离子的取代剂，在柑橘上使用铵离子也可增加脯胺酸含量来抵抗盐分逆境（Fernández-Crespo et al.，2012）。脯胺酸是有效之渗透调节剂，具有良好的抗氧化能力（Kishor et al.，2005），这可能是铵作为渗透调节剂的功能。前人研究发现，添加5mmol/L NH_4可显著抑制大麦与水稻幼苗钾离子之吸收与本研究结果相似（Ma et al.，2016；Santa-MaríA et al.，2000）。本研究也发现使用无钾养液铵离子会显著累积于植物叶片中（图5A），可能因此而维持渗透调节之功能，维持植物生长。

莴苣地上部钠离子会随着无钾处理时间增加而显著增加。许多植物中钾离子扮演保卫细胞控制气孔开阖的重要阳离子，若植物于缺钾环境气孔可能会关闭（Marschner，1995），而自由基增加，进而产生丙二醛（MDA）使植物生长减缓，低浓度的钠离子浓度可代替钾离子进入保卫细胞使气孔开放（Lv et al.，2012），在K/Na为0.03/2.97mmol/L环境的甜菜（*Beta vulgaris* L.）中也发现MDA显著低于缺钾处理组，且透过吸收钠离子可以增加渗透调节势降低游离胺基酸的含量使得植物在缺钾的环境中维持较好的生长（Pi et al.，2014）。本研究也发现在无钾的养液中莴苣会显著累积钠离子含量，可能是莴苣主动吸收钠离子维持生长的机制，美国莴苣平均每百克鲜重的钠含量在5～30mg（Kim et al.，2016），本研究结果显示并未超出此范围。Zhang et al.（2017）研究降钾养液（利用Na NO_3取代KNO_3，25%钾离子浓度）对莴苣光合作用之影响，结果显示钠离子只能部份取代钾离子之能力，光合作用效率受到抑制。

本研究中于采收前（Santa-MaríA et al.，2000）两周使用清水处理，收获时钾离子含量仍然偏高，此时清水中营养成分已不足以支持植物继续生长。要降低叶片中钾离子含量，除了利用减少养液中钾离子含量来降低植物对钾的吸收之外，可能还必须依靠植物本身成长来稀释/代谢体内已吸收的钾离子含量。本研究之4K与t.w.处理组都是在收获前2周养液中没有钾离子（前者为无钾配方，后者为清水），但是4K处理组在DAS 28～42中鲜重仍持续增加，t.w.处理则显著低于4K处理组（图2）。

本研究开发出低钾、低钠与低硝酸盐之莴苣栽培方法，可提供肾脏病人生食。辅以鲜重的考虑以4K处理组为最佳，但鲜重被牺牲了近48%。如果是熟食蔬菜由于烹调过程会降低钾离子含量为22%～90%（Martinez-Pineda et al.，2016；胡等，2009），5K处理组会更适当，因为产量高于4K处理组。

4 结论

在植物工厂中于采收前两周以清水进行栽培，可降低蔬菜的硝酸盐含量，但只降低钾离子含量约48%，且鲜重降低高达69%，此种栽培方式在商业上是无法接受的。本研究初步成果得出符合低钾、低钠与低硝酸盐要求的三低莴苣的2种栽培流程，于栽培前期使用1.5倍强度的山崎养液莴苣配方，于采收前两周改用无钾养液的栽培流程，可用于生产生食专用；采收前一周改用无钾养液，可用于生产熟食专用。由于本研究的各项处理组以一周为间隔，若改以1d为间隔的更细致的区隔应可得出符合三低要求，但鲜重较少被牺牲的更理想的栽培流程，此有待后续持续研究。

参考文献

胡怀玉、金惠民、骆菲莉. 2009. 加热前处理对蔬菜钾流失率之影响［J］. 台湾膳食营养学杂志（1）：21-28。

Agarwal，R.，A. R. Nissenson，D. Batlle，D. W. Coyne，et al. 2003. Prevalence，treatment，and control of hypertension in chronic hemodialysis patients in the United States［J］. The American Journal of Medicine 115：291-297.

Archer，M. C. 1989. Mechanisms of action of N-nitroso compounds.［J］Cancer surveys 8：241-250.

Beto，J. A.and V. K. Bansal. 2004. Medical nutrition therapy in chronic kidney failure：Integrating clinical practice guidelines. Journal of the American［J］Dietetic Association 104：404-409.

Cano，N. J. M.，M. Aparicio，G. Brunori，J. J. Carrero，et al. 2009. ESPEN guidelines on parenteral nutrition：adult renal failure［J］. Clinical Nutrition 28：401-414.

European Food Safety，A. 2008. Nitrate in vegetables - scientific opinion of the panel on contaminants in the food chain［J］. EFSA Journal 6：n/a-n/a.

Fernández-Crespo，E.，G. Camañes，P. García-Agustín. 2012. Ammonium enhances resistance to salinity stress in citrus plants［J］. Journal of plant physiology 169：1 183-1 191.

Gutiérrez，O. M.，P. Muntner，D. V. Rizk，et al. 2014. Dietary Patterns and Risk of Death and Progression to ESRD in Individuals With CKD：A Cohort Study［J］. American Journal of Kidney Diseases 64：204-213.

He，Y.，S. Terabayashi，T. Namiki. 1998. The effects of leaf position and time of sampling on nutrient concentration in the petiole sap from tomato plants cultured hydroponically［J］. Journal of the Japanese Society for Horticultural Science 67：331-336.

Hermans，C.，J. P. Hammond，P. J. White，et al. 2006. How do plants respond to nutrient shortage by biomass allocation［J］.Trends in Plant Science 11：610-617.

Hord，N. G.，Y. Tang，N. S. Bryan. 2009. Food sources of nitrates and nitrites：the physiologic context for potential health benefits［J］. The American journal of clinical nutrition 90：1-10.

Kim，M. J.，Y. Moon，J. C. Tou，B. Mou，et al. 2016. Nutritional value，bioactive compounds and health benefits of lettuce（*Lactuca sativa* L.）［J］. J. Food Compos. Anal. 49：19-34.

Kishor，P. B. K.，S. Sangam，R. N. Amrutha，P. S. Laxmi，et al. 2005. Regulation of proline biosynthesis，degradation，uptake and transport in higher plants：Its implications in plant growth and abiotic stress tolerance［J］. Current Science 88：424-438.

Lee，J. S.and Y. H. Kim. 2014. Growth and anthocyanins of lettuce grown under red or blue light-emitting diodes with distinct peak wavelength. Korean J［J］. Hortic. Sci. Technol. 32：330-339.

Lv，S.，L. Nie，P. Fan，X. Wang，et al. 2012. Sodium plays a more important role than potassium and chloride in growth of Salicornia europaea［J］. Acta Physiologiae Plantarum 34：503-513.

Ma，X. L.，C. H. Zhu，N. Yang，et al. 2016. gamma-Aminobutyric acid addition alleviates ammonium toxicity by limiting ammonium accumulation in rice（*Oryza sativa*）seedlings［J］. Physiologia plantarum 158：389-401.

Marschner，H. 1995. 2 - Ion Uptake Mechanisms of Individual Cells and Roots：Short-Distance Transport，p. 6-78，Mineral Nutrition of Higher Plants（Second Edition）［J］. Academic Press，London.

Martinez-Pineda，M.，C. Yague-Ruiz，A. Caverni-Munoz，et al. 2016. Reduction of potassium content of green bean pods and chard by culinary processing［J］. Tools for chronic kidney disease. Nefrologia 36：427-432.

Ogawa，A.，T. Eguchi，and K. Toyofuku. 2012. Cultivation methods for leafy vegetables and tomatoes with low potassium content for dialysis patients［J］. Environmental Control in Biology 50：407-414.

Ogawa，A.，S. Taguchi，C. Kawashima. 2007. A cultivation method of spinach with a low potassium content for patients on dialysis［J］. Japanese Journal of Crop Science 76：232-237.

Pi，Z.，P. Stevanato，L. H. Yv，G. Geng，X. L. et al. 2014. Effects of potassium deficiency and replacement of potassium by sodium on sugar beet plants. Russ. J［J］. Plant Physiol. 61：224-230.

Santa-MaríA，G. E.，C. H. Danna，C. Czibener. 2000. High-affinity potassium transport in barley roots. ammonium-sensitive and insensitive pathways [J] . Plant Physiology 123：297-306.

Saran，R.，B. Robinson，K. C. Abbott，L. Y. C. Agodoa，et al. 2017. US rensl data system 2016 annual data report chapter 13：international comparisons [J] . American Journal of Kidney Diseases 69：S533-S566.

Saxena，A. 2012. Nutritional problems in adult patients with chronic kidney disease [J] . Clinical Queries：Nephrology 1：222-235.

Schachtman，D.W. Liu. 1999. Molecular pieces to the puzzle of the interaction between potassium and sodium uptake in plants [J] . Trends in Plant Science 4：281-287.

Shin Jiuan，W.，P. WenHarn，Y. NaiHua，et al. 2011. Trends in nutrient and dietary intake among adults and the elderly：from NAHSIT 1993-1996 to 2005-2008 [J] . Asia Pacific journal of clinical nutrition 20：251-265.

Taneja，P.，P. Labhasetwar，P. Nagarnaik，et al. 2017. The risk of cancer as a result of elevated levels of nitrate in drinking water and vegetables in Central India [J] . Journal of Water and Health.

Tritt，L. 2004. Nutritional assessment and support of kidney transplant recipients. Journal of infusion nursing [J] ：the official publication of the Infusion Nurses Society 27：45-51.

Vinha，A. F.，R. C. Alves，S. V. P. Barreira，A. S. G. Costa，et al. 2015. Impact of boiling on phytochemicals and antioxidant activity of green vegetables consumed in the Mediterranean diet [J] . Food Funct. 6：1 157-1 163.

Wu，S. J.，W. H. Pan，N. H. Yeh，et al. 2011. Trends in nutrient and dietary intake among adults and the elderly：from NAHSIT 1993-1996 to 2005-2008 [J] . Asia Pacific journal of clinical nutrition 20：251-265.

Yoshida，T.，K. Sakuma，H. Kumagai. 2014. Nutritional and taste characteristics of low-potassium lettuce developed for patients with chronic kidney diseases [J] . Hong Kong Journal of Nephrology 16：42-45.

Zhang，G.，M. Johkan，M. Hohjo，S. Tsukagoshi，et al. 2017. Plant Growth and Photosynthesis Response to Low Potassium Conditions in Three Lettuce（*Lactuca sativa*）Types [J] . The Horticulture Journal 86：229-237.

Zhang，P.，J. Lee，G. Kang，Y. Li，et al. 2018. Disparity of nitrate and nitrite in vivo in cancer villages as compared to other areas in Huai River Basin [J] . China Science of The Total Environment 612：966-974.

纳米碳与枯草菌对黄瓜幼苗生长及土壤环境影响

周艳超，吴艳红，周海霞，韩泽宇，刘吉青，兰志谦，张雪艳*

［宁夏大学农学院，宁夏设施园艺工程技术研究中心，
宁夏设施园艺（宁夏大学）技术创新中心，银川　750021］

摘要：为筛选适宜黄瓜苗期生长的枯草芽孢杆菌与纳米碳溶胶浓度，实验以黄瓜幼苗为材料，设置纳米碳溶胶、枯草芽孢杆菌双因素实验，纳米碳溶胶施用分为0、350、650倍稀释倍数，枯草芽孢杆菌浓度为0.8×10^7cfu/ml、16×10^7cfu/ml，系统研究不同施用浓度配比对黄瓜苗期长势、根系生长状况以及土壤肥力的影响。研究结果表明，适宜浓度的枯草芽孢杆菌和纳米碳溶胶能够显著增强黄瓜幼苗的地上、地下部分生长，改善土壤状况。未施用纳米碳溶胶的情况下，16×10^7cfu/ml枯草增加根系干重、根长和表面积，显著降低土壤EC含量45.45%，提升土壤P含量12.62%，显著增强土壤磷酸酶和过氧化氢酶活性58.29%和22.92%，增加细菌、放线菌数量；未施用枯草的情况下，650倍纳米碳溶胶显著促进植株叶片生长，提升土壤有机质含量和蔗糖酶活性，增加土壤放线菌数量57.36%。8×10^7cfu/ml枯草×650倍纳米碳溶胶提高株高相对生长率6.51%，显著促进根系生长，根系干鲜重、表面积、直径和体积均为处理中最高，相对未添加菌剂与纳米碳的处理分别提高56.63%、57.14%、66.15%、56.55%、21.94%，降低土壤EC值，土壤中细菌数量是未添加菌剂与纳米碳的1.85倍，综合表现最优。

关键词：纳米碳溶胶、枯草芽孢杆菌、黄瓜、生物有机肥

The effect of Nano-carbon Sol and Bacillus Subtilis on the Cucumber Growth and Soil Environment

Zhou Yanchao，WU Yanhong，Zhou Haixia，Han Zeyu，Liu Jiqing，Lan Zhiqian，Zhang Xueyan *

［*School of Agriculture Ningxia University*，*Facility Horticulture Engineering Technique Center of Ningxia*，*Research Center for Technological Innovation of Facility Horticulture Ningxia*（*Ningxia University*），*Yinchuan* 750021］

Abstract：Aimed at screening suitable bacillus subtilis and nano-carbon sol concentration of Cucumber growth at cucumber seedling stage，this test use cucumber seedlings as materials and set two factor experiment. The application of nano-carbon sol is divided into 0，350，650 dilution multiple. The Bacillus subtilis is 0、8×10^7cfu /ml、

第一作者：周艳超（1996—），女，河北衡水人，本科生，主要从事设施蔬菜高产栽培生理生态研究，Email:905451058@qq.com

责任作者：张雪艳（1981—），女，河北保定人，博士，副教授，现主要从事设施蔬菜栽培与生理研究工作。E-mail：zhangxueyan123@sina.com

基金项目：宁夏回族自治区“十三五”重大研发项目（2016BZ0902），“十二五”国家科技支撑计划（2014BAD05B02），吴忠国家园区专项（2016BN05），宁夏回族自治区科技创新领军人才（KJT2017001）

16×10^7cfu /ml. Different concentration ratio effect on growth, soil fertility and environment of cucumber at seedling stage are researched. The results show that the suitable concentration of Bacillus subtilis and nano-carbon sol can significantly enhance the aerial part and underground part growth of cucumber seedlings and improve the soil status. Without the application of nano-carbon sol, 16×10^7cfu /ml bacillus subtilis can increase root dry weight, root length, surface area, the quantity of bacteria actinomycetes The EC of soil is reduced by 45.45%.The p content in soil is increased by 12.62%. soil phosphatase and catalase activities significantly enhance, increased by 58.29% and 22.92%. Without the application of bacillus subtilis, 650 dilution multiple of nano-carbon sol significantly promote plant leaf growth, soil organic matter content and sucrase activity, and increased the quantity of soil actinomycetes by 57.36%. The treatment of 8×10^7cfu /ml bacillus subtilis and 650 dilution multiple of nano-carbon sol is optimal. Root fresh weight, dry weight, surface area, diameter and volume are the highest in these treatments, beyond the treatment without nano-carbon sol and bacillus subtilis 56.63%、57.14%、66.15%、56.55%、21.94%.The plant height of relative growth rate is increased by 6.51%. The EC of soil is reduced. The number of bacteria in the soil is 1.85 times that of the treatment without nano-carbon sol and bacillus subtilis.

Key words: Nano-carbon sol; Bacillus subtilis; Cucumber; Bio-organic fertilizer

我国是化肥使用大国，长期、大量的施用化肥导致土壤板结，肥力下降，水体、大气污染，危害人类健康[1]。在化肥用量不断提高而成本递减，土壤质量日渐退化的状况下，减少化肥的使用量，实现设施农业可持续性的生产，已经成为一个重要的课题[2]。

生物有机肥作为无害化腐熟混合处理而成的一类具有微生物功能和有机肥效应的肥料[3]，既有利于增产增收、改善农产品品质，又可培肥土壤、减少化肥用量[4]。研究表明，生物有机肥能够调节土壤中微生物区系组成，增强土壤酶的活性，使土壤向着健康方向发展[5]。张雪艳等[6]的研究结果显示添加一定量生物质有机肥对于植株地上部形态建成以及地下部根系生长均有显著促进效果。其中，纳米碳作为一种新型有机肥料，在提高养分利用率，纳米碳材料能不同程度地促进烟草、蔬菜、水稻等作物植株生长发育，增加作物干物质积累量，提高产量[7]。刘键等[8]将纳米碳结合肥料使用，研究表明，纳米增效肥料有明显增产、促早、节肥和改善品质的作用。碳材料也能促进番茄和水稻种子对水分的吸收，促进其发芽和生长[9-10]。王艳等[11]研究表明，纳米增效肥料能促进大豆种子提早出苗，根系发达，能并且能够提高大豆抗病能力，减少大豆植株染病。

促植物生长根际细菌株在根际土壤微生态环境中能够与植物相互作用，具有改善植物根际土壤生态环境的作用[12]。枯草芽孢杆菌是目前研究较多的植物促生微生物之一，作为土壤和植物微生态区系的优势生物种群，具有较高的抗逆能力和抗菌防病作用[13]，菌株挥发物对植物有较好的促生效果[14]。有研究显示枯草芽孢杆菌不仅显著促进作物生长而且有效降低土壤线虫[15]；王翠梅等[14]的研究结果表明枯草芽孢杆菌能够防治多种蔬菜的土传病害，且对植株长有一定的促进作用，尤其是在蔬菜长期连作的地块内施用，能改善蔬菜品质和提高单位面积产量。关于其促生及生防效果研究较多，但将纳米碳溶胶与枯草有效结合，综合探究其对植株促生和土壤环境调控的研究鲜有报道。

本研究以连作黄瓜土壤为对象，设计不同浓度枯草芽孢杆菌和纳米碳溶胶，系统研究不同施用浓度配比对黄瓜苗期长势、根系生长状况以及土壤肥力的影响，确定适宜黄瓜苗期生长的枯草芽孢杆菌浓度和纳米碳溶胶的浓度及其组合，为设施黄瓜高效栽培和土壤健康保育提供理论依据。

1 材料与方法

1.1 试验材料

试验在宁夏贺兰产业园区4号温室内进行，黄瓜品种为“德尔99”，采用盆栽，花盆27cm × 20cm，每个处理10盆，每个处理重复3次，所有处理统一水肥管理。枯草芽孢杆菌选用中国农业科学院蔬菜花卉研究所的枯草芽孢杆菌剂（2×10^8活菌数/g），纳米碳溶胶购自北京风禾尽起科技有限公司，具体理化特性如下：粒径小于6nm颗粒数≥95.5%；烘干后质量百分含量：C（40.5%～41.5%）、O（51.2%～52.6%）、Na（2.1%～2.7%）、Mg（0.9%～1.3%）、Si（0.2%～0.4%）、S（0.8%～1.2%）、Cl（1.2%～1.6%）、Ca（0.9%～1.2%）；pH值（1.2～3.0）、灰分（0.05%～0.07%）。

黄瓜幼苗采用灌根法进行纳米碳溶胶、枯草芽孢杆菌双因素实验，以不添加枯草和纳米碳溶胶处理为对照（CK），以单独添加枯草芽孢杆菌、纳米碳溶胶以及枯草和纳米碳溶胶共同添加为处理。纳米碳溶胶施用分为0、350、650倍稀释倍数共3个水平，每个植株灌药量为20ml。枯草芽孢杆菌施用浓度分为0、8×10^7、16×10^7cfu/ml共3个水平，每个植株灌药量为50ml。2016年10月18日黄瓜定植，定值一周缓苗后进行灌根处理，每10d灌溉1次，连续灌溉5次（直至秧苗开始结瓜）。种植前原始土壤基本理化性质如下：速氮（24.08mg/kg）、速磷（38.52mg/kg）、速钾（99.95mg/kg）、有机质（12.93g/kg）。

1.2 植株长势测定

定植1周后，每个品种取代表植株5株，固定植株，每7d进行1次测定，测定黄瓜的株高、茎粗、叶片数及叶绿素含量，连续测定5次；株高为黄瓜生长点到根基部的垂直距离，用卷尺测量；茎粗为子叶下1cm的粗度，用游标卡尺测定；叶片数为直径大于2cm的叶片数，用目测计数法测定。

1.3 土壤指标测定

定值50d后，取根际土壤研磨过2mm筛，4℃冰箱保存，用于微生物数量的测定，剩余土壤风干后过1mm筛，用于土壤化学指标和酶活性的测定，土壤EC值采用1：5土壤悬液电导法（电导仪法）测定；土壤pH采用电位计法测定；土壤速效磷采用0.5mol/L $NaHCO_3$浸提—钼锑抗吸光光度法测定；土壤速效钾采用1mol/L NH_4AC浸提—火焰光度法测量，土壤速效氮采用1mol/L NaOH碱解—扩散法测定[16]。采用牛肉膏蛋白胨选择性培养基培养细菌，马丁孟加拉红—链霉素选择性培养基培养真菌，改良高氏一号培养基培养放线菌，均用稀释平板计数法计数[17]。脲酶活性采用靛酚蓝比色法测定；蔗糖酶的活性用3，5-二硝基水杨酸比色法测定；磷酸酶的活性采用磷酸苯二钠法测定；过氧化氢酶活性采用高锰酸钾滴定法[18]。用EPSON EXPRESSION 4990型扫描仪对根样进行扫描，用Win RHIZO根系分析软件对扫描的根系图片进行分析，得到根样的根长、根表面积、根体积和根的平均直径[19]。

1.4　数据统计与分析

测定时每个处理测量3次，结果取其平均值，数据用SPSS 17.0软件采用LSD方法在$P < 0.05$水平进行单因素显著性分析。

2　结果与分析

2.1　不同处理对黄瓜苗期植株生长的影响

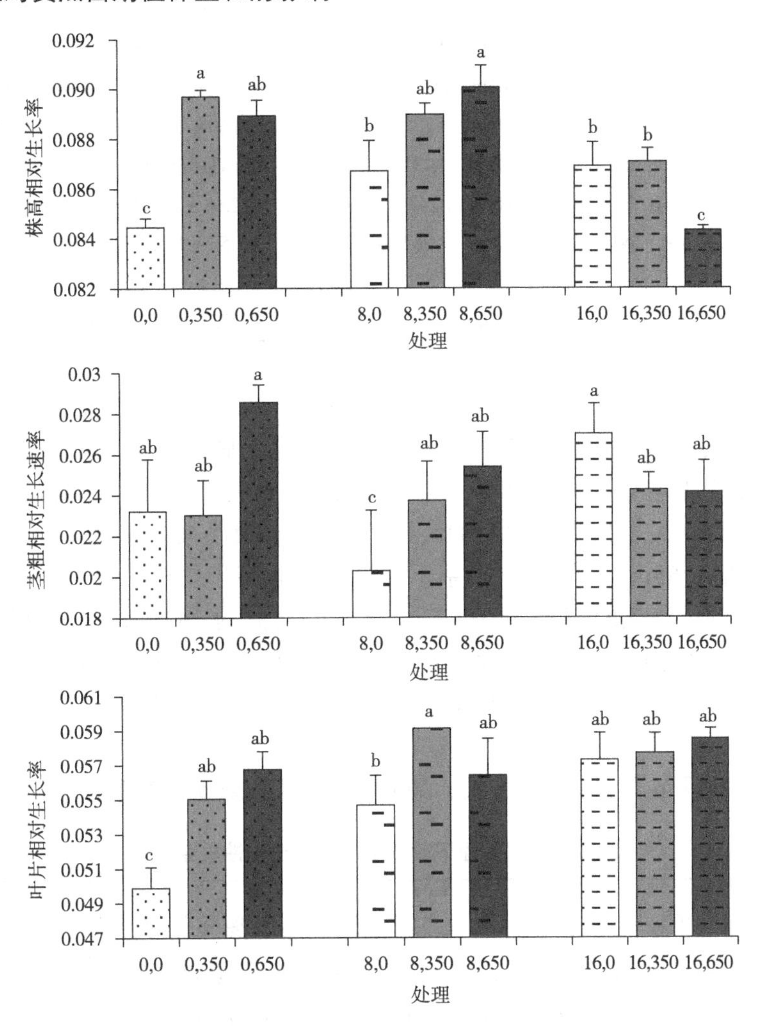

图1　不同处理下黄瓜幼苗植株长势的变化

由图1可知，在未施用枯草芽孢杆菌的处理下，纳米碳溶胶350倍、650倍稀释液的株高相对生长率相比于未施用纳米碳的处理均显著高出6.15%、5.21%；350倍、650倍稀释液的叶片相对生长率均显著高于空白对照且随稀释倍数增大呈现递增趋势，分别高出10.22%、13.83%。在施用枯草芽孢杆菌8×10^7cfu/ml的处理下，350倍、650倍稀释液的茎粗相对生长率分别显著高于空白对照16.75%、25.12%，且350倍稀释液叶片相对生长率显著高于空白、650倍稀释液株高相对生长率显著高出空白3.81%。施用枯草芽孢杆菌

16×10^7cfu/ml时，650倍稀释液株高相对生长率显著低于空白；350倍、650倍稀释液的茎粗与叶片相对生长率与空白无显著差异。

表 1　不同处理下黄瓜苗期根系生长特性变化

枯草菌浓度	溶胶稀释液	鲜重（g）	干重（g）	根长（cm）	表面积（cm^2）	直径（cm）	体积（cm^3）
0	0	8.3 ± 0.28e	0.42 ± 0.02d	794.13 ± 16.68c	254.36 ± 13.68c	1.15 ± 0.68a	9.57 ± 0.01abc
	350	9.25 ± 0.45de	0.52 ± 0.01bc	977.99 ± 11.09bc	296.95 ± 24.29bc	1.17 ± 0.44a	10.03 ± 0.15ab
	650	9.93 ± 0.35cde	0.49 ± 0.01bc	1 076.43 ± 35.91abc	334.80 ± 19.64abc	1.12 ± 0.50a	7.33 ± 0.13c
8×10^7	0	11.14 ± 0.30bc	0.51 ± 0.01bc	1 281.12 ± 12.10ab	384.60 ± 11.57ab	0.84 ± 0.02b	9.02 ± 0.78bc
	350	9.87 ± 0.118cde	0.45 ± 0.01cd	1 431.68 ± 185.32a	323.41 ± 60.40bc	0.78 ± 0.86b	8.55 ± 0.30bc
	650	13.00 ± 0.57a	0.66 ± 0.05a	1 243.22 ± 147.06ab	422.63 ± 24.77a	1.11 ± 0.12a	11.67 ± 1.22a
16×10^7	0	10.87 ± 0.82bcd	0.63 ± 0.01a	1 262.46 ± 11.64ab	361.53 ± 10.38ab	0.85 ± 0.01b	9.95 ± 0.32ab
	350	8.89 ± 0.15e	0.54 ± 0.01b	1 278.98 ± 126.91ab	337.99 ± 34.14abc	0.85 ± 0.08b	7.29 ± 1.29c
	650	11.92 ± 0.81ab	0.55 ± 0.01b	1 202.03 ± 179.81ab	356.28 ± 15.52ab	0.97 ± 0.09ab	8.58 ± 0.47bc

由表 1 可知，在未施用枯草芽孢杆菌的处理下，纳米碳溶胶350、650倍稀释液的根系鲜重、根长、表面积、直径和体积与空白对照间均无显著性差异；350倍、650倍稀释液处理下的根系干重均显著高于对照，分别高出23.81%、16.67%。在施用枯草芽孢杆菌8×10^7cfu/ml的处理下，350、650倍稀释液根系鲜重、根系根长和表面积与空白处理无显著性差异；350倍、650倍稀释液的根系干重分别显著高于空白，分别是空白的1.47倍、1.40倍，且350倍稀释液根系直径显著高于对照，是对照的1.42倍，650倍稀释液根体积最大，显著高出空白29.40%。施用枯草芽孢杆菌16×10^7cfu/ml时，350倍、650倍稀释液的根长、表面积和直径与空白无显著差异，干重显著低于对照，350倍稀释液根鲜重和根体积显著低于空白。

2.2　不同处理对黄瓜苗期土壤理化性质、酶活性及微生物数量的影响

表 2　不同处理下黄瓜苗期土壤理化性质变化

菌剂浓度	溶胶稀释液	EC（mS/cm）	pH值	N（mg/kg）	P（mg/kg）	K（mg/kg）	有机质（g/kg）
0	0	0.55 ± 0.02b	7.81 ± 0.02cd	8.68 ± 0.65c	40.98 ± 0.35b	64.93 ± 2.69cd	65.12 ± 2.29ab
	350	0.85 ± 0.01a	7.66 ± 0.00d	13.63 ± 0.57a	31 ± 1.10d	83.78 ± 0.00a	56.69 ± 2.06bc
	650	0.53 ± 0.01bc	7.70 ± 0.18d	10.64 ± 0.43b	43.27 ± 1.56b	75.70 ± 0.00b	69.86 ± 1.65a
8×10^7	0	0.37 ± 0.03f	8.24 ± 0.03a	6.72 ± 0.48d	34.58 ± 0.36c	59.54 ± 0.00e	61.77 ± 0.00abc
	350	0.44 ± 0.02de	7.99 ± 0.09bc	7.84 ± 0.32cd	40.71 ± 1.06b	67.62 ± 0.00c	57.22 ± 2.95bc
	650	0.40 ± 0.01ef	8.12 ± 0.05ab	8.21 ± .95cd	35.95 ± 0.62c	67.62 ± 0.00c	55.07 ± 3.87c
16×10^7	0	0.30 ± 0.01g	7.98 ± 0.02bc	8.68 ± 0.00c	46.15 ± 1.08a	59.54 ± 0.00e	56.29 ± 3.71c
	350	0.47 ± 0.02cd	8.00 ± 0.02bc	6.53 ± 0.52d	41.79 ± 0.61b	62.23 ± 2.69de	57.71 ± 2.84bc
	650	0.29 ± 0.02g	8.03 ± 0.02abc	7.09 ± 0.19cd	34.02 ± 0.39c	62.23 ± 2.69de	45.11 ± 2.26d

由表2可知，在未施用枯草芽孢杆菌的处理下，纳米碳溶胶350倍、650倍稀释液处理下的土壤pH值和有机质相比无显著差异；但显著增加了土壤N、K含量，N含量相比提高57.03%、22.58%，K相比提高1.29倍、1.17倍；350倍稀释液处理下的土壤P含量显著低于对照。在施用枯草芽孢杆菌8×10^7cfu/ml的处理下，纳米碳350倍稀释液与空白处理相比可显著降低土壤pH值3.03%，增加土壤P含量17.73%，增加土壤K含量14.00%；650倍稀释液增加土壤K含量14%。施用枯草芽孢杆菌16×10^7cfu/ml时，纳米碳350倍稀释液N和P含量均显著低于空白处理；650倍稀释液土壤P和有机质含量显著低于对照。

表3　不同处理下土壤酶含量的变化

枯草菌浓度	溶胶稀释倍数	过氧化氢酶（mg/kg）	脲酶（mg/kg）	磷酸酶（mg/kg）	蔗糖酶（mg/kg）
0	0	0.48 ± 0.03bc	39.44 ± 4.13b	1.75 ± 0.02d	7.05 ± 1.44cd
	350	0.42 ± 0.00c	50.82 ± 1.30a	2.00 ± 0.03c	13.32 ± 0.2ab
	650	0.55 ± 0.03ab	40.57 ± 0.37b	1.74 ± 0.02d	15.01 ± 1.10a
8×10^7	0	0.49 ± 0.02bc	42.53 ± 1.72b	1.71 ± 0.02d	5.23 ± 0.00d
	350	0.49 ± 0.02bc	38.88 ± 0.06b	2.21 ± 0.10b	11.82 ± 2.53abc
	650	0.49 ± 0.02bc	41.41 ± 0.62b	1.74 ± 0.03d	7.08 ± 0.13cd
16×10^7	0	0.59 ± 0.02a	42.39 ± 0.21b	2.77 ± 0.20a	5.91 ± 0.61cd
	350	0.49 ± 0.02bc	39.84 ± 0.33b	2.77 ± 0.05a	9.64 ± 3.97abcd
	650	0.50 ± 0.05bc	40.24 ± 0.62b	2.73 ± 0.08a	8.16 ± 0.39bcd

由表3可知，在未施用枯草芽孢杆菌的处理下，350倍、650倍稀释液处理下的土壤蔗糖酶活性均显著高于对照，分别是空白的1.89倍、2.13倍。且350倍稀释液处理下的土壤脲酶、磷酸酶活性分别显著高出对照28.85%、14.29%。在施用枯草芽孢杆菌8×10^7cfu/ml的处理下，350倍稀释液土壤磷酸酶活性显著高出空白对照29.34%，蔗糖酶活性显著高于空白且是空白处理的2.26倍；650倍稀释液与空白处理无显著性差异。施用枯草芽孢杆菌16×10^7cfu/ml时，350倍、650倍稀释液土壤脲酶、磷酸酶和蔗糖酶活性与空白处理间无显著差异；350倍、650倍稀释液处理下的土壤过氧化氢酶活性显著低于空白处理。

表4　不同处理对黄瓜苗期土壤细菌、真菌、放线菌数量的影响

菌剂浓度	溶胶稀释倍数	真菌（10^3个/g干土）	细菌（10^6个/g干土）	放线菌（10^4个/g干土）
0	0	213.00 ± 4.58ab	16.00 ± 0.00de	40.67 ± 1.86ef
	350	243.33 ± 50.44a	15.00 ± 0.58ef	37.00 ± 3.79f
	650	206.67 ± 14.53ab	13.33 ± 0.33f	64.00 ± 4.62c
8×10^7	0	227.50 ± 3.50ab	17.67 ± 1.45cde	91.33 ± 1.77a
	350	249.00 ± 8.00a	19.50 ± 0.50c	50.33 ± 0.33d
	650	230.00 ± 4.00ab	29.67 ± 0.33a	47.67 ± 3.76de

（续表）

菌剂浓度	溶胶稀释倍数	真菌（10^3个/g干土）	细菌（10^6个/g干土）	放线菌（10^4个/g干土）
16×10^7	0	160.00 ± 2.00b	23.00 ± 1.15b	69.33 ± 0.33bc
	350	177.00 ± 9.00ab	18.00 ± 1.00cd	74.67 ± 3.84b
	650	166.33 ± 10.84ab	17.50 ± 0.50cde	63.00 ± 1.53c

由表4可知，在未施用枯草芽孢杆菌的处理下，350倍稀释液处理下的土壤真菌、细菌和放线菌数与空白处理间无显著差异；650倍稀释液处理细菌数显著低于空白，但放线菌数量显著高于对照，高出空白对照57.36%。在施用枯草芽孢杆菌8×10^7cfu/ml的处理下，350倍、650倍稀释液处理的土壤放线菌数量显著低于空白，分别降低44.89%、47.80%；且650倍稀释液处理细菌显著高于对照，是对照的1.68倍。施用枯草芽孢杆菌16×10^7cfu/ml时，350倍、650倍稀释液处理下的真菌数、放线菌数与空白处理间无显著差异；350倍、650倍稀释液处理下的土壤细菌显著低于空白处理。

2.3　不同处理隶属函数值及综合排名

表5　不同处理隶属函数值及综合排名

指标	8，650	16，0	0，650	0，350	8，350	16，350	16，650	8，0	0，0
株高	0.734 4	0.393 4	0.612 0	0.695 9	0.617 3	0.4117	0.112 1	0.372 2	0.127 8
茎粗	0.726 8	0.823 4	0.916 3	0.585 0	0.627 2	0.658 0	0.651 8	0.422 0	0.595 9
叶片数	0.631 6	0.687 8	0.653 8	0.545 8	0.803 4	0.712 9	0.766 8	0.522 9	0.220 6
根长	0.516 4	0.537 2	0.336 2	0.229 8	0.720 0	0.555 0	0.471 9	0.557 3	0.031 2
表面积	0.814 9	0.586 7	0.486 8	0.345 4	0.444 2	0.498 7	0.567 1	0.672 9	0.186 3
直径	0.692 1	0.339 3	0.712 3	0.780 4	0.233 3	0.337 4	0.502 8	0.319 6	0.746 3
体积	0.803 0	0.582 4	0.245 7	0.592 6	0.401 5	0.239 8	0.405 3	0.462 4	0.533 0
鲜重	0.839 7	0.497 7	0.346 2	0.236 3	0.336 2	0.179 2	0.665 4	0.540 4	0.084 7
干重	0.734 8	0.673 1	0.268 0	0.353 6	0.171 3	0.413 4	0.432 8	0.321 8	0.080 1
EC	0.767 2	0.920 6	0.560 8	0.042 3	0.693 1	0.645 5	0.941 8	0.804 2	0.529 1
pH	0.184 0	0.333 3	0.628 5	0.666 7	0.322 9	0.312 5	0.277 8	0.059 0	0.510 4
N	0.268 8	0.322 6	0.548 4	0.892 5	0.225 8	0.075 3	0.139 8	0.096 8	0.322 6
P	0.348 7	0.888 8	0.736 1	0.086 3	0.600 7	0.657 6	0.246 0	0.275 8	0.614 8
K	0.333 3	0.000 0	0.666 7	1.000 0	0.333 3	0.111 1	0.111 1	0.000 0	0.222 2
有机质	0.420 0	0.460 0	0.905 3	0.473 3	0.490 7	0.506 7	0.093 3	0.640 0	0.750 0
过氧化氢酶	0.333 3	0.833 3	0.666 7	0.000 0	0.333 3	0.333 3	0.388 9	0.333 3	0.277 8
蔗糖酶	0.228 0	0.140 1	0.825 8	0.698 5	0.585 5	0.421 2	0.309 9	0.089 1	0.226 0
脲酶	0.358 0	0.411 3	0.312 8	0.867 5	0.221 3	0.273 2	0.294 7	0.419 1	0.251 8
磷酸酶	0.035 1	0.759 7	0.037 5	0.191 0	0.314 0	0.639 1	0.620 2	0.019 5	0.039 9
真菌	0.405 4	0.027 0	0.279 3	0.477 5	0.508 1	0.118 9	0.061 3	0.391 9	0.313 5
细菌	0.980 4	0.588 2	0.019 6	0.117 6	0.382 4	0.294 1	0.264 7	0.274 5	0.176 5

（续表）

指标	8，650	16，0	0，650	0，350	8，350	16，350	16，650	8，0	0，0
放线菌	0.276 0	0.614 6	0.531 3	0.109 4	0.317 7	0.697 9	0.515 6	0.958 3	0.166 7
平均	0.519 6	0.519 1	0.513 5	0.454 0	0.440 1	0.413 3	0.401 9	0.388 8	0.318 5
排序	1	2	3	4	5	6	7	8	9

对测定的数据进行隶属函数计算，最终得出得出，8×10^7cfu/ml枯草×650倍纳米碳处理排名第1，16×10^7cfu/ml枯草×未施用纳米碳处理排名第2，未施用枯草×650倍纳米碳处理排名第3，未施用枯草×350倍纳米碳处理排名第4，8×10^7cfu/ml枯草×350倍纳米碳处理排名第5，16×10^7cfu/ml枯草×350倍纳米碳处理排名第6，16×10^7cfu/ml枯草×650倍纳米碳处理排名第7，8×10^7cfu/ml枯草×未施用纳米碳处理排名第8，空白处理排名第9（表5）。

3 讨论

枯草芽孢杆菌作为土壤有益微生物，对土壤肥力的形成及植物养分的转化中发挥着积极的作用[20]。施入菌剂能促进植株生长、地上部分干物质的积累以及植株体对养分的吸收积累，增加植物对养分的利用效率，促进根系生长[21]。纳米碳溶胶减少肥料的挥发、淋溶等损失从而促进根系和地上部生长，并且适宜浓度下有节肥增产的功效[22]。

本研究结果表明，适宜浓度的枯草芽孢杆菌和纳米碳溶胶能够显著增强黄瓜幼苗的地上、地下部分生长，改善土壤的部分理化性质和酶活性，影响土壤菌群数量和比例；并且两因素对黄瓜幼苗生长、土壤环境的多数指标上有显著性差异。

在未施用枯草芽孢杆菌的情况下，不同纳米碳溶胶稀释液处理的株高、叶片相对生长率以及土壤N、K含量均显著高于空白对照，这可能是由于纳米碳具有表面效应和小尺寸效应，与重金属离子发生化学、物理吸附，减缓速效养分释放，增强植物对肥料的吸附功能，减少肥料的流失、淋失和固定[23]，达到提高生长势的效果，这与梁太波等[24]研究显示的纳米碳溶胶能够活化土壤磷素和钾素，增加各土层速效磷和速效钾含量的结果相一致。而且，350倍稀释液处理下的脲酶、磷酸酶和蔗糖酶活以及650倍稀释液蔗糖酶活性均有显著性提高，这是因为添加纳米碳可能增强土壤微生物的活性或植株根系活动，间接改变土壤酶活性[25]，这也与李淑敏等[7]的实验结果一致。

在未施用纳米碳溶胶的情况下，不同枯草芽孢杆菌浓度处理的株高、叶片相对生长率以及根际土壤中细菌、放线菌数量均显著优于空白对照，这可能与适量枯草芽孢杆菌的加入造成植物根际细菌群落结构改变[13]、菌株调节土壤生态环境有关，促进植物吸收营养物质的同时产生活性物质来发挥作用，直接影响到植物根际有机质的分解和养分的快速转化，达到促生的目的[12, 26]。16×10^7cfu/ml枯草芽孢杆菌处理下的根系干重、鲜重、根长和表面积均显著高于空白对照，这可能是因为枯草芽孢杆菌产生的代谢物质抑制或阻抗根部病原菌的发展，间接促进植物生长[27]。肖小露等[28]的研究结果也显示枯草芽孢杆菌发酵液处理过的辣椒株高、根长、鲜重及干重都显著高于对照组；且磷酸酶和过氧化氢酶也显著优于对照，可能是功能菌的加入，丰富了土壤中的微生物，土壤微生物的活动和代谢更加旺盛，从而提高了土壤酶的活性[29]。

通过隶属函数综合分析两因素互作处理来看，8×10^7cfu/ml枯草*650倍纳米碳处理综合所有指标表现最好，16×10^7cfu/ml枯草*未施用纳米碳处理次之，未施用枯草*650倍纳米碳处理排名第3，空白处理表现最差，这可能是因为枯草芽孢杆菌和纳米碳溶胶均有调节土壤生态环境的作用，两者互作使得促生效果更明显。

4 结论

（1）在未施用枯草芽孢杆菌的情况下，650倍纳米碳溶胶表现总体优于350倍稀释液与空白处理，能够显著促进植株叶片生长，提升土壤有机质含量和蔗糖酶活性，增加土壤放线菌数量。

（2）在未施用纳米碳溶胶的情况下，16×10^7cfu/ml枯草芽孢杆菌表现总体优于8×10^7cfu/ml枯草与空白处理，能够增加根系干重、根长和表面积，显著降低土壤EC含量，提升土壤P含量，增加细菌、放线菌的含量，显著增强土壤磷酸酶和过氧化氢酶活性。

（3）在两因素互作的情况下，8×10^7cfu/ml枯草*650倍纳米碳溶胶提高株高相对生长率，显著促进根系生长，根系干鲜重、表面积、直径和体积均为处理中最高值，一定程度上降低土壤EC值，提高土壤中细菌数量，综合排名最高。

参考文献

［1］ 张国于，董相玉. 浅谈化肥的施用所带来的危害［J］. 现代农业，2010（9）：39.

［2］ 魏彬萌. 不同种类有机肥对土壤培肥效果的研究［J］. 陕西农业科学，2017，63（10）：73-77+79.

［3］ 徐立功，徐坤，刘会诚. 生物有机肥对番茄生长发育及产量品质的影响［J］. 中国蔬菜，2006（4）：8-11.

［4］ Robert P. Larkin，Timothy S. Griffin. Control of soilborne potato diseases using Brassica green manures［J］. Crop Protection，2006，26（7）.

［5］ 李红丽，李清飞，郭夏丽，等. 调节土壤微生态防治烟草青枯病［J］. 河南农业科学，2006（2）：57-60.

［6］ 张雪艳，田蕾，高艳明，等. 生物有机肥对黄瓜幼苗生长、基质环境以及幼苗根系特征的影响［J］. 农业工程学报，2013，29（1）：117-125.

［7］ 李淑敏，马辰，李丽鹤，等. 纳米碳对玉米氮素吸收及根系活力和土壤酶活性的影响［J］. 东北农业大学学报，2014，45（7）：14-18+25.

［8］ 刘键，张阳德，张志明. 纳米生物技术促进蔬菜作物增产应用研究［J］. 湖北农业科学，2009，48（1）：123-127

［9］ Khodakovskaya M，Dervishi E，Mahmood M，et al. Carbon nanotubes are able to penetrate plant seed coat and dramatically affect seed germination and plant growth.［J］. ACS Nano，2009，3（10）：3 221-3 227.

［10］ Nair R，Mohamed M S，Gao W，et al. Effect of carbon anomaterials on the germination and growth of rice plants.［J］. Journal of Nanoscience and Nanotechnology，2012，12（3）：2 212-2 220.

［11］ 王艳，韩振，张志明，等. 纳米碳促进大豆生长发育的应用研究［J］. 腐植酸，2010（4）：17-23.

［12］ 王娟，刘东平，丁方丽，等. 促植物生长根际细菌HG28-5对黄瓜苗期生长及根际土壤微生态的影响［J］. 中国蔬菜，2016（8）：50-55.

［13］ 游偲，张立猛，计思贵，等. 枯草芽孢杆菌菌剂对烟草根际土壤细菌群落的影响［J］. 应用生态学报，2014，25（11）：3 323-3 330.

[14] 王翠梅，李剑. 枯草芽孢杆菌在有机蔬菜种植中的应用[J]. 上海农业科技，2017（4）：142+144.

[15] 张雪艳，张亚萍，许帆，等.单一或组合生防菌对田间番茄植株生长和线虫的影响[J]. 北方园艺，2016（07）：108-112.

[16] 鲍士旦. 土壤农化分析[M]. 中国农业出版社，2008.

[17] 李阜棣，喻子牛，何绍江. 农业微生物学实验技术[M]. 中国农业出版社，1996.

[18] 周礼恺，张志明. 土壤酶活性的测定方法[J]. 土壤通报，1980（5）：37-38+49.

[19] Bouma T J，Nielsen K L，Koutstaal B A S. Sample preparation and scanning protocol for computerised analysis of root length and diameter. Plant and soil，2000，218（1）：185-196.

[20] 逄焕成，李玉义，严慧峻，等. 微生物菌剂对盐碱土理化和生物性状影响的研究[J]. 农业环境科学学报，2009，28（5）：951-955.

[21] 毕延刚，田永强. 堆肥和枯草芽孢杆菌协同调控黄瓜幼苗生长的机制探究[J]. 中国农学通报，2015，31（28）：71-78.

[22] 钱银飞，邱才飞，邵彩虹，等. 纳米碳肥料增效剂对水稻产量及土壤肥力的影响[J]. 江西农业学报，2011，23（2）：125-127+139.

[23] 孙朋成. 三种环境材料对土壤铅镉固化及氮肥增效机理研究[D].北京：中国矿业大学，2016.

[24] 梁太波，赵振杰，王宝林，等. 纳米碳溶胶对碱性土壤pH和养分含量的影响[J]. 土壤，2017（5）：958-962.

[25] 马辰. 纳米碳对玉米养分利用及产量的影响[D]. 哈尔滨：东北农业大学，2014.

[26] 夏艳，徐茜，林勇，等. 植物根际促生菌（PGPR）作用机制研究进展（英文）[J]. Agricultural Science & Technology，2014，15（1）：87-90+110.

[27] 袁玉娟，徐晨伟. 枯草芽孢杆菌SQR9促进黄瓜生长的效果[J]. 中国瓜菜，2017（3）：25-28+34.

[28] 肖小露. 枯草芽孢杆菌BS193对辣椒疫病的生防作用及其抗菌机制初探[D]. 福州：福建农林大学，2017.

[29] 张静，杨江舟，胡伟，等. 生物有机肥对大豆红冠腐病及土壤酶活性的影响[J]. 农业环境科学学报，2012（3）：548-554.

红蓝白光质对番茄幼苗生长发育及光合特性的影响

文莲莲[1]，李岩[1, 2, 3*]，秦利杰[1]，周鑫[1]，倪秀男[1]，刘淑侠[1]，魏珉[1, 2, 4*]

（1. 山东农业大学园艺科学与工程学院，泰安　271018；2. 农业部黄淮海设施农业工程科学观测实验站，泰安　271018；3. 作物生物学国家重点实验室，泰安　271018；4. 山东果蔬优质高效生产协同创新中心，泰安　271018）

摘要：采用LED（发光二极管）精量调制光源，以番茄为试材，在红蓝光（R:B=3:1）基础上添加不同比例光强的白光（W，分别为0%、20%、40%、60%、80%、100%），研究不同比例红蓝白光质对番茄幼苗形态建成和光合特性的影响。结果表明: 与对照和其他处理相比，添加20%白光可显著提高番茄幼苗的茎粗、叶面积、全株鲜重、根冠比、壮苗指数、根系构型参数及根系活力，增加叶片栅栏组织、海绵组织及叶片厚度，改善光系统Ⅱ性能，提高电子传递速率、表观量子效率和叶片净光合速率。

关键词：光质；番茄；光合特性；壮苗

Effects of Different Proportions of Red，Blue and White Light on Growth and Development and Photosynthesis of Tomato Seedlings

Wen LianLian[1]，Li Yan[1, 2, 3*]，Qin LiJie[1]，Zhou Xin[1]，
Ni XiuNan[1]，Liu ShuXia[1]，Wei Min[1, 2, 4*]

（1. *College of Horticultural Science and Engineering*，*Tai'an*　271018；2. *Scientific Observing and Experimental Station of Environment Controlled Agricultural Engineering in Huang-Huai-Hai Region*，*Ministry of Agriculture*，*Tai'an*　271018；3. *State Key Laboratory of Crop Biology*，*Tai'an*　271018；4. *Shandong Collaborative Innovation Center of Fruit & Vegetable Quality and Efficient Production*，*Shangdong Agricultural University*，*Tai'an* 271018，*Tai'an*　271018）

Abstract：In this experiment，the effects of different light qualities on photosynthesis and morphogenesis of tomato seedlings were studied by adding white light at 0%、20%、40%、60%、80%、100% respevtively to red/blue （3/1） light generated by LED. The results showed that 20% treatment，compared with the control and other treatments，could significantly increase stem diameter，surface area，whole plant fresh weight，root / shoot ratio，seedling index，root configuration parameters and root activity，palisade tissue，spony tissue and vane thickness were increased.The performance of photosystem II，electron transfer rate，apparent quantum yield and net photosynthetic rate were also improved.

Key words：Light quality；Tomato；Photosynthetic characteristics；Seedling

蔬菜幼苗的健壮程度直接影响植株生长发育，并与产量和品质密切相关（汪俏梅，2000）。光质对植物形态建成、生理代谢和果实品质等方面具有明显的调控作用（李岩等，2017）。其中，红光和蓝光是植物光合作用和形态建成的主要光质（Gioil等，

2008）。研究表明，红光有利于茎粗增加、叶片扩展、生长速率提高和干物质累积（蒲高斌等，2005a，2005b），蓝光下幼苗根系活力提高，茎粗增加，光合产物分配更有利于壮苗培育（蒲高斌等，2005b）。近年来研究发现红蓝组合光明显提高番茄幼苗光合速率，碳氮代谢，加速物质积累，进而促进幼苗生长（孙娜等，2016）。但也有研究表明，与红蓝组合光相比，采用红蓝白组合光更有利于植物生长（周成波等，2017）。

LED具有体积小、寿命长、耗能低、波长固定与低发热量等优点（Guo等，2008），研究其对园艺作物生长发育的影响已成为热点（Morrow，2008）。番茄（*Solanum lycopersicum*）是我国设施主栽作物之一，有关不同光质尤其是红蓝白组合光对番茄生理生化、光合作用以及产量和品质影响的研究已有报道，且笔者所在实验室已经筛选出了利于培育番茄壮苗的适宜红蓝组合光比例（R:B=3:1），但系统研究红蓝白光质对番茄幼苗形态建成及光合特性的影响未见报道。为此，本试验通过在红蓝光（3:1）中添加白光，研究不同比例红蓝白光质对番茄幼苗形态建成、光合效率的影响，为番茄壮苗培育中的光质调控提供依据。

1 材料与方法

1.1 供试材料

供试番茄品种“SV0313TG”（*Solanum lycopersicum*，Seminis Vegetable Seeds公司，美国），经温汤浸种、催芽后，在日光温室内播于50孔育苗盘中（草炭:蛭石=2:1）。待两片子叶展平将幼苗转入到塑料钵（6.5cm × 6.5cm × 10cm）中并置于智能人工气候室，每处理80株。

1.2 试验设计

试验于2017年4—6月在山东农业大学园艺实验站日光温室及智能人工气候室内进行，LED光源均购自深圳纯英达业集团有限公司。在红蓝光（R:B=3:1）中添加不同光强比例（0%、20%、40%、60%、80%、100%）的白光（W），共6个处理，以0%（R:B=3:1）为对照。光强为300μmol·m^{-2}·s^{-1}，光周期为12h·d^{-1}，温度28℃/18℃（昼/夜），空气湿度70% ± 5%（图1）。

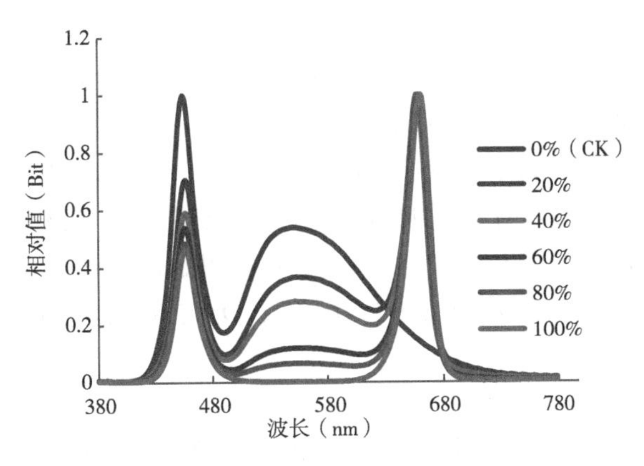

图1　不同光质的光谱

Figure 1　Spectrum figures of different light qualities

1.3 测定项目与方法

1.3.1 番茄幼苗形态生长分析

育苗第30d天，统计分析幼苗株高、茎粗、叶面积、地上部、地下部以及全株干鲜重。株高用直尺从植株子叶处到新叶最高点处测量；茎粗采用游标卡尺测量靠近子叶1cm处；叶面积使用CI-202便携式激光叶面积仪（CID Bio-science公司，美国）测定整株叶面积。将不同处理的番茄幼苗从钵中取出，先用流水冲洗干净，吸水纸吸干，测定地上部、地下部鲜重，并在烘箱中105℃杀青30min，75℃烘至恒重，称干物质。

1.3.2 根系生长发育的测定

采用Win RHIZO软件及根系扫描仪（Regent公司，加拿大）进行根系构型参数的分析，测定其总根长、总表面积、体积及根尖数等根系生长特性。并将一部分根系取出洗净吸干表面水分，称取0.5cm长根尖部分0.5g，采用氯化三苯基四氮唑（TTC）法测定根系活力。

1.3.3 石蜡切片的制作

石蜡切片的制作和数据统计参照卢福顺（2013）的方法。

1.3.4 叶片光合特性的测定

叶片色素含量的测定：参考Sartory and Grobbelaar（1984）的方法测定叶片光合色素（叶绿素a、叶绿素b和类胡萝卜素）。

光合气体交换参数的测定：采用LI-6400/XT便携式光合仪（LI-COR公司，美国），于9:00～11:30进行光合指标净光合速率（Pn）、蒸腾速率（Tr）、气孔导度（Gs）和胞间CO_2浓度（Ci）的测定。测定时使用开放气路，测定环境为内置光量子通量密度（PPFD）300$\mu mol \cdot m^{-2} \cdot s^{-1}$，$CO_2$浓度（400±10）$\mu mol \cdot mol^{-1}$，温度（26±1）℃。

光响应曲线（Pn-PFD）的测定：测定光响应曲线时采用LED光源控制光量子通量密度（PPFD），光强梯度设定为1 600、1 400、1 200、1 000、800、600、400、200、100、50、20、0$\mu mol \cdot m^{-2} \cdot s^{-1}$，测定过程中每间隔180s记录一次数据。$CO_2$浓度400$\mu mol \cdot mol^{-1}$，叶温（26±1）℃，相对湿度65%～70%。光通量密度PFD、CO_2浓度以及叶温分别由测定系统的内置可调红蓝光源、内置可调式CO_2注入系统和可调叶温热电偶监控装置控制。根据测定结果，参考Farquhar等（2001）的方法分析光合—光响应曲线，获得番茄叶片的初始斜率即表观量子效率（AQY）和最大净光合速率（Amax）。

叶绿素荧光诱导动力学曲线（OJIP）的测定：参考李鹏民等[13]的试验方法，采用Handy-PEA植物效率分析仪测量系统（Hansatech公司，英国）提供的自动程序（连续激发式荧光）进行测量。番茄叶片经过45min的充分暗适应后，用3 000$\mu mol \cdot m^{-2} \cdot s^{-1}$脉冲红光诱导，荧光信号从10μs开始记录，至1s时结束，初始记录速度为每秒钟10万次。根据Turan等（2014）的方法，将OJIP曲线进行O-P段和O点标准化，通过JIP-test对O-J-I-P曲线进行分析。

1.4 数据分析及处理

采用DPS V14.10软件进行数据统计分析、相关性分析和差异显著性检验（P<0.05）。用Microsoft Excel 2010和Sigmaplot 10.0软件进行数据统计和作图。

2　结果与分析

2.1　不同光质对番茄幼苗地上部生长的影响

从表1可以看出，不同比例红蓝白光质明显影响番茄幼苗生长。植株在处理的第30d时，对照和100%处理下株高显著低于其他处理，其他处理间无显著差异；20%处理下茎粗最大，其次是40%处理，且显著高于对照及其他处理；20%、40%处理下叶面积显著高于对照及其他处理；20%处理下幼苗根冠比明显高于对照以及除40%以外的其他处理，后者与其他处理间无显著差异；与白光相比，红蓝光有利于壮苗的培育，不同比例红蓝白光下壮苗指数差异明显，20%处理比对照和100%分别增加13.71%和24.38%。

表 1　不同光质对番茄幼苗形态的影响

Table 1　Effects of different light qualities on morphology of tomato seedlings

处理	株高（cm）	茎粗（mm）	叶面积（cm^2）	根冠比	壮苗指数
0%（CK）	30.00 ± 1.00c	3.27 ± 0.20d	273.68 ± 7.88cd	0.096 ± 0.007bc	0.175 ± 0.005b
20%	39.33 ± 0.58a	4.34 ± 0.19a	343.27 ± 11.57a	0.124 ± 0.005a	0.199 ± 0.006a
40%	40.33 ± 2.08a	3.88 ± 0.15b	349.69 ± 15.10a	0.102 ± 0.008b	0.172 ± 0.002bc
60%	41.33 ± 1.15a	3.58 ± 0.07c	297.73 ± 8.19b	0.084 ± 0.010d	0.170 ± 0.006bc
80%	41.67 ± 2.08a	3.68 ± 0.08c	277.16 ± 6.56c	0.088 ± 0.011cd	0.168 ± 0.003c
100%	33.00 ± 1.00b	3.70 ± 0.10c	255.79 ± 4.49d	0.078 ± 0.005c	0.160 ± 0.005d

不同小写字母表示处理间差异显著（$P < 0.05$）。下同。

2.2　不同光质对番茄幼苗根系发育的影响

由表2可知，20%处理下的番茄幼苗根长、总表面积、体积及根尖数都显著高于对照和其他处理，其次是处理40%和对照，对照各项指标均最低。

表 2　不同光质对番茄幼苗根系构型参数的影响

Table 2 Effects of different light qualities on the configuration parameters of root in tomato seedlings

处理	长度（cm）	总表面积（cm^2）	体积（cm^3）	根尖数
0%（CK）	754.16 ± 14.91b	115.21 ± 6.01b	2.91 ± 0.07b	1 590.00 ± 78.31b
20%	847.33 ± 13.95a	163.30 ± 9.17a	3.17 ± 0.04a	1 822.22 ± 47.30a
40%	760.79 ± 9.81b	100.68 ± 7.56c	2.47 ± 0.08c	1 576.44 ± 24.02b
60%	637.13 ± 63.82c	72.77 ± 2.92d	2.46 ± 0.07c	1 490.67 ± 38.37c
80%	543.00 ± 20.13d	64.40 ± 4.15de	1.77 ± 0.02d	1 184.50 ± 18.50d
100%	455.16 ± 50.83e	57.05 ± 0.65e	1.44 ± 0.04e	1 090.33 ± 5.69e

由图2可知，红蓝光（3:1）中添加20%的白光可显著提高番茄幼苗的根系活力，20%处理下幼苗根系活力显著高于对照和其他处理，比对照0%和100%分别增加40.68%和2.26倍。

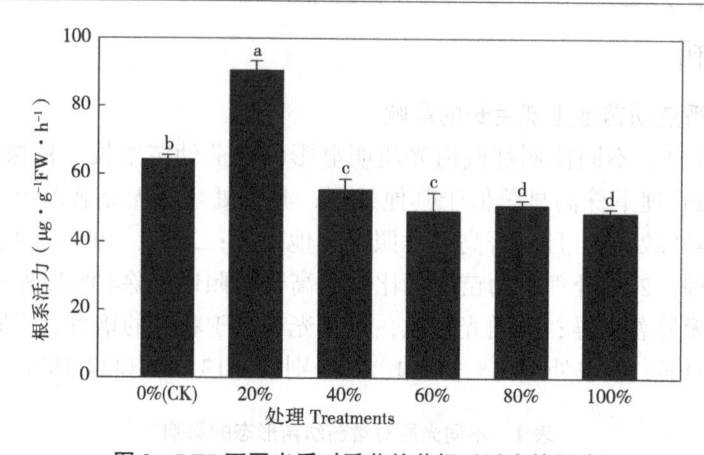

图2　LED不同光质对番茄幼苗根系活力的影响

Figure 2　Effects of different light qualities on root activity of tomato seedlings

2.3　不同光质对番茄幼苗叶片结构的影响

由图3和表3可知，光质对番茄幼苗叶片结构有显著影响。20%处理下，其栅栏细胞排列较对照和其他处理紧密，并且海绵组织排列较对照和其他处理间隙小，栅栏组织、海绵组织厚度及叶片厚度均显著高于对照和其他处理，从大到小依次为20%＞40%＞60%＞80%＞0%（CK）＞100%。

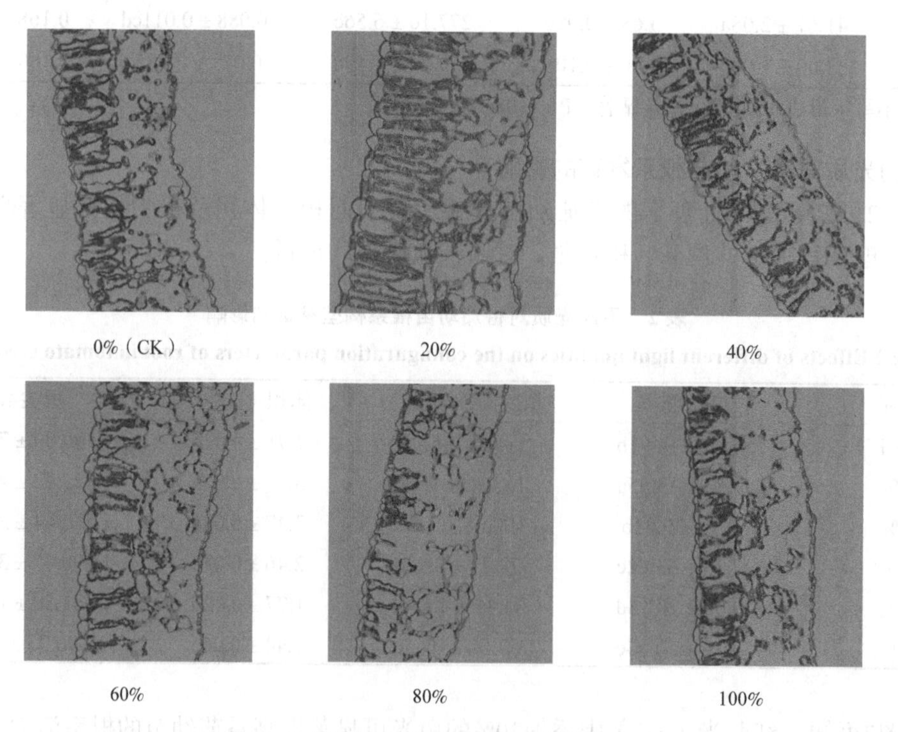

图3　不同光质对番茄幼苗叶片结构的影响

Figure 3　Effects of different light qualities on anatomical structure in tomato seedling leaves

表 3　不同光质对番茄幼苗叶片结构的影响

Table 3　Effects of different light qualities on structure in tomato seedling leaves

处理	栅栏组织厚度（μm）	海绵组织厚度（μm）	叶片厚度（μm）
0%（CK）	83.50 ± 4.54c	90.48 ± 1.19d	201.38 ± 1.22c
20%	118.48 ± 1.30a	143.45 ± 1.63a	278.62 ± 3.40a
40%	92.84 ± 3.31b	103.59 ± 2.93b	209.70 ± 0.65b
60%	83.15 ± 4.80c	96.31 ± 2.17c	210.70 ± 0.26b
80%	70.28 ± 0.61d	92.45 ± 2.37cd	202.35 ± 0.51c
100%	68.46 ± 2.28d	91.24 ± 3.01d	204.46 ± 2.34c

2.4　不同光质对番茄幼苗光合特性的影响

2.4.1　色素含量

由表4可以看出，80%处理下叶绿素a、叶绿素b、叶绿素（a+b）以及类胡萝卜素含量均高于对照和其他处理，叶绿素a/b含量则以20%最高，但除60%处理以外，各处理间差异不显著。

表 4　不同光质对番茄幼苗色素含量的影响

Table 4　Effects of different light qualities on chlorophyll of tomato seedlings

处理	叶绿素a含量	叶绿素b含量	叶绿素（a+b）	叶绿素a/b含量	类胡萝卜素
Treatment	$mg \cdot g^{-1}$（FW）	$mg \cdot g^{-1}$（FW）	$mg \cdot g^{-1}$（FW）	$mg \cdot g^{-1}$（FW）	$mg \cdot g^{-1}$（FW）
0%（CK）	1.138 ± 0.193c	0.253 ± 0.040b	1.391 ± 0.232c	0.561 ± 0.011ab	0.254 ± 0.049c
20%	1.267 ± 0.198bc	0.271 ± 0.042b	1.539 ± 0.238bc	0.584 ± 0.032a	0.300 ± 0.062bc
40%	1.447 ± 0.031abc	0.324 ± 0.003ab	1.771 ± 0.030abc	0.557 ± 0.014ab	0.323 ± 0.005bc
60%	1.351 ± 0.107bc	0.315 ± 0.027ab	1.666 ± 0.134bc	0.536 ± 0.005b	0.304 ± 0.021bc
80%	1.747 ± 0.244a	0.386 ± 0.065a	2.133 ± 0.309a	0.568 ± 0.018ab	0.417 ± 0.059a
100%	1.604 ± 0.120ab	0.360 ± 0.036a	1.964 ± 0.154ab	0.558 ± 0.024ab	0.368 ± 0.012ab

2.4.2　光合参数

不同比例红蓝白光质处理对番茄幼苗净光合速率有显著影响（表5），20%处理能显著提高番茄幼苗叶片的光合作用。20%处理下净光合速率为最高，其次是对照和40%，但三者之间无显著差异，100%明显低于对照和其他处理；20%处理下幼苗气孔导度显著高于对照和其他处理，其次是40%，100%下最低；40%处理下蒸腾速率显著高于对照和其他处理；100%处理下胞间CO_2浓度显著高于对照和其他处理，且对照和其他处理间无显著差异。

表 5　不同光质对番茄幼苗气体交换参数的影响

Table 5　Effects of different light qualities on the parameters of tomato seedlings gas exchange

处理	光合速率（$\mu mol \cdot m^{-2} \cdot s^{-1}$）	蒸腾速率（$mmol \cdot m^{-2} \cdot s^{-1}$）	气孔导度（$mmol \cdot m^{-2} \cdot s^{-1}$）	胞间CO_2浓度（$\mu mol \cdot mol^{-1}$）
0%（CK）	5.33 ± 0.06ab	3.55 ± 0.06c	0.34 ± 0.02bc	379.33 ± 9.02b
20%	5.67 ± 0.04a	3.72 ± 0.09b	0.43 ± 0.00a	377.00 ± 12.12b

（续表）

处理	光合速率（$\mu mol \cdot m^{-2} \cdot s^{-1}$）	蒸腾速率（$mmol \cdot m^{-2} \cdot s^{-1}$）	气孔导度（$mmol \cdot m^{-2} \cdot s^{-1}$）	胞间CO_2浓度（$\mu mol \cdot mol^{-1}$）
40%	5.39 ± 0.25ab	4.33 ± 0.09a	0.36 ± 0.01b	374.67 ± 7.57b
60%	4.96 ± 0.05bc	3.54 ± 0.04c	0.32 ± 0.02cd	378.67 ± 10.07b
80%	4.76 ± 0.10c	2.77 ± 0.06e	0.30 ± 0.01d	383.00 ± 11.00b
100%	4.12 ± 0.60d	2.96 ± 0.09d	0.23 ± 0.01e	402.67 ± 8.02a

2.4.3 光响应曲线

如图4所示，番茄幼苗的光合速率—光响应曲线在不同比例红蓝白光质处理下变化趋势相似。在弱光条件下，各处理间番茄幼苗的光合速率（Pn）几乎无明显差异，但随着光强的升高，光强大约在200$\mu mol \cdot m^{-2} \cdot s^{-1}$时，各处理间开始出现明显差异，在光强约1 200$\mu mol \cdot m^{-2} \cdot s^{-1}$时，各处理的光合速率基本接近光饱和（图4），最大净光合速率由大到小排列依次为20%>40%>0%（CK）>60%>80%>100%（图5-A）。此外，20%处理和80%下生长的番茄幼苗表观量子效率（AQY）均显著高于对照和其他处理（图5-B）。

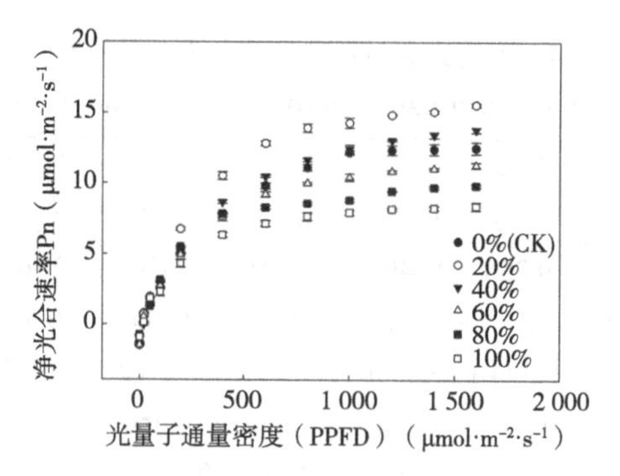

图4　不同光质对番茄幼苗光合—光响应曲线的影响

Figure 4　Effects of different light qualities on the photosynthetic lightresponse curve of tomato seedlings

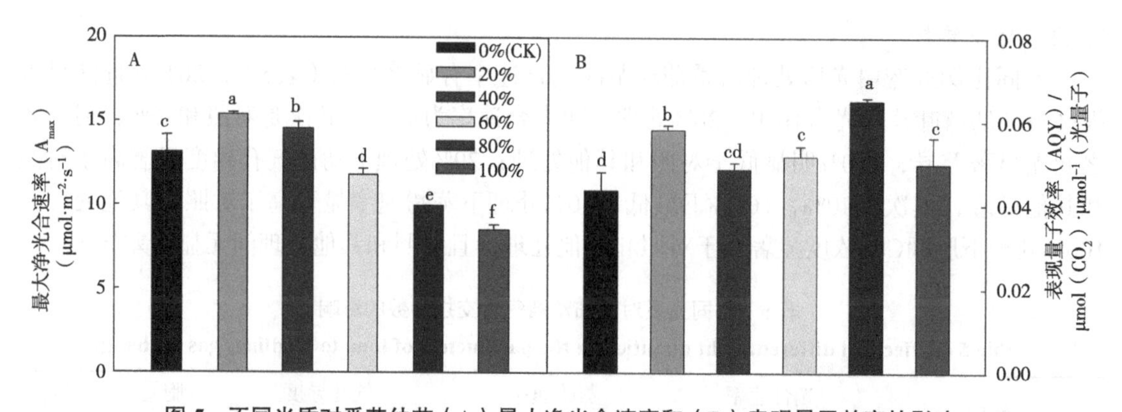

图5　不同光质对番茄幼苗（A）最大净光合速率和（B）表观量子效率的影响

Figure 5　Effects of different light qualities on（A）the maximum net photosynthetic rate（Amax）and（B）the apparent quantumyield（AQY）of tomato seedlings

2.4.4 荧光诱导动力学曲线

叶绿素荧光诱导动力学可以反映植物叶片PSII的功能以及状态变化（李鹏民等，2005）。我们为了能够更准确地分析比较不同光质对番茄幼苗叶片快速荧光诱导曲线的影响，通常将得到的快速叶绿素荧光诱导动力学曲线进行标准化，即可得到相对可变荧光（Vt）诱导动力学曲线。通过对快速叶绿素荧光诱导动力学曲线（OJIP曲线）的分析，可获得PSII原初光化学反应的大量信息。不同比例红蓝白处理对番茄幼苗叶片叶绿素的OJIP曲线有着显著的影响（图6-A，B）。在标准化的OJIP曲线中，可明显看出，各处理在O点处差异不显著，说明在O点处作用中心活性状态无差异，20%处理下幼苗在J点显著低于对照和其他处理。为进一步观察不同处理下番茄幼苗OJIP曲线，以0%（CK）作为参照，将各处理的曲线与其进行相减得到图6-B，20%处理在K、J峰均显著低于对照和其他处理，其次是60%处理在K点显著低于对照及其他处理，40%处理在I点显著低于对照及其他处理，其次是60%处理，p点除40%略高外，其他处理间无显著性差异。

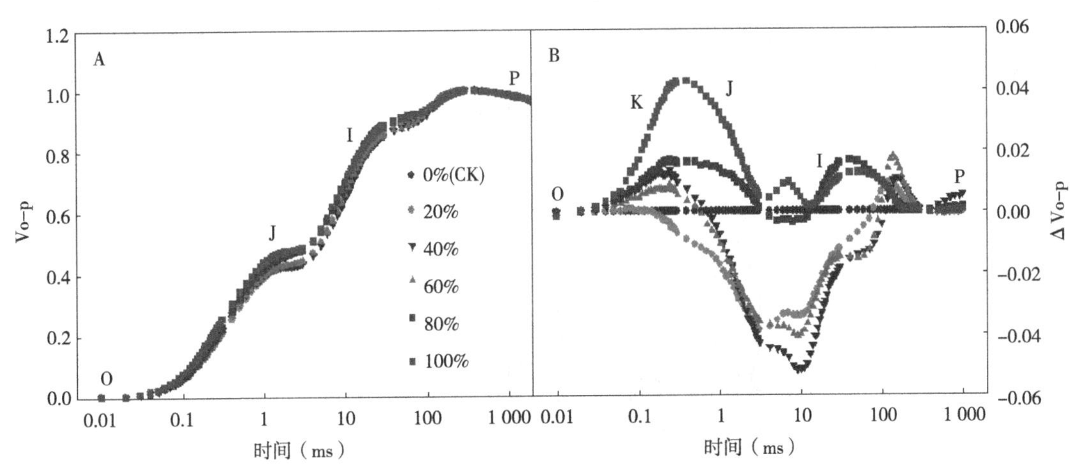

图6 不同光质对番茄幼苗光系统II（PSII）相对可变荧光强度（Vo-p）和相对可变荧光强度差值（△Vo-p）的影响

Figure 6 Effects of different light qualities on chlorophyll a relative variable fluore scence intensity（Vo-p ）and the difference of relative variable fluorescence intensity（△Vo-p ） of tomato seedlings

从图7中可知，叶片单位反应中心吸收（ABS/RC）、捕获（TRo/RC）、用于电子传递（ETo/RC）及热耗散（DIo/RC）的能量均有相同的变化趋势。20%处理下的番茄幼苗单位反应中心吸收、捕获以及用于电子传递的能量均显著高于对照和其他处理，且热耗散最低。各处理均呈现ABS/RC＞TRo/RC＞ETo/RC的趋势。这表明热耗散随着电子传递链的延伸而不断增加，从而使光能利用率降低。

由图8可以看出，20%处理下叶片单位面积吸收的光能（ABS/CS）、捕获的光能（TR/CS）以及单位面积内反应中心数目（RC/CS）均显著高于对照和其他处理，但其他处理间差异不明显；20%和40%处理下传递的光能（ET/CS）显著高于对照和其他处理；20%处理下单位面积的热耗散（DI/CS）显著低于对照和其他处理。

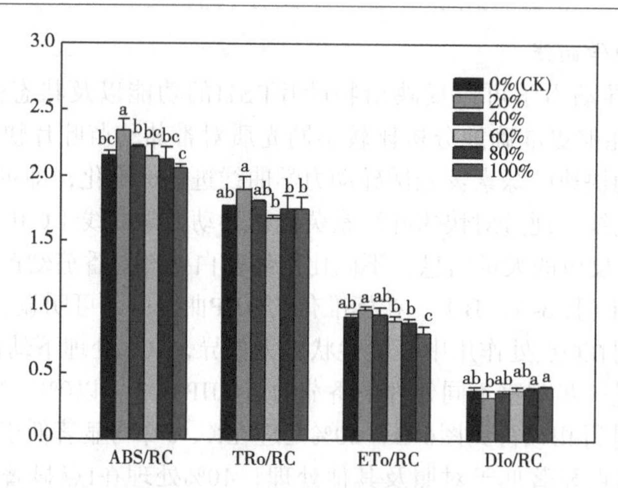

图 7　不同光质处理单位反应中心吸收（ABS/RC）、捕获（TRo/RC）、用于电子传递（ETo/RC）及热耗散掉（DIo/RC）的能量的比较

Figure 7　Relationships of the relative activities in absorption flux per reaction center（ABS/RC）, trapped energy flux per RC（TRo/RC）electron transport flux per RC（ETo/RC）and dissipated energy flux per RC（DIo/RC）of tomato seedlings in different light qualities treatment

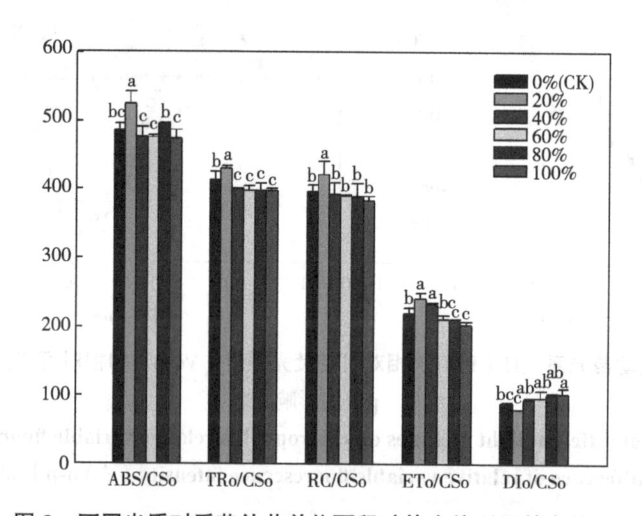

图 8　不同光质对番茄幼苗单位面积叶片光能利用效率的影响

Figure 8　Effects of different light qualities on light use efficiency per unit leaf area of tomato seedlings

3　讨论和结论

植物的生长发育受光质及其组成比例的影响（许大全等，2015）。已有研究证明，红蓝（3:1）组合光能够提高番茄幼苗光合性能，促进植株生长，利于培育壮苗（王丽伟等，2017）。在本研究中，红蓝（3:1）中添加一定的白光较红蓝（3:1）更有利于培育番茄壮苗，这与前人（周成波等，2017）的研究结果类似，其中以红蓝（3:1）光中添加20%的白光对培育番茄壮苗效果最有利。

光合色素是光合作用的基础，光质直接影响光合色素的合成，从而影响植物的光合作用。本研究表明番茄幼苗在80%处理下有效提高叶绿素和类胡萝卜素含量，可能与该处理下表观量子效率较高有关，其利用弱光的能力较强。由于复合光作用效应是一个复杂的响应过程，因此猜测引起该处理下表观量子效率较高的原因可能是光谱与植物光谱色素系统相互作用的结果（刘晓英等，2010）。研究证明光合作用受诸多因素影响，叶绿素含量的高低并不能完全反映叶片光合能力的强弱（张秋英等，2005），80%处理下幼苗叶片叶绿素和类胡萝卜素含量虽高，但净光合速率却低于对照，这与前人的研究结果类似（董飞等，2017）。另外，本研究中发现20%处理能够明显提高叶绿素a/b值，反映了植物对光能利用的程度。

典型的叶绿素荧光诱导动力学曲线是由O、J、I、P等相组成，其中O点是PSII作用中心完全开放时的荧光，此时叶片受光后发射荧光最小，J点是PSII的电子受体QA第一次处于瞬时最大程度还原态时的荧光，I点主要反映快还原型PQ库和慢还原型PQ库的大小，P点是PSII反应中心处于完全关闭状态时，此时不再接受光量子，荧光产量最高（李鹏民等，2005）。本试验中20%处理下番茄幼苗在J点显著低于对照及其他处理，表明QA被还原的速率最高，电子从QA向QB传递较顺畅，这与周成波等（2017）的研究结果相一致。此外，在本研究中发现在J点之前出现K点，K点的出现是因为水裂解系统被抑制和QA之前受体侧的部分被抑制所造成的。在抑制过程中，受伤害的是放氧复合体系统（OEC），因此K点的上升可作为OEC受伤害程度的特殊标记（Strasser等，2000）。在本研究中20%处理下幼苗在K点显著低于对照和其他处理，说明该处理OEC的活性受到的抑制最低。

通过JIP-test可分析得到光合机构的比活性，即单位反应中心（RC）或单位受光面积（CS）的各种量子效率以及单位面积的活性反应中心的数量，这些能更好的反映出植物光合器官对光能的吸收、转化以及耗散等状况（VanHeerden等，2004）。本研究结果表明，20%处理下生长的幼苗叶片对光能的吸收、转化能力最强，热耗散相对较低，PSII活性最高。

参考文献

[1] 董飞，王传增，张现征，等. 不同光质对樱桃番茄幼苗生理与光合特性的影响［J］. 植物生理学报，2017，53（7）：1208-1214

[2] 胡阳，江莎，李洁，等. 光强和光质对植物生长发育的影响［J］. 内蒙古农业大学学报，2009，30（4）：296-303

[3] 李鹏民，高辉远，Strasser RJ. 快速叶绿素荧光诱导动力学分析在光合作用研究中的应用［J］. 植物生理与分子生物学学报，2005，31（6）：559-566

[4] 李岩，王丽伟，文莲莲，等. 红蓝光质对转色期间番茄果实主要品质的影响［J］. 园艺学报，2017，44（12）：2372-2382

[5] 刘晓英，徐志刚，常涛涛，等. 不同光质LED弱光对樱桃番茄植株形态和光合性能的影响［J］. 西北植物学报，2010，30（4）：725-732

[6] 卢福顺. 水分胁迫对马铃薯生理指标和叶片结构的影响［D］.哈尔滨：东北农业大学. 2011.

[7] 蒲高斌，刘世琦，刘磊，等. 不同光质对番茄幼苗生长和生理特性的影响［J］. 园艺学报，2005，32（3）：420-425

[8] 蒲高斌，刘世琦，张珍，等. 光质对番茄幼苗生长及抗氧化酶活性的影响. 中国蔬菜，2005（9）：21-23

[9] 孙娜，魏珉，李岩，等. 光质对番茄幼苗碳氮代谢及相关酶活性的影响［J］. 园艺学报，2016，43（1）：80-88

[10] 汪俏梅. 设施栽培中培育壮苗的一些技术措施. 沈阳农业大学学报，2000，31（1），120-123

[11] 王丽伟，李岩，辛国凤，等. 不同比例红蓝光对番茄幼苗生长和光合作用的影响. 应用生态学报，2017，28（5）：1595-

1602
[12] 许大全，高伟，阮军. 光质对植物生长发育的影响［J］. 植物生理学报，2017，51（8）：1217-1234
[13] 张秋英，李发东，刘孟雨. 冬小麦叶片叶绿素含量及光合速率变化规律的研究. 中国生态农业学报，2005，13（3）：95-98
[14] 周成波，张旭，崔青青，等. LED补光光质对小白菜生长及光合作用的影响［J］.植物生理学报，2017，53（6）：1030-1038
[15] Farquhar GD，von Caemmerer S，Berry JA. Models of photosynthesis［J］. Plant Physiol，2001，125：42–45
[16] Gioia DM，Kim HH，Wheeler RM，et al. Plant Productivity in response to LED lighting［J］. Hortic Sci，2008，43：1951–1956
[17] Guo S，Liu X，Ai W，et al. Development of an improved ground-based phototype of space plant-growing facility［J］. Advances in Space Research，2008，41（5）：736–741
[18] Morrow R C. LED lighting in horticulture［J］. Hortscience，2008，43（7）：1947–1950
[19] Sartory D. P.，Grobbelaar J. U. Extraction of chlorophyll a from freshwater phytoplankton for spectrophotometric analysis. Hydrobiologia，1984，114（3）：177-187.
[20] Strasser RJ，Srivastava A，Tsimilli-Michael M. 2000. The fluorescence transient as a tool to characterize and screen photosynthetic samples. In：Yunus M，Pathre U，Mohanty P（eds）. Probing Photosynthesis：Mechanism，Regulation and Adaptation. London：Taylor and Francis Press，445–483
[21] Turan M. T. Öz. Ö.，Kayihan C.，Eyidoğan F.，et al. Evaluation of photosynthetic performance of wheat cultivars exposed to boron toxicity by the jip fluorescence test［J］. Photosynthetica，2014，52（4）：555–563
[22] Van Heerden PDR，Strasser RJ，Krüger GHJ. Reduction of dark chilling stress in N2-fixing soybean by nitrate as indicated by chlorophyll a fluorescence kinetics［J］. Physiol Plant，2004，121：239–249

锌钾营养耦合对温室无土栽培番茄产量品质的影响

侯广欣[1，2，3]，梁浩[1，2]，季延海[1，2]，刘明池[1，2]，赵敏[3]，武占会[2*]

（1. 北京市农林科学院蔬菜研究中心，北京　100097；2. 农业部华北都市农业重点实验室，北京　100097；3. 河北工程大学园林与生态工程学院，邯郸　056038）

摘要： 研究钾肥和锌肥对番茄生长发育、生理特性和光合特性的影响。以“丰收”番茄为试材，设置4处理。于第一穗花开时统一进行滴灌追肥处理，每隔14d测定株高、茎粗、叶片数等生长指标；在番茄盛果期晴天10:00左右测定植株上部第3片叶的光合速率（Pn）、蒸腾速率（Tr）、气孔导度（Gs）等光合参数；果实成熟后测定相应的品质指标以及单果重和产量。不同处理对番茄生长、光合、品质影响不同。①钾肥对番茄生长、品质和产量影响显著高于锌肥；跟单独增施钾肥相比，钾、锌同增显著提高番茄可溶性糖含量，在一定程度上提高番茄维生素C含量、可溶性固形物和糖酸比，没有显著影响；跟CK比，增施锌肥提高了番茄维生素C含量、可溶性固形物和糖酸比。②钾、锌同补对番茄光合特性指标影响显著高于其他处理，锌肥对番茄光合特性指标影响次于钾肥。③增施钾肥、锌肥对番茄有增产作用，钾、锌同施增产效果最显著，单株增产幅度为14.29%，其次是单施钾肥，增施锌肥跟CK比显著提高了单株产量。增施钾肥和锌肥对番茄光合特性、品质和产量均有提高，钾锌同增效果最显著。

关键词： 钾肥；锌肥；番茄；品质；产量；光合特性

Effects of Zinc and Potassium Nutrition on Physiology and Quality of Greenhouse Soilless Cultivated Tomato

Hou Guangxin[1，2，3]，Liang hao[1，2]，Ji Yanhai[1，2]，Liu Mingchi[1，2]，Zhao Min[3]，Wu Zhanhui[1，2*]

（[1]*College of Landscape and Ecological engineering*，*Hebei University of Enginneerring*，*Handan*　056001；[2]*National Engineering Research Center for Vegetables*，*Beijing Academy of Agriculture and Forestry Sciences*，*Beijing*　100097；[3]*Key Laboratory of North China Urban Agriculture*，*Ministry of Agriculture*，*Beijing*　100097）

Abstract: In order to study the effects of potassium fertilizer and zinc fertilizer on the growth，physiological and photosynthetic characteristics of tomato.Tomato was used as the test material，4 treatments were set. The growth indexes such as plant height，stem diameter，leaf number were determined every 14 days. Photosynthetic rate （PN），transpiration rate （Tran） and stomatal conductance （GS） of the third leaf in the upper part of the plant were measured at 10:00 in sunny days. the effects of different treatments on tomato growth，photosynthesis and quality were different. The effect of potassium fertilizer on tomato growth，quality and yield was significantly higher than that on zinc fertilizer. the effects of different treatments on tomato growth，photosynthesis and quality were significantly higher than those on zinc fertilizer. different treatments had different effects on tomato growth，photosynthesis and quality. Compared with potassium fertilizer，potassium and zinc increased the content of soluble sugar significantly，but to some extent increased the content of VC，the ratio of soluble solids and sugar to acid，but increased the content of VC in tomato compared with CK. The ratio of soluble solids to sucrose and acid was increased. 2. The effect of potassium and zinc application on photosynthetic characteristics of tomato was significantly higher than other treatments，and the effect of zinc fertilizer on photosynthesis index of tomato was lower than that of potash fertilizer. 3. Adding potassium fertilizer，zinc fertilizer could increase tomato yield，potassium，potassium，potassium，potassium，potassium，potassium，potassium，potassium，potassium，potassium，potassium and potassium. The effect of zinc fertilizer on yield was significant，the yield was increased by 14.29%. Potassium and zinc fertilizer significantly increased the photosynthetic characteristics，quality and yield of tomato.

Key words: Potash fertilizer；Zinc fertilizer；Tomato；Quality；Yield；Photosynthetic Characteristics

1　前言

钾是植物生长必需的营养元素，能够维持细胞内物质正常代谢、提高酶活性、促进光合作用，在产物的运输和蛋白质合成等方面发挥着重要作用[1-3]。番茄吸钾量居各营养元素首位[4]。缺钾会降低番茄产量，对番茄品质也有一定影响，适当增施钾肥，有助于改善品质，提高果实中维生素C含量、可溶性固形物和糖酸比。锌在作物体内参与生长素的合成，当作物缺锌时，生长素的合成受到影响，作物生长处于停滞状态，植株矮小；同时锌也是许多酶的组成成分，故锌对作物光合作用和抗逆性均有一定影响[5]。

2　材料与方法

2.1　试验场地与材料

试验于2017年9月至2018年1月在北京市农林科学院蔬菜研究中心日光连栋温室中进行，供试番茄品种为“丰收”。采用蔬菜研究中心自主研发的封闭式无土槽培系统，番茄定植到规格为长×宽×高=48cm×18cm×13cm栽培槽中，栽培基质是珍珠岩。

2.2　试验设计与方法

番茄于2017年8月14日播种，9月14日四叶一心时定植，采用双行栽培，株距35cm，小行距30cm，大行距150cm。营养液基础配方采用蔬菜研究中心番茄改良配方，共设4个处理：CK；T1—增施$ZnSO_4 \cdot 7H_2O$；T2—增施$ZnSO_4 \cdot 7H_2O$和KCl；T3—增施KCl，如表1所示，每个处理3次重复，供液方式为滴灌，番茄采用单干整枝方式整枝，第 3 穗果后摘心，于第一穗花开花时统一进行滴灌追肥处理，处理前均以蔬菜研究中心通用番茄营养液进行供液，pH值始终保持在6.3 ± 0.2之间，定期对营养液进行调整和更换。其他栽培管理方式按日常田间管理进行。

表 1　试验处理

Table 1　Experimental treatments

编号 No.	K^+浓度（mmol/L） K^+concentration（mmol / L）	Zn^{2+}浓度（μmol/L） Zn^{2+}concentration（μmol/L）
CK	8	0.77
T1	8	1.54
T2	16	1.54
T3	16	0.77

2.3 测定项目与方法

2.3.1 形态指标测定

每重复选取长势一致的6株番茄进行形态指标测定，每14d测1次，测定指标包括株高、茎粗、叶片数。番茄株高为从基部到生长点的长度，采用卷尺测量；茎粗是第一片真叶到子叶之间的横径，使用日本Mitutoyo电子数显游标卡尺测量；叶片数的计算以叶片展开（>3cm）为标准。

2.3.2 生理指标测定

光合色素含量测定：采用乙醇浸提比色法测定光合色素含量[6]。

2.3.3 光合指标测定

在番茄盛果期晴天10:00左右用6400便携式光合仪测定植株上部第4片叶的光合速率（Pn）、蒸腾速率（Tr）、气孔导度（Gs）等光合参数。每重复测定长势一致的3株植株，取其测定的平均值。

2.3.4 果实性状和品质测定

采收期采取各处理第二穗番茄果实，分别选取成熟度一致的果实进行果实性状和品质测定，果实性状指标包括：单果重、果形指数、横径、纵径；果实品质指标包括：维生素C含量[7]，可溶性固形物含量、有机酸和可溶性糖含量[8]。

2.3.5 产量指标测定

每重复选择长势一致的12株植株挂牌标记，成熟时记录采收时间和单株产量。

2.4 数据统计与分析

采用SPSS 19.0和Excel.2003软件进行数据处理、统计分析和绘图。

3 结果与分析

3.1 不同处理对番茄植株生长的影响

3.1.1 不同处理番茄株高、茎粗比较

由图1可知，处理前各处理番茄植株长势和茎粗没有显著差异；处理14d后，各处理

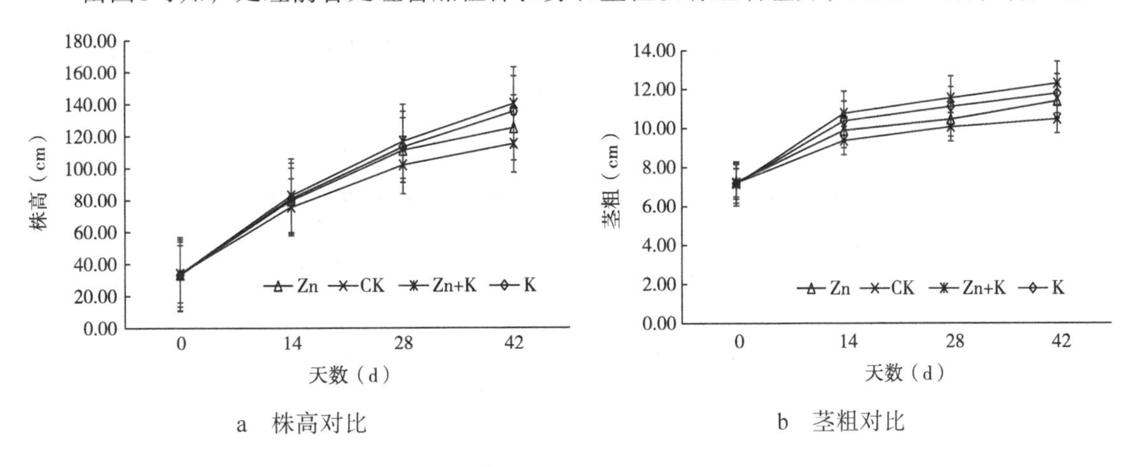

a 株高对比　　b 茎粗对比

图1 不同处理番茄株高、茎粗比较

Figure 1 Comparison of plant height and stem diameter with different treatments

株高和茎粗差异不明显，处理42d后，T2株高为140.16cm，显著高于T1和CK，较T1高12%；较CK高21.7%。处理14d前，各处理茎粗生长迅速，处理14d后，茎粗增长速度逐渐减慢，定植42d后，T3茎粗为12.29cm，显著高于其他处理，较CK高17.4%；茎粗由高到低依次为T2>T3>T1>CK。

3.1.2 不同处理番茄叶片比较

如图2所示，番茄叶片在前28d各处理间叶片数差异不显著，后期T2处理叶片数最多，显著高于其他处理；其次是T3处理；CK处理叶片数最少。

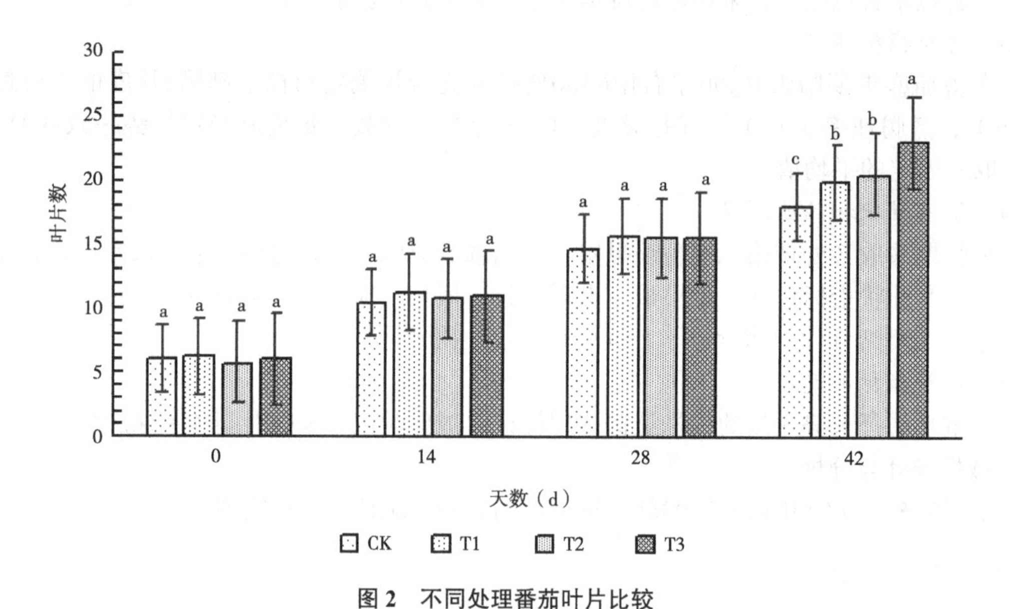

图2 不同处理番茄叶片比较

Figure 2 Comparison of tomato leaves with different treatments

3.2 不同处理对番茄叶片光合色素含量影响

由表2可以看出，叶绿素a、叶绿素a+b以及类胡萝卜素含量均以T2为最高，T2与T3没有显著差异，但显著高于T1和CK；各个处理间叶绿素b没有显著差异，但T2处理含量最高，CK最低；T2处理叶绿素a/b显著高于其他处理。

表2 不同处理对番茄叶片光合色素含量影响

Table 2 Effect of different treatments on photosynthetic pigment content in Tomato leaves

处理 treatment	叶绿素a Chlorophy Ⅱ a/（mg·g^{-1}）	叶绿素b Chlorophy Ⅱ b/（mg·g^{-1}）	叶绿素a+b Chlorophy Ⅱ a+b/（mg·g^{-1}）	叶绿素a/b Chlorophy Ⅱ a/b	类胡萝卜素 Carotenoids/（mg·g^{-1}）
CK	1.95 ± 0.09b	1.98 ± 0.05a	1.96 ± 0.02b	0.99 ± 0.07b	0.19 ± 0.003c
T1	2.01 ± 0.03b	1.95 ± 0.01a	1.99 ± 0.01b	1.02 ± 0.02ab	0.23 ± 0.010c
T2	2.62 ± 0.07a	2.11 ± 0.05a	2.37 ± 0.01a	1.24 ± 0.05a	0.33 ± 0.001a
T3	2.52 ± 0.14a	2.1 ± 0.09a	2.32 ± 0.01a	1.19 ± 0.08ab	0.28 ± 0.016b

3.3 不同处理对番茄光合特性指标影响

由表3可以看出，T2处理净光合速率、蒸腾速率、气孔导度、胞间CO_2浓度显著高于其他处理；T3处理净光合速率和蒸腾速率显著高于T1和CK；T1胞间CO_2浓度较CK显著高4.28%，较T3显著低13.32%；四个处理间荧光特性差异不显著。

表 3 不同处理对番茄光合特性指标

Table 3 Effects of different treatments on photosynthetic characteristics of Tomato

	光合特性指标				
处理 Treatment	净光合速率（$\mu molCO_2 \cdot m^{-2} \cdot s^{-1}$）	蒸腾速率（$mmolH_2O \cdot m^{-2} \cdot s^{-1}$）	气孔导度（$mol\ H_2O \cdot m^{-2} \cdot s^{-1}$）	胞间CO2浓度（$\mu mol \cdot mol^{-1}$）	荧光特性指标 Fv/Fm
CK	13.87 ± 0.1c	1.76 ± 0.03c	0.26 ± 0.02c	237.95 ± 2.97d	0.8 ± 0.001a
T1	14.54 ± 0.42c	2.07 ± 0.22c	0.28 ± 0.01bc	248.15 ± 2.23c	0.79 ± 0.002a
T2	16.98 ± 0.16a	5.85 ± 0.14a	0.47 ± 0.01a	313.21 ± 2.86a	0.79 ± 0.01a
T3	15.38 ± 0.38b	5.12 ± 0.19b	0.34 ± 0.02b	286.27 ± 2.47b	0.80 ± 0.001a

3.4 不同处理对番茄品质指标影响

由表4可以看出，第二穗果中，T2处理可溶性糖含量显著高于其他处理，较CK高32.57%，较T3高4.53%；维生素C、可溶性固形物、可滴定酸、糖酸比均以T2处理最高，T2和T3处理没有显著差异，T2处理显著高于T1和CK；T1处理可溶性固形物、可滴定酸、糖酸比均高于CK，没有显著差异；T1处理的可溶性糖含量显著高于CK，较CK高7.83%。

表 4 不同处理对番茄品质指标影响

Table 4 Effect of different treatments on Tomato quality Index

处理 treatment	维生素C含量 Vitamin C content（$mg \cdot 100g^{-1}$）	可溶性固形物 Soluble solid content（%）	可溶性糖含量 Soluble sugar content（%）	可滴定酸 Titrable acid content（%）	糖酸比 Sugar acid（%）
CK	18.88 ± 0.94b	6.75 ± 0.05b	5.74 ± 0.02d	1.11 ± 0.03c	5.14 ± 0.17b
T1	19.49 ± 0.18b	6.85 ± 0.05b	6.19 ± 0.08c	1.20 ± 0.034bc	5.16 ± 0.06b
T2	22.63 ± 0.10a	7.4 ± 0.1a	7.61 ± 0.12a	1.27 ± 0.02a	5.98 ± 0.21a
T3	22.34 ± 0.42a	7.35 ± 0.05a	7.28 ± 0.02b	1.25 ± 0.01ab	5.82 ± 004a

3.5 不同处理对番茄果实性状影响

由表5可知，T2处理单果重、果实横径、纵径和单株产量显著高于其他处理，单果重较CK、T1、T3分别高16.8%、12.29%、3.42%；T3处理单果重和单株产量显著高于T1和CK，单株产量较CK和T1高15.14%、12.75%；CK和T1果形指数显著高于其他处理，CK果形指数最高。

表 5　不同处理对番茄果实性状影响

Table 4　Effect of different treatments on Tomato shapes and properties

2穗果 2 panicle fruit	单果重 Single fruit weight（g）	果实横径 Fruit transverse diameter（mm）	果实纵径 Fruit longitudinal diameter（mm）	果形指数 Fruit shape index	单株产量 Single plant yield（g）
CK	161.40 ± 1.1d	67.37 ± 0.23c	51.31 ± 0.14d	0.76a	986.29d
T1	176.33 ± 1.12c	68.36 ± 0.08c	52.38 ± 0.04c	0.77a	1 067.2c
T2	188.52 ± 0.35a	76.48 ± 0.2a	56.52 ± 0.25a	0.74c	1 226.36a
T3	182.29 ± 0.86b	73.2 ± 1.10b	55.05 ± 0.55b	0.75b	1 135.62b

4　结论

通过研究锌钾营养耦合作用对设施无土栽培番茄产量品质的影响，结果表明，钾、锌同补显著提高了番茄株高、茎粗、光合特性和产量，净光合速率较T1和T3分别提高16.78%、10.4%；钾锌同补处理的可溶性糖含量提高32.57%；单果重量较 T1和T3 高6.91%、3.42%，较CK 高 24.34%，增产效果最明显。

参考文献

［1］　张继澍. 植物生理学［M］. 北京：高等教育出版社，2006.

［2］　谢建昌，周健民. 钾与中国农业［M］. 南京：河海大学出版社，2000.

［3］　张恩平，张淑红，李天来，等. 蔬菜钾素营养的研究现状与展望［J］. 中国农学通报，2005，21（8）：265-268.

［4］　吴国喜. 钾肥对大棚番茄品质影响及主要相关机理的研究［D］.合肥：安徽农业大学，2007.

［5］　蔡永萍. 植物生理学［M］. 北京：中国农业大学出版社，2008.

［6］　D Arnon.Copper enzymes in isolated chloroplasts，polyphenol oxidase in Brat vulgaris［J］.Plant Physiology.1949，1（24）：1-15.

［7］　蔡庆生.植物生理学实验［M］.北京：中国农业大学出版社，2013.

［8］　张以顺，黄霞，陈云凤.植物生理学实验［M］.北京：高等教育出版社，2009.